Intermediate Algebra

Intermediate Algebra

J. Louis Nanney
Miami-Dade Commumity College

John L. Cable
Miami-Dade Commumity College

 Wm. C. Brown Publishers

Book Team

Editor *Earl McPeek*
Developmental Editor *Theresa Grutz*
Production Editor *Eugenia M. Collins*
Designer *K. Wayne Harms*
Art Editor *Janice M. Roerig*
Photo Editor *Michelle Oberhoffer*

 Wm. C. Brown Publishers

President *G. Franklin Lewis*
Vice President, Publisher *George Wm. Bergquist*
Vice President, Publisher *Thomas E. Doran*
Vice President, Operations and Production *Beverly Kolz*
National Sales Manager *Virginia S. Moffat*
Advertising Manager *Ann M. Knepper*
Marketing Manager *John W. Calhoun*
Editor in Chief *Edward G. Jaffe*
Managing Editor, Production *Colleen A. Yonda*
Production Editorial Manager *Julie A. Kennedy*
Production Editorial Manager *Ann Fuerste*
Publishing Services Manager *Karen J. Slaght*
Manager of Visuals and Design *Faye M. Schilling*

Cover credit Robert Phillips

One: AP/Wide World Photos; **Two:** Courtesy Ron Johnson/Iowa Department of Natural Resources; **Three:** NASA; **Four:** © James L. Shaffer; **Five:** © Bob Clay/Clay Images; **Six:** © Keith R. Porter/Photo Researchers, Inc.; **Seven:** © Jim Cronk/Lightwave; **Eight:** © Jim Cronk/Lightwave; **Nine:** © Dan Chidester/The Image Works, Inc.; **Ten:** © Bob Clay/Clay Images; **Eleven:** © James L. Shaffer.

Library of Congress Catalog Card Number: 90-55918

ISBN 0-697-11653-0

Printed in the United States of America by Wm. C. Brown Publishers, 2460 Kerper Boulevard, Dubuque, IA 52001

10 9 8 7 6 5 4 3 2 1

CONTENTS

CHAPTER
3   **Solving Word Problems**

CHAPTER
4   **Polynomials: Products and Factoring**

CHAPTER
5   **Algebraic Fractions**

Exponents and Radicals

Quadratic Equations

CHAPTER
8   **Graphing: Functions, Relations, and Inequalities**

CHAPTER
9   **Systems of Equations and Inequalities**

CHAPTER

10   Logarithms

CHAPTER

11   Sequences, Series, and the Binomial Theorem

PREFACE

A course in intermediate algebra may be the final course in mathematics for some students with certain majors. For others it is a course that "bridges the gap" between elementary or beginning algebra and college algebra. The student who has only one course in secondary school algebra or whose background in algebra is not strong will benefit from an intermediate course. This text is "zero based" providing a thorough review of elementary algebra topics for those who need it.

Pedagogical Features

Format

Each chapter begins with an introductory paragraph previewing the chapter coverage. To maintain a flow of topics each section has an introductory paragraph which gives a purpose for that section and, where possible, relates the new material to previous sections or chapters.

Readability

The authors have made a special effort to produce a text for the student. One new concept at a time is introduced with clear, concise, mathematically correct explanations and definitions. These are followed by many detailed examples and an abundance of exercises.

Objectives

Each section of the text contains a statement of the objectives. Thus a clearly stated purpose for that section is available to both student and instructor.

Exercises

Careful attention has been given to each set of exercises so they properly reinforce the material presented and fulfill the stated objectives. They also provide review, and where possible lead into the next concept to be presented. The abundance of exercises are graded in difficulty and odd–even problems are similar.

Word Problems

Mastering the techniques of solving word problems may well be the most demanding task for students at this level. Hence a separate chapter is devoted to this topic. The authors have carefully presented methods of classifying and solving different types of problems. Every effort has been made to assure that most of the word problems have practical applications to the real world. Also, whenever possible, exercise sets throughout the text include applications of the topics discussed.

Margins

A unique feature of this text is the judicious use of margins. They contain objectives, helpful notes, hints, warnings against common errors, references to properties, and other comments. The language in the margins is less formal than that in the text. In short, the margins are designed to compliment the text with a self-contained student study guide.

Warnings

The authors have drawn from their many years of teaching at this level to identify common errors made by students. These warnings are contained in the margins and refer directly to the textual material being presented.

Pretest/Chapter Test

Each chapter begins with a pretest and ends with a chapter test. These tests are similarly constructed so the student may compare results and evaluate progress. Answers to these tests are keyed to sections within the chapter as an aid to any necessary review.

Chapter Summary

Each chapter is followed by a summary that lists key words and procedures covered in the chapter. These are referenced to the sections that discuss them.

Chapter Review

After the chapter summary, an abundance of review exercises is provided. Answers to these exercises are keyed to sections within the chapter so topics needing further study can easily be identified.

Cumulative Tests

Cumulative tests are provided after chapter 3, chapter 6, and chapter 9. Answers to these tests are keyed to the chapter and section for easy reference.

End of Book Test

A test covering the entire text is provided after chapter 11. Answers to this test are keyed to the chapter and section for easy reference.

Answers to the Exercises

Answers to the odd-numbered exercises and all answers to the pretests, chapter tests, and cumulative tests are provided in appendix B at the end of the book. Since the odd and even questions in the exercise sets and chapter reviews are similar, the even-numbered answers have been omitted as some instructors use them as hand-in exercises.

Ancillary Package

For the Instructor

The *Annotated Instructor's Edition* provides a convenient source for answers to all exercise and margin problems. Each answer is placed directly below the problem in boldface type. The *Annotated Instructor's Edition* eliminates the need to search through a separate answer key.

The *Test Item File/Quiz Item File* is a printed version of the computerized *TestPak* and *QuizPak* which allows you to choose test items based on chapter, section, or objective. The objectives are taken directly from each section of *Intermediate Algebra.*

WCB *TestPak,* our computerized testing service, provides you with a mail-in/call-in testing program and the complete test item file on diskette for use with IBM PC, Apple or Macintosh computers. **WCB** *TestPak* requires no programming experience. Tests can be generated randomly, by selecting specific test items or the *Intermediate Algebra* section objectives. In addition, new test items can be added and existing test items can be edited.

WCB *QuizPak,* a part of *TestPak 3.0,* provides students with true/false, multiple choice, and matching questions from the *Quiz Item File* for each chapter in the text. Using this portion of the program will help your students prepare for examinations. Also included with the **WCB** *QuizPak* is an on-line testing option that allows professors to prepare tests for students to take using the computer. The computer will automatically grade the test and update the gradebook file.

WCB *GradePak,* also a part of *TestPak 3.0,* is a computerized grade management system for instructors. This program tracks student performance on examinations and assignments. It will compute each student's percentage and corresponding letter grade, as well as the class average. Printouts can be made utilizing both text and graphics.

Diagnostic Masters to Accompany Intermediate Algebra are reproducible masters keyed to each section objective in the text, and reproducible chapter tests. *Diagnostic Masters* can be used for placement, additional assignments, instructional examples, quizzes, reviews, or exams. A complete answer key for all masters is included.

For the Student

The *Student's Solutions Manual* contains helpful hints, warnings, and solutions to selected exercises in *Intermediate Algebra* with clear explanations of each step. It is available for student purchase.

On the **Videotapes** the instructor introduces a concept, provides detailed explanations of example problems that illustrate the concept, including applications, and concludes with a summary. The tapes are available free to qualified adopters.

The **Audiotapes** have been developed specifically to accompany *Intermediate Algebra.* They begin with a brief synopsis of the section, followed by clear discussions of examples with warnings and hints where appropriate. Exercises are solved for each section of the text: students are directed to turn off the tape and solve a specific problem and turn the tape on again for a complete explanation of the correct solution.

Algebra Problem Solver Software is available to reinforce the key concepts of each chapter. Students can practice problems generated by the computer or create their own practice problems. Step-by-step solutions with complete explanations guide students to clear understanding mastery of the major concepts and skills of *Intermediate Algebra.*

Acknowledgments

A text of this caliber cannot be produced without the assistance of many people. The authors wish to express their gratitude to the editors and staff at Wm. C. Brown Publishers for their constant support. We also wish to thank the following reviewers for their constructive comments.

Janet G. Melancon
Loyola University

Michael J. Bonanno
Suffolk County Community College

Sue L. Korsak
New Mexico State University

William A. Neal
Fresno City College

G. Don Benson
Virginia Western Community College

PRETEST

Before beginning this chapter, answer as many of the following questions as you can. When you have finished the chapter, take the practice test at the end of the chapter and compare the scores of the two tests to see how much you have learned.

A N S W E R S

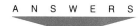

1. Name the property that the following statement illustrates:
$(4 + 1) + 7 = 4 + (1 + 7)$

2. Find the negative of $-\dfrac{1}{2}$.

1. _____

3. Find the negative of 3.5.

2. _____

4. Identify the signed number represented by point A.

3. _____

5. Identify the signed number represented by point B.

4. _____

6. Evaluate: $-|9|$

5. _____

6. _____

A N S W E R S

7. Evaluate: $|-(-3)|$

8. If the temperature was 15° and then fell twenty degrees, express the new temperature using a signed number.

7. _____

8. _____

9. Combine: $(-21) + (+14)$

10. Combine: $(-17) - (-32)$

9. _____

10. _____

11. Evaluate: $(-12)(-4)$

12. Evaluate: $\dfrac{-30}{+5}$

11. _____

12. _____

13. Find the product: $(-2)(4)(-3)(-1)$

14. Simplify: $(2x^2y)(-6x^2y^3)(-x^3)$

13. _____

14. _____

15. Remove parentheses and simplify: $16 - (-3 - 8) + 4$

16. Combine like terms: $24xy - 5x^2 - 9xy + 7x^2$

15. _____

16. _____

17. Find the product: $2xy(3x^2 - 4xy + 5y^3)$

18. Remove the grouping symbols and simplify: $14x - 3\{4x - 6[2x - 3(5x - 2)]\}$

17. _____

18. _____

19. _____

19. Evaluate: $3x^2 - 2xy - 5y^2$ if $x = -2, y = 3$.

20. A distance formula from physics is $d = rt$. Find d when $r = 46$ and $t = 3.5$.

20. _____

SCORE: _____

2

1

Basic Concepts—A Review

After losing nine yards on the first down, the Browns completed a pass for a twelve-yard gain on the second down. What was the net gain or loss?

Certain fundamental concepts form the foundations of algebra and algebraic manipulations. These fundamentals were studied in elementary algebra but are of such basic importance that a review is in order. Topics in this chapter include the operations on signed numbers, use of grouping symbols, and simple operations on monomials and other polynomials. Mastering these skills is necessary to the further study of algebra.

1-1 REAL NUMBERS AND THEIR PROPERTIES

O B J E C T I V E S

▼

Upon completing this section you should be able to:
1. Classify any real number.
2. Identify the properties of real numbers.

Braces { } are used to enclose the elements of a set.

Do you see that mixed numbers and decimals are also rational numbers?

Note that a number can not be both rational and irrational.

The set of counting numbers is a *subset* of the set of whole numbers, which in turn is a subset of the set of integers, and so on.

Elementary and intermediate algebra deal almost exclusively with a set of numbers called the *real numbers*. In this section we will give a brief outline of the set of real numbers and list their properties.

The numbers first encountered in arithmetic are those used in counting, or the set of **counting numbers** or **natural numbers.**

$$\{1,2,3,4,5, . . .\}$$

This set is next extended to include zero and we have the set of **whole numbers.**

$$\{0,1,2,3,4, . . .\}$$

We can extend the set even further to include the negatives or opposites of the counting numbers. This produces the set of **integers.**

$$\{. . .,-4,-3,-2,-1,0,1,2,3,4, . . .\}$$

Another set of numbers that is commonly used is the set of **rational numbers,** those numbers that can be expressed as a ratio of two integers and are often referred to as **fractions.** This set includes the integers since, for example, 3 can be expressed as $\frac{3}{1}, \frac{6}{2}$, and so on.

The set of numbers that cannot be expressed as a ratio of two integers is called the set of **irrational numbers.** This set includes such numbers as π, $\sqrt{5}$, $\sqrt[3]{7}$, and so on.

The set of rational and irrational numbers together make up a set called the **real numbers.**

The following chart shows the different subsets of the set of real numbers.

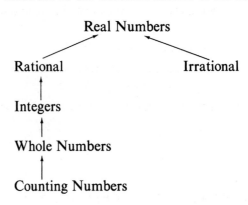

Example 1   To which of these sets, counting numbers, whole numbers, integers, rational numbers, irrational numbers, do the following real numbers belong?

$$\textbf{a. } 3 \qquad \textbf{b. } -5 \qquad \textbf{c. } \frac{2}{3} \qquad \textbf{d. } 0$$

Answers   **a.** 3 is a counting number, a whole number, an integer, and a rational number.

Some numbers are in more than one set.

b. -5 is an integer and a rational number.

c. $\frac{2}{3}$ is a rational number.

d. 0 is a whole number, an integer, and a rational number.

0 is rational since it can be expressed as a ratio of 0 over any other integer.

The **properties** (called axioms) of the set of real numbers can be used (along with definitions) to justify all of the manipulations used in algebra. If this set of properties is accepted then all other properties of real numbers can be proved as theorems. Although in this text an intuitive approach, rather than one of rigorous proof, is used, mention shall often be made of these properties. They are listed here for easy reference.

A *theorem* is a statement that can be mathematically proved using the properties and definitions.

Properties (Axioms) of the Real Numbers

1. *The real numbers are closed under addition.* If a and b are real numbers, then $(a + b)$ is also a real number.

2. *Addition of real numbers is commutative.* If a and b are real numbers, then $a + b = b + a$.

3. *Addition of real numbers is associative.* If a, b, and c are real numbers, then
$$(a + b) + c = a + (b + c).$$

4. There is a real number that is the *additive identity.* This number is zero. For all real numbers a, $a + 0 = a$.

5. Every real number has a *negative (additive inverse).* If a is a real number, then there is a real number $(-a)$ such that $a + (-a) = 0$.

6. *The real numbers are closed under multiplication.* If a and b are real numbers, then $a \times b$ is also a real number.

7. *Multiplication of real numbers is commutative.* If a and b are real numbers, then $a \times b = b \times a$.

8. *Multiplication of real numbers is associative.* If a, b, and c are real numbers, then
$$(a \times b) \times c = a \times (b \times c).$$

9. There is a real number that is the *multiplicative identity.* This number is 1. For all real numbers a, $a \times 1 = a$.

10. Each nonzero number has a *reciprocal (multiplicative inverse).* If a is a nonzero real number, then there is a real number $\frac{1}{a}$ such that
$$a \times \frac{1}{a} = 1.$$

11. The real numbers obey the *distributive property of multiplication over addition.* If a, b, and c are real numbers, then
$$a(b + c) = (a \times b) + (a \times c).$$

If we add two numbers, the result is a real number (sometimes called *closure*).

Does $3 + 6 = 6 + 3$?

When adding a series of numbers, it doesn't matter which two are added first.

Note that 0 is its own negative.

When we multiply two real numbers, the result is a real number.

Again, it doesn't matter which two are multiplied first.

$$3 \times \frac{1}{3} = 1$$

Evaluate each side of
$$3(2 + 4) = (3 \times 2) + (3 \times 4).$$

Example 2   What property is illustrated by

a. $3 + (-3) = 0$

b. $2 \times 5 = 5 \times 2$

c. $4 \times \dfrac{1}{4} = 1$

Answers

The negative of 3 is -3.

a. additive inverse

b. commutative property of multiplication

c. multiplicative inverse

▼ **EXERCISE 1-1-1**

To which of these sets, (a) counting numbers, (b) whole numbers, (c) integers, (d) rational numbers, (e) irrational numbers, does each of the following numbers belong?

1. 5

2. 8

3. -4

4. -6

5. 0

6. 1

7. 23

8. -42

9. $\sqrt{2}$

10. $\sqrt{3}$

11. $\dfrac{1}{2}$

12. $\dfrac{2}{5}$

13. $-\dfrac{9}{4}$

14. $-\dfrac{7}{2}$

15. π

16. $\sqrt{11}$

17. $3\dfrac{1}{2}$

18. $5\dfrac{2}{3}$

19. 1.6

20. .01

Name the property that each of the following illustrates.

21. $6 + 8$ is a real number.

22. 3×7 is a real number.

23. $12 + 0 = 12$

24. $4 + (-4) = 0$

25. $(2 \times 3) \times 8 = 2 \times (3 \times 8)$

26. $6 \times \dfrac{1}{6} = 1$

27. $3 + 7 = 7 + 3$

28. $(2 + 5) + 9 = 2 + (5 + 9)$

29. $25 \times 1 = 25$

30. $1 + 0 = 1$

31. $3(5 + 4) = (3 \times 5) + (3 \times 4)$

32. $3 \times 5 = 5 \times 3$

 1-2 THE MEANING OF SIGNED NUMBERS

In this section we will establish the meaning of a set of numbers called **signed numbers.** Signed numbers are numbers preceded by a plus (+) or minus (−) sign. A nonzero number without a sign is assumed to be preceded by a plus sign. We will also see that the minus sign has more than one meaning. Examples of signed numbers are $+5$, -6, $+\dfrac{2}{3}$, $-\dfrac{4}{5}$, -100, $+7$, 8, 12, -4, and so on.

> Numbers preceded by a plus sign are called **positive numbers** and numbers preceded by a minus sign are called **negative numbers.**

One useful way of representing signed numbers is on a **number line.**

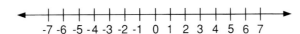

Note that the positive numbers are placed in order to the right of the zero and the negative numbers are placed in order to the left of zero. The arrows indicate the line continues indefinitely in both directions.

Zero is neither positive nor negative. We will make use of the number line in the next section.

To be successful in using operations on signed numbers, a very important distinction must now be made. We must distinguish between a *negative number* and the *negative of a number*. We have already noted that a negative number is a number preceded by a minus sign. On the number line we noted that negative numbers are to the left of zero. Pay special attention to the following definitions.

OBJECTIVES

Upon completing this section you should be able to:

1. Distinguish between positive and negative numbers and represent them on a number line.
2. Find the negative of a given number.
3. Find the absolute value of a given number.
4. Remove grouping symbols preceded by a minus sign.

What is the only unsigned number? (Answer: 0)

We will agree that the negative of zero is zero.

> The **negative of a number** is the opposite of that number, or that number with the opposite sign.

Example 1   The negative of $+5$ is -5.

Example 2   The negative of $+\dfrac{2}{3}$ is $-\dfrac{2}{3}$.

x stands for any number.

Example 3   The negative of $+x$ is $-x$.

Example 4   The negative of -5 is $+5$.

Example 5   The negative of $-x$ is $+x$.

Example 6   If $+7$ represents seven steps north, then -7 represents seven steps south.

Can you find examples of signed numbers? Check the newspaper for signed numbers in the stock market report, weather report, and so forth.

Example 7   If $-3°$ represents three degrees below zero, then $+3°$ represents three degrees above zero.

Example 8   What is the negative of -10?

   Answer   Since "the negative of" means "opposite of," the negative of -10 is $+10$.

The minus sign is often used as a symbol for "the negative of" or "opposite of." Therefore $-(-10)$ means "the opposite of" -10.

Example 9 $-(+7) = -7$

Example 10 $-(-2) = +2$

The distance from any number to zero on the number line is defined to be the **absolute value** of that number.

Note that if x is negative, then $-x$ is positive.

> The **absolute value** of a number x is written $|x|$, where
> $$|x| = x \text{ if } x \text{ is zero or positive,}$$
> $$|x| = -x \text{ if } x \text{ is negative.}$$

The absolute value of a number is always positive or zero.

Example 11 $|3| = 3$

Example 12 $|-3| = 3$

▼ EXERCISE 1-2-1

Find the negative of each number.

1. $+3$

2. $+8$

3. -4

4. -15

5. $+\dfrac{1}{2}$

6. $-\dfrac{3}{5}$

7. -5.3

8. $+13.1$

9. $+y$

10. $+a$

11. $-b$

12. $-x$

13. 7

14. 23

15. $-4\dfrac{1}{3}$

16. $+1.12$

17. 0

18. $.81$

19. 100

20. $5\dfrac{3}{4}$

21. -7.38

Simplify.

22. $-(+5)$

23. $-(+4)$

24. $-(-3)$

25. $-(+2.05)$

26. $-(9)$

27. $-\left(\dfrac{1}{2}\right)$

28. $-(-11)$

29. $-\left(-\dfrac{3}{4}\right)$

30. $-(-a)$

Identify the signed numbers represented by the following points.

31.

32.

Represent the signed numbers on a number line.

33. -3

34. -2

35. $-(-1)$

36. $-(-2)$

37. -4

38. -5

39. $-(+2)$

40. $-(+4)$

41. If $+20$ represents a temperature of twenty degrees above zero, use a signed number to represent a temperature of twenty degrees below zero.

42. If $+400$ represents four hundred feet above sea level, use a signed number to represent four hundred feet below sea level.

43. If -10 represents a drop of ten points on the stock market, use a signed number to represent a gain of ten points.

44. During a particular day, the temperature rose twelve degrees. Indicate this rise in temperature using a signed number.

45. A football team loses five yards on a play. Use a signed number to indicate this loss.

46. During the night, the temperature fell to fifteen degrees below zero. Use a signed number to indicate this temperature.

47. A news reporter stated that the Dow-Jones Average dropped thirteen points. Indicate this drop using a signed number.

48. Inflation increased one and a half percent last month. Indicate this increase using a signed number.

49. A person's weight was recorded two weeks ago at 165 pounds. This week the person's weight is 159 pounds. Indicate the loss of weight using a signed number.

50. The balance in a savings account was $1,000 last year and is $1,264 this year. Use a signed number to indicate the change from last year to this year.

Evaluate each of the following.

51. $|5|$

52. $|13|$

53. $|-4|$

54. $|-17|$

55. $|0|$

56. $|2 - 2|$

57. $|21 - 21|$

58. $|-(-7)|$

59. $|-3.14|$

60. $\left|-\dfrac{3}{8}\right|$

61. $-|9|$

62. $-|-14|$

If we enclose a number expression in grouping symbols and place a minus sign before it, we are indicating the "negative" of the entire expression within the grouping symbols.

If we want to write the negative of the expression $a + b + 5$, we place the expression in parentheses and place a minus sign in front of it obtaining $-(a + b + 5)$.

Example 13 $-(a + b)$ means "the negative of, the sum of a and b."

Example 14 $-(x + 3 - b)$ means "the negative of, x plus 3 minus b."

> *To find the negative of an expression enclosed in grouping symbols* find the negative of each term within the symbols.

Terms are separated by addition and subtraction signs.

Example 15 $-(a + b) = -a - b$

▽ **WARNING** A very common error in finding the negative of an expression is not changing *every* sign.	**Example 16** $-(x - y + 4) = -x + y - 4$ **Example 17** $-(a - b) = -a + b$ **Example 18** $-(a - 3 + 4 - x) = -a + 3 - 4 + x$

Let us now restate the rule in more general terms.

> *To remove grouping symbols preceded by a minus sign* change the sign of every term that is within the symbols. *To remove grouping symbols preceded by a plus sign* do not change the sign of any term that was within the symbols.

We will remove parentheses in each of the following.

Example 19 $-(a + b - 3) = -a - b + 3$

This could also be written as $a + b - 3$.

Example 20 $+(a + b - 3) = +a + b - 3$

A plus sign is understood to precede the parentheses.

Example 21 $(x + y - a + b) = x + y - a + b$

▼ **EXERCISE 1–2–2**

Remove the grouping symbols.

1. $-(x + y)$ **2.** $-(a + 4)$ **3.** $+(x - 5)$

4. $+(a - b)$ **5.** $-(a - b - c)$ **6.** $-(a + b - c)$

7. $-(x + y - 3)$ **8.** $+(x - y + 3)$ **9.** $(a - b + 8)$

10. $-(x + y - 9)$ **11.** $+(a - c - 4)$ **12.** $+(a + b - 6)$

13. $-(a + b - c)$ **14.** $-(x - y + z)$ **15.** $-(x - y - 15)$

16. $(-x + 4 - z)$ **17.** $-(x + 4 - z)$ **18.** $-(a - b + c - 5)$

19. $(a - b)$ **20.** $-(b - a)$ **21.** $(a - 2b + c)$

22. $(-a + b - 4c)$ **23.** $-(-x + y)$ **24.** $-(-2x - y)$

25. $-(-a - 6b + 4)$ **26.** $-(-3a + b - 16)$ **27.** $-(5x - y + 4z - 3)$

28. $(x - 8y - 2z - 10)$ **29.** $-(a + 2b - 5c + 6)$ **30.** $-(-2a + 3b - 4c - 9)$

◥◣ 1-3 COMBINING SIGNED NUMBERS

In the preceding section we established the meaning of the set of signed numbers. We are now ready to begin exploration of basic operations on this new set of numbers.

Example 1   If you start at a point and move six steps north, then move three more steps north, where will you be in relation to your starting point?

     Answer   Nine steps north

Example 2   If the temperature at 6:00 P.M. is zero degrees and from 6:00 P.M. until midnight the temperature falls three degrees, then from midnight until 4:00 A.M. it falls seven more degrees, what is the temperature at 4:00 A.M.?

     Answer   Ten degrees below zero or $-10°$

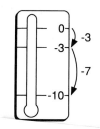

O B J E C T I V E S

Upon completing this section you should be able to:

1. Determine the result of combining two or more signed numbers.
2. Use the number line to combine signed numbers.

If we start at zero, a fall of three degrees takes the temperature to $-3°$. Another fall of seven degrees takes it to $-10°$.

Example 3   If at 9:00 A.M. the temperature is $+20°$ and from 9:00 A.M. until 1:00 P.M. it rises fourteen degrees, then from 1:00 P.M. until 5:00 P.M. it falls eighteen degrees, what is the temperature at 5:00 P.M.?

Answer $+16°$

Starting at $+20°$, a rise of $14°$ takes the temperature to $+34°$. A fall of $18°$ takes it to $+16°$.

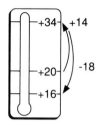

These are examples of combining signed numbers. If "steps north" is represented as positive, then six steps north followed by three steps north equals nine steps north could be written as

$$+6 + 3 = +9.$$

If a fall in temperature is represented by a minus sign, then a fall of three degrees followed by a fall of seven degrees can be written as

$$-3 - 7 = -10.$$

Note: -10 means a net fall of ten degrees.

If a rise in temperature is plus and a fall in temperature is minus, then example 3 can be written as

$$+20 + 14 - 18 = +16.$$

▼ EXERCISE 1-3-1

Use a signed number to indicate the result of each of the following.

1. If we consider north as positive and south as negative, then ten steps north followed by three steps south equals _____ .

2. Five steps south followed by two steps north equals _____ .

3. Four steps north followed by six steps south equals _____ .

4. Three steps south followed by six steps south equals _____ .

5. Eight steps south followed by ten steps north equals _____ .

6. A profit of $10 combined with a profit of $12 equals _____ .

7. A profit of $16 combined with a loss of $9 equals _____ .

8. A loss of $20 combined with a profit of $12 equals _____ .

9. A loss of $6 combined with a loss of $18 equals _____ .

10. A profit of $30 combined with a loss of $36 equals _____ .

11. A gain of two yards in a football game followed by a loss of six yards equals _____ .

12. A loss of two yards followed by a loss of five yards equals _____ .

13. A loss of eight yards followed by a gain of twelve yards equals _____ .

14. A gain of twelve yards followed by a loss of fifteen yards equals _____ .

15. A fall of six degrees in temperature followed by a fall of two degrees equals _____ .

16. A rise of eight degrees in temperature followed by a fall of two degrees equals _____ .

17. If at 10:00 P.M. the temperature was 65° and from 10:00 P.M. until 4:00 A.M. it fell twelve degrees, what was the temperature at 4:00 A.M.?

18. If the temperature was 50° at 6:00 A.M. and from 6:00 A.M. to 2:00 P.M. it rose fifteen degrees, then from 2:00 P.M. to 7:00 P.M. it fell nine degrees, what was the temperature at 7:00 P.M.?

19. Ten steps north, followed by six steps south, followed by nine steps north, followed by twelve steps south equals _____ .

20. A temperature starts at 10°. It falls fifteen degrees, then rises eight degrees, then falls three degrees. What is the final temperature?

In questions 21–40 write each of the previous questions (1–20) as a number statement.

21. (1) _____

22. (2) _____

23. (3) _____

24. (4) _____

25. (5) _____   **26.** (6) _____

27. (7) _____   **28.** (8) _____

29. (9) _____   **30.** (10) _____

31. (11) _____   **32.** (12) _____

33. (13) _____   **34.** (14) _____

35. (15) _____   **36.** (16) _____

37. (17) _____   **38.** (18) _____

39. (19) _____   **40.** (20) _____

The number line mentioned in the previous section is a useful tool in combining signed numbers.

Recall that the arrows indicate that the line continues indefinitely.

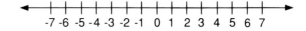

Note again that the positive numbers are placed in order to the right of zero and the negative numbers are placed in order to the left of zero.

If a movement to the right is considered $(+)$ and a movement to the left $(-)$, then consider the following examples.

Example 4 $+3 + 4 = +7$ indicates "start at zero, move three units to the right, then move four units to the right."

Remember, $+$ means move to the right.

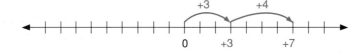

Example 5 $-2 - 4 = -6$ indicates "start at zero, move two units to the left, then move four units to the left."

Remember, $-$ means move to the left.

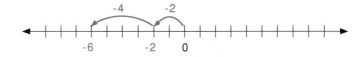

Example 6 $-6 + 8 = +2$ indicates "start at zero, move six units to the left, then move eight units to the right."

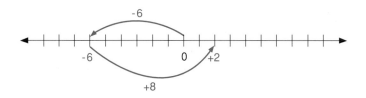

Example 7   Combine: $3 - 6 + 4 - 5 =$

Solution   Using the number line and starting at zero, first move three units to the right.

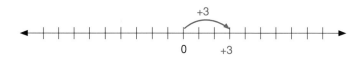

This example combines four numbers, but we can use the number line to combine as many numbers as we wish.

Then move six units to the left.

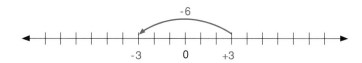

Then move four units to the right.

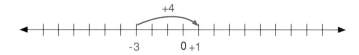

Notice that each step starts from the previous position.

Finally, move five units to the left.

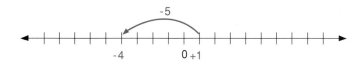

The final location is at -4. Therefore $3 - 6 + 4 - 5 = -4$.

The answer completes the number statement.

Example 8   After losing nine yards on the first down, the Browns completed a pass for a twelve-yard gain on the second down. What was the net gain or loss?

Solution   First write a number statement.

$$-9 + 12 =$$

Recall that in the previous exercise set you practiced writing number statements.

Using the number line, we obtain

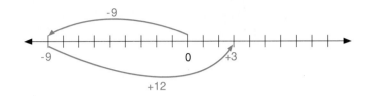

Therefore $-9 + 12 = +3$ (a three-yard gain).

▼ **EXERCISE 1-3-2**
___

Use a number line to combine each of the following.

1. $+3 + 5 =$

2. $+4 + 2 =$

3. $-3 - 4 =$

4. $-6 - 3 =$

5. $-4 + 10 =$

6. $-3 + 8 =$

7. $9 - 6 =$

8. $4 - 8 =$

9. $-2 + 7 =$

10. $-9 + 4 =$

11. $-11 + 5 =$

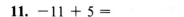

12. $-10 + 14 =$

13. $+8 - 8 =$

14. $-11 + 11 =$

15. $+13 - 20 =$

16. $9 - 15 =$

17. $-11 + 18 =$

18. $-16 + 25 =$

19. $8 - 17 =$

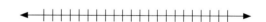

20. $+10 - 21 =$

21. $4 - 2 + 7 =$

22. $3 - 1 + 5 =$

23. $-6 + 13 - 7 =$

24. $-4 + 13 - 9 =$

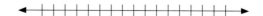

25. $-5 + 8 - 4 + 1 =$

26. $3 - 8 + 5 - 4 =$

27. $3 - 9 + 6 - 14 =$

28. $-14 + 3 - 9 + 6 =$

29. $-8 + 6 - 5 + 7 =$

30. $-5 + 6 - 8 + 7 =$

31. The stock market gained ten points in the morning, then lost six points in the afternoon. What was the net gain or loss for the day?

32. The stock market lost eight points on Thursday and gained five points on Friday. What was the net gain or loss for the two days?

33. The temperature rose nine degrees in the morning and dropped twelve degrees in the afternoon. What was the net gain or loss of temperature?

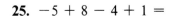

34. The temperature dropped three degrees the first hour, then rose five degrees the second hour. What was the net gain or loss of temperature?

35. A worker earned $25 in the morning and spent $18 in the afternoon. What was the net gain or loss for the day?

36. A poker player lost $46 the first hour of a game and won $38 the second hour. What was the net gain or loss at the end of the second hour?

37. A weight-watcher lost five pounds the first week, gained two pounds the second week, lost three pounds the third week, and lost two pounds the fourth week. What was the net gain or loss for the four weeks?

38. The stock market gained six points on Monday, lost eight points on Tuesday, lost three points on Wednesday, and gained nine points on Thursday. If the market was closed on Friday, what was the net gain or loss for the week?

39. The temperature in Denver was 28° C at 1:00 P.M. During the next hour, the temperature rose four degrees. During the second hour, the temperature dropped three degrees and during the third hour, it dropped another six degrees. What was the temperature at 4:00 P.M.?

40. In American football a team has four plays (downs) during which the ball must be moved ten yards forward or given up to the other team. On a series of four downs the Miami Dolphins did the following:

first down	gained five yards
second down	lost eight yards
third down	gained twelve yards
fourth down	gained two yards

Did the team gain enough yards to keep the ball? How many yards did they gain in the four downs?

1-4 RULES FOR COMBINING SIGNED NUMBERS

In the preceding section we used the meaning of a signed number and the number line to work problems combining signed numbers. We are now ready to establish rules for performing this operation. First we make the following observations from the discussion and examples of the preceding sections.

> **1.** A sign (plus or minus) affects only the number expression to its right.

Example 1 In $+6 - 7$ the plus sign $(+)$ affects only the 6 and the minus sign $(-)$ affects only the 7.

> **2.** A sign preceding parentheses affects all terms inside the parentheses.

Example 2 In $-(+6 - 8)$ the $(-)$ before the parentheses affects both the $+6$ and -8.

> **3.** We combine numbers only two at a time.

OBJECTIVES

Upon completing this section you should be able to:
1. Combine numbers having like signs.
2. Combine numbers having unlike signs.
3. Combine a series of several signed numbers.

Recall that if we removed parentheses we would have
$$-(+6 - 8) = -6 + 8.$$

In example 3 we could also first combine the $+4 + 3$ to get $+7$, then combine $+6 + 7$ to get $+13$. Remember the associative property of addition?

Example 3   Combine: $+6 + 4 + 3$

Solution   We first combine $+6 + 4$ to get $+10$, then combine $+10 + 3$ to get $+13$.

Observation 3 makes it clear that we only need rules for combining two signed numbers.

To combine two numbers with the same sign add the numbers and attach the common sign.

Example 4 $+6 + 8 = +14$. We add 6 and 8 to get 14 and then use the $+$ sign that is common to both 6 and 8.

Example 5 $-3 - 9 = -12$. We add 3 and 9 to get 12 and then attach the $-$ sign that is common to both 3 and 9.

▼ EXERCISE 1–4–1

Combine.

1. $+5 + 7 =$

2. $+12 + 8 =$

3. $+13 + 28 =$

4. $+24 + 31 =$

5. $8 + 9 =$

6. $16 + 37 =$

7. $-2 - 6 =$

8. $-9 - 5 =$

9. $-8 - 15 =$

10. $-11 - 39 =$

11. $-3 - 18 =$

12. $-1 - 14 =$

13. $12 + 49 =$

14. $6 + 25 =$

15. $-18 - 14 =$

16. $-11 - 25 =$

17. $-15 - 16 =$

18. $-19 - 34 =$

19. $17 + 18 =$

20. $24 + 96 =$

21. $+8 + 13 + 4 =$

22. $+10 + 6 + 14 =$

23. $5 + 7 + 11 =$

24. $7 + 21 + 9 =$

25. $-4 - 8 - 16 =$

26. $-3 - 1 - 12 =$

27. $-6 - 15 - 9 =$

28. $-4 - 28 - 15 =$

29. $2 + 14 + 1 =$

30. $21 + 16 + 18 =$

31. $-18 - 1 - 5 =$

32. $-5 - 1 - 6 =$

33. $-9 - 15 - 3 =$

34. $-10 - 54 - 26 =$

35. $26 + 35 + 13 =$

36. $18 + 12 + 23 =$

37. $8 + 4 + 15 + 2 =$

38. $14 + 3 + 29 + 8 =$

39. $-6 - 27 - 4 - 1 =$

40. $-6 - 21 - 9 - 4 =$

> *To combine two numbers with unlike signs* subtract the smaller number from the larger (without regard to sign) and attach the sign of the larger number.

Example 6 $-7 + 11 = +4$. We subtract 7 from 11 and use the $+$ sign because 11 is the larger of the two numbers.

We can also think of it as subtracting the absolute values of the two numbers.

Example 7 $+7 - 11 = -4$. We subtract 7 from 11 and use the $-$ sign because 11 is the larger of the two numbers (without regard to sign).

Try working these examples using the number line.

Example 8 $-14 + 8 = -6$. We subtract 8 from 14 and use the $-$ sign of 14.

▼ EXERCISE 1-4-2

Combine each of the following pairs of numbers.

1. $+8 - 3 =$

2. $+16 - 13 =$

3. $11 - 7 =$

4. $11 - 5 =$

5. $-6 + 15 =$

6. $-4 + 12 =$

7. $+14 - 20 =$

8. $9 - 21 =$

9. $6 - 8 =$

10. $5 - 11 =$

11. $14 - 28 =$

12. $12 - 31 =$

13. $+18 - 18 =$

14. $-21 + 21 =$

15. $+21 - 20 =$

16. $+15 - 16 =$

17. $+1 - 1 =$

18. $-1 + 1 =$

19. $-2 + 5 =$

20. $-13 + 4 =$

21. $+11 - 19 =$

22. $+51 - 60 =$

23. $18 - 20 =$

24. $16 - 24 =$

25. $10 - 11 =$

26. $22 - 32 =$

27. $-19 + 10 =$

28. $-12 + 10 =$

29. $-35 + 49 =$

30. $-21 + 10 =$

31. $68 - 101 =$

32. $49 - 98 =$

33. $+17 - 17 =$

34. $-43 + 19 =$

35. $-7 + 32 =$

36. $-12 + 51 =$

37. $-23 + 22 =$ **38.** $-5 + 5 =$ **39.** $-9 + 35 =$ **40.** $51 - 64 =$

Note again that rules for combining signed numbers apply to *only two* numbers at a time. If one expression contains several numbers, we must apply the rules more than once.

Example 9 Combine: $-6 + 4 + 8 - 3 =$

Solution First combine -6 and $+4$ to obtain -2.

In this example we combine the numbers from left to right.

$$\boxed{-6 + 4} + 8 - 3$$
$$= \quad -2 + 8 - 3$$

Next combine -2 and $+8$ to obtain $+6$.

$$\boxed{-2 + 8} - 3$$
$$= \quad +6 - 3$$

Then combine $+6$ and -3 to obtain the final result.

$$+6 - 3 = +3$$

Since the order of combining numbers will not change the answer, we can, in the previous example, proceed in a different way.

The commutative and associative properties of addition allow this.

Example 10 Combine: $-6 + 4 + 8 - 3 =$

Solution Combine -6 and -3 to obtain -9.

In a problem such as this many students like to combine all the negative numbers and all the positive numbers, and then combine the two resulting numbers with unlike signs.

$$\boxed{-6} + 4 + 8 \boxed{-3}$$
$$= \quad -9 + 4 + 8$$

Next combine $+4$ and $+8$ to obtain $+12$.

$$-9 \boxed{+ 4 + 8}$$
$$= -9 + 12$$

The final answer is obtained by combining -9 and $+12$.

$$-9 + 12 = +3$$

Both of these approaches should be practiced so that you can choose whichever method is easier for a particular problem.

Example 11 Combine: $+6 - 3 - 8 - 7 + 4 =$

Work this example by combining the numbers in order from left to right.

Solution At a glance we notice that $+6$ and $+4$ give us $+10$. $-3 - 7 - 8$ gives -18. $+10 - 18 = -8$. This order gives easier combinations than combining left to right. The choice of order is left to the student, since the answer will be the same either way.

▼ **EXERCISE 1-4-3**

Combine.

1. $3 - 6 + 9 =$
$12 - 6 \quad = 6$

2. $5 - 8 + 10 =$

3. $2 - 7 + 5 =$
$7 - 7 \quad = 0$

4. $5 - 12 + 1 =$

5. $-5 + 8 - 2 =$
$8 - 7 \quad = 1$

6. $-3 + 10 - 4 =$

7. $-11 + 3 - 2 =$
$3 - 13 = -10$

8. $-14 + 9 - 1 =$

9. $5 - 6 + 4 =$
$9 - 6 \quad = 3$

10. $8 - 10 + 2 =$

11. $6 - 8 - 4 =$
$-12 + 6 \quad = ^-6$

12. $9 - 7 - 6 =$

13. $-3 + 9 - 5 =$
$-8 + 9 \quad = 1$

14. $-5 + 12 - 7 =$

15. $3 - 9 + 7 =$
$-9 + 10 \quad = 1$

16. $4 - 7 + 6 =$

17. $5 + 9 - 5 =$
$14 - 5 \quad = 9$

18. $14 + 3 - 18 =$

19. $21 - 10 - 12 =$
$-22 + 21 = -1$

20. $7 - 3 - 5 =$

21. $13 - 17 + 5 - 4 =$
$-21 + 18 \quad = ^-3$

22. $15 - 20 + 6 - 8 =$

23. $-9 + 10 - 6 + 3 =$
$-15 + 13 \quad = ^-2$

24. $-4 + 5 - 12 + 6 =$

25. $-7 + 5 + 7 + 9 =$
$21 - 7 \quad = 14$

26. $-16 + 3 + 16 + 4 =$

27. $9 + 20 - 9 + 5 =$
$34 - 9 = 25$

28. $15 + 37 - 15 + 3 =$

29. $27 - 59 - 27 + 50 =$

$-59 + 50 \qquad = -9$

30. $13 - 31 - 13 + 3 =$

31. $12 - 7 + 4 - 5 =$

$16 - 12 = +4$

32. $14 - 27 - 13 - 8 =$

33. $-20 + 17 - 5 + 25 = \quad 17$

34. $-19 + 7 - 22 + 16 =$

35. $8 - 7 + 2 - 4 + 9 =$

$-11 + 19 \qquad = 8$

36. $11 - 5 + 3 - 1 + 3 =$

$17 - 6 \qquad = 11$

37. $34 - 16 + 9 + 16 - 3 =$

$43 - 3 \qquad = 40$

38. $22 - 19 + 3 - 14 + 8 =$

39. $6 - 4 + 8 + 16 - 9 - 8 =$

$30 - 21 \qquad = 9$

40. $9 - 5 + 13 + 6 - 7 - 20 =$

In section 1–2 we learned a rule for removing parentheses. We recognize that in some instances the sign of the term within the parentheses is changed when the parentheses are removed. For this reason we never combine a term outside parentheses with one inside parentheses without removing the parentheses first.

This will occur when the parentheses are preceded by a negative sign.

Example 12 Evaluate: $5 - (4 - 6)$

Solution We first remove the parentheses.

$$5 - (4 - 6) = 5 - 4 + 6$$

Recall how to remove parentheses.

Then combine obtaining

$$5 - 4 + 6 = 7.$$

Example 13 $12 - (-3) = 12 + 3 = 15$

Example 14 $7 + (-6 - 4) = 7 - 6 - 4 = -3$

Always remove parentheses before combining!

Example 15 $12 + (-3) = 12 - 3 = 9$

Example 16 $-8 - (-10 - 7) = -8 + 10 + 7 = 9$

Example 17 $10 - (-13 + 10 - 4) = 10 + 13 - 10 + 4$

$$= 17$$

▼ EXERCISE 1-4-4

Combine by removing parentheses first.

1. $8 + (-5) =$

2. $7 + (-8) =$

3. $10 - (+4) =$

4. $6 - (+11) =$

5. $3 - (-8) =$

6. $5 - (-3) =$

7. $6 + (8 - 2) =$

8. $9 + (6 - 3) =$

9. $5 + (9 - 12) =$

10. $7 + (3 - 8) =$

11. $10 - (2 + 5) =$

12. $18 - (6 + 4) =$

13. $13 - (2 - 1) =$

14. $11 - (8 - 10) =$

15. $21 - (4 - 10) =$

16. $16 - (7 - 11) =$

17. $8 + (-5 - 4) =$

18. $4 + (-6 - 2) =$

19. $4 - (-5 + 9) =$

20. $9 - (-10 + 12) =$

21. $4 + (13 - 6 - 4) =$

22. $8 + (10 - 3 - 4) =$

23. $11 + (-6 - 3 - 8) =$

24. $14 + (-8 - 4 - 7) =$

25. $8 - (4 - 3 + 2) =$

26. $9 - (5 - 4 + 7) =$

27. $-3 - (7 + 4 - 8) =$

28. $-6 - (4 - 9 + 4) =$

29. $10 - (8 + 5 - 9 + 2) =$

30. $12 - (10 - 4 + 5 - 9) =$

We have used the word "combine" to include both addition and subtraction. The reason becomes clear when we realize that the sign ($-$) is used to indicate both *subtraction* and *the negative of*. The expression $5 - (-4)$ can be thought of as

Recall the distinction between a negative number and the negative of a number.

    1. "Subtract -4 from 5."

or

    2. "Add the negative of -4 to 5."

Thus far we have combined these numbers by thinking of this in the second way. We changed $5 - (-4)$ to $5 + 4$ and obtained 9.

The following definition should make it clear that both statements are the same.

Subtraction is adding the negative. In symbols: $a - b$ (b subtracted from a) is the same as $a + (-b)$ (the negative of b added to a).

▼ WARNING

Don't forget, you can never write two signs together without using parentheses.

Example 18 $8 - (+6) = 8 + (-6) = 2$

Example 19 $3 - (+7) = 3 + (-7) = -4$

Example 20 $8 - 5 = 8 + (-5) = 3$

Notice here that $8 - 5$ means $8 - (+5)$.

Example 21 $4 - 10 = 4 + (-10) = -6$

Example 22 $6 - (-2) = 6 + 2 = 8$

Example 23 $-14 - (-9) = -14 + 9 = -5$

Example 24 Subtract -8 from 5.

    Solution This should be expressed as $5 - (-8)$. Then $5 - (-8) = 5 + 8 = 13$.

This symbol means "subtract."

$$5 \overset{\downarrow}{-} (\underset{\uparrow}{-} 8)$$

This symbol means "the negative of."

▼ EXERCISE 1-4-5

Rewrite each subtraction problem as an addition problem and determine the answer.

    1. $6 - (+4) =$ **2.** $10 - (+5) =$ **3.** $9 - (+12) =$

4. $4 - (+16) =$

5. $13 - (-2) =$

6. $18 - (-5) =$

7. $-16 - (-9) =$

8. $2 - (-20) =$

9. $22 - (-6) =$

10. $-31 - (-7) =$

11. $3 - (+5) =$

12. $15 - (+18) =$

13. $-11 - (-1) =$

14. $8 - (-8) =$

15. $24 - (+24) =$

16. $-16 - (-4) =$

17. $-25 - (-3) =$

18. $3 - (+9) =$

19. $18 - (+18) =$

20. $32 - (-32) =$

21. $8 - 3 =$

22. $9 - 4 =$

23. $8 - 21 =$

24. $10 - 39 =$

25. Subtract 12 from 8.

26. Subtract 17 from 9.

27. Subtract -5 from 7.

28. Subtract -4 from 9.

29. Subtract 14 from -9.

30. Subtract 18 from -16.

31. Subtract -6 from -13.

32. Subtract -14 from -8.

◣◥ 1–5 MULTIPLICATION AND DIVISION OF SIGNED NUMBERS

In section 1–4 we learned how to add and subtract (combine) signed numbers. We will now proceed to the other two basic operations, multiplication and division.

Multiplication can be thought of as "shortcut addition." For instance, if we wish to multiply 7×5, we can think of this as "seven fives" or $5 + 5 + 5 + 5 + 5 + 5 + 5$ and obtain the result, 35. From now on we will write $(7)(5)$ or $7(5)$ to indicate the product instead of 7×5, since we will be using letters as well as numbers and do not want to confuse the letter x with the operation of multiplication. Applying this same technique to signed numbers will lead to one of the rules for their multiplication. Think of $(7)(-5)$ as seven (-5)s or $(-5) + (-5) + (-5) + (-5) + (-5) + (-5) + (-5)$. Using our rules for combining signed numbers, we see that $(7)(-5) = -35$.

The product of a positive number and a negative number yields a negative number.

Example 1 $(-2)(5) = -10$

Example 2 $(3)(-12) = -36$

Example 3 $(-238)(147) = -34{,}986$

Example 4 $\left(\dfrac{2}{3}\right)\left(-\dfrac{3}{4}\right) = -\dfrac{1}{2}$

From arithmetic we know the following rule.

The product of two positive numbers is positive.

Be careful to note that in each of these rules the word "two" is very important. Since we can multiply only two numbers at a time, we only need rules for two numbers.

Example 5   Find the product of $(3)(-2)(7)$.

Solution   We can first multiply $(3)(-2)$ and get -6, then multiply $(-6)(7)$ and get -42.

We could instead first multiply $(3)(7)$ and get 21, then multiply $(-2)(21)$ and get -42.

We could also first multiply $(-2)(7)$ and get -14, then multiply $(3)(-14)$ and get -42.

So $(3)(-2)(7) = -42$.

OBJECTIVES

Upon completing this section you should be able to:

1. Apply the rules for multiplying signed numbers.
2. Apply the distributive property of multiplication over addition.
3. Divide signed numbers.

Remember that multiplying any number by zero always gives a product of zero.
$$(5)(0) = 0$$
$$(-4)(0) = 0$$

The word binary means two numbers. Multiplication is called a *binary* operation.

Instead of $(3)(-2)(7)$ we could write $(3)(7)(-2)$. This is because multiplication is commutative.

▼ EXERCISE 1–5–1

Find the products.

1. $(+8)(-5) =$ **2.** $(+9)(-4) =$ **3.** $(-5)(+8) =$ **4.** $(-4)(+9) =$

5. $(7)(-3) =$ **6.** $(6)(-7) =$ **7.** $(-4)(6) =$ **8.** $(-8)(0) =$

9. $14(-3) =$ **10.** $6(-11) =$ **11.** $\left(-\dfrac{1}{2}\right)(8) =$ **12.** $\left(-\dfrac{1}{3}\right)(12) =$

13. $\left(\dfrac{2}{3}\right)\left(-\dfrac{9}{10}\right) =$ **14.** $\left(\dfrac{4}{5}\right)\left(-\dfrac{3}{4}\right) =$ **15.** $(-3)(2.5) =$ **16.** $(-4)(4.2) =$

17. $(5.1)(-2.3) =$ **18.** $(4.0)(-6.2) =$ **19.** $\left(\dfrac{2}{3}\right)\left(-\dfrac{3}{8}\right) =$ **20.** $\left(-\dfrac{4}{5}\right)\left(\dfrac{15}{16}\right) =$

21. $\left(-\dfrac{1}{3}\right)(6.3) =$ **22.** $\left(-\dfrac{1}{2}\right)(4.6) =$ **23.** $(5)(-2)(4) =$ **24.** $(2)(-4)(3) =$

25. $(-3)(2)(8) =$ **26.** $(-6)(3)(10) =$ **27.** $(5)(3)(-4) =$ **28.** $(6)(2)(-1) =$

29. $(-10)\left(\dfrac{1}{2}\right)(7) =$ **30.** $(-25)(4)\left(\dfrac{1}{5}\right) =$ **31.** $(-16)(121)(0) =$ **32.** $(14)(0)(-15) =$

33. $(4)\left(-\dfrac{1}{2}\right)(3) =$ **34.** $\left(\dfrac{2}{3}\right)\left(-\dfrac{1}{4}\right)(6) =$ **35.** $\left(-\dfrac{2}{7}\right)(14)\left(2\dfrac{3}{8}\right) =$ **36.** $(1.2)(-3.1)(5.0) =$

37. $(18)(2.3)(-4) =$ **38.** $\left(-\dfrac{3}{4}\right)(18)\left(1\dfrac{5}{9}\right) =$ **39.** $(5.4)\left(-3\dfrac{4}{5}\right)(0) =$ **40.** $\left(\dfrac{1}{2}\right)\left(\dfrac{1}{4}\right)\left(-\dfrac{1}{16}\right) =$

> When an algebraic expression is composed of parts *connected by addition signs*, these parts are called the **terms** of the expression.

Example 6 $a + b$ has two terms: a and b.

$\boxed{a} + \boxed{b}$
two terms

Example 7 $2x + 5y + 3$ has three terms: $2x$, $5y$, and 3.

> When an algebraic expression is composed of parts *to be multiplied*, these parts are called the **factors** of the expression.

Example 8 ab has two factors: a and b.

$\boxed{a}\,\boxed{b}$
two factors

Example 9 $3(a + b)$ is one term having two factors.

Note that the factor $(a + b)$ has two terms, but the entire expression is made up of factors.

ab means $a \times b$. Also, $3(a + b)$ means $3 \times (a + b)$.

To estabish a rule for the product or quotient of two negative numbers we will make use of the distributive property of multiplication over addition.

$$a(b + c) = ab + ac$$

We may express this property in words by saying "if terms enclosed in parentheses are to be multiplied by a number, multiply each term in the parentheses by that number."

Notice that a is a factor. That is, it is multiplying the quantity $(b + c)$.

Example 10 $5(4 + 3) = 5(4) + 5(3) = 20 + 15 = 35$.

In this problem we could, of course, first combine the $4 + 3$ getting $5(4 + 3) = 5(7) = 35$.

When we indicate the product of a number and a letter such as $4(x)$, we usually do not use parentheses. We simply write $4x$.

Example 11 $4(x + 6) = 4(x) + 4(6) = 4x + 24.$

Example 12 $3(a + 2b + 4) = 3(a) + 3(2b) + 3(4)$
$$= 3a + 6b + 12$$

Example 13 $-5(2x + 3y + 4) = -5(2x) + (-5)(3y) + (-5)(4)$
$$= -10x - 15y - 20$$

▼ **EXERCISE 1–5–2**

Remove parentheses.

1. $5(x + 4) =$

2. $3(x + 6) =$

3. $6(a + 5) =$

4. $7(a + 1) =$

5. $3(2a + 4) =$

6. $4(3a + 5) =$

7. $2(x - 9) =$

8. $5(x - 4) =$

9. $2(3x + 4) =$

10. $8(2x + 3) =$

11. $-3(a + 7) =$

12. $-4(a + 2) =$

13. $2(x - 3) =$

14. $2(x - 5) =$

15. $-3(2x + 5) =$

16. $-6(3x + 8) =$

17. $14(3x - 2) =$

18. $13(2a - 5) =$

19. $-8(2x + 5) =$

20. $-5(3x + 4) =$

21. $4(a + b + 4) =$

22. $3(a + b + 6) =$

23. $2(x + 3y + 4) =$

24. $2(x + 4y + 3) =$

25. $3(2a + 4b + 2) =$

26. $4(2a + 5b + 11) =$

27. $5(2x - y + 3) =$

28. $3(5x - 4y - 1) =$

29. $-2(x + 3y + 11) =$

30. $-8(3x + 5y + 3) =$

31. $-4(3a + 2b + c) =$

32. $-5(2x + y + 4) =$

We now wish to establish a rule for multiplying two negative numbers.

By using our rules for adding signed numbers we have worked problems such as $-35 + 35 = 0$. We also know that multiplying by zero gives zero. For instance, $(-7)(0) = 0$.

Recall the additive inverse.

Now consider this problem.

Find the result of $(-7)[(5) + (-5)]$. If we note that $(5) + (-5) = 0$, then we have $(-7)(0) = 0$ and thus $(-7)[(5) + (-5)] = 0$. However if we use the distributive property, we get

The brackets [ ] serve the same function as parentheses.

$$(-7)[(5) + (-5)] = (-7)(5) + (-7)(-5).$$

We now have $(-35) + (-7)(-5)$ and we know the result must be zero. If $(-35) + (-7)(-5) = 0$, it follows that $(-7)(-5) = +35$ since $+35$ is the only number that can be added to -35 and give zero as the result.

Each number has only one additive inverse.

The choice of numbers in the discussion would not change the conclusion. Hence, we have the following rule.

> *The product of two negative numbers* is positive.

Do not confuse the rule for adding two negative numbers with the rule for multiplying them.

Example 14 $(-2)(-5) = 10$

Example 15 $(-3)(-15) = 45$

Example 16 $\left(-\dfrac{2}{3}\right)\left(-\dfrac{3}{4}\right) = \dfrac{1}{2}$

Once more we must realize the importance of the word *two*. This rule, as all others, must be applied to only two numbers at a time.

Example 17 Find the product of $(-3)(-2)(-5)$.

Find this product by first multiplying $(-2)(-5)$.

Solution If we first multiply $(-3)(-2)$, we get $+6$. Then we multiply $(+6)(-5)$ and get -30. So $(-3)(-2)(-5) = -30$.

Example 18 Find the product of $(-3)(4)(-5)$.

Find this product by first multiplying $(-3)(-5)$.

Solution First if we multiply $(-3)(4)$, we get -12. Now we multiply $(-12)(-5)$ and get $+60$. So $(-3)(4)(-5) = 60$.

Example 19 Find the product of $(-2)(5)(-3)(-1)(4)$.

How many factors are negative? If there are an odd number of negative factors, what sign will the product have?

$$\begin{aligned} \textit{Solution} \quad (-2)(5)(-3)(-1)(4) &= -10(-3)(-1)(4) \\ &= 30(-1)(4) \\ &= -30(4) \\ &= -120 \end{aligned}$$

Example 20 Find the product of $(-4)(-1)(-2)(3)(-3)$.

How many factors are negative? If there are an even number of negative factors, what sign will the product have?

$$\begin{aligned} \textit{Solution} \quad (-4)(-1)(-2)(3)(-3) &= 4(-2)(3)(-3) \\ &= -8(3)(-3) \\ &= -24(-3) \\ &= 72 \end{aligned}$$

◤ EXERCISE 1–5–3

Find the products.

1. $(-2)(-3) =$

2. $(-4)(-3) =$

3. $(-1)(-8) =$

4. $(-1)(-5) =$

5. $(-3)(-7) =$

6. $(-8)(-10) =$

7. $(-5)(-9) =$

8. $(-6)(-4) =$

9. $\left(-\dfrac{1}{2}\right)(-30) =$

10. $\left(-\dfrac{1}{3}\right)(-9) =$

11. $\left(-\dfrac{1}{3}\right)\left(-\dfrac{3}{5}\right) =$

12. $\left(-\dfrac{2}{3}\right)\left(-\dfrac{1}{4}\right) =$

13. $(-5)(+8) =$

14. $(-3)(+5) =$

15. $7(-3) =$

16. $9(-4) =$

17. $(-1.2)(-5.0) =$

18. $(-3.2)(-1.5) =$

19. $(6)(-2)(-5) =$

20. $(4)(-9)(-3) =$

21. $(-1)(-3)(-4) =$

22. $(-2)(-8)(5) =$

23. $(-5)(6)(-2) =$

24. $(-3)(4)(-5) =$

25. $(-1)(-5)\left(-\dfrac{1}{5}\right) =$

26. $\left(-\dfrac{1}{3}\right)(-12)\left(-\dfrac{1}{2}\right) =$

27. $(-3)(5)(0) =$

28. $(-7)(-14)(0) =$

29. $(3)(-1)(-5)(10) =$

30. $(-4)(3)(-11)(2) =$

31. $(-1)(-5)(-2)(-3) =$ **32.** $(-3)(-1)(-4)(-6) =$ **33.** $(-2)(-7)(3)(4) =$

34. $(3)(-2)(-1)(-7) =$ **35.** $(-10)(10)(-1)(-3) =$ **36.** $(-8)(4)(-2)(10) =$

37. $(-2)(-1)(-10)(4)(1) =$ **38.** $\left(-\dfrac{1}{2}\right)(-14)\left(-\dfrac{3}{7}\right)(-2) =$

39. $(27)\left(-\dfrac{1}{9}\right)\left(-\dfrac{1}{3}\right)(-37) =$ **40.** $(49)(-72)(-104)(0)(-23) =$

The multiplicative inverse of a number is often referred to as the *reciprocal* of the number.

Examples:

The multiplicative inverse of $\dfrac{2}{3}$ is $\dfrac{3}{2}$. The multiplicative inverse of 8 is $\dfrac{1}{8}$. The multiplicative inverse of $-\dfrac{3}{5}$ is $-\dfrac{5}{3}$.

Division is defined as "multiplication by the inverse." In a problem such as $12 \div 6$ we are dividing by 6. Using the definition, the problem is the same as multiplying by the multiplicative inverse of 6, which is $\dfrac{1}{6}$.

Thus
$$12 \div 6 = 12 \times \dfrac{1}{6}.$$

This relationship ties division and multiplication together so that the rules for division are the same as the rules for multiplication. For convenience, we will now restate the rules as a single rule.

Notice that the sign of a number and the sign of its multiplicative inverse are the same.

> In multiplying or dividing signed numbers *the product or quotient of two numbers with like signs* is positive, and *the product or quotient of two numbers with unlike signs* is negative.

Example 21 $-12 \div (-6) = +2$

Example 22 $-12 \div (+6) = -2$

Example 23 $+12 \div (-6) = -2$

Example 24 $\dfrac{-8}{-2} = +4$

Example 25 $\dfrac{-8}{+2} = -4$

At times it may be more convenient to write a division problem in fraction form. Thus $-12 \div (+6)$ could be written as $\dfrac{-12}{+6}$.

Example 26 $\left(-\dfrac{2}{3}\right) \div \left(-\dfrac{3}{4}\right) = \left(-\dfrac{2}{3}\right)\left(-\dfrac{4}{3}\right) = +\dfrac{8}{9}$

Example 27 $\dfrac{(-6)(-7)}{(-3)} = -14$

▼ **EXERCISE 1–5–4**

Find the quotients.

1. $\dfrac{-6}{-2} =$

2. $\dfrac{-14}{-2} =$

3. $\dfrac{-6}{+2} =$

4. $\dfrac{-14}{+2} =$

5. $\dfrac{-15}{-3} =$

6. $\dfrac{-8}{-4} =$

7. $\dfrac{+15}{-3} =$

8. $\dfrac{-18}{-3} =$

9. $\dfrac{-20}{+5} =$

10. $\dfrac{+18}{-3} =$

11. $\dfrac{-42}{-6} =$

12. $\dfrac{+24}{-6} =$

13. $(-20) \div (-5) =$

14. $(-56) \div (-7) =$

15. $(+81) \div (-3) =$

16. $(+20) \div (-4) =$

17. $(-28) \div (+4) =$

18. $(-100) \div (+20) =$

19. $\left(-\dfrac{1}{2}\right) \div \left(-\dfrac{1}{8}\right) =$

20. $\left(\dfrac{3}{4}\right) \div \left(-\dfrac{9}{16}\right) =$

21. $\left(-\dfrac{2}{5}\right) \div \left(\dfrac{3}{7}\right) =$ **22.** $\left(-\dfrac{4}{5}\right) \div \left(\dfrac{5}{8}\right) =$ **23.** $\left(\dfrac{3}{8}\right) \div \left(-\dfrac{3}{16}\right) =$ **24.** $\left(\dfrac{2}{7}\right) \div \left(-\dfrac{4}{5}\right) =$

25. $(-3) \div \left(-\dfrac{6}{7}\right) =$ **26.** $-9 \div \left(-\dfrac{3}{4}\right) =$ **27.** $\left(-\dfrac{5}{7}\right) \div 15 =$ **28.** $\left(-\dfrac{3}{8}\right) \div 12 =$

29. $\dfrac{(-5)(-3)}{(-6)} =$ **30.** $\dfrac{(-2)(-5)}{(-10)} =$ **31.** $\dfrac{(-3)(-6)}{-9} =$ **32.** $\dfrac{(-2)(-9)}{-3} =$

33. $\dfrac{(+4)(-6)}{-2} =$ **34.** $\dfrac{3(-10)}{-6} =$ **35.** $\dfrac{(-6)(+4)}{12} =$ **36.** $\dfrac{(-5)(+4)}{-10} =$

37. $\dfrac{3-15}{-6} =$ **38.** $\dfrac{5-33}{-7} =$ **39.** $\dfrac{9-27}{6} =$ **40.** $\dfrac{18-4}{-7} =$

◣◤ 1–6 POSITIVE WHOLE NUMBER EXPONENTS

O B J E C T I V E S

Upon completing this section you should be able to:

1. Evaluate expressions having exponents.
2. Apply the laws of exponents.

In this section we wish to define and give the rules for multiplying and dividing with positive whole number exponents. In a later chapter we will explore zero, negative, and rational exponents.

First we must introduce some terminology. In the expression $5x^3$, 5 is called a **numerical coefficient,** x is a **base,** and 3 (written half size and one-half space above) is called an **exponent.**

A **positive whole number exponent** indicates the number of times a base is to be used as a factor.

5 is used as a factor three times.

Example 1 $5^3 = (5)(5)(5) = 125$

3 is used as a factor twice.

Example 2 $3^2 = (3)(3) = 9$

How many times is 2 used as a factor?

Example 3 $2^5 = (2)(2)(2)(2)(2) = 32$

> If x is a real number and n is a positive whole number, then
>
> $$x^n = \underbrace{(x)(x)(x) \cdots (x)}_{n \text{ times}}$$

x is used as a factor n times.

Care must always be taken to identify the base in an expression involving exponents because the exponent never affects anything except the base.

An exponent is sometimes referred to as a *power*. For example, 5^3 could be read as "five to the third power."

Example 4   In the expression ax^3, a is the coefficient, x is the base, and 3 is the exponent.

$$ax^3 \text{ means } (a)(x)(x)(x).$$

Example 5   In the expression $(ax)^3$, 1 is the coefficient (understood), ax is the base, and 3 is the exponent.

Carefully note the difference in the bases in examples 4 and 5.

$$(ax)^3 \text{ means } (ax)(ax)(ax).$$

When we write a literal number such as x, it will be understood that the coefficient is 1 and the exponent is 1. This can be very important in many operations.

$$x \text{ means } 1x^1$$

◥ EXERCISE 1–6–1

Evaluate.

1. 3^2

2. 5^2

3. 2^3

4. 7^3

5. 5^4

6. 2^4

7. 3^5

8. 4^5

9. $\left(\dfrac{1}{2}\right)^3$

10. $\left(\dfrac{1}{3}\right)^4$

11. $2(5)^2$

12. $[2(5)]^2$

13. $[2(3)]^5$

14. $2(3)^5$

15. $5(3)^4$

16. $[5(3)]^4$

17. $10 + 5^2$ **18.** $3^4 - 2^5$ **19.** $3(5)^2 + 2(5) - 8$ **20.** $35 + 3^2 - 5(2)^3$

The laws of positive whole number exponents are derived directly from the definition.

This means three factors of x times two factors of x.

Example 6   Find the product of x^3x^2.

Solution   With nothing but the definition of an exponent we would approach this problem by noting that

$$x^3 = (x)(x)(x) \text{ and } x^2 = (x)(x), \text{ so}$$
$$x^3x^2 = [(x)(x)(x)][(x)(x)] = x^5.$$

This and similar examples would lead us to the following law.

To multiply like bases add the exponents.

> **First Law of Exponents** $x^a x^b = x^{a+b}$

Be careful to note that the bases must be identical before this law can be used.

Example 7 $x^4x^2 = x^6$

Example 8 $y^3y^4 = y^7$

Example 9 $a^2a^5 = a^7$

The bases are not the same so we leave it as is.

Example 10 $x^3y^4 = x^3y^4$ (Rule does not apply.)

Recall $x = x^1$.

Example 11 $x^2x = x^3$

Example 12 $x^4xx^2 = x^7$

If an expression contains the product of different bases, we apply the law to those bases that are alike.

We can rewrite this as $x^4x^3y^2y^4$ by the commutative property.

Example 13 $x^4y^2x^3y^4 = x^7y^6$

Example 14 $x^2y^2z^2x^4y = x^6y^3z^2$

As we multiply expressions involving exponents, we cannot forget the rules for operations on signed numbers.

Example 15 $(2x^5)(5x^2) = (2)(5)(x^5)(x^2) = 10x^7$

Note carefully the distinction between coefficient and base.

Example 16 $(-3x^4)(4x^2) = (-3)(4)(x^4)(x^2) = -12x^6$

Example 17 $(-5x^3)(-2y^2) = (-5)(-2)(x^3)(y^2) = 10x^3y^2$

> Notice that the coefficients are multiplied even though the bases are not alike and thus the exponents cannot be added.

Example 18 $(2x^2y)(-6xy^3) = -12x^3y^4$

$$(2)(-6) = -12$$
$$(x^2)(x) = x^3$$
$$(y)(y^3) = y^4$$

▼ **EXERCISE 1-6-2**

Apply the first law of exponents to simplify.

1. x^2x^5 x^7 **2.** a^3a^4 **3.** y^3y^8 y^{11}

4. b^4b^5 **5.** aa^4 A^5 **6.** c^5c

7. a^3b^4 $= a^3b^4$ **8.** x^2y **9.** $x^4xy^3x^8$ $x^{12} xy^3$ $x^{13}y^3$

10. $x^3y^5x^2y^4$ **11.** $a^2bc^4ab^2a^3$ **12.** $x^5yz^2xy^4z$

                       $a^5 bc^4 ab^2$ $a^6b^3c^4$

13. $(4x^3)(3x^5)$ **14.** $(2b^3)(5b^2)$ **15.** $(-3x^2)(7x^3)$

     $12x^8$ $-21x^5$

16. $(-4a^3)(3a^5)$ **17.** $(11a^3)(-6a^7)$ **18.** $(9x^2)(-5x)$

                         $-66a^{10}$

19. $(-8a^2)(-4b^3)$ A^2 b^7 **20.** $(-3x^4)(-x^3)$ **21.** $(7xy)(5xy)$

     $32a^2b^3 =$ $35x^2y^2$

$(-8(2)^2) = -32 \cdot \dfrac{(-4(7)^3)}{343} =$

$[-32 \cdot -1372]$ -1372 (43904)

22. $(4ab)(6ab)$

23. $(-2x^2y)(3xy^3)$

$-6x^3y^4$

24. $(3a^3b)(7ab^5)$

25. $(6a^2b)(5a^3c)$

$30a^5bc$

26. $(8xy^4)(6y^3z^3)$

27. $(2x^2)(-5x^5)(3x^4)$

$(-10x^7)(3x^4) = -30x^{11}$

28. $(3a^4)(-4a^3)(a^5)$

29. $(-7a^2b^2)(-2ab)(3ab)$

$(+14a^3b^3)(3ab)$

$+42a^4b^4$

30. $(-5x^3y)(3xy^2)(-2xy)$

31. $(-6xy)(-2x^2)(-4y^5)$

$(12x^3y)(-4y^5) = -48x^3y^6$

32. $(-2a^2b)(-a^3b)(-3b^4)$

33. $(11ab)(3b^2c)(-5a^2c^3)$

$(33Ab^3c)(-5A^2c^3)$

$-165a^3b^3c^4$

34. $(6xy^2)(2y^3z)(-4x^3z^2)$

35. $(-3x^2y^3)(-8xy^5z^2)(5yz)$

$(24x^3y^8z^2)(5yz)$

$120x^3y^9z^3$

36. $(-2a^3b)(-4ab^2c^2)(3c^4)$

37. $(6x^2)(-5y^2)(2z)$

$(-30x^2y^2)(2z)$

$-60x^2y^2z$

38. $(-3a^3)(-2b^2)(5c^2)$

39. $(-2a^2b)(-5b^2c^2)(6c^3d^2)$

$(10A^2b^3c^2)(6c^3d^2)$

$60A^2b^3c^5d^2$

40. $(5x^4z)(-x^5y^4z^2)(-9yz^4)$

CLASS

$(-5x^9y^4z^3)(-9yz^4)$

$45x^9y^5z^7$

We now wish to establish a second law of exponents. Note in the following example how this law is derived from the definition of an exponent and the first law of exponents.

This says "raise x^2 to the third power."

Example 19 Simplify $(x^2)^3$.

Solution $(x^2)^3 = (x^2)(x^2)(x^2)$ by the meaning of the exponent 3.

Now by the first law of exponents we have

We add exponents.

$$(x^2)(x^2)(x^2) = x^6$$
so
$$(x^2)^3 = x^6.$$

This leads us to the following law.

Second Law of Exponents $(x^a)^b = x^{ab}$	To raise a power of a base to a power multiply the exponents.

Example 20 $(x^4)^2 = x^{(4)(2)} = x^8$

Example 21 $(x^3y^2)^4 = x^{(3)(4)}y^{(2)(4)} = x^{12}y^8$

Note that when factors are grouped in parentheses, each factor is affected by the exponent.

Example 22 $(2x^3y^8)^3 = (2)^3(x^3)^3(y^8)^3$
$$= 8x^9y^{24}$$

Again, distinguish between the base and the coefficient.

Example 23 $(3xy^3)^2 = 9x^2y^6$

▼ EXERCISE 1-6-3

Simplify the following using the second law of exponents.

1. $(x^3)^2$ x^6

2. $(a^2)^2$ A^4

3. $(a^4)^3$ a^{12}

4. $(x^3)^5$ x^{15}

5. $(x^2)^4$ x^8

6. $(a^3)^6$ a^{18}

7. $(x^3)^3$ x^9

8. $(x^4)^8$ x^{32}

9. $(w^4)^4$ w^{16}

10. $(z^{12})^3$ z^{36}

11. $(x^2y^3)^3$ x^6y^9

12. $(a^4b^3)^2$ a^8b^6

13. $(a^4b)^5$ $a^{20}b^5$

14. $(xy^5)^3$

15. $(x^3y^6)^2$ x^6y^{12}

16. $(a^2b^4)^4$

17. $(xy^2z^3)^5$ $x^5y^{10}z^{15}$

18. $(x^8yz^3)^5$

19. $(a^3b^5c^4)^9$ $a^{27}b^{45}c^{36}$

20. $(x^2y^9z^3)^8$

21. $(-2x^3y^4)^3$ $-8x^9y^{12}$

22. $(3a^2b^4)^3$ $27a^6b^{12}$ class

23. $(5ab^3)^4$ $625a^4b^{12}$

24. $(-2x^3y^4)^5$

25. $(2x^3y^5)^5$

$(2x^3y^5)^5$
$32x^{15}y^{25}$

26. $(4x^2y^9)^3$

27. $(3a^2bc^4)^4$

$81A^8b^4c^{16}$

28. $(2a^2b^5c)^6$

29. $(6x^2y^8z^3)^2$

$36x^4y^{16}z^6$

30. $(2a^3b^2c^8)^7$

In the following examples we will again use the definition of an exponent.

Example 24 Simplify: $\dfrac{x^5}{x^3}$

Solution $\dfrac{x^5}{x^3} = \dfrac{(x)(x)(x)(x)(x)}{(x)(x)(x)}$

$= \left(\dfrac{x}{x}\right)\left(\dfrac{x}{x}\right)\left(\dfrac{x}{x}\right)(x)(x)$

$= (1)(1)(1)(x)(x)$

$= x^2$

Any nonzero real number divided by itself is 1, so $\dfrac{x}{x} = 1$.

In such an example we do not have to separate the $\left(\dfrac{x}{x}\right)$ quantities if we remember that any nonzero quantity divided by itself is equal to 1. In example 26 we could write

This is sometimes referred to as "cancelling."

$$\frac{x^5}{x^3} = \frac{\overset{1}{(\cancel{x})}\overset{1}{(\cancel{x})}\overset{1}{(\cancel{x})}(x)(x)}{\underset{1}{(\cancel{x})}\underset{1}{(\cancel{x})}\underset{1}{(\cancel{x})}} = x^2.$$

Example 25 $\dfrac{x^2}{x^6} = \dfrac{\overset{1}{(\cancel{x})}\overset{1}{(\cancel{x})}}{\underset{1}{(\cancel{x})}\underset{1}{(\cancel{x})}(x)(x)(x)(x)} = \dfrac{1}{x^4}$

These examples lead us to another law.

When we later introduce negative exponents, we will only need the first part of this law.

> **Third Law of Exponents** $\dfrac{x^a}{x^b} = x^{a-b}$ if a is larger than b.
>
> $\dfrac{x^a}{x^b} = \dfrac{1}{x^{b-a}}$ if a is smaller than b.

Example 26 $\dfrac{x^7}{x^3} = x^{7-3} = x^4$

Here a is larger than b.

Example 27 $\dfrac{x^4}{x^{12}} = \dfrac{1}{x^{12-4}} = \dfrac{1}{x^8}$

Here a is smaller than b.

We also need to be aware of the laws for signed numbers when working with coefficients.

Example 28 $\dfrac{12x^3}{6x^2} = 2x$

Reduce this type of fraction in two steps: 1. Reduce the coefficients. 2. Use the third law of exponents.

Example 29 $\dfrac{-3x^4}{18x^6} = -\dfrac{1}{6x^2}$

Example 30 $\dfrac{5x^4}{12x^{10}} = \dfrac{5}{12x^6}$

Notice in example 32 that the fraction $\dfrac{5}{12}$ will not reduce, but the exponents can be subtracted because the bases are the same.

We can only use the third law of exponents here.

Example 31 $\dfrac{6x^3}{15y^2} = \dfrac{2x^3}{5y^2}$

Notice in example 33 that $\dfrac{6}{15}$ is reduced to $\dfrac{2}{5}$ even though the bases are not alike and thus the exponents cannot be subtracted.

Always check the bases carefully.

Example 32 $\dfrac{6xy^2}{-15x^3y} = -\dfrac{2y}{5x^2}$

▼ **EXERCISE 1-6-4**

Simplify the following using the third law of exponents.

1. $\dfrac{x^4}{x^3}$ ✕

2. $\dfrac{a^5}{a^2}$

3. $\dfrac{a^2}{a^7}$ $\dfrac{1}{a5}$

4. $\dfrac{x^4}{x^9}$

5. $\dfrac{x^4y^4}{x^2y^8}$ $\dfrac{x^2}{y^4}$

6. $\dfrac{a^2b^9}{a^5b^{12}}$ $\dfrac{1}{a^3b^3}$

7. $\dfrac{x^4yz^3}{x^2y^4z}$ $\dfrac{x^2z^2}{y^3}$

8. $\dfrac{x^2y^9z^4}{xy^{10}z^7}$

9. $\dfrac{x^2yz^3}{xz^5}$ $\dfrac{xy}{z^2}$

10. $\dfrac{a^4b^2c^5}{a^3bc^4}$

11. $\dfrac{4x^5}{2x^3}$ $2x^2$

12. $\dfrac{6a^8}{2a^6}$

13. $\dfrac{10x^9}{5x^3}$ $2x^6$

14. $\dfrac{8a^{10}}{2a^5}$

15. $\dfrac{4b^7}{12b^2}$ $\dfrac{b^5}{3}$

16. $\dfrac{18x^4}{6x^8}$

17. $\dfrac{9a^2}{15a^{10}}$ $\dfrac{3}{5a^8}$

18. $\dfrac{18w^{16}}{32w^4}$

19. $\dfrac{10x^3y^2}{2x^2y}$ $5xy$

20. $\dfrac{50a^3b^4}{15a^6b^8}$

21. $\dfrac{-8x^3}{2x^5}$ $\dfrac{-4}{x^2}$

$-\dfrac{4}{x^2}$

22. $\dfrac{10a^2}{-5a^3}$

23. $\dfrac{-4a^3b}{-12ab^4}$ $\dfrac{a^2}{3b^3}$

24. $\dfrac{-3w^2z^3}{15wz^9}$

25. $\dfrac{18x^3y^5}{24xy^7}$ $\dfrac{3x^2}{4y^2}$

26. $\dfrac{34a^3b^4}{17a^3b}$

27. $\dfrac{-a^3}{-a^4}$ $\dfrac{1}{a}$

28. $\dfrac{-14x^5}{2x^3}$

29. $\dfrac{x^{14}}{4x^7}$ $\dfrac{x^7}{4}$

30. $\dfrac{-15x^4}{-3xy}$

31. $\dfrac{21x^2y^7}{-6x^5y^3}$ $-\dfrac{7y^4}{2x^3}$

32. $\dfrac{12x^5y^6}{-4x^2y}$

33. $\dfrac{-36x^3y^5}{9x^6y}$ $-\dfrac{4y^4}{x^3}$

34. $\dfrac{-3x^4y^3}{27x^7y^5}$ $-\dfrac{1}{9x^3y^5}$

35. $\dfrac{-x^2y^2}{-x^2y^2}$ $=1$

36. $\dfrac{12ab}{-24a^3b^2}$ $-\dfrac{1}{2a^2b}$

37. $\dfrac{-15xy}{21yz}$ $-\dfrac{5x}{7z}$

38. $\dfrac{7a^2bc}{84ab^2c}$

39. $\dfrac{27w^3x^2y^4}{-36w^2y^4}$ $-\dfrac{3wx^2}{4}$

40. $\dfrac{-132a^3bc^4}{77a^5c}$

◣ 1-7 COMBINING LIKE TERMS

An expression in which letters are used to represent numbers is called a **literal expression,** or sometimes an **algebraic expression.**

It is important that we learn to operate on such expressions and here we wish to concern ourselves with adding and subtracting them.

First we must remember that **factors** are indicated products and **terms** are indicated sums.

Example 1 $3xy$ has three factors: 3, x, and y.

Example 2 $3x + y$ has no factors (except itself and 1) but has two terms: $3x$ and y. Notice that an individual term such as $3x$ may have factors but the entire expression is made up of terms.

Example 3   Name the terms in $x + y - 3xy$.

   Answer   The terms are x, y, and $-3xy$.

Example 4   Name the factors in xyz.

   Answer   The factors are x, y, and z. We could say 1, x, y, and z because 1 is a factor of every expression.

OBJECTIVES

▼

Upon completing this section you should be able to:

1. Identify like terms.
2. Combine like terms.

Factors are multiplied together.

Terms are added together.

 $x + y - 3xy = x + y + (-3xy)$

Example 5 Name the factors in $3(x + y)$.

The factor $(x + y)$ has two terms.

Answer The factors are 3 and $(x + y)$. Note that the entire expression is a product.

In an expression such as $3xy$ the number 3 is a numerical factor or coefficient and the x and y are literal factors.

The word *identical* means *exactly* the same.

> Two terms are **like terms** if their literal factors are identical.

Example 6 $3x$ and $5x$ are like terms.

Note that having the same letters is not enough. They must have the same exponents as well.

Example 7 $3x^2$ and $5x$ are *not* like terms.

Example 8 $3xy$ and $4xy$ are like terms.

Example 9 $2x^2y$ and $3xy$ are *not* like terms.

It is very important to note that like terms are absolutely identical except for the coefficients. Only like terms can be combined.

> *To combine like terms* combine the numerical coefficients and use this result as the coefficient of the common literal factors.

2 and 3 are the coefficients.

Example 10 $2xy + 3xy = (2 + 3)xy = 5xy$

Example 11 $7x + 2y - 3x + 4y = 4x + 6y$

Example 12 $3x^2y + 2xy - x^2y = 2x^2y + 2xy$

An expression should always be left with the fewest possible terms. This is one meaning of the word *simplify*.

Example 13 Simplify: $2x + 3y - 5x^2 + 4x + 12x^2$

The commutative property of addition allows this.

Answer The simplified form would be $7x^2 + 6x + 3y$. The correctness of this answer does not depend on the order in which the terms are written.

Example 14 Simplify: $5xy - 2x - 3y + 2xy - 3x - 7y$

Answer The simplified form is $7xy - 5x - 10y$.

▼ **EXERCISE 1-7-1**

Simplify by combining like terms.

1. $4x + 7x =$ $11x$

2. $5a + 8a =$ $13a$

3. $9x - 5x =$ $4x$

4. $11a - 6a =$ $5a$

5. $3a + a =$ $4A$

6. $x + 5x =$ $6x$

7. $5x^2 + 3x^2 =$ $8x^2$

8. $3x^2 + x^2 =$ $4x^2$

9. $7a^3 - a^3 =$ $6a^3$

10. $10a^3 - 5a^3 =$ $5a^3$

11. $4xy + 12xy =$ $16xy$

12. $12ab - 5ab =$ $7ab$

13. $8xy - 7xy =$ xy

14. $3a^2b + 8a^2b =$ $11a^2b$

15. $9a + 6b - 5a =$ $4a + 6b$

16. $5x - 4y + 6x =$ $11x - 4y$

17. $6ab^2 + 9a^2b + 5ab^2 =$ $11ab^2 + 9a^2b$

18. $13xy^2 - 7x^2y - 5xy^2 =$ $8xy^2 - 7x^2y$

19. $14abc + 3ab + 8abc - 2ab =$ $22abc + ab$

20. $4xy + 9x^2y + 5xy - 3x^2y =$ $9xy + 6x^2y$

21. $5ab + 11ac - 4ab - 10ac =$

22. $6x^2y + 21xy^2 + 4x^2y - 17xy^2 =$

23. $13a^2 + 14a - 11a^2 - 6a + a^2 =$

24. $8xyz + 3xz - 5xyz - 2xz - 3xyz =$

25. $14x^3y + 3x^3y^2 - 9x^3y + 5x^3y^2 - 5x^3y =$

26. $5a^2b + 10ab - 4a^2b - 4ab + 2ab^2 =$

27. $4x^2y - 5xy^2 + 15x^2y^2 + 3x^2y - 6x^2y^2 =$

28. $15ab^2 + 9a^2b - 5ab + 4a^2b - 6ab^2 =$

29. $20m^2n + 18n^2m - 17nm^2 - 12mn^2 =$

30. $7xyz - 2yxz - 3zxy - zyx - xzy =$

 1-8 PRODUCT OF A MONOMIAL AND OTHER POLYNOMIALS

O B J E C T I V E S

Upon completing this section you should be able to:

1. Recognize polynomials.
2. Identify monomials, binomials, and trinomials.
3. Find the product of a monomial and a polynomial.

In the previous section we defined literal expressions and how to combine like terms. In this section we introduce a special type of literal expression known as a polynomial.

> A literal expression having one or more terms that contain only nonnegative whole number exponents is called a **polynomial.**

Special names are given to polynomials with one, two, or three terms since these are so commonly used. A polynomial with one term is called a **monomial.** A polynomial with two terms is called a **binomial.** A polynomial with three terms is called a **trinomial.**

This is a single term.

Example 1 $5xy$ is a monomial.

Two terms.

Example 2 $5x + y$ is a binomial.

One term again.

Example 3 $3x^2y^4z$ is a monomial.

Three terms.

Example 4 $3x + 2y - 4z$ is a trinomial.

In a later chapter we will introduce negative and fractional exponents. At this point we simply wish to state that expressions containing terms such as x^{-3}, $\sqrt{x}$, or $x^{2/3}$ are *not* polynomials.

The distributive property, $a(b + c) = ab + ac$, gives us a way of changing from a product of factors, a and $(b + c)$, to a sum of terms, ab and ac. The distributive property can be generalized to include multiplying a monomial by another polynomial having any number of terms.

$$a(b + c + d + \cdots) = ab + ac + ad + \cdots$$

See section 1–6.

We have already multiplied monomials by monomials. We now state a rule for multiplying a monomial by any other polynomial.

> To multiply a monomial and another polynomial multiply each term of the other polynomial by the monomial.

Example 5   Multiply: $3x(5x + 2)$

Solution $3x(5x + 2) = 3x(5x) + 3x(2)$
$$= 15x^2 + 6x$$

$3x$ is the monomial.

Example 6   Multiply: $4x^2(2x - 3y + 2x^4)$

Solution $4x^2(2x - 3y + 2x^4) = 4x^2(2x) - 4x^2(3y) + 4x^2(2x^4)$
$$= 8x^3 - 12x^2y + 8x^6$$

Notice how the laws of exponents and the rules for signed numbers are used here.

Example 7   Multiply: $(3x + 4y^2)3xy$

Solution $(3x + 4y^2)3xy = 3xy(3x) + 3xy(4y^2)$
$$= 9x^2y + 12xy^3$$

You may instead write $3xy(3x + 4y^2)$. The commutative property allows this.

Example 8   Multiply: $-3x(x - y)$

Solution $-3x(x - y) = -3x(x) + 3x(y)$
$$= -3x^2 + 3xy$$

This answer could also be written as $3xy - 3x^2$.

We learned earlier to remove parentheses preceded by a negative sign by taking the negative of each term within the parentheses. Notice that if the number 1 is understood to precede the parentheses, then the same result would be accomplished.

We have said before that 1 is always a factor.

Example 9   Remove parentheses: $-(2x - y + 3z)$

Solution $-(2x - y + 3z) = -1(2x - y + 3z)$
$$= -1(2x) + 1(y) - 1(3z)$$
$$= -2x + y - 3z$$

Could we write this as $y - 2x - 3z$?

Using the distributive property to clear parentheses is often needed to simplify an expression.

Example 10   Simplify: $2(3x + 2y) + x$

Solution   Note that as written, this expression has two terms, $2(3x + 2y)$ and x, and since these are not like terms they cannot be combined. However, if we remove parentheses, we have

Only like terms can be combined.

$$2(3x + 2y) + x = 6x + 4y + x$$
$$= 7x + 4y.$$

$6x$ and x are like terms.

Example 11   Simplify: $2x(3x + 4) - x(2x - 1)$

Solution $2x(3x + 4) - x(2x - 1) = 2x(3x) + 2x(4) - x(2x) + x(1)$
$$= 6x^2 + 8x - 2x^2 + x$$
$$= 4x^2 + 9x$$

Combine $6x^2 - 2x^2$ and also $8x + x$.

▼ **EXERCISE 1-8-1**

Find the products and simplify where possible.

1. $3(x + 2y)$ **2.** $5(a - b)$

3. $-2(2x + y)$ **4.** $-4(2a + 3b)$

5. $x(x + y)$ **6.** $a(a + b)$

7. $3x(2x + 5y)$ **8.** $2a(3a - 4b)$

9. $-(2x - 7y)$ **10.** $-(4x + y)$

11. $(x - 5y)2x$ **12.** $4a(2a - 3b)$

13. $5(2x + y - 3z)$ **14.** $8(2x - 3y + 2z)$

15. $-6x(x - 3y)$ **16.** $(2x + 5y)(-4x)$

17. $-x(2x - 4y)$ **18.** $-a(5a - 2b)$

19. $(3x - 2y - z)(-2x)$ **20.** $-7a(3a - 2b + 5c)$

21. $-(3x + 5y - z)$ **22.** $-(a - 4b + 3)$

23. $3xy(2x - 3y - 5z)$

24. $(3a + 4b - c)(2ac)$

25. $-7y(x - 2y - 3z)$

26. $-8x(2x - y + 4)$

27. $2x(-x^2 + 3x + 1)$

28. $(-a^2 + 4b - c)5ab$

29. $(2 + 5x - 4x^2)3x$

30. $2xz(3x + 5z^2)$

31. $-5y(-4y^2 - 2y + 1)$

32. $-3a(-a^2 + 2ab - 4c)$

33. $-7xy(2x^2y^3 + 3x^2y - 5xy^2)$

34. $5ac(4a^2 + 3bc - 5c)$

35. $2xy(3x^2y - 2xy^2 + xy)$

36. $-4bc(2ab - b^2c + 5ab^2)$

37. $2xyz(x^2y - 5y^2z + 9xz)$

38. $2abc(3a^2b + 5b^2c^2 - bc^3)$

39. $-xyz(x^2y^2z - 10xz + 3y^2z^3)$

40. $-3a^2c(a^3bc - 3ab^3 + 8a^2bc)$

41. $3(x + y) - 2x$

42. $6(2x - 1) + 3x$

43. $7(x + 2) + 3(x - 4)$

44. $9(x + 3) - 4(x - 2)$

45. $2x(x - 3) + 4x(2x + 1)$

46. $5x(2x + 4) - 6x(x - 3)$

47. $3x(2x^2 + x) - 2x(2x^2 + 4x) + 5x^2$

48. $4x(3x^2 - 8x) + 2x(4x^2 + 5x) - 6x^2$

49. Express the area of this rectangle as a polynomial.

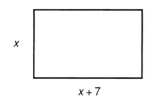

x

$x + 7$

50. Express the area of this rectangle as a polynomial.

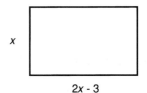

x

$2x - 3$

 **1–9 ORDER OF OPERATIONS AND GROUPING SYMBOLS**

OBJECTIVES

Upon completing this section you should be able to:

1. Perform operations in a specific order when grouping symbols are not used.
2. Perform operations in the correct order indicated by grouping symbols.

Scientific calculators have this "logic" built into them.

In the previous sections of this chapter we have used parentheses as grouping symbols and have used the distributive property to remove parentheses. We now need to look at two situations not discussed thus far. First, how do we handle the situation when there are no grouping symbols and operations other than addition are involved? Second, how do we handle the situation when we have grouping symbols within grouping symbols?

Consider the number expression $2 + 3 \times 5$. Notice that no grouping symbols are present in the expression. Should the answer be $2 + 3 \times 5 = 5 \times 5 = 25$? Or should it be $2 + 3 \times 5 = 2 + 15 = 17$? Obviously if we had no rule or convention to follow, we could claim either answer correct. For this reason mathematicians have agreed to the following rule.

> *If no grouping symbols occur in an expression to be evaluated*, the operations follow this order: First exponents; next multiplication and division as they occur left to right; then addition and subtraction as they occur left to right.

Try this problem on a scientific calculator.

This rule gives the answer to the previous expression as $2 + 3 \times 5 = 2 + 15 = 17$.

Example 1   Evaluate: $2 \times 5^2 + 3$

Solution   We must first evaluate 5^2, then multiply by 2, and then add 3.

$$\begin{aligned} 2 \times 5^2 + 3 &= 2 \times 25 + 3 \\ &= 50 + 3 \\ &= 53 \end{aligned}$$

First exponents; then multiplication; then addition.

Example 2   Evaluate: $6^2 \div 2 \times 3 + 6$

Solution $\begin{aligned} 6^2 \div 2 \times 3 + 6 &= 36 \div 2 \times 3 + 6 \\ &= 18 \times 3 + 6 \\ &= 54 + 6 \\ &= 60 \end{aligned}$

Notice that division is performed before multiplication because it is to the left of multiplication.

Example 3   Evaluate: $5 + 3 \times 10 \div 2 + 1$

Solution $\begin{aligned} 5 + 3 \times 10 \div 2 + 1 &= 5 + 30 \div 2 + 1 \\ &= 5 + 15 + 1 \\ &= 21 \end{aligned}$

◢ **EXERCISE 1-9-1**

Evaluate the following expressions.

1. $5 - 3 + 7 =$

2. $8 - 4 + 6 =$

3. $2 \times 4 - 6 =$

4. $3 \times 5 - 11 =$

5. $3 + 4 \times 2 =$

6. $4 + 3 \times 7 =$

7. $8 \div 2 + 4 =$

8. $9 \div 3 + 6 =$

9. $4 + 6 \div 2 =$

10. $12 - 8 \div 4 =$

11. $4 \times 3 + 6 =$

12. $8 \times 3 + 5 =$

13. $12 + 6 \times 2 =$

14. $15 - 2 \times 4 =$

15. $7 + 3 \times 5 - 2 =$

16. $3 + 4 \times 3 + 2 =$

17. $3 \times 2^3 + 5 =$

18. $4 \times 3^2 - 6 =$

19. $15 + 10 \div 5 - 3 =$

20. $6 + 21 \div 3 - 5 =$

21. $4^2 \div 2 + 4 \times 3 =$

22. $2^6 \div 4 - 3 \times 2 =$

23. $4 \div 2 + 8 \times 3 =$

24. $40 \div 2 - 5 \times 4 =$

25. $16 \div 2^2 \times 3 - 5 =$

26. $32 \div 4^2 \times 2 - 21 =$

27. $3 \times 5 - 9 \div 3 + 5 =$

28. $12 \div 2 - 2 \times 3 + 1 =$

29. $6 + 4 \times 5 - 3 =$

30. $3^2 + 2 \times 7 + 5 =$

31. $4 \times 6^2 \div 9 - 25 =$

32. $3 \times 6^2 \div 4 + 17 =$

33. $2 \times 5 + 7 \times 3 - 5 \times 2 =$

34. $2 \times 8 \div 4 \times 3 \div 2 - 6 =$ **35.** $3 + 2 \times 8 \div 4 - 5 =$ **36.** $4 + 27 \div 3 \times 2 - 6 =$

37. $7^2 - 3 \times 2^3 - 5 =$ **38.** $5^3 - 3^2 \times 2 + 6 =$ **39.** $6 \times 5 \div 3 + 12 =$

40. $24 \div 6 \times 2 - 8 =$

In evaluating a number expression we have already seen that parentheses group numbers together as a single element. We found, for instance, that $-(a + b + c)$ means the negative of each term within the parentheses and that $2(a + b + c)$ means to multiply each term in the parentheses by 2. Let us now consider simplifying this expression.

$$2[3x + 5(2x + 1)]$$

First we must pause to note that brackets [ ], braces { }, and parentheses ( ) are all grouping symbols and all have the same function. The different symbols are used simply for ease of determining where an expression starts and stops. In the expression $2[3x + 5(2x + 1)]$ we know that all terms within the brackets must be multiplied by 2. We note that the terms within the brackets are $3x$ and $5(2x + 1)$. So we could simplify by multiplying these by 2 and then simplifying further. However, to avoid confusion a good (but not necessary) rule to follow is:

Recall that $-(a + b + c)$ can also be thought of as $-1(a + b + c)$.

It would be helpful to pause here and learn the names of these symbols.

Errors can easily be made doing this.

When simplifying an expression that has grouping symbols within grouping symbols, remove the *innermost* set of symbols first.

It is important to perform only one step at a time.

Applying this rule will always avoid confusion and will eliminate many careless errors. In the expression $2[3x + 5(2x + 1)]$ we first simplify $5(2x + 1)$, obtaining $2[3x + 10x + 5]$. Now multiplying each term within the brackets by 2, we get $6x + 20x + 10$ or $26x + 10$ as the simplified expression.

Example 4 Simplify: $5 - 2[3x - 4(2x + 3)]$

Remove parentheses.

Remove brackets.

Combine like terms.

Solution $5 - 2[3x - 4(2x + 3)] = 5 - 2[3x - 8x - 12]$
$$= 5 - 6x + 16x + 24$$
$$= 10x + 29$$

It is suggested that only one operation be performed in each step. Combining steps may shorten the work, but quick wrong answers are not an advantage. Also, like terms can be combined within the same grouping symbol, but the authors have found that fewer mistakes are made if all grouping symbols are removed *before* any terms are combined.

Example 5 Simplify: $3x\{x + 4[2x - (x^2 - 3)]\} + 5x^2$

Remove parentheses.

Remove brackets.

Remove braces.

Combine like terms.

Solution $3x\{x + 4[2x - (x^2 - 3)]\} + 5x^2$
$$= 3x\{x + 4[2x - x^2 + 3]\} + 5x^2$$
$$= 3x\{x + 8x - 4x^2 + 12\} + 5x^2$$
$$= 3x^2 + 24x^2 - 12x^3 + 36x + 5x^2$$
$$= 32x^2 - 12x^3 + 36x$$

The order in which the final answer is written does not affect its correctness.

▼ **EXERCISE 1–9–2**

Simplify.

1. $4 + [9 - (3 + 4)] =$

2. $5 + [10 - (4 + 3)] =$

3. $6 + 2[8 + 3(6 - 3)] =$

4. $19 - 3[18 - 4(7 - 4)] =$

5. $2[2x + 3(x + 4)] =$

6. $3[2x + 5(x - 3)] =$

7. $3[4x - 2(3x - 1)] =$

8. $4[3x - 2(3x - 4)] =$

9. $8 - 3[2x - 3(2x + 4)] =$

10. $11 - 4[2x - 4(3x + 2)] =$

11. $8x^2 + 2x[3x - (x + y)] =$

12. $7xy - [2xy - (xy + 1)] =$

13. $3x[x^2 - 2(2x - 1)] =$

14. $2x[3x - 4(x^2 - 3)] =$

15. $-\{7a - 2[a - (4a + 1)]\} =$

16. $-2\{3x - 5[2x - (3x - 2)]\} =$

17. $3x\{x - 3[2x - 3(2x + 1)]\} =$

18. $-5x\{3x - 2[x + 2(3x - 4)]\} =$

19. $2x\{3x + 2[3x - (2x^2 - 3)]\} - 4x^2 =$

20. $3x\{x - 4[x - (3x^2 + 5)]\} + 6x^2 =$

21. $6x + [x^2 - 3(2x - 1)] - 3 =$

22. $5x + 3[2x^2 - 3(4x + 3)] - 6x^2 =$

23. $5x^2 + [2x + 3x(x - 4)] =$

24. $6x^2 - 2[5x - 2x(3x + 1)] =$

25. $-\{5a - 3a[2a - 4(a^2 - 7a)]\} =$

26. $6x - \{4y - 3[2x + 5(x - y)]\} =$

27. $x\{10x^2 - 3x[4 - 2x(x - 5)]\} + 20x^3 =$

28. $2x\{5x - x[3 + 2x(x - 6)]\} + 4x^4 =$

29. $2xy^2 + 5\{6xy - 2x[2x - y(y - 3)]\} =$

30. $5a^2b - 6\{45ab + 3b[4b - 3a(2a + 5)]\} =$

◢◣ 1–10 EVALUATING LITERAL EXPRESSIONS

<table>
<tr><td>

O B J E C T I V E S

▼

Upon completing this section you should be able to:

1. Substitute numbers for letters in literal expressions.
2. Evaluate the expression once the substitutions have been made.

</td><td>

The **substitution principle** used in mathematics says simply that "a quantity may be substituted for its equal in any process." We use this when working with formulas such as those from physics, geometry, or business.

The order of operations and the proper use of grouping symbols as studied in the previous section must be followed.

Example 1   Evaluate the following if $x = -5$.
      a. x^2 **b.** $-x^2$ **c.** $(-x)^2$

</td></tr>
</table>

Solutions

a. $x^2 = (-5)^2 = 25$ Substitute -5 for x.
b. $-x^2 = -(-5)^2 = -25$ Here we square only x.
c. $(-x)^2 = [-(-5)]^2 = 5^2 = 25$ Here we square $-x$.

Example 2   Evaluate the following if $x = 3$.

    a. $2x^2$ **b.** $(2x)^2$

Solutions

    a. $2x^2 = 2(3)^2 = 2(9) = 18$ Here we square only x.
    b. $(2x)^2 = [2(3)]^2 = 6^2 = 36$ Here we square $2x$.

Notice, in example 2, that the order of operations dictates that in $2x^2$ the value of x^2 must be found before multiplying by 2 and that parentheses dictate that in $(2x)^2$ the value of $2x$ must be found before squaring occurs.

Example 3   Evaluate $x^2 - 3x + y^2 - 4$ if $x = -2$ and $y = 3$.

    Solution $\begin{aligned}x^2 - 3x + y^2 - 4 &= (-2)^2 - 3(-2) + (3)^2 - 4 \\ &= 4 + 6 + 9 - 4 \\ &= 15\end{aligned}$

> Remember that in a literal expression the letters are merely holding a place for various numbers that may be assigned to them. For that reason these letters are sometimes called *placeholders* or *variables*.

A good way to avoid errors, especially with signs, is to first use parentheses for each literal number and then go back and fill in its value.

Example 4   Evaluate $a - 3ab + b^2 - abc$ if $a = 1, b = -1, c = 2$.

    Solution   The previous suggestion would have us write

$$a - 3ab + b^2 - abc = (\ ) - 3(\ )(\ ) + (\ )^2 - (\ )(\ )(\ ).$$

Place parentheses for each literal number.

Now we go back and place 1 for each a, -1 for each b, and 2 for each c, giving

$$\begin{aligned} &= (1) - 3(1)(-1) + (-1)^2 - (1)(-1)(2) \\ &= 1 + 3 + 1 + 2 \\ &= 7 \end{aligned}$$

Example 5   The formula for finding the area of a trapezoid is $A = \dfrac{1}{2}h(b_1 + b_2)$. Find A if $h = 5, b_1 = 3,$ and $b_2 = 9$.

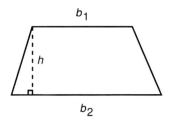

h is the height and b_1 and b_2 are the bases of the trapezoid.

Solution $A = \dfrac{1}{2}h(b_1 + b_2) = \dfrac{1}{2}(5)(3 + 9)$

$$= \dfrac{1}{2}(5)(12)$$

$$= 30$$

Example 6   *I = Prt* is the formula for simple interest where *I* is the interest, *P* represents the principal, *r* the rate of interest per unit of time, and *t* the units of time. Find the interest on $10,000 invested at 9% interest per year for two years.

Recall 9% = .09.

Solution   *I = Prt*
$$= (10,000)(.09)(2)$$
$$= \$1,800$$

◥ EXERCISE 1–10–1

Evaluate the expressions in questions 1–26.

1. $2x + 5$ if $x = 5$

2. $3x - 2$ if $x = 4$

3. $3x + 5$ if $x = -7$

4. $16 - 2x$ if $x = -3$

5. x^2 if $x = -3$

6. x^2 if $x = -2$

7. $-x^2$ if $x = -3$

8. $-x^2$ if $x = -2$

9. $(-x)^2$ if $x = -3$

10. $(-x)^2$ if $x = -2$

11. $3x^3$ if $x = -2$

12. $2x^3$ if $x - -5$

13. $(3x)^3$ if $x = -2$

14. $(2x)^3$ if $x = -5$

15. $x^2 + 3x - 1$ if $x = -3$

16. $a^2 - 2a + 4$ if $a = -4$

17. $4x^2 - 5x + 2$ if $x = -2$

18. $2x^2 - 3x + 1$ if $x = -3$

19. $2x - 3y$ if $x = 4, y = -2$

20. $2a - 5b$ if $a = 6, b = -3$

21. $3x^2 + xy - 6$ if $x = -2$, $y = -1$

22. $3xy - 2y^2$ if $x = 3$, $y = -4$

23. $5x^2 - (5x)^2$ if $x = -2$

24. $3x^2 - (3x)^2$ if $x = -4$

25. $x^3 - 2xz - 3y$ if $x = -2$, $y = -3$, $z = 4$

26. $2x^3 + 3xyz - 2z^2$ if $x = 2$, $y = 5$, $z = -1$

27. The perimeter of a rectangle is given by $P = 2\ell + 2w$. Find P when $\ell = 12$ and $w = 7$.

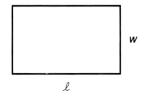

28. The area of a rectangle is given by $A = bh$. Find A when $b = 9$ and $h = 3$.

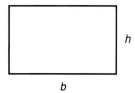

29. The perimeter of a square is given by $P = 4s$, where s represents the length of one side. Find P when $s = 8$.

30. The area of a square is given by $A = s^2$. Find A when $s = 3.5$.

31. The perimeter of a triangle is given by the formula $P = a + b + c$. Find P when $a = 14.5$, $b = 17.8$, and $c = 10.6$.

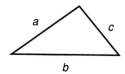

32. The area of a triangle is given by $A = \dfrac{1}{2}bh$. Find A when $b = 7$ and $h = 6$.

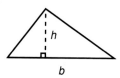

33. A distance formula from physics is $d = rt$. Find d when $r = 55$ and $t = 4$.

34. A force formula from physics is $F = ma$. Find F when $m = 120$ and $a = 32$.

35. A formula from business for finding interest is $I = Prt$. Find I when $P = \$8,000$; $r = .18$; and $t = 3$.

36. The circumference of a circle is given by $C = \pi d$, where d represents the diameter of the circle and π is a constant number that is approximately equal to $\dfrac{22}{7}$. Find C when $d = 14$ and $\pi = \dfrac{22}{7}$.

37. The area of a circle is given by $A = \pi r^2$, where r represents the radius of the circle. Find A when $\pi = \dfrac{22}{7}$ and $r = 7$.

38. A formula for changing Fahrenheit temperature to Celsius is given by $C = \dfrac{5}{9}(F - 32)$. Find C when $F = 68$.

39. A formula for changing Celsius temperature to Fahrenheit is $F = \dfrac{9}{5}C + 32$. Find F when $C = -10$.

40. The volume of a rectangular solid is given by $V = xyz$. Find V when $x = 3$, $y = 1.5$, and $z = 4$.

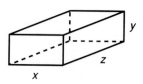

41. The volume of a cube is given by $V = s^3$. Find V when $s = 3.5$.

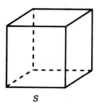

42. The volume of a cylinder of height h having a circular base with radius r is given by $V = \pi r^2 h$. Find V when $\pi = \frac{22}{7}$, $r = 1.4$, and $h = 3$.

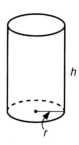

43. The area of a trapezoid is given by $A = \frac{1}{2}h(b_1 + b_2)$. Find A when $h = 9$, $b_1 = 7$, and $b_2 = 13$.

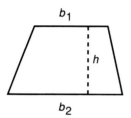

44. The volume of a right circular cone is given by $V = \frac{1}{3}\pi r^2 h$. Find V when $\pi = \frac{22}{7}$, $r = 7$, and $h = 9$.

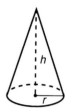

45. A distance formula for a free-falling object is given by $s = \frac{1}{2}gt^2$. Find s when $g = 32$ and $t = 5$.

46. The volume of a sphere is given by $V = \frac{4}{3}\pi r^3$. Find V when $\pi = \frac{22}{7}$ and $r = 2.1$.

47. A formula from physics is $s = h + vt - 490t^2$. Find s when $h = 5,500$; $v = 100$; and $t = 3$.

48. A formula from business is $T = P(1 + rt)$. Find T when $P = \$12,000$; $r = 0.15$; and $t = 5$.

49. A formula from physics is $H = .4nd^2$. Find H when $n = 2$ and $d = 10$.

50. A formula from physics is $K = \frac{1}{2}mv^2$. Find K when $m = 90$ and $v = 40$.

1 SUMMARY

Key Words

Section 1-1

- The set of **real numbers** is composed of the sets of **rational** and **irrational** numbers.

- The **properties** of the real numbers can be used to justify all manipulations used in algebra.

Section 1-2

- **Signed numbers** are preceded by a plus or minus sign.

- **Positive numbers** are numbers preceded by a plus sign.

- **Negative numbers** are numbers preceded by a minus sign.

- The **negative of a number** is that number with the opposite sign.

- A **number line** is useful for visualizing the relative positions of numbers.

- The **absolute value** of a number is its distance from zero on the number line.

Section 1-4

- **Combining** numbers involves the operations of addition and subtraction.

- **Subtraction** is adding the opposite.

Section 1-5

- **Terms** are connected by addition signs.

- **Factors** are parts to be multiplied.

Section 1-6

- In the expression $5x^4$
  5 is called the **numerical coefficient,**
  x is called the **base,**
  4 is called the **exponent.**

- A **positive whole number exponent** indicates the number of times a base is to be used as a factor.

Section 1-7

- A **literal** or **algebraic expression** is one in which letters are used to represent numbers.

- **Like terms** have exactly the same literal factors.

Section 1-8

- A **polynomial** is a literal expression having one or more terms that contain only nonnegative whole number exponents.

- A **monomial** is a polynomial having one term.

- A **binomial** is a polynomial having two terms.

- A **trinomial** is a polynomial having three terms.

Section 1-9

- **Brackets, braces,** and **parentheses** are all grouping symbols having the same function.

Procedures

Section 1-2

- To find the negative of a number change its sign.

- To find the absolute value of a number x:
  $|x| = x$, if x is zero or positive,
  $|x| = -x$, if x is negative.

- To remove parentheses preceded by a minus sign find the negative of each term within the parentheses.

- To remove parentheses preceded by a plus sign do not change the sign of any term within the parentheses.

Section 1-3

- If a movement to the right is considered positive and a movement to the left is negative, the number line can be used to combine signed numbers.

Section 1-4

- To combine two numbers with like signs add the numbers and use the common sign.

- To combine two numbers with unlike signs subtract the smaller from the larger (without regard to sign) and attach the sign of the larger.

Section 1-5

- To multiply or divide signed numbers two like signs give a positive product or quotient and two unlike signs give a negative product or quotient.

Section 1-6

- The laws of exponents for positive whole number exponents are:

  I. $x^a x^b = x^{a+b}$

  II. $(x^a)^b = x^{ab}$

  III. $\dfrac{x^a}{x^b} = x^{a-b}$ is a is larger than b.

  $\dfrac{x^a}{x^b} = \dfrac{1}{x^{b-a}}$ if a is smaller than b.

Section 1-7

- Only like terms can be combined.

- To combine like terms combine the numerical coefficients and use this result as the coefficient of the common literal factors.

Section 1-8

- To multiply a monomial and another polynomial multiply each term of the polynomial by the monomial.

Section 1-9

- If no parentheses occur in an expression to be evaluated, the operations should be performed in the following order:
  1. exponents,
  2. multiplication and division from left to right,
  3. addition and subtraction from left to right.

- When removing grouping symbols within grouping symbols, remove the innermost set first.

Section 1-10

- To evaluate a literal expression substitute the given values for the literal numbers and perform the indicated operations.

NAME: _____

CLASS / SECTION: _____   DATE: _____

A   N   S   W   E   R   S

1. _____

2. _____

3. _____

4. _____

5. _____

6. _____

7. _____

8. _____

9. _____

10. _____

11. _____

12. _____

13. _____

14. _____

15. _____

16. _____

17. _____

18. _____

19. _____

20. _____

State which set(s), (a) counting numbers, (b) whole numbers, (c) integers, (d) rational numbers, (e) irrational numbers, each of the given numbers belongs in.

1. 3

2. -100

3. $-\dfrac{2}{5}$

4. $\dfrac{1}{4}$

5. π

6. $\sqrt{7}$

7. 3.6

8. -11.4

Name the property that each illustrates.

9. $3(6) = 6(3)$.

10. $5 + 0 = 5$

11. $(3 + 5) + 6 = 6 + (3 + 5)$

12. $3\left(\dfrac{1}{3}\right) = 1$

13. $4(2 + 3) = 4(2) + 4(3)$

14. $2 + (3 + 4) = (2 + 3) + 4$

15. $8 + (-8) = 0$

16. $15(1) = 15$

Find the negative of each number.

17. $+14$

18. $+\dfrac{3}{5}$

19. $-\dfrac{1}{3}$

20. 0

Represent the signed numbers on a number line.

21. −4

21. _____

22. −(+3)

22. _____

23. −(−5)

23. _____

24. _____

24. −(−2)

25. _____

Evaluate.

26. _____

25. $|-8|$ **26.** $|+21|$ **27.** $-|-4|$ **28.** $|-(-5)|$

27. _____

28. _____

Remove the parentheses.

29. $-(x - 4)$ **30.** $-(x + y - 1)$

29. _____

30. _____

31. $-(3x - 2y + 5)$ **32.** $+(2x - 3y + 4)$

31. _____

Use a number line to combine the numbers.

32. _____

33. $-15 + 7$

33. _____

34. $3 - 11$

34. _____

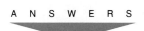

A N S W E R S

35. _____

36. _____

37. _____

38. _____

39. _____

40. _____

41. _____

42. _____

43. _____

44. _____

45. _____

46. _____

47. _____

48. _____

49. _____

50. _____

35. $-2 + 9 - 5 + 4$

$\longleftarrow \! +\! \longrightarrow$

36. $3 - 8 + 10 - 3$

$\longleftarrow \! +\! \longrightarrow$

Combine the numbers.

37. $(+7) + (-9)$ **38.** $(-17) + (-5)$

39. $\left(+\dfrac{5}{9}\right) + \left(-\dfrac{2}{9}\right)$ **40.** $\left(-\dfrac{1}{2}\right) + \left(+\dfrac{2}{3}\right)$

41. $(-5) + (+8) + (-10)$ **42.** $(+6) + (-15) + (+4)$

43. $5 - 8 + 2$ **44.** $4 - (-6) - 1$

45. $-3 + (-4) + 9$ **46.** $-13 + 5 - (-16) - 2$

47. $-4 + 9 + 2 - 8 - 7 + 6$ **48.** $4 - (9 - 2)$

49. $5 - (-6 - 9) + 2$ **50.** $7 - (5 - 16) - 4$

Evaluate.

51. $(-4)(+12)$

52. $(-6)\left(+\dfrac{1}{3}\right)$

51. _____

52. _____

53. $\left(-\dfrac{3}{4}\right)\left(-\dfrac{2}{7}\right)$

54. $(-2.3)(-3.5)$

53. _____

54. _____

55. $(8)(-5)(-7)$

56. $(-4)(3)(10)$

55. _____

57. $\dfrac{-36}{-3}$

58. $\dfrac{-32}{+8}$

56. _____

57. _____

59. $\left(\dfrac{4}{9}\right) \div \left(-\dfrac{2}{3}\right)$

60. $(-15) \div \left(-\dfrac{3}{5}\right)$

58. _____

59. _____

61. $\dfrac{(-6)(5)}{-10}$

62. $\dfrac{(-8)(-6)}{+12}$

60. _____

61. _____

63. $(-3)(-4)(-5)$

64. $(-3)(-1)(5)(-10)\left(-\dfrac{2}{5}\right)(3)$

62. _____

63. _____

65. $\dfrac{(-4)(-9)}{-6}$

66. $\dfrac{-18 + 3}{-5}$

64. _____

65. _____

66. _____

Simplify.

67. x^3x^5

68. a^3a^6

67. _____

68. _____

69. $(-2a^2)(4a^4)$

70. $(-5x)(-3x^2)$

69. _____

71. $(3x^2y)(-4xy^2)(-x^3y^4)$

72. $(-3x^2y)(-2x^4)(-6x^2y^3)$

70. _____

71. _____

73. $(x^4)^2$

74. $(a^3)^3$

72. _____

73. _____

75. $(2ab^3c^2)^4$

76. $(3x^2y^3z)^2$

74. _____

77. $\dfrac{10x^4}{-2x}$

78. $\dfrac{-16x^2}{2x^5}$

75. _____

76. _____

77. _____

79. $\dfrac{13a^3b^4c^2}{-39ab^6c}$

80. $\dfrac{7ab^3c}{-21a^5b}$

78. _____

79. _____

Simplify by combining like terms.

80. _____

81. $4x + 3y - 2x$

82. $7a^2b - 4ab + 3a^2b$

81. _____

83. $3x^2y + 4xy - 7xy^2 + 9x^2y - 6xy$

82. _____

83. _____

84. $5x^2y^2 - 4x^2y + 6x^2y - 9x^2y^2 - 5x^2y$

84. _____

Find the products.

85. $-3(2x - y)$

86. $-5(3a - 4b)$

85. _____

87. $2xy(3x^2 - 2xy + 4xy^3)$

88. $3x^2y(2x^2y + 3x^3 - 4xy)$

86. _____

87. _____

89. $-3a^2b^3(5a^4b - 7abc + 3a^3c^2)$

90. $-2abc^2(3a^2b - 4a^2bc^4 - 11ac)$

88. _____

89. _____

90. _____

Evaluate the expressions.

91. $3 + 4 \times 6$

92. $10 - 2 \times 7$

91. _____

92. _____

93. $2^3 - 12 \div 4 + 6$

94. $4 - 3^2 \times 6 \div 9$

93. _____

94. _____

Remove grouping symbols and combine like terms.

95. _____

95. $3x - (2x + 5) - 7$

96. $10a - (9 - 4a) + 16$

96. _____

97. $8 - (3x + 5) + 12x$

98. $15 + (5x - 7) - (9x - 2)$

97. _____

98. _____

99. $14 - 3(2x - 1)$

100. $5(3a - 2) - 4(3a + 1)$

99. _____

100. _____

101. $3(a - 2) - 5(2a - 4)$

102. $8(3x - 4) + 3x - 2(6x + 5)$

101. _____

102. _____

103. _____

104. _____

105. _____

106. _____

107. _____

108. _____

109. _____

110. _____

111. _____

112. _____

113. _____

114. _____

115. _____

116. _____

103. $13x - 5[x - (3x + 4)]$

104. $5a - [-3 - 2(3a - 2)]$

105. $2[3a - 2(a + 3) + a]$

106. $10x - 3[2x + 5(x - 3)]$

107. $a + 4\{3a - [5(a + 1) - a]\}$

108. $15x - 2\{2x - 4[3 - 4(x - 5)]\}$

Evaluate each expression.

109. $3x^3$ if $x = -5$

110. $-x^3$ if $x = -2$

111. $2x^2 - 3x + 1$ if $x = -4$

112. $x^2 - y^3$ if $x = -3$ and $y = -4$

113. $5x^3 - 2xy^2 - 3y$ if $x = -3$ and $y = 4$

114. $5a^3 - 3a^2b^3 + ab$ if $a = -1$ and $b = -3$

115. A formula from mathematics is $L = a + (n - 1)d$. Find L when $a = 3, n = 9, d = \dfrac{1}{2}$.

116. A formula from physics is $v = \dfrac{d}{t}$. Find v when $d = 1,200; t = 2.5$.

117. A formula from mathematics is $S = \dfrac{a}{1 - r}$. Find S when $a = 12$,
$r = \dfrac{1}{4}$.

117. _____

118. A formula from geometry is $V = \dfrac{h}{6}(B + 4M + b)$. Find V when
$h = 15, B = 12, M = 8, b = 4$.

118. _____

119. A formula from physics is $I = V\left(\dfrac{1}{R} + \dfrac{1}{r}\right)$. Find I when $V = 30$,
$R = 2, r = 5$.

119. _____

120. _____

120. A formula from mathematics is $S = \dfrac{a - r\ell}{1 - r}$. Find S when $a = 4$,
$r = \dfrac{1}{3}, \ell = 6$.

SCORE: _____

NAME:

CLASS / SECTION: DATE:

A N S W E R S

1. Name the property that the following statement illustrates:

$5 \times 2 = 2 \times 5$

2. Find the negative of $+\dfrac{4}{5}$

1. _____

3. Find the negative of -8.3.

2. _____

4. Identify the signed number represented by point A.

3. _____

5. Identify the signed number represented by point B.

4. _____

5. _____

6. Evaluate: $-|-5|$

6. _____

7. Evaluate: $|-(+11)|$

8. Use a signed number to indicate the result of combining a loss of $54 with a gain of $39.

7. _____

8. _____

9. Combine: $(-43) + (+52)$

10. Combine: $(-13) - (-8)$

9. _____

10. _____

11. Evaluate: $(-9)(+4)$

12. Evaluate: $\dfrac{-48}{-4}$

11. _____

13. Find the product:
$(-3)(-5)(-2)(-4)$

14. Simplify:
$(-3a^3b)(-2ab^4)(-5a^2b^2)$

12. _____

13. _____

15. Remove parentheses and simplify: $18 - (5 - 11) - 6$

16. Combine like terms:
$15ab + 6a^2 - 11ab - 9a^2$

14. _____

15. _____

17. Find the product:
$4x^2y(3xy^2 - 5x^3 + 4x^3y)$

18. Remove the grouping symbols and simplify:
$12x +$
$4\{3x - 2[4x - 3(3x - 2)]\}$

16. _____

17. _____

19. Evaluate: $2x^2 + 5xy - 3y^3$ if $x = 5, y = -1$

20. The area of a triangle is given by $A = \dfrac{1}{2}bh$. Find A when $b = 12.5$ and $h = 14$.

18. _____

19. _____

20. _____

SCORE: _____

PRETEST

Before beginning this chapter, answer as many of the following questions as you can. When you have finished the chapter, take the practice test at the end of the chapter and compare the scores of the two tests to see how much you have learned.

A N S W E R S

1. Classify $3(x + 4)$ $= 5x - (2x - 12)$ as conditional or as an identity.

2. Are the two equations $x = 5$ and $\frac{1}{2}x + 3 = 2x - \frac{9}{2}$ equivalent?

1. _____

2. _____

Solve for x in each of the equations in questions 3–8.

3. _____

3. $4x = -28$

4. $3x - 4 = 10 - x$

4. _____

5. $3(2x - 5) = 2(3 - x) - 1$

6. $\frac{1}{2}\left(x + \frac{1}{3}\right)$ $= \frac{1}{3}\left(x + \frac{1}{2}\right) - 1$

5. _____

6. _____

7. $\frac{5}{8}x + \frac{3}{4} = \frac{1}{2}x - 2$

8. $x - (3x + 4)$ $= \frac{1}{2}(x + 1) + \frac{11}{2}$

7. _____

8. _____

9. _____

10. _____

11. _____

12. _____

13. _____

14. _____

SCORE: _____

9. The ratio of boys to girls in grade 4 of a certain school is 6 to 5. If there are 15 girls in the class, how many boys are in the class?

10. The area of a triangle is 48 square inches and the base is 12 in. long. Find the height.

$$\left(\text{Use the formula } A = \frac{1}{2}bh. \right)$$

11. Solve for x if $3abx = 2a(bx + c) + 3ac$.

12. Solve for x and graph the results on the number line:

$3x + 1 \geq 5x - 7$

0

13. Solve for x: $|2x - 3| = 15$

14. Solve for x and graph the solution on the number line:

$|5x + 3| \geq 12$

0

C H A P T E R

2

Equations and Inequalities

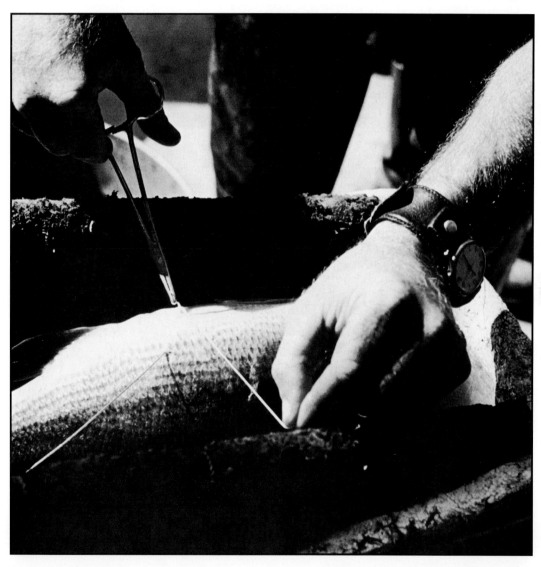

A wildlife management team, conducting a study on a particular lake, caught, tagged, and released 75 black bass. Several weeks later, they caught 125 black bass and observed that 3 of them were tagged. What is the total number of black bass they could expect to be present in the lake?

The solution of equations is the central theme of algebra. Skills learned manipulating the numbers and symbols of algebra allow us to find answers to problems by solving equations. In this chapter we will study techniques for solving first-degree equations with one variable. We will also study techniques for solving and graphing first-degree inequalities with one variable.

2–1 CONDITIONAL AND EQUIVALENT EQUATIONS

OBJECTIVES

Upon completing this section you should be able to:
1. Classify an equation as conditional or an identity.
2. Solve simple equations mentally.
3. Determine if certain equations are equivalent.

In this section we will present the mathematical definition of "equation." We will also identify the two main types of equations and discuss what is meant by a "solution" or "root."

> An **equation** is a statement in symbols that two number expressions are equal.

Equations can be classified in two main types:

    1. An **identity** is true for all values of the literal and arithmetical numbers in it.

Example 1 $5 \times 4 = 20$

Example 2 $3 + 6 = 9$

Example 3 $2x + 3x = 5x$

    2. A **conditional** equation is true for only certain values of the literal numbers in it.

Example 4 $x + 3 = 9$ is true only if $x = 6$.

Example 5 $3x - 4 = 11$ is true only if $x = 5$.

> A **solution** or **root** of an equation is the value of the variable or variables that make the equation a true statement.

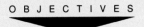

$$x = \circled{6} \leftarrow \text{solution or root}$$

The solution or root is said to *satisfy the equation*.

Solving an equation means finding the solution or root.

Many equations can be solved mentally. Ability to solve an equation mentally will depend on your ability to manipulate the numbers of arithmetic. The better

To solve the equation

    $x + 2 = 7$

means to find the value of x that makes the statement true. What is the solution (root) of this equation?

you know the facts of multiplication and addition, the more adept you will be at mentally solving equations.

Example 6   Solve for x: $x + 3 = 7$

Solution   To have a true statement we need a value for x that, when added to 3, will yield 7. Our knowledge of arithmetic indicates that 4 is the needed value. Therefore the solution to the equation is $x = 4$.

> Is there any other number besides $x = 4$ that would satisfy the equation?

Example 7   Solve for x: $x - 5 = 3$

Solution   What number do we subtract 5 from to obtain 3? Again, our experience with arithmetic tells us that $8 - 5 = 3$. Therefore the solution is $x = 8$.

> The variable is sometimes referred to as a *placeholder*. That is, it is holding the place for a value that will make the statement either true or false.

Example 8   Solve for x: $3x = 15$

Solution   What number must be multiplied by 3 to obtain 15? Our answer is $x = 5$.

Example 9   Solve for x: $2x - 1 = 5$

Solution   We would subtract 1 from 6 to obtain 5. Thus $2x = 6$. Then $x = 3$.

Regardless of how an equation is solved, the solution should *always* be checked for correctness.

> **WARNING**
>
> Many students think that when they have found the solution to an equation, the problem is finished. *Not so!* The final step should *always* be to check the solution.

Example 10   A student solved the equation $5x - 3 = 4x + 2$ and found an answer of $x = 6$. Was this right or wrong?

Solution   Does $x = 6$ satisfy the equation $5x - 3 = 4x + 2$? To check we substitute 6 for x in the equation to see if we obtain a true statement.

$$5x - 3 = 4x + 2$$
$$5(6) - 3 = 4(6) + 2$$
$$30 - 3 = 24 + 2$$
$$27 = 26$$

The answer $x = 6$ is wrong.

Another student solved the same equation and found $x = 5$.

$$5x - 3 = 4x + 2$$
$$5(5) - 3 = 4(5) + 2$$
$$25 - 3 = 20 + 2$$
$$22 = 22$$

The answer $x = 5$ is correct.

Is $x = 3$ a solution of
$x - 1 = 2$?
Is $x = 3$ a solution of
$2x + 1 = 7$?

What can be said about the
equations $x - 1 = 2$ and
$2x + 1 = 7$?

> Two equations are **equivalent** if they have the same solution or
> solutions. $3x = 6$ and $2x + 1 = 5$ are equivalent because in both
> cases $x = 2$ is a solution.

Techniques for solving equations will involve processes for changing an equation
to an equivalent equation. If a complicated equation such as $2x - 4 + 3x$
$= x + 2 + 3x$ can be changed to a simple equation $x = 6$, and the equation
$x = 6$ is equivalent to the original equation, then we have solved the equation.

Two things now become very important.

1. Are two equations equivalent?
2. How can we change an equation to another equation that is
equivalent to it?

The answer to the first question is found by using the substitution principle.

Example 11 Are $5x + 2 = 6x - 1$ and $x = 3$ equivalent equations?

To use the substitution principle
correctly, substitute the numeral 3
for x wherever x appears in the
equation.

$$5\,\widehat{x} + 2 = 6\,\widehat{x} - 1$$
$$\uparrow \qquad\quad \uparrow$$
$$\textcircled{3} \qquad\quad \textcircled{3}$$

Solution The question becomes, does $5(3) + 2 = 6(3) - 1$?

$$15 + 2 = 18 - 1$$
$$17 = 17$$

Answer: Yes, they are equivalent.

The answer to the second question involves the techniques for solving equations
that will be discussed in the next few sections.

▼ **EXERCISE 2–1–1**

Classify each equation as conditional or an identity.

1. $3x + x = 4x$ **2.** $2x + 1 = 5$ **3.** $x - 2 = 11$ **4.** $6x = 4x + 2x$

5. $7x = 10x - 3x$ **6.** $6x + 9x = 15x$ **7.** $20x - 8x = 7x + 5x$ **8.** $x + 5 = 2x - 1$

9. $4x - 3 = 17$ **10.** $3x + 2x = 4x + 9$

Solve the equations mentally and check by substitution.

11. $x + 4 = 7$　　　**12.** $2x = 6$　　　**13.** $x - 3 = 8$　　　**14.** $x + 1 = 10$

15. $3x = 12$　　　**16.** $2x = 1$　　　**17.** $x - 7 = 3$　　　**18.** $10 - x = 6$

19. $5x = 30$　　　**20.** $x + 9 = 11$　　　**21.** $5x = 3$　　　**22.** $2x + 1 = 11$

23. $9x = 3$　　　**24.** $12 + x = 25$　　　**25.** $x + 5 = 5$　　　**26.** $3x - 5 = 22$

27. $2x - 1 = 13$　　　**28.** $x - 3 = 0$　　　**29.** $3x + 4 = 22$　　　**30.** $3x + 2 = 2x + 7$

Determine which of the following pairs of equations are equivalent.

31. $x + 7 = 2x + 6$　　and　　$x = 1$　　　　**32.** $3x - 5 = 2x + 1$　　and　　$x = 5$

33. $2x + 1 = x$　　and　　$x = 1$　　　　**34.** $4x - 3 = x + 6$　　and　　$x = 3$

35. $5x - 4 = 3x + 2$　　and　　$x = 2$　　　　**36.** $11 - 2x = 7 - x$　　and　　$x = 4$

37. $2x - 2 + x = 6 + x$　　and　　$x = 5$　　　　**38.** $3x + 1 = x - 5 + 4x$　　and　　$x = 6$

39. $3x + 1 = x + 5$　　and　　$x = 2$　　　　**40.** $2x + 17 = 31 - 5x$　　and　　$x = 2$

2-2 THE DIVISION RULE

As stated in the previous section, we wish to establish ways of changing a given equation to an equivalent equation. In this section we will use division as one way of accomplishing this.

> *If each term of an equation is divided by the same nonzero number,* the resulting equation is equivalent to the original equation.

To prepare to use the division rule for solving equations we must make note of the following process:

Just as 18 meters $\div$ 2 = 9 meters,

$$18x \div 2 = 9x.$$

And as 18 meters $\div$ 6 = 3 meters,

$$18x \div 6 = 3x.$$

Also as 18 meters $\div$ 18 = 1 meter,

$$18x \div 18 = 1x.$$

We usually write $1x$ as x, with the 1 understood.

Example 1 Solve for x if $3x = 10$.

We want this quantity to be $1x$ so we divide by 3, since

$$3x \div 3 = 1x, \text{ or } x.$$

In the equation $5x = 15$ we would divide both sides by 5.

In the equation $7x = 21$ we would divide both sides by 7.

Solution The division rule allows us to divide both sides of $3x = 10$ by the same numbers and our goal of finding a value of x would indicate that we divide by 3.

$$3x = 10$$

$$\frac{3x}{3} = \frac{10}{3}$$

$$x = \frac{10}{3} \text{ or } 3\frac{1}{3}$$

Check: $3x = 10$ and $x = \frac{10}{3}$. Are these equivalent equations?

$$3\left(\frac{10}{3}\right) = 10$$

$$10 = 10$$

The solution is correct.

Example 2   Solve for x if $5x = 20$.

Solution
$$5x = 20$$
$$\frac{5x}{5} = \frac{20}{5} \quad \text{(division rule)}$$
$$x = 4$$

Check:
$$5x = 20$$
$$5(4) = 20$$
$$20 = 20$$

Example 3   Solve for x if $8x = 4$.

Solution
$$8x = 4$$
$$\frac{8x}{8} = \frac{4}{8} \quad \text{(division rule)}$$
$$x = \frac{4}{8} \text{ or } \frac{1}{2}$$

Check:
$$8x = 4$$
$$8\left(\frac{1}{2}\right) = 4$$
$$4 = 4$$

Note that the division rule allows us to divide each term of an equation by *any nonzero* number and the resulting equation is equivalent to the original equation.

Therefore we could divide each side of the equation
$$4x = 12$$
by 5 and obtain $\frac{4}{5}x = \frac{12}{5}$, which is equivalent to the original equation.

Dividing by 5 does not help find the solution, however. What number should we divide by to find the solution?

Example 4   Solve for x if $2x = -6$.

Solution
$$2x = -6$$
$$\frac{2x}{2} = \frac{-6}{2} \quad \text{(division rule)}$$
$$x = -3$$

Check:
$$2x = -6$$
$$2(-3) = -6$$
$$-6 = -6$$

Example 5   Solve for x if $-3x = 12$.

Solution
$$-3x = 12$$
$$\frac{-3x}{-3} = \frac{12}{-3}$$
$$x = -4$$

Check:
$$-3x = 12$$
$$-3(-4) = 12$$
$$12 = 12$$

Notice also that the division rule does *not* allow us to divide by zero.

Since dividing by zero is not allowed in mathematics, expressions such as $\frac{5x}{0}$ and $\frac{8}{0}$ are meaningless.

Example 6   Solve for x if $-7x = -28$.

Solution
$$-7x = -28$$
$$\frac{-7x}{-7} = \frac{-28}{-7}$$
$$x = 4$$

Check:
$$-7x = -28$$
$$-7(+4) = -28$$
$$-28 = -28$$

Example 7   Solve for x if $-x = 27$.

Solution
$$-x = 27$$
$$\frac{-x}{-1} = \frac{27}{-1}$$
$$x = -27$$

The term $-x$ is the same as $(-1)x = -1x$.

Check:
$$-x = 27$$
$$-(-27) = 27$$
$$27 = 27$$

Example 8   Solve for x if $\frac{3}{4}x = -9$.

Solution
$$\frac{3}{4}x = -9$$

Recall from arithmetic that dividing by $\frac{3}{4}$ is the same as multiplying by $\frac{4}{3}$.

$$\frac{\frac{3}{4}x}{\frac{3}{4}} = \frac{-9}{\frac{3}{4}}$$
$$x = -9\left(\frac{4}{3}\right)$$
$$x = -12$$

Check:
$$\frac{3}{4}x = -9$$
$$\frac{3}{4}(-12) = -9$$
$$-9 = -9$$

Example 9   The formula for the circumference of a circle is $C = 2\pi r$, where C represents circumference, r represents the radius, and π is approximately 3.14. Find the radius of a circle if the circumference is measured to be 40.72 cm. (Answer correct to two decimal places.)

The circumference of a circle is the distance around the circle.

Solution   To solve a problem involving a formula we first use the substitution principle.

$$C = 2\pi r$$
$$40.72 = 2(3.14)r$$
$$40.72 = 6.28r$$
$$\frac{40.72}{6.28} = r \qquad \text{(division rule)}$$
$$r = 6.48$$

Notice at this point we now have only *one* unknown, r, in the equation.

Note that $6.48 = r$ is the same as $r = 6.48$. This is called the *reflexive property of equality*.

Check: $C = 2\pi r$
$$40.72 = 2(3.14)(6.48)$$
$$40.72 = 40.72$$

Example 10 18 is what percent of 30?

Solution The equation is

$$18 = (x)(30)$$
$$\frac{18}{30} = \frac{x(30)}{30} \qquad \text{(division rule)}$$
$$x = .6$$
$$x = 60\%.$$

Note that $(x)(30)$ and $(30)(x)$ would result in the same product. This is an example of the *commutative property* of multiplication. Thus the equation $18 = (x)(30)$ could be written as $18 = 30x$, which could also be written as $30x = 18$ (reflexive property).

Check: $18 = (x)(30)$
$$18 = (.6)(30)$$
$$18 = 18$$

▼ **EXERCISE 2-2-1**

Solve for x and check.

1. $2x = 24$

$X = \dfrac{24}{2}$

$X = 12$

2. $3x = 12$

$X = \dfrac{12}{3}$

$X = 4$

3. $-4x = 24$

$X = \dfrac{24}{-4}$

$X = -6$

4. $3x = -12$

$X = \dfrac{-12}{3}$

$X = -4$

5. $-5x = -15$

$X = \dfrac{-15}{-5}$

$X = 3$

6. $5x = 15$

$X = \dfrac{15}{5}$

$X = 3$

7. $3x = 1$

$X = \dfrac{1}{3}$

8. $15x = 15$

$X = 1$

9. $4x = -2$

$X = \dfrac{-2}{4}$

$X = \dfrac{-1}{2}$

10. $-7x = 21$

$X = \dfrac{21}{-7}$

$X = -3$

11. $3x = 9$

$X = \dfrac{9}{3}$

$X = 3$

12. $9x = 3$

$X = \dfrac{3}{9}$

$X = \dfrac{1}{3}$

13. $-3x = 9$

14. $9x = -3$

15. $10x = 4$

16. $12x = 180$

17. $52x = -13$

18. $\dfrac{2}{3}x = 10$

19. $-\dfrac{1}{2}x = 7$ **20.** $\dfrac{3}{5}x = \dfrac{9}{20}$ **21.** $-\dfrac{2}{3}x = -8$

22. $-5x = 1$ **23.** $13x = 26$ **24.** $26x = 13$

25. The formula for the area of a rectangle is $A = \ell w$, where A is area, ℓ is length, and w is width. If the area of a rectangle is 391 square meters and the length is 23 meters, find the width.

26. The formula for the area of a triangle is $A = \dfrac{1}{2}bh$, where A is area, b is the base, and h is the altitude. If the area of the triangle is 279 cm² and the altitude is 18 cm, find the length of the base.

27. The formula $d = rt$ gives the distance d traveled at a constant rate r in the time t. If a man drives 347 kilometers in five hours, what was his average rate?

28. $I = prt$ is the formula from business. I represents interest, p represents the principal or amount invested, r represents the rate given in percent, and t represents the time. If an investment of $10,000 returns as interest $6,180 at the annual rate of 10.3%, for how many years was the money invested?

29. A loan of $8,500 was made for four years. The total interest paid was $4,760. What annual rate of interest does this represent? (Use the formula in problem 28.)

30. 621 is 11.5% of what number?

31. 336.6 is what percent of 748?

32. A car was purchased at 23% off the list price. If this sale price was $9,694.01, what was the list price? (Round off the answer to the nearest cent.)

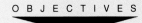 **2–3 THE ADDITION RULE**

OBJECTIVES

Upon completing this section you should be able to:
1. Use the addition rule to solve equations.
2. Solve some basic applied problems whose solutions involve using the addition rule.

In section 2–2 we used the operation of division to change some given equations to equivalent equations. The next operation we explore for this purpose is addition.

> *If the same quantity is added to both sides of an equation*, the resulting equation will be equivalent to the original equation.

Example 1 Solve for x if $x + 7 = 12$.

Solution Even though this equation can easily be solved mentally, we wish to use the addition rule to show a technique. Your thinking should be in this manner. "Since my goal is to solve for x, I must have x by itself on one side of the equation. I now have $x + 7$ and must find a way to get x alone in an equation equivalent to the given equation." We know that if we add (-7) to 7 the result is zero and also that $x + 0$ is x. So, we proceed as follows:

We add (-7) to both sides of the equation.

$$x + 7 = 12$$
$$x + 7 + (-7) = 12 + (-7) \qquad \text{(addition rule)}$$
$$x + 0 = 12 - 7$$
$$x = 5$$

$$Check:\quad x + 7 = 12$$
$$5 + 7 = 12$$
$$12 = 12$$

Remember that "adding the negative" is the definition of subtraction. So if it is easier, subtract the same quantity from both sides.

Example 2   Solve for x: $x - 8 = -4$

Solution
$$x - 8 = -4$$
$$x - 8 + 8 = -4 + 8 \quad \text{(adding 8 to both sides)}$$
$$x + 0 = 4$$
$$x = 4$$

Note that $x + 0$ may be written simply as x since zero added to any quantity equals the quantity itself.

Check: $x - 8 = -4$
$$4 - 8 = -4$$
$$-4 = -4$$

Example 3   Solve for x: $3x = 2x - 3$

Solution   Note that here the variable occurs on both sides of the equation. Our goal is to arrive at an equation of the form $x = $ (some number), so we elect to subtract $2x$ from both sides [or add $(-2x)$ to both sides].

$$3x = 2x - 3$$
$$3x - 2x = 2x - 2x - 3$$
$$x = -3$$

Check: $3x = 2x - 2$
$$3(-3) = 2(-3) - 3$$
$$-9 = -6 - 3$$
$$-9 = -9$$

Remember that checking your solution is an important step in solving equations.

Example 4   Solve for x: $2x + 6 = 3x - 9$

Solution
$$2x + 6 = 3x - 9$$
$$2x - 3x + 6 = 3x - 3x - 9 \quad \text{(add } -3x \text{ to both sides)}$$
$$-x + 6 = -9$$
$$-x + 6 - 6 = -9 - 6 \quad \text{(add } -6 \text{ to both sides)}$$
$$-x = -15$$
$$\frac{-x}{-1} = \frac{-15}{-1} \quad \text{(division rule)}$$
$$x = 15$$

Note that we could also add $(-2x)$ to both sides of the equation obtaining
$$6 = x - 9.$$
Then adding $(+9)$ to both sides we have
$$15 = x$$
or
$$x = 15.$$

Check: $2x + 6 = 3x - 9$
$$2(15) + 6 = 3(15) - 9$$
$$30 + 6 = 45 - 9$$
$$36 = 36$$

Note in example 4 that we needed to use the division rule from the previous section. Almost all previous knowledge of arithmetic and algebra will be needed in solving equations.

Example 5   Solve for x: $2(x - 3) = x + 3(x + 4)$

Solution   First eliminate parentheses.

$$2(x - 3) = x + 3(x + 4)$$
$$2x - 6 = x + 3x + 12$$
$$2x - 6 = 4x + 12$$

Always combine similar terms when possible.

We now wish to get all variables on one side and numbers on the other of the equation.

Again, complete the problem by adding $(-2x)$ to both sides of
$$2x - 6 = 4x + 12.$$

$$2x - 4x - 6 = 4x - 4x + 12 \quad \text{(subtract } 4x\text{)}$$
$$-2x - 6 = 12$$
$$-2x - 6 + 6 = 12 + 6 \quad \text{(add 6)}$$
$$-2x = 18$$
$$\frac{-2x}{-2} = \frac{18}{-2} \quad \text{(division rule)}$$
$$x = -9$$

Note that this is the original equation.

When checking, always substitute into the original equation.

Check:
$$2(x - 3) = x + 3(x + 4)$$
$$2(-9 - 3) = -9 + 3(-9 + 4)$$
$$2(-12) = -9 + 3(-5)$$
$$-24 = -9 - 15$$
$$-24 = -24$$

Example 6   If the perimeter of a rectangle is 54 cm and the length is 15 cm, what is the width?

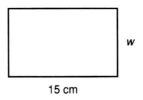

15 cm

We will use (w) to represent the unknown width.

Solution   Using the formula $P = 2\ell + 2w$, we have

Be very careful to substitute the correct quantity for each letter.

$$54 = 2(15) + 2w$$
$$54 = 30 + 2w$$
$$54 - 30 = 30 - 30 + 2w \quad \text{(addition rule)}$$
$$24 = 2w$$
$$\frac{24}{2} = \frac{2w}{2} \quad \text{(division rule)}$$
$$w = 12$$

Note that you can rewrite
$$24 = 2w$$
as $2w = 24$ if you prefer (reflexive property).

Check:
$$54 = 2(15) + 2(12)$$
$$54 = 30 + 24$$
$$54 = 54$$

▼ EXERCISE 2-3-1

Solve for x and check.

1. $x + 3 = 5$

$X = 5 - 3$
$X = 2$
$2 + 3 = 5$

2. $x - 4 = 6$

$X = 6 + 4$
$X = 10$
$10 - 4 = 6$

3. $x + 7 = 2$

$X = 2 - 7$
$X = -5$
$-5 + 7 = 2$

4. $x + 4 = 11$

$X = 11 - 4$
$X = 7$
$7 + 4 = 11$

5. $x - 7 = 2$

$X = 2 + 7$
$X = 9$
$9 - 7 = 2$

6. $4 - x = -5$

7. $x + 1 = 9$

$X = 9 - 1$
$X = 8$
$8 + 1 = 9$

8. $x - 5 = 1$

9. $4 + x = -5$

$X = -5 - 4$
$X = -9$
$4 - 9 = -5$

10. $x + 3.4 = 8.1$

11. $x - 8 = 8$

$X = 8 + 8$
$X = 16$
$16 - 8 = 8$

12. $2x = 3x - 5$

13. $x + \dfrac{3}{4} = 4\dfrac{2}{5}$

14. $x - \dfrac{1}{2} = 4$

15. $7x + 3 = 6x$

16. $x - 15 = 52$

17. $x - \dfrac{2}{3} = \dfrac{3}{5}$

18. $2x = 5 - 3x$

19. $x + 5 = 5$

20. $x - 3.6 = 2.7$

21. $3 - 7x = 6x$

22. $2x = x + 1$

23. $x - 11.5 = 4.6$

24. $x + 12 = 10$

25. $3x = 2x + 7$

26. $2x - 3 = 5$

27. $3x - 4 = 2x + 5$

28. $7x = 6x + 1$

29. $5x - 9 = 31$

30. $3x + 4 = 2x - 5$

31. $8 + x = 2x$

32. $3x - 1 = 1$

33. $3x + 4 = 4x - 5$

34. $5x + 2 = 4x + 10$

35. $14x - 5 = 9$

36. $x + 1 = 2x + 3$

37. $7x + 5 = 6x + 14$

38. $8x - 3 = 1$

39. $x + 5 = 2x + 6$

40. $11x = 7x + 20$

41. $3x = 10 - 2x$

42. $x - 1 = 2x - 3$

43. $8x = 3x + 5$

44. $6x = 21 - x$

45. $x + 8 = 2$

46. $4x + 3 = x + 18$

47. $4x = 5 - x$

48. $x - 8 = 2$

49. $5x + 4 = 2x + 25$

50. $2x = 5 - 4x$

51. $12x + 4 = x - 3$

52. $8x + 5 = 2x + 5$

53. $12x - 4 = x + 3$

54. $5x = 2 - 3x$

55. $17x + 15 = 20x + 3$

56. $3x - 4 = 8 - x$

57. $5x + 7 = 4x + 1$

58. $x + 9 = 5x + 1$

59. $3x - 3 = 12 - 3x$

$3x + 3x = 12 + 3$

$6x = 15$

$x = 15/6$

$x = 2\frac{1}{3}$

60. $x + 2 = 10 - 3x$

61. $5x - 7 = 4x + 1$

62. $3(x + 2) = 3$

63. $2(x + 4) = 3(x - 1)$

64. $3(x - 2) = 3$

65. $2(4 - x) = 3(1 - x)$

66. Two sides of a triangle are 10.1 m and 13.5 m. If the perimeter of the triangle is 47 m, find the length of the third side. (Use $P = a + b + c$.)

67. If the perimeter of a rectangle is 39.8 cm and the width is 7.5 cm, what is the length? (Use $P = 2\ell + 2w$.)

68. What was the cost of an item that sold for $12.95 if the margin was $3.05? (Use $S = C + M$.)

69. The total earnings (E) of a waitress are equal to the sum of the tips (t) and the product of the hourly rate (r) and the number of hours (h) worked ($E = rh + t$). If Jane worked eight hours at $1.50 per hour, and her total earnings were $50.75, what did she make in tips?

70. Using the formula $M = S - C$, determine the selling price of an article that costs \$38.75 and is sold with a margin of \$5.65.

71. $5F = 9C + 160$ is a relationship between Fahrenheit (F) and Celsius (C) temperatures. Find the Celsius temperature if the temperature is 86° F.

72. A relationship between Fahrenheit (F) and Celsius (C) temperature is given by $5F - 160 = 9C$. Find the Fahrenheit temperature if the Celsius temperature is 32° C.

73. Find the margin on the sale of an item if the profit was \$10.35 and the overhead was \$3.54. (Use $P = M - O$.)

74. The distance (s) in meters above the ground at a given time (t) of an object dropped from a height (h) is given by the formula $s = h - 4.9t^2$. Find the height that an object was dropped from if it strikes the ground in three seconds. (Hint: Use $s = 0$.)

 **2-4 THE MULTIPLICATION RULE**

O B J E C T I V E S

Upon completing this section you should be able to:

1. Use the multiplication rule to solve equations.
2. Solve proportions.
3. Solve some basic applied problems using the multiplication rule.

In the two previous sections we have solved equations using division and addition. Since subtraction is adding the negative, we have actually used three of the four basic operations to change a given equation to an equivalent equation. We now proceed to multiplication.

> *If each side of an equation is multiplied by the same nonzero number,* the resulting equation is equivalent to the original equation.

In elementary arithmetic the most difficult operations are those involving fractions. The multiplication rule allows us to avoid these operations when solving equations by finding an equivalent equation that contains only whole numbers.

Remember that to multiply a whole number by a fraction use the rule

$$A \cdot \frac{B}{C} = \frac{AB}{C}.$$

Example 1 $10 \times \dfrac{2}{3} = \dfrac{10(2)}{3} = \dfrac{20}{3}$

Note that in each case only the numerator of the fraction is multiplied by the whole number.

Example 2 $5 \times \dfrac{3}{5} = \dfrac{15}{5}$ or 3

Example 3 $6 \times \dfrac{2}{3} = \dfrac{12}{3}$ or 4

When multiplying a whole number by a fraction, some answers are whole numbers and some are still fractions. Do you see why?

Example 4   Solve for x if $\dfrac{2}{3}x = 18$.

Solution   Keep in mind that we wish to obtain x alone on one side of the equation. Also keep in mind that we would like to get an equation in whole numbers that is equivalent to the equation given. With these things in mind we multiply both sides of the equation by 3.

$$\frac{2}{3}x = 18$$

$$(3)\left(\frac{2}{3}x\right) = 3(18)$$

$$2x = 54$$

To eliminate the fraction we need to multiply by a number that is divisible by 3.

We now have an equation that contains only whole numbers but is equivalent to the original fractional equation. So we proceed to solve for x.

We could have multiplied both sides by 6, 9, 12, and so on, but the equation is simpler and easier to work with if we use the smallest multiple.

$$2x = 54$$

$$\frac{2x}{2} = \frac{54}{2} \qquad \text{(division rule)}$$

$$x = 27$$

Check: $\dfrac{2}{3}x = 18$

$$\frac{2}{3}(27) = 18$$

$$18 = 18$$

Example 5   Solve for x if $\dfrac{5}{8}x = 12$.

Solution

$$\frac{5}{8}x = 12$$

$$8\left(\frac{5}{8}x\right) = (8)(12) \qquad \text{(multiplication rule)}$$

$$5x = 96$$

$$\frac{5x}{5} = \frac{96}{5} \qquad \text{(division rule)}$$

$$x = 19\frac{1}{5}$$

See if you obtain the same solution by multiplying each side of the original equation by 16.

Check:

$$\frac{5}{8}x = 12$$

$$\left(\frac{5}{8}\right)\left(19\frac{1}{5}\right) = 12$$

$$12 = 12$$

Example 6   Solve for x if $\dfrac{2}{3}x = \dfrac{3}{5}$.

Solution   Here the task is the same but a little more complex. We have two fractions to eliminate. We must multiply both sides of the equation by a number that is divisible by both 3 and 5. It is best to use the smallest of such numbers (called the *least common multiple*). We will multiply by 15.

What is the least common denominator for $\dfrac{2}{3}$ and $\dfrac{3}{5}$?

$$\frac{2}{3}x = \frac{3}{5}$$

$$(15)\left(\frac{2}{3}x\right) = (15)\left(\frac{3}{5}\right)$$

$$10x = 9$$

$$\frac{10x}{10} = \frac{9}{10} \qquad \text{(division rule)}$$

$$x = \frac{9}{10}$$

Would you get the same solution if you multiplied both sides by 30 or 45 or 60?

Check:

$$\frac{2}{3}x = \frac{3}{5}$$

$$\left(\frac{2}{3}\right)\left(\frac{9}{10}\right) = \frac{3}{5}$$

$$\frac{3}{5} = \frac{3}{5}$$

Example 7 Solve for x if $\frac{5}{8}x + 3 = 5\frac{1}{2}$.

Solution The least common multiple for 8 and 2 is 8, so we multiply each side of the equation by 8.

$$\frac{5}{8}x + 3 = 5\frac{1}{2}$$

$$(8)\left(\frac{5}{8}x\right) + 8(3) = 8\left(\frac{11}{2}\right) \quad \text{(multiplication rule)}$$

$$5x + 24 = 44$$

Before multiplying, change any mixed numbers to improper fractions. In this example change $5\frac{1}{2}$ to $\frac{11}{2}$.

Be careful that *each* term on both sides of the equation is multiplied by the same number.

$$5x + 24 = 44$$

$$5x + 24 - 24 = 44 - 24 \quad \text{(addition rule)}$$

$$5x = 20$$

$$\frac{5x}{5} = \frac{20}{5} \quad \text{(division rule)}$$

$$x = 4$$

Note that in this example we used three rules to find the solution.

Check: $\quad \frac{5}{8}x + 3 = 5\frac{1}{2}$

$$\frac{5}{8}(4) + 3 = 5\frac{1}{2}$$

$$\frac{5}{2} + 3 = 5\frac{1}{2}$$

$$\frac{11}{2} = 5\frac{1}{2}$$

$$5\frac{1}{2} = 5\frac{1}{2}$$

Example 8 Solve for x if $\frac{2}{3}x + x = \frac{5}{8}$.

Solution The least common multiple for 3 and 8 is 24, so we multiply each side of the equation by 24.

$$24\left(\frac{2}{3}x\right) + 24(x) = 24\left(\frac{5}{8}\right)$$

$$16x + 24x = 15$$

$$40x = 15$$

$$\frac{40x}{40} = \frac{15}{40} \quad \text{(division rule)}$$

$$x = \frac{15}{40}$$

$$x = \frac{3}{8}$$

Remember that *each term* must be multiplied by 24.

All answers should always be given in the simplest form.

Remember, always check the answer by substituting it in the *original* equation.

Check:

$$\frac{2}{3}x + x = \frac{5}{8}$$

$$\frac{2}{3}\left(\frac{3}{8}\right) + \frac{3}{8} = \frac{5}{8}$$

$$\frac{2}{8} + \frac{3}{8} = \frac{5}{8}$$

$$\frac{5}{8} = \frac{5}{8}$$

Example 9 Solve for x if $\dfrac{3}{2}\left(x + \dfrac{1}{5}\right) = \dfrac{2}{3}$.

Solution Be very careful when solving a problem that contains parentheses. Remember that parentheses group elements together as a single element. When multiplying both sides of an equation by a number, be careful to multiply each *term*. As long as parentheses are present it is easy to become confused as to which part is a term and which part is perhaps only part of a factor. To avoid confusion *always* remove all parentheses first.

Recall that to remove parentheses multiply each term inside the parentheses by the number immediately preceding the parentheses.

$$\frac{3}{2}\left(x + \frac{1}{5}\right) = \frac{2}{3}$$

$$\frac{3}{2}x + \frac{3}{10} = \frac{2}{3}$$

In this form we immediately recognize terms and see that the least common multiple is 30. Multiply each term of each side of the equation by 30.

$$30\left(\frac{3}{2}x\right) + 30\left(\frac{3}{10}\right) = 30\left(\frac{2}{3}\right)$$

$$45x + 9 = 20$$

$$45x + 9 - 9 = 20 - 9 \qquad \text{(addition rule)}$$

$$45x = 11$$

$$\frac{45x}{45} = \frac{11}{45} \qquad \text{(division rule)}$$

$$x = \frac{11}{45}$$

Check:

$$\frac{3}{2}\left(\frac{11}{45} + \frac{1}{5}\right) = \frac{2}{3}$$

$$\frac{3}{2}\left(\frac{20}{45}\right) = \frac{2}{3}$$

$$\frac{2}{3} = \frac{2}{3}$$

Example 10 Solve for x if $5x - \dfrac{2}{3}(x - 6) = \dfrac{1}{2}$.

Solution Again, first remove parentheses.

$$5x - \frac{2}{3}x + 4 = \frac{1}{2}$$

When removing parentheses, be careful to obtain the correct signs.

The least common multiple is 6, so multiply each term of the equation by 6.

$$6(5x) - 6\left(\frac{2}{3}x\right) + (6)(4) = (6)\frac{1}{2}$$

$$30x - 4x + 24 = 3$$

$$26x + 24 = 3 \qquad \text{(combining like terms)} \qquad 30x - 4x = 26x$$

$$26x + 24 - 24 = 3 - 24 \qquad \text{(addition rule)}$$

$$26x = -21$$

$$\frac{26x}{26} = \frac{-21}{26} \qquad \text{(division rule)}$$

$$x = -\frac{21}{26}$$

Check:
$$5x - \frac{2}{3}(x - 6) = \frac{1}{2}$$

$$5\left(-\frac{21}{26}\right) - \frac{2}{3}\left(-\frac{21}{26} - 6\right) = \frac{1}{2}$$

Again, be careful of signs.

$$-\frac{105}{26} - \frac{2}{3}\left(-\frac{177}{26}\right) = \frac{1}{2}$$

$$-\frac{105}{26} + \frac{118}{26} = \frac{1}{2}$$

$$\frac{13}{26} = \frac{1}{2}$$

$$\frac{1}{2} = \frac{1}{2}$$

▼ **EXERCISE 2–4–1**

Solve for x.

1. $\dfrac{1}{2}x = 6$

2. $\dfrac{1}{3}x = 12$

3. $\dfrac{2}{3}x = -8$

4. $-\dfrac{5}{8}x = -14$ **5.** $\dfrac{2}{3}x = 10$ **6.** $\dfrac{3}{5}x = 9$

7. $\dfrac{3}{5}x = -15$ **8.** $\dfrac{3}{4}x = 8$ **9.** $\dfrac{2}{3}x = 5$

10. $-\dfrac{3}{4}x = 8$ **11.** $\dfrac{7}{8}x = 42$ **12.** $\dfrac{3}{7}x = 14$

13. $-\dfrac{3}{4}x = -8$ **14.** $\dfrac{1}{2}x = \dfrac{3}{4}$ **15.** $\dfrac{5}{7}x = -12$

16. $\dfrac{7}{8}x = -42$

17. $\dfrac{2}{3}x = \dfrac{5}{8}$

18. $\dfrac{3}{5}x = \dfrac{5}{6}$

19. $\dfrac{5}{8}x = \dfrac{1}{4}$

20. $-\dfrac{2}{3}x = \dfrac{2}{3}$

21. $\dfrac{5}{6}x = \dfrac{2}{3}$

22. $\dfrac{2}{5}x = \dfrac{4}{15}$

23. $-\dfrac{5}{8}x = \dfrac{1}{2}$

24. $\dfrac{7}{3}x = -\dfrac{5}{8}$

25. $\dfrac{1}{2}x + 5 = .12$

26. $\dfrac{4}{5}x - 5 = \dfrac{1}{2}$

27. $\dfrac{1}{4}x - \dfrac{2}{3} = \dfrac{1}{2}$

28. $\dfrac{2}{3}x + x + 18 = 0$

29. $\dfrac{7}{8}x + 4 = \dfrac{21}{4}$

30. $\dfrac{3}{4}x - \dfrac{1}{2} = x + \dfrac{1}{3}$

31. $\dfrac{1}{2}x + 12 = \dfrac{3}{4}x$

32. $\dfrac{2}{3}x - 8 = 14 - x$

33. $2x - \dfrac{3}{5} = \dfrac{3}{10}x + 8$

34. $\dfrac{4}{9}x - 2 = \dfrac{1}{3}x + \dfrac{1}{2}$

35. $\dfrac{5}{6}x + \dfrac{2}{3} = \dfrac{1}{2}x + \dfrac{3}{4}$

36. $\dfrac{1}{2}(x + 4) = 5$

37. $\dfrac{2}{3}\left(x - \dfrac{3}{8}\right) = \dfrac{1}{2}$

38. $6 - \dfrac{3}{8}(x - 2) = 14$

39. Find the altitude h of a triangle if the area A is 12 cm² and the base b is 3 cm.

$$\left(\text{Use } A = \frac{1}{2}bh.\right)$$

40. The distance (s) in feet that a falling object travels in t seconds is given by $s = \frac{1}{2}gt^2$, where g is the acceleration due to gravity. Find g if an object fell 144 feet in 3 seconds.

41. A relationship between Fahrenheit (F) and Celsius (C) temperature is given by $F = \frac{9}{5}C + 32$. Find C if $F = 212°$.

42. The volume of a rectangular solid is given by the formula $V = \ell wh$, where ℓ is length, w is width, and h is height. If a room is $16\frac{1}{2}$ feet long and 15 feet wide, how high must it be to have a volume of 2,970 cubic feet?

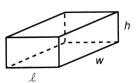

Solving simple equations by multiplying both sides by the same number occurs frequently in the study of *ratio* and *proportion*.

The **ratio** of a number x to a number y can be written as $x:y$ or $\dfrac{x}{y}$. In general, the fractional form is more meaningful and useful. Thus we will write the ratio of 3 to 4 as $\dfrac{3}{4}$.

A *ratio* is the quotient of two numbers.

A **proportion** is a statement that two ratios are equal.

Example 11 Solve for x: $\dfrac{x}{15} = \dfrac{2}{5}$

Solution We need to find a value of x such that the ratio of x to 15 is equal to the ratio of 2 to 5.

Why do we multiply both sides by 15?

Multiplying each side of the equation by 15, we obtain

$$\boxed{15}\left(\frac{x}{15}\right) = \boxed{15}\left(\frac{2}{5}\right)$$
$$x = 6.$$

Check this solution in the original equation.

Thus $\frac{6}{15} = \frac{2}{5}$.

Example 12 What number x has the same ratio to 3 as 6 has to 9?

Solution To solve for x we first write the proportion: $\frac{x}{3} = \frac{6}{9}$.

Make sure that you write the ratio in the correct order. "x is to 3" should be written as $\frac{x}{3}$ and "6 is to 9" should be written as $\frac{6}{9}$.

Next we multiply each side of the equation by 9.

$$\boxed{9}\left(\frac{x}{3}\right) = \boxed{9}\left(\frac{6}{9}\right)$$
$$3x = 6$$
$$x = 2$$

Check the solution.

Example 13 $\frac{2}{5}$ of a meter is how many 10ths of a meter?

Solution We set up the equation $\frac{x}{10} = \frac{2}{5}$. Multiply each side by 10.

Say to yourself, "2 is to 5 as x is to 10."

$$\boxed{10}\left(\frac{x}{10}\right) = \boxed{10}\left(\frac{2}{5}\right)$$
$$x = 4$$

Check!

Example 14 What number has the same ratio to 6 as 5 has to 8?

Solution The proportion is $\frac{x}{6} = \frac{5}{8}$.

Multiplying each side by 24 gives

$$\boxed{24}\left(\frac{x}{6}\right) = \boxed{24}\left(\frac{5}{8}\right)$$
$$4x = 15$$

Check!

$$x = \frac{15}{4}.$$

Example 15 Two sons were to divide an inheritance in the ratio of 3 to 5. If the son who received the larger portion got $20,000, what was the total amount of the inheritance?

Again, be careful in setting up the proportion. In the ratio $\frac{3}{5}$, 5 is the larger portion. Therefore since $20,000 is the larger portion, it must also appear in the denominator.

Solution In the proportion $\frac{3}{5} = \frac{x}{20,000}$, x represents the amount received by the other son.

$$\frac{3}{5} = \frac{x}{20,000}$$

$$\frac{3}{5}(20,000) = \frac{x}{20,000}(20,000)$$

$$x = 12,000$$

We now add \$20,000 + \$12,000 to obtain the total amount of \$32,000.

Check!

Example 16 If the legal requirements for room capacity require 3 cubic meters of air space per person, how many people can legally occupy a room that measures 6 meters wide, 8 meters long, and 3 meters high?

Solution We first use $V = \ell wh$ to find the volume of the room.

$$V = 8 \times 6 \times 3$$
$$V = 144 \text{ cubic meters.}$$

We now have the proportion $\dfrac{1}{3} = \dfrac{x}{144}$

This means "1 person is to 3 cubic meters as x people are to 144 cubic meters."

$$(144)\frac{1}{3} = \frac{x}{144}(144)$$

$$x = 48$$

So 48 people would be the legal room capacity.

Check the solution.

▼ EXERCISE 2-4-2

Solve for x.

1. $\dfrac{x}{12} = \dfrac{3}{4}$

2. $\dfrac{x}{9} = \dfrac{2}{3}$

3. $\dfrac{x}{10} = \dfrac{2}{5}$

4. $\dfrac{x}{22} = \dfrac{5}{11}$

5. $\dfrac{x}{8} = \dfrac{3}{4}$

6. $\dfrac{x}{16} = \dfrac{1}{2}$

7. $\dfrac{x}{36} = \dfrac{5}{3}$

8. $\dfrac{x}{128} = \dfrac{7}{4}$

9. $\dfrac{x}{3} = \dfrac{5}{6}$

10. $\dfrac{x}{4} = \dfrac{3}{20}$

11. What number has the same ratio to 15 as 2 has to 3?

12. What number has the same ratio to 5 as 6 has to 30?

13. What number has the same ratio to 49 as 2 has to 7?

14. What number has the same ratio to 8 as 2 has to 3?

15. The ratio of miles to kilometers is 5 to 8. How many miles are there in 100 kilometers?

16. The ratio of teeth on a gear A to those on gear B is 5 to 7. If gear B has 35 teeth, how many teeth does gear A have?

17. The ratio of kilograms to pounds is 5 to 11. Find your weight in kilograms.

18. The ratio of the production of two machines is 5 to 8. How many items will the slower machine produce while the faster one produces 2,400?

19. If an automobile uses three gallons of gasoline to travel 69 miles, how many gallons are necessary to travel 161 miles?

20. A company manufactured 1,200 calculators. They tested 100 calculators at random and found 2 were defective. What is the total number of defective calculators they could expect?

21. If two teaspoons of an instant coffee are used for three cups of coffee, how many are needed for eighteen cups?

22. In a certain population three out of five people have brown eyes. If seventy-five people are selected at random, how many could be expected to have brown eyes?

23. Hockey player Phil Esposito played 1,241 games in 17 seasons. How many games did he average in 3 seasons?

24. A wildlife management team, conducting a study on a particular lake, caught, tagged, and released 75 black bass. Several weeks later, they caught 125 black bass and observed that 3 of them were tagged. What is the total number of black bass they could expect to be present in the lake?

25. The "golden ratio" is 1 to 1.618. This ratio is believed by some to be the width to length ratio of a rectangle most pleasing to the eye. Assuming this to be true, what width would a bookcover be if the length is 27 cm and it is to be most pleasing to the eye? (correct to two decimal places)

◣◤ 2–5 COMBINING RULES FOR SOLVING EQUATIONS

In the previous sections we have introduced and used the division, addition, and multiplication rules. In many equations you have seen that more than one rule must be used to arrive at a solution. In fact, many equations can require all the rules plus almost every other manipulation of algebra that we have learned.

In this section a step-by-step method is introduced. This method is not the only approach to equations involving more than one rule, but experience has shown that following this order gives a smoother, more mistake-free procedure.

> **Step 1** Remove all parentheses.
> **Step 2** Eliminate fractions by multiplying each term of the equation by the least common multiple of all denominators.
> **Step 3** Simplify by combining like terms on each side of the equation.

OBJECTIVES

▼

Upon completing this section you should be able to:
1. Use combinations of the various rules to solve more complex equations.
2. Apply the orderly steps established in this section to systematically solve equations.

Consider these six steps a guide to solving equations.

Step 4 Add (or subtract) the necessary terms to obtain an equation with the numbers of arithmetic on one side and the unknown on the other.

Step 5 Divide both sides of the equation by the coefficient of the unknown.

Remember, the coefficient is the number being multiplied by the letter. (That is, in the expression $5x$, the coefficient is 5.)

Step 6 Check your solution.

It is suggested that the student think each step in order. If a particular step does not apply, then skip it and go to the next one.

Example 1   Solve for x if $3x - 4 = 5x + 2$.

Solution   **Step 1** No parentheses, so skip this step.

Step 2 No fractions, so skip this step.

Step 3 No terms can be combined.

Step 4 This is our starting point.

Solve the equation in this example by first subtracting $3x$ from each side.

$$3x - 4 = 5x + 2$$
$$3x - 4 + 4 = 5x + 2 + 4 \qquad \text{(addition rule)}$$
$$3x = 5x + 6$$
$$3x - 5x = 5x - 5x + 6 \qquad \text{(addition rule)}$$

Step 5
$$-2x = 6$$
$$\frac{-2x}{-2} = \frac{6}{-2} \qquad \text{(division rule)}$$
$$x = -3$$

Step 6 *Check:*
$$3x - 4 = 5x + 2$$
$$3(-3) - 4 = 5(-3) + 2$$
$$-9 - 4 = -15 + 2$$
$$-13 = -13$$

Example 2   Solve for x if $\frac{2}{3}x + 7 = 2x - 1$.

Solution   **Step 1** Skip.

Step 2 Starting point.

> **WARNING**
>
> Many students obtain wrong solutions because they fail to multiply *every* term of the equation by the same factor.

$$\frac{2}{3}x + 7 = 2x - 1$$
$$3\left(\frac{2}{3}x\right) + 3(7) = 3(2x) - 3(1) \qquad \text{(multiplication rule)}$$
$$2x + 21 = 6x - 3$$

Step 3 Skip.

Solve $2x + 21 = 6x - 3$ by first subtracting $2x$ from each side.

Step 4
$$2x + 21 - 21 = 6x - 3 - 21 \qquad \text{(addition rule)}$$
$$2x = 6x - 24$$
$$2x - 6x = 6x - 6x - 24 \qquad \text{(addition rule)}$$
$$-4x = -24$$

Step 5 $$\frac{-4x}{-4} = \frac{-24}{-4}$$ (divison rule)

$$x = 6$$

Step 6 *Check:* $\frac{2}{3}x + 7 = 2x - 1$

$$\frac{2}{3}(6) + 7 = 2(6) - 1$$

$$4 + 7 = 12 - 1$$

$$11 = 11$$

Note that the check uses the original equation.

Example 3   Solve for x if $3(x - 4) + 5 = 4x - 11$.

Solution   First remove parentheses.

$$3(x - 4) + 5 = 4x - 11$$
$$3x - 12 + 5 = 4x - 11$$

Since there are no fractions, we will now combine terms.

$$3x - 7 = 4x - 11$$

Try solving this equation by subtracting $3x$ from each side.

Now, adding 7 and subtracting $4x$ from both sides gives

$$-x = -4$$
$$\frac{-x}{-1} = \frac{-4}{-1}$$ (division rule)
$$x = 4.$$

Check: $3(x - 4) + 5 = 4x - 11$
$$3(4 - 4) + 5 = 4(4) - 11$$
$$3(0) + 5 = 16 - 11$$
$$5 = 5$$

Example 4   Solve for x if $\frac{1}{2}x + 2\left(x + \frac{2}{3}\right) = \frac{3}{4}\left(x - \frac{1}{2}\right) + 2$.

Solution $$\frac{1}{2}x + 2x + \frac{4}{3} = \frac{3}{4}x - \frac{3}{8} + 2$$ (removing parentheses)

The LCM for 2, 3, 4, and 8 is 24.

$$24\left(\frac{1}{2}x\right) + 24(2x) + 24\left(\frac{4}{3}\right) = 24\left(\frac{3}{4}x\right) - 24\left(\frac{3}{8}\right) + 24(2)$$

Again, notice *every* term is multiplied by 24.

$$12x + 48x + 32 = 18x - 9 + 48$$
$$60x + 32 = 18x + 39$$ (combining terms)
$$42x = 7$$ (addition rule)
$$\frac{42x}{42} = \frac{7}{42}$$ (division rule)
$$x = \frac{1}{6}$$

Some students hesitate to check an answer when it looks complicated. The successful student will check the answer since it will tell if the solution is correct.

Check:

$$\frac{1}{2}x + 2\left(x + \frac{2}{3}\right) = \frac{3}{4}\left(x - \frac{1}{2}\right) + 2$$

$$\frac{1}{2}\left(\frac{1}{6}\right) + 2\left(\frac{1}{6} + \frac{2}{3}\right) = \frac{3}{4}\left(\frac{1}{6} - \frac{1}{2}\right) + 2$$

$$\frac{1}{12} + 2\left(\frac{5}{6}\right) = \frac{3}{4}\left(-\frac{1}{3}\right) + 2$$

$$\frac{1}{12} + \frac{5}{3} = -\frac{1}{4} + 2$$

$$\frac{1}{12} + \frac{20}{12} = -\frac{1}{4} + \frac{8}{4}$$

$$\frac{21}{12} = \frac{7}{4}$$

$$\frac{7}{4} = \frac{7}{4}$$

Example 5   Solve for x if $\frac{2}{3}\left(x + \frac{3}{4}\right) = \frac{5}{6}$.

Remember to multiply every term inside the parentheses by the number preceding the parentheses.

Solution

$$\frac{2}{3}x + \frac{1}{2} = \frac{5}{6} \qquad \text{(removing parentheses)}$$

$$6\left(\frac{2}{3}x\right) + 6\left(\frac{1}{2}\right) = 6\left(\frac{5}{6}\right) \qquad \text{(multiplication rule)}$$

$$4x + 3 = 5$$

$$4x = 2 \qquad \text{(addition rule)}$$

$$\frac{4x}{4} = \frac{2}{4} \qquad \text{(division rule)}$$

$$x = \frac{1}{2}$$

Check:

$$\frac{2}{3}\left(x + \frac{3}{4}\right) = \frac{5}{6}$$

$$\frac{2}{3}\left(\frac{1}{2} + \frac{3}{4}\right) = \frac{5}{6}$$

$$\frac{2}{3}\left(\frac{5}{4}\right) = \frac{5}{6}$$

$$\frac{5}{6} = \frac{5}{6}$$

Even a complicated looking equation such as this can be solved by taking each of the six steps *one at a time.*

Example 6   Solve for x if $\frac{3}{7}(2x + 14) + 3 = \frac{1}{2}(5x + 18)$.

Solution

$$\frac{6x}{7} + 6 + 3 = \frac{5x}{2} + 9 \qquad \text{(removing parentheses)}$$

$$12x + 84 + 42 = 35x + 126 \qquad \text{(multiplication rule)}$$

$$12x + 126 = 35x + 126 \qquad \text{(combining terms)}$$

$$-23x = 0 \qquad \text{(addition rule)}$$

$$\frac{-23x}{-23} = \frac{0}{-23} \qquad \text{(division rule)}$$

$$x = 0$$

Recall that zero divided by any nonzero number is zero.

Check:

$$\frac{3}{7}(2x + 14) + 3 = \frac{1}{2}(5x + 18)$$

$$\frac{3}{7}[2(0) + 14] + 3 = \frac{1}{2}[5(0) + 18]$$

$$\frac{3}{7}(14) + 3 = \frac{1}{2}(18)$$

$$6 + 3 = 9$$

$$9 = 9$$

Example 7   Solve for x if $\dfrac{x + 6}{5} = \dfrac{2}{3}x + 4$.

Solution   The LCM for 5 and 3 is 15, so multiply both sides by 15.

$$15\left(\frac{x + 6}{5}\right) = 15\left(\frac{2}{3}x\right) + 15(4) \qquad \text{(multiplication rule)}$$

$$3(x + 6) = 10x + 60$$

$$3x + 18 = 10x + 60 \qquad \text{(remove parentheses)}$$

$$-7x = 42 \qquad \text{(addition rule)}$$

$$x = -6 \qquad \text{(division rule)}$$

Many students would forget to multiply the 4 by 15.

Notice that parentheses were introduced when we multiplied the left side by the LCM. Why?

Check:

$$\frac{x + 6}{5} = \frac{2}{3}x + 4$$

$$\frac{-6 + 6}{5} = \frac{2}{3}(-6) + 4$$

$$\frac{0}{5} = -4 + 4$$

$$0 = 0$$

Example 8   Mrs. Jones collected $960.00 principal and interest from an investment. If the interest was one-half of the original investment, what was the amount of the original investment?

The "principal" is the original amount invested.

Solution   Let $x =$ original investment, then $\dfrac{1}{2}x =$ amount of interest.

$$\text{So } x + \frac{1}{2}x = 960$$

$$2x + x = 1{,}920 \qquad \text{(multiplication rule)}$$

$$3x = 1{,}920$$

$$x = \$640$$

Check: $\qquad x + \dfrac{1}{2}x = 960$

$$640 + \dfrac{1}{2}(640) = 960$$

$$640 + 320 = 960$$

$$960 = 960$$

▼ **EXERCISE 2-5-1**

Solve for x.

✓ **1.** $\dfrac{1}{2}x + \dfrac{3}{4} = 5$

$$X + \frac{6}{4} = 10$$
$$X = 10 - 1.5$$
$$X = 8.5$$

2. $3x + 5 = x + 7$

✓ **3.** $\dfrac{1}{3}(x + 6) = x + 4$

$$\frac{1}{3}X + \frac{6}{3} = X + 4$$
$$X + 6 = 3X + 12$$
$$X - 3X = 12 - 6$$
$$-2X = 6$$
$$X = {}^-3$$

4. $x = \dfrac{1}{2} - \dfrac{2}{3}x$

5. $4x - 3 = x - 9$

$$4X - X = -9 + 3$$
$$3X = -6$$
$$X = -2$$

6. $3 - x = 2(1 - x)$

✓ **7.** $\dfrac{2}{3}x + \dfrac{1}{2} = \dfrac{5}{6}$

$$2X + \frac{3}{2} = \frac{15}{6}$$
$$2X = \frac{15}{6} - \frac{9}{6}$$
$$2X = 1$$
$$X = .5$$

8. $3x - 1 = 2(x - 5)$

✓ **9.** $6x + 2 = 2\left(x - \dfrac{1}{3}\right)$

$$6X + 2 = 2X - \frac{2}{3}$$
$$6X - 2X = -\frac{6}{3} - \frac{2}{3}$$
$$4X = -\frac{8}{3} \qquad X = -\frac{2}{3}$$
$$12X = -8$$
$$X = -8/12$$

10. $2x - \dfrac{3}{5} = \dfrac{2}{3}x$

11. $3x + 2 = 5x - 8$

$3x + 2 = 5x - 8$

$3x - 5x = -8 - 2$

$-2x = -10$

$-x = -5$

$x = 5$

12. $3(x + 4) + 1 = 9 - x$

13. $2x - \dfrac{2}{3} = \dfrac{3}{5}x + 2$

$2x - \dfrac{2}{3} = \dfrac{3}{5}x + 2$

$\dfrac{30}{15}x - \dfrac{10}{15} = \dfrac{9}{15}x + \dfrac{30}{15}$

$\dfrac{30}{15}x - \dfrac{9}{15}x = \dfrac{30}{15} + \dfrac{10}{15}$

$\dfrac{21}{15}x = \dfrac{40}{15}$ $21x = 40$

$x = 1\,^{19}/_{21}$

14. $x - 7 = 4x + 5$

15. $\dfrac{5}{8}(2x + 4) = \dfrac{2}{3}(x - 5)$

$\dfrac{5}{8}(2x + 4 = \dfrac{2}{3}(x - 5)$ $\dfrac{30}{24}x - \dfrac{16}{24}x = \dfrac{14}{24}x$

$\dfrac{10}{8}x + \dfrac{20}{8} = \dfrac{2}{3}x - \dfrac{16}{3}$ $\dfrac{-140}{24} = \dfrac{14}{24}x$

$\dfrac{30}{24}x + \dfrac{60}{24} = \dfrac{16}{24}x - \dfrac{80}{24}$ $14 = -140$

$x = -10$

16. $2x + 3 = 3x + 5$

17. $4(x + 6) = 9x + 16$

$4x + 24 = 9x + 16$

$4x - 9x = 16 - 24$

$-5x = -8$

$x = 8/5 \ \text{a} \ 1\,^3/_5$

18. $4x - \dfrac{3}{8} = \dfrac{3}{4}x + 3$

19. $x - 6 = 5x - 14$

$x + 5x = -14 + 6$

$-4x = -8$

$x = 2$

20. $\dfrac{3}{4}x + \dfrac{2}{3} = \dfrac{5}{6} + \dfrac{1}{2}x$

21. $3(x - 4) = 5 + 2(x + 1)$

$3x - 12 = 5 + 2x + 2$

$3x - 12 = 2x + 7$

$3x - 2x = 7 + 12$

$x = 19$

22. $3x - \dfrac{5}{7} = \dfrac{3}{4}x + 4$

✓ 23. $5x - 4 = 2x + 6$

$5x - 2x = 6 + 4$

$3x = 10$

$X = 10/3$

$X = 3\frac{1}{3}$

24. $\frac{3}{5}(2x - 1) + 4 = 5 - (x + 7)$

✓ 25. $\frac{2}{3}x - 5 = \frac{2}{9}x + 1$

$\frac{6}{9}x - \frac{45}{9} = \frac{2}{9}x + 1$

$\frac{6}{9}x - \frac{2}{9}x = \frac{45}{9} + \frac{9}{9}$ $X = 13\frac{1}{2}$

$\frac{4}{9}x = \frac{54}{9}$ $4X = 54$

26. $3x + 7 = 2x + \frac{1}{3}(2x + 1)$

✓ 27. $5x + \frac{3}{4} = 5 - \frac{1}{3}x$

$5x + 3/4 = 5 - \frac{1}{3}x$

$\frac{60}{12}x + \frac{9}{12} = \frac{60}{12} - \frac{4}{12}x$ $X = \frac{51}{64}$

$\frac{60}{12}x + \frac{4}{12}x = \frac{60}{12} - \frac{9}{12} = 64X = 51$

28. $3x + 7 = 5x - 4$

29. $2(x + 3) + 4x = 4\left(x - \frac{1}{2}\right)$

30. $\frac{3}{8} + 5x = \frac{3}{4}x + 2$

31. $3x + 2(x - 5) = 7 - (x + 3)$

32. $3 - x = \frac{1}{2}(7 + 2x)$

33. $\frac{5}{8}x - \frac{1}{2} = \frac{3}{5}x + 3$

34. $5x - 3(x + 1) = 5$

35. $x + \dfrac{3}{8} = 3x + \dfrac{3}{4}$

$8x + 3 = 24x + 6$

$3 - 6 = 24x - 8x$

$\dfrac{-3}{16} = 16x \qquad -\dfrac{3}{16} = X$

36. $\dfrac{4}{5}x + 6 - \dfrac{1}{3}x = \dfrac{2x}{5} + 9$

37. $7x + 5 = 2(x - 1) - 21$

$7x + 5 = 2x - 23$

$5x = -28$

$X = -\dfrac{28}{5} = -5\dfrac{3}{5}$

38. $\dfrac{1}{3}(x + 3) + 8 = \dfrac{1}{2}x + 11$

39. $x + \dfrac{2}{3}x = \dfrac{3}{5}x + 32$

$15x + 10x = 9x + 480$

$25x = 9x + 480$

$25x - 9x = 480$

$16x = 480 \qquad X = 30$

40. $5 + 6x - 3 = 2 + 4x$

41. $\dfrac{2}{3}(2x - 7) + 2 = \dfrac{1}{2}(4x - 13)$

$\dfrac{4x}{3} - \dfrac{14}{3} + 2 = \dfrac{4x}{2} - \dfrac{13}{2} \qquad 23 - 4x$

$8x - 28 + 12 = 12x - 39 \qquad \dfrac{23}{4} = X$

$8x - 28 + 12 = 12x - 39 \qquad 5\dfrac{3}{4} = X$

$39 + 16 = 4x$

42. $\dfrac{2}{3}x = 11 + \dfrac{1}{2}x - \dfrac{3}{5}$

43. $3(x + 1) + 3 = 7(x - 2)$

44. $x + \dfrac{19}{6} = \dfrac{2}{3} - \dfrac{1}{2}(2x - 5)$ $\qquad \boxed{X = 0}$

$x + \dfrac{19}{6} = \dfrac{2}{3} - X - \dfrac{5}{2}$

$2x = \dfrac{2}{3} + \dfrac{19}{6} + \dfrac{5}{2} \qquad 2x = \dfrac{-5}{6}$

$2x = \dfrac{4}{6} - \dfrac{19}{6} + \dfrac{15}{6} \qquad 12X = -10$

$2x = \dfrac{19}{6} - \dfrac{19}{6} = \qquad X = -\dfrac{5}{6}$

45. $\dfrac{4}{5}x - 10 = \dfrac{2}{3}x + 2 - \dfrac{1}{5}x$

46. $5x - (2x - 3) = 4(x + 9)$

47. $\dfrac{5}{6}(x + 6) = \dfrac{1}{2}(10 - x)$

48. $6x - \dfrac{2}{3} = \dfrac{5}{8}x + 4 + \dfrac{2}{3}x$

49. $\dfrac{x}{2} = \dfrac{1}{5} - x$

50. $\dfrac{3}{5}(4 - x) - 6 = x + 6$

51. $\dfrac{4}{7}x = 28 - \dfrac{2}{3}x$

52. $4 - \dfrac{x + 2}{3} = 1$

53. $\dfrac{1}{3}(x + 6) + 5 = 2x + 28$

54. $2x + \dfrac{1}{3} = \dfrac{5}{8}x + 1$

55. $13 - \dfrac{2}{5}x = 25 - \dfrac{3}{4}x$

56. $x - \dfrac{1}{2} = \dfrac{x}{3} + 7$

57. $\dfrac{x - 3}{4} - 9 = 11$

58. $\dfrac{5}{8}(x + 16) = \dfrac{2}{3}(2x + 15)$

59. $\dfrac{2}{3}x - 1 = \dfrac{3}{5}x + \dfrac{1}{6}$

60. $x - \dfrac{5}{9} = \dfrac{5}{12}x + \dfrac{3}{4}$

61. $5(x + 4) = \dfrac{1}{3}(x - 10)$

62. $\dfrac{2x + 5}{3} = 0$

63. $\dfrac{2}{3} - 3(x - 1) = \dfrac{3}{5}$

64. $2x - 3(x - 2) = \dfrac{1}{2}(x + 1)$

65. $\dfrac{2}{3}x + 1 = x - \dfrac{5}{2}$

66. $4x - 2(x - 3) = \dfrac{1}{4}(x + 3)$

67. $x - \dfrac{1}{5}x + 1 = \dfrac{1}{3}(x - 5)$

68. $\dfrac{2}{3}x + \dfrac{x}{2} - \dfrac{1}{2} = \dfrac{1}{3}(x - 14)$

69. $\dfrac{2}{3}x - \dfrac{1}{2}x = x + \dfrac{1}{6}$

70. $\dfrac{x}{5} - \dfrac{2}{3}x + \dfrac{1}{2} = \dfrac{1}{3}(x - 4)$

71. The width of a rectangle is two-thirds the length. If the perimeter of the rectangle is 250 feet, find the length. (The formula for the perimeter is $P = 2\ell + 2w$.)

72. An article cost a dealer $65.00. If the margin is to be one-fourth of the selling price, what should the article sell for? (Cost + margin = selling price.)

73. The selling price of a car is $8,952.30. If the dealer has set the margin to be one-twentieth of the cost, what was the cost of the car? (Use the formula in problem 72.)

74. A certain handyman estimates the cost of small jobs by assuming the amount for materials to be one-third the amount for labor. He estimated repairing Mr. Smith's porch would amount to $64.00 (labor plus materials). What did he expect to get for his labor?

75. A sale sign states "Ladies' Dress Shoes, $\frac{1}{3}$ Off." If the sale price of a certain pair of shoes is $45.00, what was the original price?

 2-6 LITERAL EQUATIONS

OBJECTIVES

Upon completing this section you should be able to:

1. Identify literal equations.
2. Apply the rules previously learned in order to solve literal equations.

Think of c as the unknown and all other letters as known quantities.

An equation having more than one letter is sometimes called a **literal** equation. It is occasionally necessary to solve such an equation for one of the letters in terms of the others. The step-by-step procedure discussed and used in the previous section is still valid.

Example 1 Solve for c: $3(x + c) - 4y = 2x - 5c$

Solution First remove parentheses.

$$3x + 3c - 4y = 2x - 5c$$

At this point we should note that since we are solving for c, we wish to get c on one side and all other terms on the other side of the equation. By adding terms to both sides of the equation, we get

$$3c + 5c = 2x - 3x + 4y$$
$$8c = -x + 4y \qquad \text{(combining terms)}$$
$$c = \frac{-x + 4y}{8} \qquad \text{(division rule)}$$

Example 2   Solve for x: $3abx + cy = 2abx + 4cy$

Solution   First subtract $2abx$ from both sides.

We want all terms involving x on one side and all other terms on the opposite side.

$$3abx + cy - 2abx = 2abx + 4cy - 2abx$$
$$\text{or}$$
$$abx + cy = 4cy$$

Remember, abx is the same as $1abx$.

Subtract cy from both sides.

$$abx + cy - cy = 4cy - cy$$
$$\text{or}$$
$$abx = 3cy$$

Divide both sides by ab.

Divide by the coefficient of x, which in this case is ab.

$$\frac{abx}{ab} = \frac{3cy}{ab}$$
$$\text{or}$$
$$x = \frac{3cy}{ab}$$

Example 3   Solve for x: $\dfrac{2}{3}(x + y) = 3(x + a)$

Solution   First remove parentheses.

$$\frac{2}{3}x + \frac{2}{3}y = 3x + 3a$$

Multiply both sides by 3.

Remember, you must multiply every term by 3.

$$2x + 2y = 9x + 9a$$

Subtract $9x$ from both sides.

$$-7x + 2y = 9a$$

Subtract $2y$ from both sides.

$$-7x = 9a - 2y$$

Divide both sides by -7.

$$x = \frac{9a - 2y}{-7}$$

Let's solve the equation
$$2x + 2y = 9x + 9a$$
by first subtracting $2x$ from both sides obtaining
$$2y = 7x + 9a.$$
Now subtract $9a$ from both sides getting
$$2y - 9a = 7x.$$
Finally dividing both sides by 7 gives
$$\frac{2y - 9a}{7} = x.$$
Is this solution equivalent to
$$x = \frac{9a - 2y}{-7}\ ?$$

Sometimes the form of an answer may be changed. In this example we could multiply both numerator and denominator by (-1) (this does not change the value of the answer) and obtain

$$x = \frac{9a - 2y}{-7} = \frac{(-1)(9a - 2y)}{(-1)(-7)} = \frac{-9a + 2y}{7} = \frac{2y - 9a}{7}.$$

The advantage of this last expression over the first is that there are not so many negative signs in the answer.

The most commonly used literal expressions are formulas from geometry, physics, business, electronics, and so on.

A trapezoid is a four-sided figure with two parallel sides.

Example 4 $A = \frac{1}{2}h(b + c)$ is the formula for the area of a trapezoid.

Solve for c.

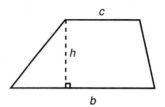

Removing parentheses does not mean merely to erase them. We must multiply each term inside the parentheses by the factor preceding the parentheses.

Solution   First remove the parentheses.

$$A = \frac{1}{2}hb + \frac{1}{2}hc$$

Multiply both sides by 2.

$$2A = hb + hc$$

Subtract hc from both sides.

$$2A - hc = hb$$

Subtract $2A$ from both sides.

$$-hc = hb - 2A$$

Divide both sides by $-h$.

$$c = \frac{hb - 2A}{-h}$$

It is often best to write the final answer using the least number of negative signs.

We may change the form of the answer.

$$c = \frac{hb - 2A}{-h} = \frac{(-1)(hb - 2A)}{(-1)(-h)}$$

$$= \frac{-hb + 2A}{h} = \frac{2A - hb}{h}$$

This is a variation of the interest formula $I = prt$ where t represents time in years. Notice that $\frac{D}{365}$ is time in years.

Example 5 $I = \frac{prD}{365}$ is a formula giving interest earned for a period of D days when the principal and yearly rate are known. Find the yearly rate when the amount of interest, the principal, and the number of days are all known.

Solution   The problem requires solving $I = \frac{prD}{365}$ for r.

$$365I = prD \qquad \text{(multiplication rule)}$$

$$\frac{365I}{pD} = r \qquad \text{(division rule)}$$

Notice that pD is the coefficient of r.

Notice that in this example r was left on the right side and thus the computation was simpler. We can rewrite the answer another way if we wish.

$$r = \frac{365I}{pD}$$

▼ EXERCISE 2-6-1

Solve for x.

1. $2x + y = x + 3y$

$2x - x = 3y - y$

$x = 2y$

2. $3x - 2y = x + 4y$

3. $3x + 2y = 6y + x$

$3x - x = 6y - 2y$

$2x = 4y$

$x = 2y$

4. $2x + 8y = 5x - y$

5. $2(x + a) = x - 4a$

$2x + 2a = x - 4a$

$2x - x = -4a - 2a$

$x = -6a$

6. $3x + a = 7(x - a)$

7. $4(x - 2y) + y = 3x - 5y$

$4x - 8y + y = 3x - 5y$

$4x - 7y = 3x - 5y$

$4x - 3x = 7y - 5y$

$x = 2y$

8. $3(2x + y) = 19y + 4x$

9. $4a - 2x = 3(4x - 8a)$

10. $11x + 15y = 5(3x - y)$ **11.** $3ax + 2bc = 4(ax - 5bc)$ **12.** $2ax + bc = 2(3ax + bc)$

13. $7abx - 3y = 4abx + 5a$ **14.** $3(2a - 5x) = 2(a + 3x)$ **15.** $4x + 3a = 5(2b - x)$

16. $5(x - a) = \dfrac{2}{3}(x + 1)$ **17.** $3(2x + 5) = \dfrac{1}{5}(x + a)$ **18.** $\dfrac{2}{5}(x - a) = 4(x + a)$

19. $7x + a - 3b = 5x + 3a$ **20.** $2x + 3(x - a) = 7(x - a)$

21. $A = \frac{1}{2}bh$ is the formula for the area of a triangle. Solve for h.

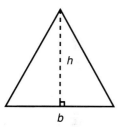

22. The formula for simple interest is $I = prt$. Solve for r.

23. A distance formula from physics is $s = \frac{1}{2}gt^2$. Solve for g.

24. $F = \frac{9}{5}C + 32$ is a formula for changing temperature from Celsius to Fahrenheit degrees. Solve for C.

25. First solve the equation $I = \frac{prD}{365}$ for p (principal) (see example 5), then find how much should be invested for 90 days at 18% annual rate of interest to yield $200.00 interest.

◣◤ 2-7 FIRST-DEGREE INEQUALITIES

We have already discussed the set of **rational numbers** as those numbers that can be expressed as a ratio of two integers. There is also a set of numbers, called the **irrational numbers,** that cannot be expressed as the ratio of integers. This set includes such numbers as π, $\sqrt{5}$, $\sqrt[3]{7}$, and so on. The set composed of rational and irrational numbers together is called the **real numbers.**

Given any two real numbers a and b, it is always possible to state $a = b$ or $a \neq b$. Many times we are only interested in whether or not two numbers are equal, but there are situations in which we also wish to represent the relative size of numbers that are not equal.

The symbols $<$ and $>$ are **inequality symbols** or **order relations** and are used to show the relative sizes of the values of two numbers. We usually read the symbol $<$ as "less than." For instance, $a < b$ is read as "a is less than b." We usually read the symbol $>$ as "greater than." For instance, $a > b$ is read as "a is greater than b." Notice that we have stated that we usually read $a < b$ as a is less than

OBJECTIVES

Upon completing this section you should be able to:

1. Use the inequality symbol to represent the relative positions of two numbers on the number line.
2. Graph inequalities on the number line.
3. Solve first-degree inequalities.

The statement $2 < 5$ can be read as "two is less than five" or "five is greater than two."

b. But this is only because we read from left to right. In other words, "*a* is less than *b*" is the same as saying "*b* is greater than *a*." Actually then, we have just one symbol that is written two ways only for convenience of reading. One way to remember the meaning of the symbol is that the pointed end is toward the *lesser* of the two numbers.

 < 5

What positive number can be added to 2 to give 5?

> **$a < b$,** "*a* is less than *b*," if and only if there is a positive number *c* that can be added to *a* to give $a + c = b$.

What positive number can be added to -6 to give 2?

In simpler words this definition states that *a* is less than *b* if we must add something to *a* to get *b*. Of course, the "something" must be positive.

If you think of the number line, you know that adding a positive number is equivalent to moving to the right on the number line. This gives rise to the following alternative definition, which may be easier to visualize.

This also means that *b* is to the right of *a*.

> **$a < b$** means that *a* is to the left of *b* on the real number line.

Example 1

We could also write $6 > 3$.

$3 < 6$ because 3 is to the left of 6.

Example 2

We could also write $0 > -4$.

$-4 < 0$ because -4 is to the left of zero.

Example 3

$4 > -2$ because 4 is to the right of -2

or

$-2 < 4$ since -2 is to the left of 4.

Example 4

$-6 < -2$ since -6 is to the left of -2.

EXERCISE 2-7-1

Locate the following numbers on the number line and replace the question mark with > or <.

1. 6 ? 10   6 < 10

```
-3 -2 -1  0  1  2  3  4  5  6  7  8  9 10
```

2. −6 ? −10   −6 > −10

```
-10 -9 -8 -7 -6 -5 -4 -3 -2 -1  0  1  2  3
```

3. −3 ? 3   −3 < 3

```
-4 -3 -2 -1  0  1  2  3  4
```

4. −4 ? −1   −4 < −1

```
-5 -4 -3 -2 -1  0  1
```

5. 4 ? 1   4 > 1

```
-1  0  1  2  3  4  5
```

6. 1 ? 4   1 < 4

```
-1  0  1  2  3  4  5  6  7
```

7. −2 ? −3   −2 > −3

```
-5 -4 -3 -2 -1  0  1  2  3
```

8. −5 ? −3   −5 < −3

```
-6 -5 -4 -3 -2 -1  0  1  2
```

9. 0 ? 7   0 < 7

```
-1  0  1  2  3  4  5  6  7
```

10. 0 ? −3   0 > −3

```
-4 -3 -2 -1  0  1  2  3  4
```

The mathematical statement $x < 3$ (read as "x is less than 3") indicates that the variable x can be any number less than (or to the left of) 3. Remember, we are considering the real numbers and not just integers, so do not think of the values for x for $x < 3$ as only $2, 1, 0, -1$, and so on.

As a matter of fact, to name the number x that is the largest number less than 3 is an impossible task. It can be indicated on the number line, however. To do this we need a symbol to represent the meaning of a statement such as $x < 3$.

The symbols **(** and **)** used on the number line will indicate the end point is not included in the set.

Example 5   Graph $x < 3$ on the number line.

This graph represents every real number less than 3.

Solution

Note that the graph has an arrow indicating that the line continues without end to the left.

Example 6   Graph $x > 4$ on the number line.

This graph represents every real number greater than 4.

Solution

Example 7   Graph $x > -5$ on the number line.

This graph represents every real number greater than -5.

Solution

Example 8   Make a number line graph showing that $x > -1$ and $x < 5$. (Remember that "and" indicates both conditions must apply.)

This graph represents all real numbers that are between -1 and 5.

Solution

The statement in the last example, $x > -1$ and $x < 5$, can be condensed to read $-1 < x < 5$.

Example 9   Graph $-3 < x < 3$.

Solution

If we wish to include the endpoint in the set, we use a different symbol, $\leq$ or $\geq$. We read these symbols as "equal to or less than" and "equal to or greater than."

Example 10 $x \geq 4$ indicates the number 4 *and* all real numbers to the right of 4 on the number line.

What does $x < 4$ represent?

> The symbols **[** and **]** used on the number line will indicate the end point is included in the set.

Example 11 Graph $x \geq 1$ on the number line.

Solution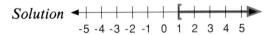

This graph represents the number 1 and all real numbers greater than 1.

Example 12 Graph $x \leq -3$ on the number line.

Solution

This graph represents all real numbers less than or equal to -3.

Example 13 Write an algebraic statement represented by the following graph.

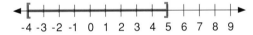

Solution $x \geq -2$

Example 14 Write an algebraic statement for the following graph.

![number line from -4 to 9 with bracket at -4 and at 5]

Solution $x \geq -4$ and $x \leq 5$, which may be written as $-4 \leq x \leq 5$

This graph represents all real numbers between -4 and 5 *including* -4 and 5.

Example 15 Write an algebraic statement for the following graph.

![number line from -4 to 9 with open paren at -2 and bracket at 4]

Solution $x > -2$ and $x \leq 4$, which may be written as $-2 < x \leq 4$

This graph includes 4 but *not* -2.

Example 16 Graph $x > 2\frac{1}{2}$ on the number line.

Solution This example presents a small problem. How can we indicate $2\frac{1}{2}$ on the line? If we estimate the point, then another person might misread the statement. Could you possibly tell if the point represents $2\frac{1}{2}$ or maybe $2\frac{7}{16}$? Since the purpose of a graph is to clarify, always name the end point.

Where would you label the number π on the number line?

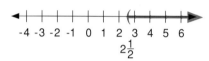

```
◄─┼──┼──┼──┼──┼──┼──┼─(┼──┼──┼──►
 -4 -3 -2 -1  0  1  2  3  4  5  6
                     2½
```


▼ EXERCISE 2-7-2

Construct a graph on the number line for each of the following:

1. $x > 7$

```
◄─┼──┼──┼──┼──┼──┼──┼──┼─╫─┼──┼──►
 -1  0  1  2  3  4  5  6  7  8  9
```

2. $x < 5$

```
◄─┼──┼──┼──┼──┼──┼──┼──┼──┼──►
 -1  0  1  2  3  4  5  6  7  8
```

3. $x < -1$

```
◄─┼──┼─╫─┼──┼──┼──┼──┼──┼──►
 -3 -2 -1  0  1  2  3  4  5  6
```

4. $x > -3$

```
◄─┼──┼──┼──┼──┼──┼──┼──┼──┼──►
 -4 -3 -2 -1  0  1  2  3  4  5
```

5. $x \geq 1$

```
◄─┼──┼──┼──┼──┼─╫─┼──┼──┼──►
 -4 -3 -2 -1  0  1  2  3  4  5
```

6. $x \leq 1$

```
◄─┼──┼──┼──┼──┼──┼──┼──┼──┼──►
 -4 -3 -2 -1  0  1  2  3  4  5
```

7. $x \geq -4$

```
◄─┼──┼──┼──┼──┼──┼──┼──┼──►
 -4 -3 -2 -1  0  1  2  3  4  5
```

8. $x \leq -2$

```
◄─┼──┼──┼──┼──┼──┼──┼──┼──┼──►
 -4 -3 -2 -1  0  1  2  3  4  5
```

9. $3 < x < 7$

```
◄─┼──┼──┼──┼──┼─(┼──┼──┼─)┼──►
 -2 -1  0  1  2  3  4  5  6  7  8
```

10. $-2 \leq x \leq 5$

```
◄─┼──┼──┼──┼──┼──┼──┼──┼──┼──►
 -4 -3 -2 -1  0  1  2  3  4  5
```

11. $-5 \leq x < 0$

```
◄─┼──┼──┼──┼──┼──┼─)┼──┼──┼──►
 -6 -5 -4 -3 -2 -1  0  1  2  3
```

12. $x > 2\frac{3}{10}$

```
◄─┼──┼──┼──┼──┼──┼──┼──┼──┼──►
 -2 -1  0  1  2  3  4  5  6
```

13. $x < 5.4$

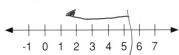

14. $-2\frac{1}{2} \le x < 3\frac{1}{2}$

15. $6 < x < 7$

In each of the following write an algebraic statement for the graph.

16.

$x > 0$

17.

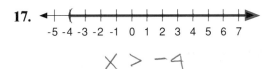

$x > -4$

18.

$x < -4$

19.

$-4 \ge x \le 4$

20.

$x \le 2$

The solutions for first-degree inequalities involve the same basic rules as equations—but with one exception.

> *If the same quantity is added to each side of an inequality,* the results are unequal in the same order.

Example 17 If $5 < 8$, then $5 + 2 < 8 + 2$.

Example 18 If $7 < 10$, then $7 - 3 < 10 - 3$.

This becomes $x < 4$.

Example 19 If $x + 6 < 10$, then $x + 6 - 6 < 10 - 6$.

This last example indicates how we would solve the inequality $x + 6 < 10$.

Example 20 Solve the inequality and graph the solution on a number line.

Solution

Note that the procedure is the same as in solving equalities.

$$2(x + 2) < x - 5$$
$$2x + 4 < x - 5$$
$$2x + 4 - 4 < x - 5 - 4 \qquad \text{(adding } -4 \text{ to both sides)}$$
$$2x - x < -9 \qquad \text{(adding } -x \text{ to both sides)}$$
$$x < -9$$

(number line from -10 to 4)

▼ EXERCISE 2-7-3
_____

Use the addition rule to solve the following and graph the solutions.

1. $x + 3 < 7$

2. $x - 5 < 0$

3. $3x + 4 < 2x - 1$

4. $5x < 4x + 1$

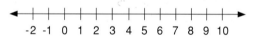

5. $2(x + 3) \leq x + 9$

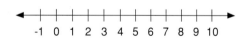

6. $-3(x - 1) \geq 2(1 - 2x)$

7. $5(x + 3) > 2(2x - 1)$

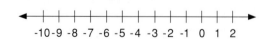

8. $2x + 3(x + 2) \leq 2(2x + 1) - 3$

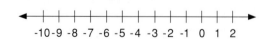

9. $3x + 5 \geq 2(x + 5) - 10$

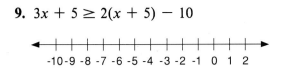

10. $3x + 7 - 2x \leq 7$

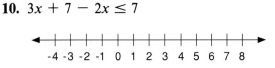

We will now use the addition rule to show an important concept concerning multiplication or division of inequalities.

Suppose $x > a$.

Now add $-x$ to both sides by the addition rule.

$$x - x > a - x$$
$$0 > a - x$$

Now add $-a$ to both sides by the addition rule.

$$0 - a > a - x - a$$
$$-a > -x$$

The last statement, $-a > -x$, can be rewritten as $-x < -a$.

We have thus established the following:

"If $x > a$, then $-x < -a$." This translates into the following rule.

> Remember, *adding* the same quantity to both sides of an inequality does *not* change its direction.

> For example:
> If $5 > 3$ then $-5 < -3$.

If an inequality is multiplied or divided by a negative number the results will be unequal in the *opposite* order.

Example 21 Solve for x if $-2x > 6$.

Solution To obtain x on the left side we must divide by -2.

$$\frac{-2x}{-2} < \frac{6}{-2}$$

Note the change from $>$ to $<$.

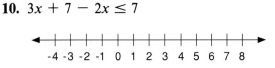

$$x < -3.$$

> Notice that as soon as we divide by a negative quantity, we must change the direction of the inequality.

Take special note of this rule. Each time you divide by a negative number, you must change the direction of the inequality symbol. This is the *only* difference between solving equations (first-degree, one variable) and solving inequalities.

Division or multiplication by a positive number will *not* change the sense of the inequality, *but* the same operations by a negative number *will* change it. This one point is a source of many errors—so be very careful!

Let us now review the step-by-step method from the previous section and note the differences when solving inequalities.

Again, these five steps should be memorized.

Step 1 Remove all parentheses. (No change)

Step 2 Eliminate fractions by multiplying all terms by the least common denominator of all fractions. (No change, since multiplying by a positive number)

Step 3 Simplify by combining like terms on each side of the inequality. (No change)

Step 4 Add or subtract quantities to get the unknown on one side, the numbers on the other. (No change)

Step 5 Divide both sides of the inequality by the coefficient of the unknown. If the coefficient is positive, the inequality will remain the same. If the coefficient is negative, the inequality will be reversed. (This is the important difference.)

Example 22 Solve for x and graph the results.

$$2(x - 3) \geq 3x - 4$$

Solution

$$2x - 6 \geq 3x - 4 \qquad \text{(removing parentheses)}$$
$$2x - 6 + 6 \geq 3x - 4 + 6 \qquad \text{(adding terms)}$$
$$2x \geq 3x + 2$$
$$2x - 3x \geq 3x + 2 - 3x \qquad \text{(adding terms)}$$
$$-x \geq 2$$
$$\frac{-x}{-1} \leq \frac{2}{-1} \qquad \text{(dividing by the coefficient of } x\text{)}$$
$$x \leq -2 \qquad \text{(Note the change of the symbol.)}$$

What must be done when dividing by a negative?

Example 23 Solve for x and graph the results.

$$\frac{2}{3}x + 2 > 3(x + 1)$$

Solution $\frac{2}{3}x + 2 > 3x + 3$ (removing parentheses)

$2x + 6 > 9x + 9$ (multiply by 3)

$-7x > 3$ (adding like terms)

$\frac{-7x}{-7} < \frac{3}{-7}$ (divide by -7)

$x < -\frac{3}{7}$

Notice again, the direction of the inequality reverses.

Example 24   Solve for x and graph the results.

$$5(x + 3) + 2x \geq \frac{3}{4}(x - 7) + 5x - 1$$

Solution $5x + 15 + 2x \geq \frac{3}{4}x - \frac{21}{4} + 5x - 1$ (removing parentheses)

$20x + 60 + 8x \geq 3x - 21 + 20x - 4$ (multiplying by 4)

$28x + 60 \geq 23x - 25$ (combining like terms)

$5x \geq -85$ (adding terms to both sides)

$x \geq -17$ (dividing by the coefficient of x)

Remember to multiply *every* term by 4.

Dividing by a positive quantity does not change the direction of the inequality.

Since we must graph $x \geq -17$, we can allow each unit to represent 5.

Example 25 Solve for x and graph the results.

$$\frac{1}{3}x + \frac{2}{3} < \frac{5}{6}x - 1$$

Solution $2x + 4 < 5x - 6$ (multiply by 6)

$-3x < -10$ (add terms to both sides)

$\frac{-3x}{-3} > \frac{-10}{-3}$ (divide by -3)

$x > +\frac{10}{3}$

Don't forget to label the endpoint.

EXERCISE 2-7-4

Solve the following inequalities and graph the results on the number line.

1. $3x < 6$

$3x < 6$
$x < 2$

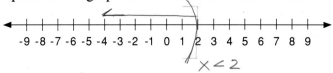

$x < 2$

2. $3x \geq 15$

$x \geq 5$

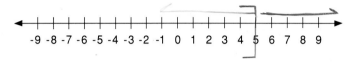

3. $2x \leq -8$

$x \leq -4$

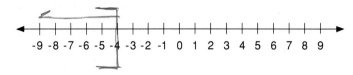

4. $-3x > 9$

$-3x > 9$
$x < -3$

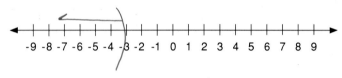

5. $-5x < -10$

$-5x < -10$
$x > 2$

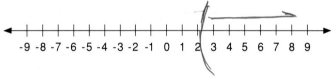

6. $-x \geq 3$

$-x \geq 3$
$x \leq -3$

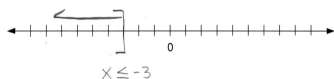

$x \leq -3$

7. $\frac{1}{2}x < 2$

$x < 4$

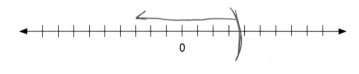

8. $\frac{1}{5}x \leq -1$

$x \leq -5$

9. $\frac{2}{3}x > 4$

$2x > 12$
$x > 6$

10. $-\frac{1}{2}x < 3$

$-2x < 6$
$x > -3$

11. $-\frac{1}{3}x \geq -1$

$-1x \geq -3$

$x \leq 3$

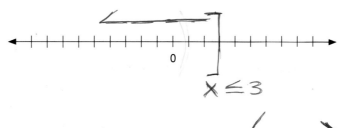

$x \leq 3$

12. $-\frac{2}{5}x < -2$

$-2x < -10$

$x > 5$

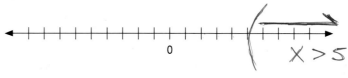

$x > 5$

13. $5x - 1 > 3(x + 1)$

$5x - 1 > 3x + 3$

$5x - 3x > 3 + 1$

$2x > 4 \quad x > 2$

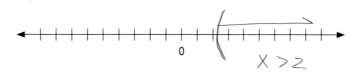

$x > 2$

14. $3(x + 3) + 2x > 3 + 7x$

$3x + 9 + 2x > 3 + 7x$

$5x + 9 > 3 + 7x$

$-2x > -6 \qquad x < 3$

15. $2(x + 3) \leq 7(x + 2) + 2$

$2x + 6 \leq 7x + 14 + 2$

$2x + 6 \leq 7x + 16$

$6 - 16 \leq 7x - 2x \quad \frac{-10 \leq 5x}{5}$

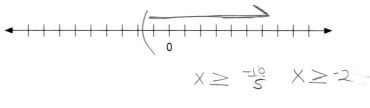

$x \geq \frac{-10}{5} \quad x \geq -2$

16. $x > 4\left(\frac{1}{3}x + \frac{1}{2}\right)$

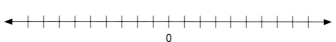

17. $\frac{2}{3}x + 1 \geq 2x - \left(\frac{x}{2} - 6\right)$

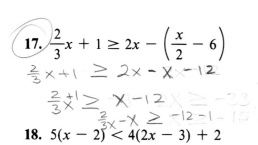

$\frac{2}{3}x + 1 \geq 2x - x - 12$

$\frac{2}{3}x + 1 \geq x - 12$

$\frac{2}{3}x - x \geq -12 - 1$

$\frac{2}{3}x - 1x \geq -13$

$\frac{1}{3}x \geq -13$

$x \geq -39 \; ?$

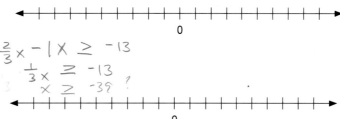

18. $5(x - 2) < 4(2x - 3) + 2$

19. $6x + 2\left(\frac{1}{3}x - 10\right) > 0$

$6x + \frac{2}{3}x - 20 > 0$

$18x + 2x - 60 > 0$

$20x > 60$

$x > 3$

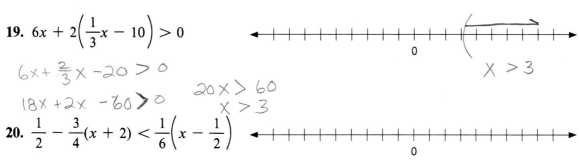

$x > 3$

20. $\frac{1}{2} - \frac{3}{4}(x + 2) < \frac{1}{6}\left(x - \frac{1}{2}\right)$

21. $\frac{4}{5}(x - 2) \le \frac{1}{2}(x - 1) + \frac{1}{4}$

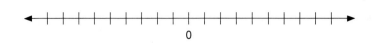

22. $\frac{2}{3}(x - 5) + 1 \ge \frac{1}{5}(2x - 3) + \frac{2}{3}$

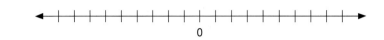

23. $3(x - 4) + 5 \le 5x - 11$

24. $\frac{1}{2}x + 2\left(x + \frac{2}{3}\right) \ge \frac{3}{4}\left(x - \frac{1}{2}\right) + 2$

25. $\frac{2}{3}\left(x + \frac{3}{4}\right) < \frac{5}{6}$

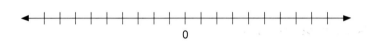

26. $\frac{x + 6}{5} > \frac{2}{3}x + 4$

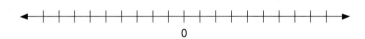

27. $3x + 7 \le 2x + \frac{1}{3}(2x + 1)$

28. $3x + 7 < 5x - 4$

29. $2(x + 3) + 4x > 4\left(x - \frac{1}{2}\right)$

30. $\dfrac{3}{8} + 5x < \dfrac{3}{4}x + 2$

31. $3x + 2(x - 5) \le 7 - (x + 3)$

32. $3 - x \ge \dfrac{1}{2}(7 + 2x)$

33. $\dfrac{5}{8}x - \dfrac{1}{2} < \dfrac{3}{5}x + 3$

34. $5x - 3(x + 1) < 5$

35. $x + \dfrac{3}{8} \ge 3x + \dfrac{3}{4}$

36. $\dfrac{4}{5}x + 6 - \dfrac{1}{3}x \le \dfrac{2x}{5} + 9$

37. $7x + 5 > 2(x - 1) - 21$

38. $\dfrac{1}{3}(x + 3) + 8 > \dfrac{1}{2}x + 11$

39. $x + \dfrac{6}{7} + \dfrac{2}{3}x < 3x - \dfrac{22}{7}$

40. $5 + 6x - 3 \le 2 + 4x$

41. $\dfrac{2}{3}(2x - 7) + 2 \ge \dfrac{1}{2}(4x - 13)$

42. $\dfrac{2}{3}x \le 10 + \dfrac{1}{2}x$

43. $3(x + 1) + 3 > 7(x - 2)$

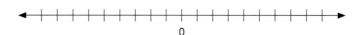

44. $x + \dfrac{19}{6} \le \dfrac{2}{3} + \dfrac{1}{2}(4x + 5)$

45. $\dfrac{4}{5}x - 10 > \dfrac{2}{3}x + 2 - \dfrac{1}{5}x$

46. $5x - (2x - 3) > 4(x + 9)$

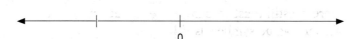

47. $\dfrac{5}{6}(x + 6) < \dfrac{1}{2}(10 - x)$

48. $\dfrac{x}{2} \ge \dfrac{1}{5} - x$

49. $\dfrac{3}{5}(4 - x) - 6 > x + 6$

0

50. $\dfrac{1}{3}(x + 6) + 5 \geq 2x + 28$

0

◣◥ 2–8 ABSOLUTE VALUE EQUATIONS AND INEQUALITIES

In section 1–2 we defined absolute value as the distance of a number from zero on the number line. We gave the formal definition, which we will repeat here.

> $|x| = x$ if x is zero or positive, $|x| = -x$ if x is negative.

O B J E C T I V E S

Upon completing this section you should be able to:

1. Solve equations involving absolute values.
2. Solve inequalities involving absolute values.

Absolute value is actually a measure of distance without regard to direction. This concept becomes important in situations where a measure of "nearness" is needed.

Example 1 Solve for x if $|x| = 3$.

Solution From the definition of absolute value we know the quantity within the absolute value symbols could either be 3 or -3 since $|3| = 3$ and $|-3| = 3$. So we have two solutions: $x = 3$ or $x = -3$.

You might want to briefly review that part of section 1–2 dealing with absolute value.

Example 2 Solve for x if $|x + 5| = 3$.

Solution Here we still must recognize that the quantity within the absolute value symbols is either 3 or -3.

So $|x + 5| = 3$ means that $x + 5 = 3$ or $x + 5 = -3$. We solve both equations and find $x = -2$ or $x = -8$.

Check: If $x = -2$, then

$$|x + 5| = |-2 + 5| = |3| = 3.$$

If $x = -8$, then

$$|x + 5| = |-8 + 5| = |-3| = 3.$$

Remember to always check your solutions in the original equation.

Example 3   Solve for x if $|3x + 1| = 7$.

Solution $|3x + 1| = 7$ means that

$$3x + 1 = 7 \qquad \text{or} \qquad 3x + 1 = -7$$
$$3x = 6 \qquad\qquad\qquad 3x = -8$$
$$x = 2 \qquad\qquad\qquad x = -\frac{8}{3}$$

Check: If $x = 2$, then

We substitute 2 for x.

$$|3x + 1| = |3(2) + 1| = |7| = 7.$$

If $x = -\frac{8}{3}$, then

We replace x with $-\frac{8}{3}$.

$$|3x + 1| = \left|3\left(-\frac{8}{3}\right) + 1\right| = |-7| = 7.$$

Example 4   Solve for x if $\left|\frac{2}{3}x + 4\right| = 10$.

Solution $\left|\frac{2}{3}x + 4\right| = 10$ means that

$$\frac{2}{3}x + 4 = 10 \qquad \text{or} \qquad \frac{2}{3}x + 4 = -10$$
$$2x + 12 = 30 \qquad\qquad 2x + 12 = -30$$
$$2x = 18 \qquad\qquad\qquad 2x = -42$$
$$x = 9 \qquad\qquad\qquad x = -21$$

Check these answers.

▼   **EXERCISE 2-8-1**

Solve for each of the following.

1. $|x| = 2$

    2 or -2

2. $|x| = 5$

3. $|x| = 9$

    9 -9

4. $|x| = 4$

5. $|x + 1| = 3$

    2 -4

6. $|x + 3| = 6$

7. $|x - 4| = 1$

$5 \quad 3$

8. $|x - 2| = 3$

9. $|x| = 0$

0

10. $|x + 4| = 0$

11. $|2x + 1| = 5$

$2x + 1 = 5$

$2x = 5 - 1$

$2x = 4$

$x = 2$

$2x + 1 = -5$

$2x = -5 - 1 \quad x = -3$

$2x = -6$

12. $|3x + 2| = 7$

13. $|3x - 2| = 4$

$3x - 2 = 4$

$3x = 4 + 2$

$3x = 6 \quad x = 2$

$3x - 2 = -4$

$3x = -4 + 2$

$3x = -2 \quad x = -\frac{2}{3}$

14. $|2x - 5| = 11$

15. $\left|\frac{1}{2}x + 4\right| = 6$ (ok)

$|x + 8 = 12 \qquad |x + 8 = -12$

$x = 12 - 8 \qquad x = -8 - 12$

$x = 4 \qquad\qquad x = -20$

16. $\left|\frac{1}{3}x + 1\right| = 8$

17. $\left|\frac{2}{5}x - 2\right| = 0$

$2x - 10 = 0$

$2x = 10$

$x = 5$

18. $\left|\frac{4}{5}x - 1\right| = 3$

19. $\left|\frac{3}{4}x + 7\right| = 2$ (ok)

$3x + 28 = 8$

$3x = 8 - 28$

$3x = -20$

$x = -\frac{20}{3}$

$x = 12 \quad (-8)$

20. $\left|\frac{2}{3}x + 3\right| = 1$

The statement $|x| = 5$ means that x is 5 units from zero giving us $x = 5$ or $x = -5$. Suppose we have the inequality $|x| < 5$? This expression means that x is *within* 5 units of zero and hence has all values between -5 and 5. As before, we can write $-5 < x < 5$, which can be read as "x is greater than -5 and less than 5."

Any number between -5 and 5 will have an absolute value of less than 5.

Now look at $|x| > 5$. This means that x must be more than 5 units away from zero. We can write this as $x < -5$ or $x > 5$. Note that we must use the word *or* since a number cannot be less than -5 and greater than 5 at the same time.

For any positive number a $|x| < a$ means $x > -a$ **and** $x < a$, which can be written as $-a < x < a$. $|x| > a$ means $x < -a$ **or** $x > a$.

"Less than" gives "between." "Greater than" gives an "excluded middle."

Example 5   Use the real number line to graph the solution of $|x + 1| < 10$.

Solution $|x + 1| < 10$ means that

This is from the rule.

$$x + 1 > -10 \quad \textit{and} \quad x + 1 < 10.$$

Solving, we have $x > -11$ and $x < 9$, which can be written as $-11 < x < 9$. The graph of this solution is

Note that -11 and 9 are *not* included in the solution.

Example 6   Graph $|x + 1| > 10$.

Solution $|x + 1| > 10$ means that

Again, this is from the rule.

$$x + 1 < -10 \quad \textit{or} \quad x + 1 > 10$$
$$x < -11 \quad \text{or} \quad x > 9.$$

The graph of this solution is

The numbers -11 through 9 are the "excluded middle." We sometimes refer to "$x < -11$ or $x > 9$" as being disjoint.

Example 7   Graph $|2x - 3| < 4$.

Solution $|2x - 3| < 4$ means that

$$2x - 3 > -4 \quad \textit{and} \quad 2x - 3 < 4$$
$$2x > -1 \qquad\qquad 2x < 7$$
$$x > -\frac{1}{2} \qquad\qquad x < \frac{7}{2}.$$

Of course, solutions very often involve fractions.

The graph of this solution is

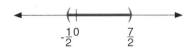

The symbols $\geq$ or $\leq$ will give solutions that include the endpoints.

Don't forget to use brackets when endpoints are included.

Example 8   Solve for x and graph the solution of $\left|\dfrac{1}{2}x - 3\right| \geq 5$.

Solution $\left|\dfrac{1}{2}x - 3\right| \geq 5$ means that

$$\dfrac{1}{2}x - 3 \leq -5 \qquad or \qquad \dfrac{1}{2}x - 3 \geq 5$$

$$x - 6 \leq -10 \qquad\qquad x - 6 \geq 10$$

$$x \leq -4 \qquad\qquad\qquad x \geq 16.$$

To clear fractions multiply *each* term by 2.

The graph of this solution is

Note the use of brackets.

Example 9   Solve for x and graph the solution of $\left|\dfrac{2}{3}x + 3\right| \leq 11$.

Solution $\left|\dfrac{2}{3}x + 3\right| \leq 11$ means that

$$\dfrac{2}{3}x + 3 \geq -11 \qquad and \qquad \dfrac{2}{3}x + 3 \leq 11$$

$$2x + 9 \geq -33 \qquad\qquad 2x + 9 \leq 33$$

$$2x \geq -42 \qquad\qquad\qquad 2x \leq 24$$

$$x \geq -21 \qquad\qquad\qquad x \leq 12.$$

Now multiply each term by 3.

The graph of this solution is

▼ **EXERCISE 2-8-2**

Solve and graph the solution for each of the following.

1. $|x + 2| < 5$

2. $|x + 3| < 2$

3. $|x - 3| \leq 1$

4. $|x - 4| \leq 4$

5. $|x + 1| > 3$

6. $|x + 3| > 5$

7. $|x + 4| < 10$

8. $|x + 2| < 6$

9. $|x + 7| > 3$

10. $|x + 5| > 2$

11. $|x + 2| \leq 3$

12. $|x + 1| \leq 6$

13. $\left| \dfrac{1}{2}x - 3 \right| > 5$

14. $\left| \dfrac{1}{3}x - 2 \right| > 7$

15. $|2x - 3| \geq 11$

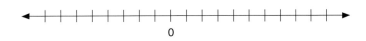

16. $|3x - 2| \geq 5$

17. $\left| \dfrac{2}{3}x + 1 \right| \leq 4$

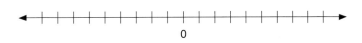

18. $\left| \dfrac{3}{4}x + 2 \right| \leq 8$

19. $\left| \dfrac{1}{3}x - 5 \right| > 1$

20. $\left| \dfrac{2}{3}x - 10 \right| > 4$

Key Words

Section 2–1

- An **equation** is a statement in symbols that two number expressions are equal.

- An **identity** is true for all values of the literal and arithmetic numbers in it.

- A **conditional equation** is true for only certain values of the literal numbers in it.

- A **solution** or **root** of an equation is the value of the variable that makes the equation a true statement.

- Two equations are **equivalent** if they have the same solution set.

Section 2–4

- A **ratio** is the quotient of two numbers.

- A **proportion** is a statement that two ratios are equal.

Section 2–6

- A **literal equation** is an equation involving more than one letter.

Section 2–7

- The symbols $<$ and $>$ are **inequality symbols** or **order relations.**

- $a < b$ means that a is to the left of b on the real number line.

- The double symbols $\leq$ or $\geq$ indicate that the endpoints are included in the solution set.

Section 2–8

- $|x| = a$ means $x = a$ **or** $x = -a$.

- $|x| < a$ means $x > -a$ **and** $x < a$.

- $|x| > a$ means $x < -a$ **or** $x > a$.

Procedures

Section 2–2

- If each term of an equation is divided by the same nonzero number, the resulting equation is equivalent to the original equation.

Section 2–3

- If the same quantity is added to both sides of an equation, the resulting equation is equivalent to the original equation.

Section 2–4

- If each side of an equation is multiplied by the same nonzero number, the resulting equation is equivalent to the original equation.

Section 2–5

- To solve an equation follow these steps:

  Step 1 Remove all parentheses.

  Step 2 Eliminate fractions by multiplying each term by the least common multiple of all denominators in the equation.

  Step 3 Combine like terms on each side of the equation.

  Step 4 Add or subtract terms to obtain the unknown quantity on one side and all other terms on the other side.

  Step 5 Divide each term by the coefficient of the unknown quantity.

  Step 6 Check your answer(s).

Section 2–6

- To solve a literal equation for one letter in terms of the others follow the same steps as indicated in section 2–5.

Section 2–7

- To solve an inequality follow the same steps as when solving an equation. The *only* difference may arise in step 5 (division). When an inequality is divided by a negative number, the results will be unequal in the *opposite* order.

Classify the equations as conditional or as an identity.

1. $2x = x + 3$

2. $5x - x = 4x$

3. $6x - 2x = 3x + x$

4. $7x - 3 = 4x$

5. $6x = 3x + 3$

6. $2x + 3 - x = x + 3$

Determine which pairs of equations are equivalent.

7. $3x - 4 = 2x$ and $x = 4$

8. $5x + 3 = 18$ and $x = 3$

9. $2x - 1 = 11$ and $x = 5$

10. $4x + 3 = 2x + 15$ and $x = 9$

11. $3x + 1 = x + 2$ and $x = \dfrac{1}{2}$

12. $6x + 1 - 2x = x + 1$ and $x = 0$

Solve for x.

13. $7x = 56$

14. $-5x = 20$

15. $x + 6 = 25$

16. $x - 12 = 36$

A N S W E R S

1. _____

2. _____

3. _____

4. _____

5. _____

6. _____

7. _____

8. _____

9. _____

10. _____

11. _____

12. _____

13. _____

14. _____

15. _____

16. _____

A N S W E R S

17. _____

18. _____

19. _____

20. _____

21. _____

22. _____

23. _____

24. _____

25. _____

26. _____

27. _____

28. _____

29. _____

30. _____

17. $4x = 6x + 10$

18. $2x = 104$

19. $x + 13 = 52$

20. $3(2x + 1) = 5(x + 1)$

21. $x - 7 = 68$

22. $3x = 25$

23. $2x + 3(x + 1) = 6x$

24. $3x = x + 8$

25. $x - (4x + 3) = 3(x - 9)$

26. $7x + 3 = \dfrac{1}{2}(x + 25) + 7$

27. $\dfrac{2}{3}x = 4$

28. $\dfrac{1}{3}(2x + 1) = \dfrac{1}{2}(x - 1)$

29. $5x = 41$

30. $\dfrac{2}{3}(x + 2) = \dfrac{1}{4}x - 2$

31. $7x = 5x + 9$

32. $\dfrac{1}{5}x = 20$

31. _____

33. $\dfrac{1}{3} + 2x = \dfrac{1}{2}x + 4$

34. $6x = 78$

32. _____

33. _____

34. _____

35. $3(x + 2) - 4 = \dfrac{2}{3}x + 7$

36. $4x + 5 = 2x + 17$

35. _____

36. _____

37. $5 - x + \dfrac{2}{3}x = \dfrac{1}{2}x$

38. $\dfrac{2}{5}x = \dfrac{4}{7}$

37. _____

38. _____

39. $x - \dfrac{2}{5} = \dfrac{1}{7}$

40. $4x + \dfrac{3}{4} = \dfrac{1}{2}x + 3$

39. _____

40. _____

41. $\dfrac{5}{6}x - 3 = \dfrac{1}{2}$

42. $14x = 168$

41. _____

42. _____

43. _____

44. _____

45. _____

46. _____

47. _____

48. _____

49. _____

50. _____

51. _____

52. _____

53. _____

54. _____

43. $3(2x - 1) + 2 = 5x + 11$

44. $\dfrac{2}{3}x - 18 = \dfrac{1}{2}x + 5$

45. $5x + 2 = x + 19$

46. $3x - 4 = 19$

47. $\dfrac{3}{8}x = \dfrac{5}{6}$

48. $\dfrac{4}{7}x - 5 = \dfrac{1}{2}x$

49. $7x - \dfrac{1}{2} = \dfrac{4}{5}x + 4$

50. $11x = 166$

51. $\dfrac{1}{3}(5x - 7) = \dfrac{3}{8}(x + 6) - 2$

52. $4x - 1 = 3x + 5$

53. $\dfrac{2}{3} + 5x = \dfrac{3}{2} + 3x$

54. $8x - 5 = x + 37$

55. $3x + 5 = 2x + 5$

56. $\dfrac{5}{6}x - \dfrac{2}{3} = \dfrac{1}{2}x + 1$

57. $3\dfrac{1}{2}x = 4\dfrac{1}{3}$

58. $x - 6 = 14 - 3x$

59. $\dfrac{3}{2}x - 2 = \dfrac{1}{5} + \dfrac{1}{3}x$

60. $\dfrac{3}{5}x + 1 = \dfrac{1}{2}(x + 3)$

61. $\dfrac{1}{4}x - \dfrac{1}{2} = \dfrac{2}{3} - \dfrac{1}{2}x$

62. $\dfrac{3}{4} + \dfrac{2}{3}x = \dfrac{1}{2}x + 2$

63. $\dfrac{x}{4} + 9 = \dfrac{3x}{2} - \dfrac{3}{2}$

64. $x - \dfrac{7}{8} = \dfrac{3}{5}x + \dfrac{3}{4}$

65. $2(xy + 3) = 5xy + z$

66. $3a(x - b) = 2ax - bc$

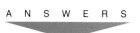

ANSWERS

55. _____

56. _____

57. _____

58. _____

59. _____

60. _____

61. _____

62. _____

63. _____

64. _____

65. _____

66. _____

67. _____

68. _____

69. _____

70. _____

71. _____

72. _____

73. _____

74. _____

75. _____

76. _____

77. _____

78. _____

67. $3(x + y) = 5x - 4y + 7$

68. $\dfrac{2}{3}x + \dfrac{5}{8}y = 4$

69. $7x + 2y = 6$

70. $\dfrac{1}{2}x + 4y = 12$

71. $3(2x - 1) + y = 5x + 3$

72. If $P = 2\ell + 2w$, solve for ℓ.

73. If $A = \dfrac{1}{2}h(a + b)$, solve for b.

74. If $V = \pi r^2 h$, solve for h.

75. What number x has the same ratio to 100 as 2 to 3?

76. Seven has the same ratio to some number x as 4 has to 12. Find x.

77. Gear A has 30 teeth and gear B has 35. If this same ratio is to be maintained and gear A is replaced by a gear with 12 teeth, how many teeth must be on the gear replacing B?

78. Graph the statement $x > -2$ on the number line.

0

79. Graph the statement $x \leq 5$ on the number line.

79. _____

80. Graph the statement $-3 < x \leq 4$ on the number line.

80. _____

Solve for x.

81. _____

81. $|x + 5| = 6$ **82.** $|x + 3| = 8$ **83.** $|3x - 1| = 10$

82. _____

84. $|2x - 5| = 3$ **85.** $\left|\dfrac{2}{3}x + 5\right| = 1$ **86.** $\left|\dfrac{3}{4}x + 12\right| = 9$

83. _____

84. _____

Solve for x and **graph the result on the number line.**

85. _____

87. $2x - 5 < x - 1$

86. _____

88. $3x - 2 < 2x + 3$

87. _____

88. _____

89. $5x - 6 \leq 7x + 4$

89. _____

A N S W E R S

90. _____

91. _____

92. _____

93. _____

94. _____

95. _____

96. _____

90. $x + 2 \geq 3x - 6$

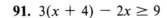

91. $3(x + 4) - 2x \geq 9$

92. $\dfrac{2}{3}(x - 4) + \dfrac{1}{2} < \dfrac{9}{2}$

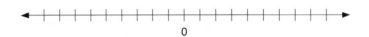

93. $5(x + 1) - 7 \leq 3(x - 2)$

94. $\dfrac{x}{4} + 2x \geq 7x - 3$

95. $\dfrac{1}{2}(4x - 3) + \dfrac{2}{3} > x + 1$

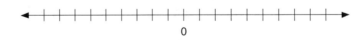

96. $3(x + 2) + \dfrac{x}{5} \leq$

$\dfrac{1}{2}(x - 6) + 36$

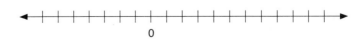

97. $|x - 3| < 4$

97. _____

98. $|x - 5| < 7$

98. _____

99. $|2x + 5| \leq 3$

99. _____

100. $|3x + 8| \leq 2$

100. _____

101. $\left|\dfrac{2}{3}x + 2\right| \geq 6$

101. _____

102. $|3x + 8| \geq 5$

102. _____

103. _____

104. _____

105. _____

106. _____

107. _____

108. _____

109. _____

110. _____

111. _____

112. _____

SCORE: _____

103. The area of a rectangle is 225 square meters. Find the length if the width is 12.5 meters.

104. A car traveled 315.7 kilometers in $5\frac{1}{2}$ hours. Find the average rate of speed of the car.

105. The volume of a rectangular solid is 60 cubic meters. Find the length of the solid if the width is 5 meters and the height is $\frac{2}{3}$ meters.

106. The wages (w) made by a worker is equal to the product of the hourly rate (r) and the number of hours worked (t). The formula is given by $w = rt$. Find the hourly rate of a worker earning \$217.50 in $37\frac{1}{2}$ hours.

107. If the property tax rate is \$12.00 per \$1,000 assessed valuation, what is the tax on property assessed at \$39,500?

108. 12.5% of a number is 1,312.5. Find the number.

109. Find the amount of money borrowed for five years at an annual rate of $12\frac{1}{2}$% if the total interest paid was \$3,437.50.

110. A relationship between Fahrenheit (F) and Celsius (C) temperature is given by $9C = 5F - 160$. Find the Fahrenheit temperature if the Celsius temperature is 23.5 degrees.

111. Find the width of a rectangle if the perimeter is 400 cm and the length is 115.8 cm.

112. The perimeter of a certain rectangle is $27\frac{1}{2}$ meters. Find the width of the rectangle if the width is two-thirds the length.

1. Classify $3(x + 4)$
$= 5x - (x - 4)$ as a
conditional equation or as an
identity.

2. Are the two equations

$3\left(\dfrac{x}{2} + 1\right) - 2 = 7$ and

$x = 4$ equivalent equations?

A   N   S   W   E   R   S

1. _____

2. _____

Solve for x in each of the equations 3 through 8.

3. _____

3. $-7x = 28$

4. $3x + 4 = 10 - x$

4. _____

5. $2(x + 3) = 7(x + 2) + 2$

6. $\dfrac{4}{9}x - 2 = \dfrac{1}{3}x + \dfrac{1}{2}$

5. _____

6. _____

7. $\dfrac{3}{5}(2x - 1) + 4$
$= 5 - (x + 7)$

8. $\dfrac{5}{8}x - \dfrac{1}{2} = \dfrac{3}{5}x + 3$

7. _____

8. _____

9. If the ratio of passing grades to
failing grades is seven to two,
how many students would fail if
twenty-eight passed?

10. The length of a certain
rectangle is $2\dfrac{1}{2}$ times the
width. If the perimeter of the
rectangle is 35, find the length
and width. Use the formula
$P = 2\ell + 2w$.

9. _____

10. _____

11. _____

11. If $A = \dfrac{1}{2}h(a + b)$, solve for b.

12. Solve for x and graph the results on the number line: $\dfrac{1}{2}x + 2 \geq x - \dfrac{1}{3}$

<--+-->
                                    0

12. _____

13. Solve for x: $|5x - 1| = 9$

13. _____

14. Solve for x and graph the solution on the number line: $|3x - 4| \geq 5$

<--+-->
                                0

14. _____

SCORE: _____

C H A P T E R

3

PRETEST

Before beginning this chapter, answer as many of the following questions as you can. When you have finished the chapter, take the practice test at the end of the chapter and compare the scores of the two tests to see how much you have learned.

A N S W E R S

1. Write an algebraic expression for each of the following:

    a. Twice a number x, decreased by nine

    b. One-third a number y, increased by one

    c. 17.3% of a number x

    d. The value in cents of d dimes

1a. _____

1b. _____

1c. _____

2. The second side of a triangle is twice the first, and the third side is two units more than the first. If the perimeter of the triangle is 26 units, find each side.

3. A beginning tennis class has twice as many students as the advanced class. The intermediate class has three more students than the advanced class. How many students are in the advanced class if the total enrollment for the three tennis classes is 43?

1d. _____

2. _____

3. _____

4. Two towns are 677 kilometers apart. A car leaves town A bound for town B at 75 kilometers per hour. Twenty minutes later a car leaves town B bound for town A at 88 kilometers per hour. In how many hours after the car leaves town B will they meet?

5. How many liters of pure dye must be added to 32 liters of a 6% dye solution to obtain a 10% dye solution? (Answer correct to nearest tenth of a liter.)

3

Solving Word Problems

A spacecraft leaves the moon for earth (a distance of 238,250 miles) traveling at 6,500 mph. Three hours later a spacecraft leaves earth for the moon traveling at 6,000 mph. In how many hours after the moonbound spacecraft departs will they pass each other?

In chapter 2 we developed the techniques for solving equations. Equations arise as a means of solving verbal or word problems. Since word problems are a large part of algebra, it is necessary to develop techniques of solving them. In this chapter we will concentrate on ways of outlining and solving verbal problems.

 **3-1 FROM WORDS TO ALGEBRA**

<table>
<tr><td>

O B J E C T I V E S

Upon completing this section you should be able to:

1. Change a word phrase to an algebraic expression.
2. Express a relationship between two or more unknowns in a given statement by using one unknown.

</td><td>

The primary task when attempting to solve a word problem is one of translation. The problem is written in one language and must be translated into another— the language of algebra. This translation process must be precise if we are to be successful in solving the problem.

Example 1   Statement: A number increased by seven is twelve.
Algebraic translation: $x + 7 = 12$

Do these two statements have the same meaning? Does the algebraic equation make the same statement as the English sentence? If so, we have correctly translated from one language to another and can easily solve for the missing number.

Before we begin to outline completed sentences or solve problems we need to review the meaning of certain phrases.

</td></tr>
</table>

Note that a phrase will yield only an algebraic expression and *not* a complete equation. It is impossible to solve for the unknown without a complete equation.

Example 2 Write an algebraic expression for: A certain number decreased by four.

Solution If we let x represent "a certain number," and recognize that "decreased by" is subtraction, then our answer is $x - 4$.

There are key words that give clues to the operations to be used.

Addition—words such as "increased by," "sum," "more than," "greater than," "total"

Subtraction—words such as "decreased by," "less than," "difference," "diminished by"

Multiplication—words such as "times," "of," "product," "twice"

Division—words such as "quotient," "divided by"

Example 3 Write an algebraic expression for: Five times a certain number.

Solution If x represents "a certain number," then the expression would be $5x$.

Example 4 Write an algebraic expression for: Five more than a certain number.

Solution If x represents "a certain number," and "more than" means addition, the expression is $x + 5$.

Example 5 Write an algebraic expression for: Seven more than twice a certain number.

Solution Again, allowing x to represent the unknown number, we have $2x + 7$.

Example 6 Write an algebraic expression for: 5% of a given number.

Solution First write 5% as the decimal .05. If x represents the "given number," then the expression would be $.05x$.

Example 7   Write an algebraic expression for: The value in cents of d dimes.

Solution   The value of a dime is 10 cents. Therefore to indicate the value of d dimes multiply 10 by d, obtaining $10d$.

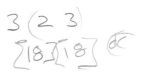

▼ EXERCISE 3-1-1

Write an algebraic expression for each.

1. A number increased by five

$$x + 5$$

2. A number decreased by three

$$x - 3$$

3. Five more than a given number

$$x + 5$$

4. Eight less than a given number

$$x - 8$$

5. A given number, less nine

$$x - 9$$

6. Nine, less a given number

$$9 - x$$

7. Ten greater than a given number

$$x + 10$$

8. The sum of x and 4

$$x + 4$$

9. Twice a certain number

$$2x$$

10. Eight times a certain number

$$8x$$

11. Twice a certain number, increased by two

$$2x + 2$$

12. Three times a certain number, increased by six

$$3x + 6$$

13. Six times a certain number, decreased by four

$$6x - 4$$

14. Nine times a certain number, decreased by one

$$9x - 1$$

15. One-fifth a certain number

$$\frac{1}{5}x$$

16. Half a certain number

$$\frac{1}{2}x \quad \text{or} \quad \frac{x}{2}$$

17. Three less than a number

$$x - 3$$

18. Four less than twice a number

$$4 - 2x$$

19. The product of a and 8

$$(a)(8) = x$$

20. The product of x and y

$$(x)(y) = x$$

21. Half the product of x and 5

$$\frac{(x)(5)}{2} \quad \text{or} \quad \frac{1}{2}(x5)$$

22. Three times the product of a and b

$$3(Ab) = x$$

23. Half the sum of b and 7

$$\frac{(b+7)}{2} =$$

$$\frac{1}{2}(b+7) =$$

24. Twice the sum of x and 3

$$2(x+3)$$

25. Three times the difference of x and 9

$(x-9)3$

26. 25% of a number

$.25x$

$\frac{1}{4}x$

27. 15% of a number

$.15x$

28. 7% of a number

$.07x$

29. 10% of a number

$x \cdot .10$

$.1x$

30. 16.8% of a number

$.168x$

31. 9.5% of a number

$.095x$

32. The value in cents of n nickels

$5n = x$

33. The value in cents of d dimes

$10d = \text{CENTS}$

34. The value in cents of q quarters

$25q = x$

35. The total value in cents of d dimes and q quarters

$10d + 25q = \text{CENTS}$

36. The number of days in w weeks

$7w$

37. The number of months in y years

$12y =$

38. The number of weeks in y years

$52y$

39. The number of weeks in d days

$\frac{d}{7}$

40. The number of years in m months

$\frac{m}{12} = \#\text{YRS}$

It may not always be possible to relate the unknowns in a problem using only one unknown. You will be able to do so, however, with all problems in this chapter.

When more than one unknown number is involved in a problem, we try to outline it in such a way that all unknowns are expressed in terms of one unknown.

Example 8   The Sears Tower in Chicago is eight stories taller than the Empire State Building in New York. Write algebraic expressions for the height of each building using one unknown.

Solution   From the information given we do not know how many stories either building has. We will choose one of the buildings and represent the number of stories by x.

We could also choose x to represent the height of the Sears Tower. Then $x - 8$ would represent the height of the Empire State Building.

> Let $x =$ number of stories in the Empire State Building.
> Then $x + 8 =$ number of stories in the Sears Tower.

Example 9   The length of a rectangle is three meters more than the width. Write expressions for the length and width using one unknown.

Solution
Let x = width of the rectangle.
Then $x + 3$ = length of the rectangle.

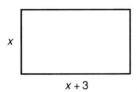

It is sometimes helpful to use a diagram to see the relationships between unknowns.

Example 10 The width of a rectangle is one-fourth the length. Write expressions for the length and width using one unknown.

Solution
Let x = width.
Then $4x$ = length.

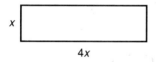

Note that if we were to let x represent the length, we would have a fraction representing the width. This certainly would not be wrong but we can avoid the use of fractions by careful selection.

If we let x = length
then $\frac{1}{4}x$ = width.

Example 11 The sum of two numbers is 20. Write expressions for both numbers using one unknown.

Solution
Let x = first number.
Then $20 - x$ = second number.

We could also let
x = second number.
What expression would represent the first number?

Example 12 Express algebraically the relationship of the unknowns in this sentence: A certain number is four more than a second number and is three less than a third number.

Solution To express these three numbers algebraically first decide which will be represented by x.

For instance, if x represents the first number, we have the following:

$$x = \text{first number}$$
$$x - 4 = \text{second number}$$
$$x + 3 = \text{third number.}$$

Ask yourself, "Is the first number four more than the second?" and "Is the first number three less than the third?" If the answers are "yes," you have correctly outlined the sentence.

Now suppose, in the same example, we decide to allow x to represent the second number.

Which number to represent by x is arbitrary, but re-read the relationships several times to see if one option is preferable over another.

We now have $x + 4 =$ first number

$x =$ second number

$x + 7 =$ third number.

Represent the three numbers if x represents the third number.

Ask the same questions again, "Is the first number four more than the second?" and "Is the first number three less than the third?" This outline is also correct.

▶   **EXERCISE 3–1–2**

Express the unknowns in terms of x.

1. A certain number is eight more than a second number. Write expressions for the numbers.

$$X$$
$$X - 8 = Y$$

2. A certain number is five less than a second number. Write expressions for the numbers.

$$X$$
$$X - 5 = Y$$

3. The Sears Tower in Chicago is 104 feet taller than the World Trade Center in New York. Write expressions for the height of each building.

if $X = WTC$
than $X + 104' = ST$

4. The height of the Empire State Building in New York is 204 feet less than that of the Sears Tower in Chicago. Write expressions for the height of each building.

$ESB = X$
$ST = X + 204'$

5. During the second week of production of a new car, General Motors produced 612 units more than it did during the first week. Write expressions for the number of cars produced each week.

$1^{st} wK = X$
$2^{nd} wK = X + 612$

6. A student purchased a biology book and a math book. The biology book cost five dollars more than the math book. Write expressions for the cost of each book.

$X = math book$
$X + 5 = Bio Book$

7. A 7:00 A.M. math class has eight fewer students than a 9:00 A.M. class. Write expressions for the number of students in each class.

8. The capacity of computer A is 32 kilobytes more than computer B. Write expressions for the capacity of each computer.

9. The population of San Francisco is twice that of Long Beach. Write expressions for the population of each city.

10. The population of El Paso is half that of Boston. Write expressions for the population of each city.

11. The length of a rectangle is five meters more than the width. Write expressions for the length and width.

$w = x$

$L = x + 5m$

12. The width of a rectangle is one-third the length. Write expressions for the length and width.

$L = x$

$w = \frac{1}{3} x$

13. The length of a rectangle is three centimeters more than twice the width. Write expressions for the length and width.

LENGTH $= 2w + 3$ cm

14. The width of a rectangle is three inches more than half the length. Write expressions for the length and width.

$w = \frac{1}{2} x + 3$

$L = x$

15. The sum of two numbers is 40. Write expressions for the numbers.

NO $\checkmark$ $x + y = 40$ x

NO $\checkmark$ $x + x = 40$ $40 - x$

16. The total income of a married couple is $50,000. Write expressions for the income of each person.

17. The difference of two numbers is 18. Write expressions for the two numbers.

18. Computer A is more expensive than computer B. If the difference in their cost is $250, write expressions for the cost of each.

$A - B = ^{\$}250$

19. An individual makes two investments that total $10,000. Write expressions to represent the two investments.

20. A person has $8,000 less invested in stocks than in bonds. Write expressions for the amount invested in each.

21. The price of this year's Chevrolet Celebrity is 5% higher than last year's. Write expressions for last year's price and this year's price.

22. This year's enrollment at a college is 7% less than last year's. Write expressions for the enrollment for last year and this year.

23. One number is four more than the second and nine less than the third. Write expressions for the three numbers.

24. The price of a certain model of a Texas Instruments calculator is six dollars more than a Sharp calculator and three dollars less than one made by Hewlett-Packard. Write expressions for the three prices.

25. One number is half the second and three times the third. Write expressions for the three numbers.

26. On a particular flight the cost of a coach ticket is half the cost of a first-class ticket and thirty dollars more than an economy ticket. Write expressions for the cost of the three tickets.

27. Write expressions for three consecutive even integers. (Examples of consecutive even integers are 2, 4, 6, 8, 10, and so on.)

$$X$$
$$X+2$$
$$X+4$$

28. Write expressions for three consecutive odd integers. (Examples of consecutive odd integers are 1, 3, 5, 7, 9, and so on.)

29. The cost of a house today is three times its cost in 1970 and twice its cost in 1976. Write expressions for the cost in 1970, 1976, and now.

30. The capacity of computer A is twice that of computer B and 32 kilobytes less than computer C. Write expressions for the capacity of the three computers.

31. Jane has five dollars more than Bob and thirteen dollars less than Jim. Write expressions for the numbers of dollars each person has.

32. The load capacity of a Ford pick-up truck is eight cubic feet more than a Nissan and nine cubic feet more than a Toyota. Write expressions for the load capacity of the three trucks.

33. A Buick Century obtains four miles per gallon more than a Chevrolet Impala and eight miles per gallon less than a Toyota Corolla. Write expressions for the mileage for the three cars.

Buick = X
Chevy = X - 4
Toyota X + 8

34. The state of Ohio has half the population of California and twice the population of Indiana. Write expressions for the population of the three states.

35. A certain amount is invested in an account for one year at 14% interest. Write expressions for the original amount, the interest earned, and the total amount in the account at the end of the year.

36. The length of one side of a triangle is twice the length of the second side and four less than the length of the third side. Write expressions for the lengths of the three sides.

37. A meter stick is cut into two pieces. Write expressions for the length of each piece in centimeters. (Recall that 1 m = 100 cm.)

38. The atomic weight of sulfur is twice that of oxygen and eight times that of helium. Write expressions for the atomic weight of each element.

39. A person has twice as many dimes as nickels and two more quarters than dimes. Write expressions for the number of each kind of coin the person has.

40. Write an expression for the total value, in cents, of the coins discussed in problem 39.

 **3-2 SOLVING WORD PROBLEMS**

In the preceding exercises we outlined relationships between unknowns within a statement. If we are to solve a problem and find the unknown numbers, there must be within the problem a sentence that yields an equation.

Example 1   The Sears Tower in Chicago is eight stories taller than the Empire State Building in New York. If the total number of stories in both buildings is 212, find the number of stories in each building.

The word "total" implies addition.

Solution   From example 8 in the previous section we have

$$x = \text{number of stories in the Empire State Building}$$
$$x + 8 = \text{number of stories in the Sears Tower.}$$

This time, however, we have the added statement that the total number of stories is 212. Thus, we can write the equation

$$x + (x + 8) = 212.$$

Solving this equation gives

$$2x = 204$$
$$x = 102.$$

What does x represent?

Always re-read the problem to make sure you have answered the question.

Notice that we have found only the number of stories in the Empire State Building. The question asked us to find the number of stories in each building. Thus, to find the number of stories in the Sears Tower we must substitute 102 for x in the expression $x + 8$, obtaining

$$x + 8 = (102) + 8$$
$$= 110.$$

The answers are

Get in the habit of always summarizing your answers.

$$\text{number of stories in the Empire State Building} = 102$$
$$\text{number of stories in the Sears Tower} = 110.$$

Checking your answers is one of the most important parts of the solution.

To check the answers, 110 is eight more than 102 and the sum of 110 and 102 is 212.

Example 2   The length of a rectangle is three meters more than the width. Find the length and width if the perimeter of the rectangle is 26 meters.

Solution   Let x = width
$x + 3$ = length.

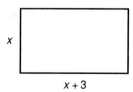

Since the perimeter is equal to the sum of twice the length and twice the width, or $P = 2\ell + 2w$, we can write the equation

$$2(x) + 2(x + 3) = 26.$$

Solving, we obtain

$$2x + 2x + 6 = 26$$
$$4x = 20$$
$$x = 5.$$
$$\text{Also } x + 3 = 8.$$

Thus, the width is 5 meters and the length is 8 meters.

Check: The length (8) is three more than the width (5), and the perimeter is $2(8) + 2(5) = 16 + 10 = 26$.

The perimeter is the distance around the figure. Thus, two sides of the rectangle measure x and the other two sides measure $x + 3$.

Again, the check is very important.

Example 3 The width of a rectangle is one-fourth the length. If the perimeter is 200 centimeters, find the length.

Solution
$$\text{Let } x = \text{width}$$
$$4x = \text{length}.$$

This particular way of stating the unknowns avoids fractions.

Again, the perimeter is twice the width plus twice the length. Therefore,

$$2(x) + 2(4x) = 200$$
$$10x = 200$$
$$x = 20.$$

The problem asks for the length only. Thus,

$$4x = 4(20) = 80.$$

The length is 80 cm.

Check: The width (20) is one fourth the length (80). Also, the perimeter is $2(20) + 2(80) = 40 + 160 = 200$.

What does x represent?

Try this problem again using
$$x = \text{length}$$
$$\frac{1}{4}x = \text{width}.$$

Which way is easier?

Example 4 The sum of two numbers is 20. Their difference is 4. Find the numbers.

Solution
$$\text{Let } \quad x = \text{first number}$$
$$20 - x = \text{second number}.$$

If the difference of the two numbers is 4, write the equation

$$x - (20 - x) = 4.$$

We could also write the equation $(20 - x) - x = 4$ since we don't know which number is larger. Solve the above equation.

Solving, we obtain

$$2x = 24$$
$$x = 12.$$
$$\text{Also } 20 - x = 8.$$

The two numbers are 8 and 12.

Check: The sum of 8 and 12 is 20. The difference of 8 and 12 is 4.

Example 5   A certain number is four more than a second number and three less than a third number. Find the numbers if their sum is 23.

Solution   In this problem we are asked to find the numbers and are given a statement about their sum.

If we let x represent the first number, we have

$$x = \text{first number}$$
$$x - 4 = \text{second number}$$
$$x + 3 = \text{third number.}$$

The statement "their sum is 23" gives the equation

$$x + (x - 4) + (x + 3) = 23.$$

If we now solve the equation, we obtain

$$3x - 1 = 23$$
$$3x = 24$$
$$x = 8.$$

Leaving the answer as $x = 8$ would not be a solution to the problem. We are asked to "find the numbers." The answer must be

$$\text{first number} = 8$$
$$\text{second number} = 4$$
$$\text{third number} = 11.$$

We check the problem by noting that the first number (8) is four more than the second (4) and is three less than the third (11) and that their sum is 23.

We have followed five basic steps in solving the above examples. These steps should be observed when solving any word problem.

Again, we could just as well let x represent the second or third number. The important thing to remember is to label each algebraic expression with the number that it represents.

▼ WARNING

Do not just solve the equation for x and think you have solved the problem. Check to see what x represents and re-read the problem to see what it is asking for.

1. Write expressions for the unknowns.
2. Write an equation that relates the unknowns to each other.
3. Solve the equation.
4. Make sure you have answered the question.
5. Check your answers to make sure they agree with the original problem.

Make sure you know these five important steps before working the following exercise set.

▼ EXERCISE 3-2-1

Solve each of the following using the five steps discussed in this section.

1. A certain number is eight more than a second number. The sum of the two numbers is 22. Find the numbers.

$$x + 8 = 22$$

2. A certain number is five less than a second number. The sum of the two numbers is 33. Find the numbers.

3. The Sears Tower in Chicago is 104 feet taller than the World Trade Center in New York. If the sum of their heights is 2,804 feet, find the height of each building.

$$(x) + (x + 109) = 2804$$
$$2x + 109 = 2804$$
$$2x = 2804 - 109$$
$$2x = 2700$$
$$x = 1350$$

1350'
1454'

4. The height of the Empire State Building in New York is 204 feet less than the height of the Sears Tower in Chicago. If the sum of their heights is 2,704 feet, find the height of the Empire State Building.

5. During the second week of production of a new car, General Motors produced 612 units more than it did during the first week. If the total production for the two weeks was 18,416, find the production for each week.

$$18416 = x + x + 612$$
$$18416 - 612 = 2x$$
$$17804 = 2x$$

8902 1ST
9514 2ND

6. A student purchased a biology book and a math book. The biology book cost five dollars more than the math book and the total cost of both books was $37.00. Find the cost of the biology book.

7. A 7:00 A.M. math class has eight fewer students than a 9:00 A.M. class. Find the number of students in each class if the total number of students in both classes is 84.

$$84 = x + x - 8$$
$$84 = 2x - 8$$
$$92 = 2x$$

x = 46 —
x - 8 = 38 —

8. The capacity of computer A is 32 kilobytes more than the capacity of computer B. The total capacity of both computers is 96 kilobytes. Find the capacity of each computer.

9. The population of San Francisco is twice that of Long Beach. If the total population of both cities is 1,074,000, find the population of San Francisco.

$1,074,000 = X + 2X$

$1,074,000 = 3X$

$358000 = X$ L.B.

$716000 = 2X$ SAN FRAN

10. The population of El Paso is half that of Boston. Find the population of both cities if the total population for both cities is 960,000.

11. The length of a rectangle is five meters more than the width. Find the length and width if the perimeter is 46 meters.

$46 = L \times W$ $5 + W = L$

$46 = (L+5) \times (W)$

12. The width of a rectangle is one-third the length. The perimeter is 384 feet. Find the length and width.

$X = width$ $384 = 2(3X) + 2(X)$

$3X = length$ $384 = 6X + 2X$

$L = 144$ $384 = 8X$

$W = 48$ $48 = X$

13. The length of a rectangle is three centimeters more than twice the width. If the perimeter is 192 centimeters, find the length.

14. The width of a rectangle is three inches more than half the length. If the perimeter is 108 inches, find the width.

15. The sum of two numbers is 40. Their difference is 12. Find the numbers.

$X + Y = 40$ $(X+Y)(X-Y) = 52$

$X - Y = 12$ $2X = 52$

$\quad 26 + 14 = 40$ $X = 26$

$\quad 26 - 14 = 12$

16. The total income of a married couple is $50,000. The wife earns $5,800 more than the husband. Find the earnings of each.

17. The difference of two numbers is 18. Their sum is 82. Find the numbers.

$X - Y = 18$ $50 - 32 = 18$

$X + Y = 82$ $50 + 32 = 82$

$(X-Y) + (X+Y) = 100$

$\quad 2X = 100$

$\quad X = 50$

18. Computer A is $250 more expensive than computer B. If the total cost of both computers is $11,090, find the cost of each.

19. An individual makes two investments that total $10,000. If the first investment is $3,256 more than the second, find the amount of each investment.

$$X + X + 3256 = 10,000 \qquad 3372 \text{ Sec}$$
$$2X + 3256 = 10000 \qquad 6628 \text{ First}$$
$$2X = 6744 \qquad \overline{ 10,000}$$
$$X = 3372$$

20. A person has $8,000 less invested in stocks than in bonds. If the total invested in both is $86,500, find the amount invested in stocks.

21. A certain number is 37 more than a second number. If the sum of the two numbers is 15, find the numbers.

22. A certain number is 25 less than a second number. Find the numbers if their sum is -13.

23. One number is four more than a second number and nine less than a third number. Find the three numbers if their sum is 50.

$$(X) + (X-4) + (X+9) = 50$$
$$3X + 5 = 50 \qquad 15$$
$$3X = 45 \qquad 11$$
$$X = 15 \qquad \underline{24}$$
$$50$$

24. The price of a certain model of a Texas Instruments calculator is six dollars more than a Sharp calculator and three dollars less than one made by Hewlett-Packard. If the total price of the three calculators is $46.50, find the price of each.

25. On a particular flight the cost of a coach ticket is half the cost of a first-class ticket and thirty dollars more than an economy ticket. If the total cost of the three tickets is $406, find the cost of each.

$$\text{Coach } (X) = \tfrac{1}{2} \text{ First } +30 \text{ Econ}$$
$$406 = (X+30) + X + (2X+30) \qquad \begin{array}{ll} \text{First} & 233 \\ \text{Coach} & 116.50 \\ \text{Econ} & 86.50 \end{array}$$
$$406 = 4X + 60$$
$$346 = 4X \qquad X = 86.50$$

26. Jane has five dollars more than Bob and thirteen dollars less than Jim. If they have a total of $197, find the amount each person has.

$$\text{Jane} = X \qquad \text{Bob} = X-5 \qquad \text{Jim} = X+13$$
$$197 = (X-5) + (X+13) + (X)$$
$$197 = 3X + 13 - 5$$
$$197 = 3X + 8 \qquad (63) + (63+5) + (63-13)$$
$$197 - 8 = 3X$$
$$189 = 3X \qquad X = 63$$

27. A Buick Century obtains four miles per gallon better gas mileage than a Chevrolet Impala and eight miles per gallon less than a Toyota Corolla. If the average mileage for the three cars is $23\tfrac{1}{3}$ mpg, find the mileage of each.

28. The state of Ohio has half the population of California and twice the population of Indiana. If the total population of the three states is 38,500,000, find the population of each state.

29. One side of a triangle is twice the length of the second side and four centimeters less than that of the third side. If the perimeter of the triangle is 84 centimeters, find the length of each side.

30. The atomic weight of sulfur is twice that of oxygen and eight times that of helium. If the total atomic weights of the three elements is 52, find the atomic weight of each.

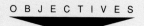 3–3 NUMBER RELATION PROBLEMS

O B J E C T I V E S

Upon completing this section you should be able to:
1. Identify a problem as being a number relation problem.
2. Express relationships between unknowns and determine an equation to solve a number relation problem.
3. Apply formulas to solve certain types of number relation problems.

Read the following three problems carefully.

1. Jim is two years older than Sue and four years younger than Hugh. The sum of their ages is 59. Find the age of each.
2. The length of rope is 59 meters. It is cut in three pieces such that the first is two meters longer than the second and four meters shorter than the third. Find the length of each piece.
3. There are three numbers such that the first is two more than the second and four less than the third. If their sum is 59, find the numbers.

At first glance these may seem to be three different problems. However, closer examination shows that they are actually the same. For all three a possible outline is

$(x + 2)$, (x), and $(x + 6)$ represent the three unknowns.

The equation $(x + 2) + (x) + (x + 6) = 59$ will lead to the solutions.

Problems such as these are classified as **number relation problems.** If you recognize a problem in this class, then the outline for solving it must show a relation between the numbers. The equation will then be based on a statement about the sum, difference, and so on.

The problems you solved in the last section were all number relation problems. They all involved comparing two or more numbers.

Example 1   If we wish to solve problem 3 above, we could let

$$x + 2 = \text{first number}$$
$$x = \text{second number}$$
$$x + 6 = \text{third number.}$$

Again, x does not have to be used to represent the second number. It could be used to represent one of the other numbers. We would then have a different set of expressions for the three numbers.

Then

$$(x + 2) + x + (x + 6) = 59.$$

Solving, we obtain

$$x = 17$$
$$x + 2 = 19$$
$$x + 6 = 23.$$

Our answers are

$$\text{first number} = 19$$
$$\text{second number} = 17$$
$$\text{third number} = 23.$$

Check: The first number (19) is two more than the second (17) and four less than the third (23). Also, the sum $19 + 17 + 23 = 59$.

Problems comparing age, height, weight, and so forth are number relation problems.

Example 2 Janet is three years older than Maria. The sum of their ages is 39. Find the age of each.

Solution Let
$$x = \text{Maria's age}$$
$$x + 3 = \text{Janet's age}.$$
Then $x + (x + 3) = 39.$

Solving, we obtain

$$x = 18$$
$$x + 3 = 21.$$

The answers are

$$\text{Maria's age} = 18 \text{ years}$$
$$\text{Janet's age} = 21 \text{ years}.$$

Check: 21 is three more than 18 and $21 + 18 = 39$.

Example 3 Pat weighs ten pounds more than Paul and three pounds less than Pete. If the total weight of all three is 518 pounds, find the weight of each.

Solution Let
$$x + 10 = \text{Pat's weight}$$
$$x = \text{Paul's weight}$$
$$x + 13 = \text{Pete's weight}$$
$\left.\right\}$ Step 1

Then $(x + 10) + x + (x + 13) = 518.\big\}$ Step 2

Solving, we obtain

$$x = 165$$
$$x + 10 = 175$$
$$x + 13 = 178.$$
$\left.\right\}$ Step 3

The answers are

$$\text{Pat's weight} = 175 \text{ pounds}$$
$$\text{Paul's weight} = 165 \text{ pounds}$$
$$\text{Pete's weight} = 178 \text{ pounds}.$$
$\left.\right\}$ Step 4

Work this problem if x = first number.

Work this problem if x = Janet's age.

Remember the five steps in solving a verbal problem.

Check: Pat's weight (175 pounds) is ten more than
Paul's (165 pounds) and three less than Pete's (178
pounds). The sum is 175 + 165 + 178 = 518.   Step 5

The field of business contains number relation problems.

Example 4   A customer purchases a stereo radio for $306.80, including a 4%
sales tax. Find the price of the radio and the amount of sales tax.

Note the tax is on the price of the
radio and *not* on $306.80.

Solution   If x represents the price of the radio, then the sales tax is found
by multiplying the price of the radio by 4%.

Thus, we have

$$x = \text{price of the radio}$$
$$.04x = \text{sales tax.}$$

Remember, to change a percent to
a decimal divide by 100.

Then the total sale is the sum of the price of the radio and the
sales tax.

$$x + .04x = 306.80$$

Solving the equation, we obtain

Solve the equation
   $x + .04x = 306.80$
here in the margin. It will give
you practice in solving equations
involving decimals.

$$x = 295$$
$$.04x = 11.80.$$

The answers are

$$\text{price of the radio} = \$295.00$$
$$\text{sales tax} = \$11.80.$$

Check: The total of $295.00 and $11.80 is $306.80.

Some number relation problems require prior knowledge of a formula to set up
the equation.

Example 5   In a certain rectangle the length is two more than the width. If
the length is increased by three and the width by two, the perimeter of the
new rectangle will be twice that of the original. Find the length and width of
the original rectangle.

The perimeter of a rectangle is
twice the length added to twice
the width.

Solution   Notice that to solve the problem, you must know that

$$P = 2\ell + 2w$$

is the formula for the perimeter of a rectangle. We outline this as

original rectangle

$\ell = x + 2$

$w = x$

$P_o = 2(x + 2) + 2x$

new rectangle

$\ell = (x + 2) + 3$

$w = x + 2$

$P_n = 2[(x + 2) + 3] + 2(x + 2).$

Re-read the original problem to see if the expression for the dimensions of the two rectangles make sense.

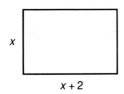

x

$x + 2$

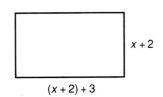

$x + 2$

$(x + 2) + 3$

Notice that we used *subscripts o* and *n* to distinguish the original perimeter from the new perimeter.

P_o represents the original perimeter.

P_n represents the new perimeter.

The equation comes from the statement that $P_n = 2P_o$.

$$2[(x + 2) + 3] + 2(x + 2) = 2[2(x + 2) + 2x]$$
$$2x + 10 + 2x + 4 = 2(4x + 4)$$
$$4x + 14 = 8x + 8$$
$$-4x = -6$$
$$x = \frac{-6}{-4} = \frac{3}{2}$$

Recall the order of operations in removing grouping symbols.

Since x represents the width, we have

$$w = \frac{3}{2}.$$

Again, just solving for x does not solve the problem.

Also

$$\ell = x + 2$$
$$= \frac{3}{2} + 2$$
$$= \frac{7}{2}.$$

Checking, we find

$$P_o = 2\ell + 2w$$
$$= 2\left(\frac{7}{2}\right) + 2\left(\frac{3}{2}\right)$$
$$= 10$$

$$P_n = 2\left(\frac{7}{2} + 3\right) + 2\left(\frac{3}{2} + 2\right)$$
$$= 13 + 7$$
$$= 20.$$

Once the length and width are determined, the check will tell if they are correct.

Thus, the perimeter of the new rectangle is twice that of the original rectangle, and the answer checks.

Example 6   An individual invests $8,000 for one year. Part is invested at 6% and the rest at 5%. If the total interest at the end of the year is $460, how much was invested at each rate?

Solution   Here we must use the interest formula

$$I = prt,$$

where I represents the interest, r represents the rate of interest, p represents the principal amount invested, and t is the time in years. We let x = amount invested at 5% and $8,000 - x$ = amount invested at 6%. Then, the interest earned at 5% is

$$I = prt$$
$$= (.05)(x)(1).$$

The interest earned at 6% is

$$I = prt$$
$$= (.06)(8,000 - x)(1).$$

The equation representing the total interest is

$$.05x + .06(8,000 - x) = 460.$$

Solving the equation gives

$$x = 2,000$$
$$8,000 - x = 6,000.$$

Thus, $2,000 was invested at 5% and $6,000 at 6%.

Check: 5% of $2,000 is $100 and 6% of $6,000 is $360. The two interest amounts total $460.

The rate of interest r is expressed as a decimal number. Change 5% to a decimal number.

$$5\% = \frac{5}{100} = .05$$
$$6\% = \frac{}{100} =$$

Notice that it is important that $(8,000 - x)$ be enclosed in parentheses since 6% is multiplying the entire expression.

$$.05x + .06(8,000 - x) = 460$$
$$.05x + 480 - .06x = 460$$
$$-.01x = -20$$
$$x = 2,000$$

Example 7   The interest due on a two-year $300 loan amounted to $90. What annual rate of interest was charged?

Solution   Again, the formula $I = prt$ is useful. Here $I = \$90$, $p = \$300$, and $t = 2$ years. Therefore,

$$I = prt$$
$$90 = (300)r(2)$$
$$\frac{90}{600} = \frac{600r}{600}$$
$$.15 = r.$$

Thus, the rate of interest was 15%.

Check: $300 at 15% for 2 years yields $90 interest.

Percent problems come in three basic forms. Each is easily translated to algebraic equations as shown below:

1. What is 10% of 30?
$$x = .10 \cdot 30$$
2. 3 is 10% of what number?
$$3 = .10 \cdot x$$
3. 3 is what percent of 30?
$$3 = x \cdot 30$$
Solve each of the above.

▼ **EXERCISE 3-3-1**

Solve each of the problems.

1. A second number is five more than the first number. The sum of the two numbers is 83. Find the two numbers.

2. One number is twice another number. The sum of the two numbers is 42. Find the two numbers.

3. Jane is five years older than Louis. The sum of their ages is 41. Find the age of each.

4. Jose is twice as old as Steve. The sum of their ages is 63. Find the age of each.

5. One number is three more than five times another. The sum of the two numbers is 75. Find the numbers.

6. One number is five more than three times another. The difference of the two numbers is 53. Find the numbers.

7. Mark is three years younger than twice Mario's age. The difference of their ages is 14. Find their ages.

8. A Toyota obtains twice the mileage of a Buick. If the average mileage for the two cars is 27 mpg, find the mileage of each.

9. In 1979 Roger Staubach of the Dallas Cowboys completed 36 more passes than he did in 1978. If his total number of completed passes for both years was 498, how many passes did he complete in each year?

10. In 1980 tennis player Bjorn Borg earned $49,000 more than John McEnroe. If their total earnings were $895,000, find the earnings of each.

11. The length of a rectangle is 20 centimeters more than its width. Find the length and width if the perimeter is 224 centimeters.

12. The length of a rectangle is twice the width. Find the length and width if the perimeter is 420 centimeters.

13. A certain number is three times another number. Their difference is 16. Find the numbers.

14. The Amazon River is three times as long as the Ohio-Allegheny River. Find the length of each river if the difference in their lengths is 2,610 miles.

15. The sum of two numbers is 233. Their difference is 9. Find the two numbers.

16. The combined areas of the states of Connecticut and Massachusetts is 13,266 square miles. If Massachusetts is 3,248 square miles larger than Connecticut, find the area of each state.

17. The sum of two numbers is 40. Twice the first number added to the second number is 51. Find the numbers.

18. A student purchases a notebook and a calculator for $14.45. Another student purchases two notebooks and a calculator for $18.95. Find the price of a calculator and the price of a notebook.

19. A certain number is twelve more than a second number and twenty less than a third number. Find the numbers if their sum is 152.

20. The World Trade Center is 100 feet taller than the Empire State Building and 104 feet shorter than the Sears Tower. Laid end to end, the three buildings would measure a total of 4,054 feet (over 3/4 mile). Find the height of each building.

21. Three consecutive even integers have a sum of 90. Find the integers.

22. The sum of three consecutive odd integers is 117. Find the integers.

23. Jim is two years older than Sue and four years younger than Hugh. The sum of their ages is 59. Find the age of each.

24. Lake Erie is three times as deep as Lake-of-the-Woods and 565 feet shallower than Lake Ontario. If the total depth of the three lakes is 1,055 feet, find the depth of each.

25. An item sells for a certain price. If a 5% sales tax is added to the price, the total amount of the sale is $28.14. Find the original selling price of the item and the sales tax.

26. The total cost of an item, including 7% sales tax, is $196.88. Find the price of the item excluding sales tax.

27. Town A has a population ten percent greater than town B. The total population of the two towns is 49,245. Find the population of each.

28. $56.25 is to be divided between two people in such a way that one person receives three and one-half times as much as the other. Find the amount each is to receive.

29. Alice is 15 centimeters taller than Jane and 3 centimeters shorter than Diane. The sum of their heights is 489 centimeters. How tall is each?

30. The selling price of car B is $1,411 more than car A and $1,922 more than car C. The total sales price of all three cars is $20,163. Find the price of each.

31. Jim weighs five kilograms less than Tom and two kilograms less than Bob. Find the weight of each person if their total weight is 223 kilograms.

32. During a certain month of sales, salesperson A earned $23 more commission than salesperson B. Salesperson C earned twice as much as salesperson B. Find the commission for each salesperson if their total commissions were $3,515.

33. The Saxon SX-20S copier produces eight more copies per minute than the Canon NP-120. If together they produce 32 copies per minute, find the number each produces.

34. The Royal 130R copier produces ten more copies per minute than the Saxon SX-20S. The Xerox 8200 produces ten more copies per minute than twice the production of the Royal 130R. If the three machines together can produce 120 copies per minute, find the number each produces.

35. Three cyclists traveled a total distance of 223 kilometers. Cyclist A traveled five kilometers farther than cyclist B and three kilometers farther than cyclist C. Find the distance traveled by each cyclist.

36. The length of a rectangle is three meters more than twice the width. The perimeter of the rectangle is 48 meters. Find the length and width.

37. In a given rectangle the length is twice the width. If the length is increased by seven and the width is increased by eight, the perimeter is doubled. Find the dimensions of the original rectangle.

38. An individual makes two investments that total $10,000. One investment is at 11% and the other at 13%. If the total interest for one year is $1,190, find the amount invested at each rate.

39. The base of a given triangle is ten. If the length of the altitude is increased so that it is four more than twice the original, and the base is decreased by two, then the area of the new triangle is twice the area of the original. Find the altitude of the original triangle.

$$\left(\text{Hint: Use } A = \frac{1}{2}bh.\right)$$

40. A dealer has been selling an item for a profit of half the cost of the item. The cost of the item is increased by ten dollars and the profit is increased by two dollars. The new selling price is twice the original selling price. Find the original cost of the item.

41. 621 is 11.5% of what number?

42. 2.6 is 20% of what number?

43. 207 is what percent of 1,656?

44. 9 is what percent of 5?

45. A loan of $8,500 was made for four years. The total interest paid was $4,760. What annual rate of interest does this represent?

46. A car was purchased for $9,694.01. This is 23% off the list price. What was the list price?

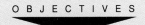 **3–4 DISTANCE-RATE-TIME PROBLEMS**

Another formula often found in verbal problems is

$$d = rt \text{ (distance} = \text{rate} \times \text{time)},$$

which is the distance formula for constant rate. Given the rate at which an object is moving and the time that it moves at this rate, we can find the distance the object moves.

> **OBJECTIVES**
>
> Upon completing this section you should be able to:
> 1. Identify distance-rate-time problems.
> 2. Apply the distance formula to solve problems in this group.

Example 1   A car travels at a constant rate of 55 mph for four hours. How far does it travel?

How far would the car travel in one hour? In two hours? In three hours?

Solution   Using the formula $d = rt$, substitute $r = 55$ and $t = 4$, obtaining

$$d = (55)(4)$$
$$= 220.$$

Thus, the distance traveled is 220 miles.

If the distance and rate are both given, we can find the time.

Example 2   How long will it take a plane whose average ground speed is 530 mph to travel 2,120 miles?

Always check to see that the units agree. That is, if the speed is given in *miles* per hour, then the distance must also be in *miles*.

Solution   In this case, let $d = 2,120$ and $r = 530$ in the formula

$$d = rt$$
$$2,120 = 530t$$
$$t = 4.$$

It will therefore take four hours to travel the distance.

We can also solve for the rate if we are given the distance and time.

Example 3   A person walked 21 miles in $3\frac{1}{2}$ hours. What was the person's average rate?

Solution   We substitute $d = 21$ and $t = 3.5$ in the formula

$$d = rt$$
$$21 = 3.5r$$
$$r = 6.$$

You could also leave the time in fraction form.

Then

$$21 = 3\frac{1}{2}r$$

$$r = 21\left(\frac{2}{7}\right) = 6.$$

The person's average rate was six miles per hour.

One type of distance problem involves two objects leaving from the same point and traveling in the same direction.

Example 4   A bank robber leaves town heading north at an average speed of 120 kilometers per hour. The sheriff leaves two hours later in a plane that travels at 200 kilometers per hour. How long will it take the sheriff to catch the robber?

Solution   We outline as follows:

	d	r	t
Robber	120(2 + x)	120	2 + x
Sheriff	200x	200	x

In this table x represents the time the sheriff takes to catch the robber. Notice that the distance $120(x + 2)$ and $200x$ come from the formula $d = rt$.

The equation comes from the fact that the sheriff and the robber will have traveled the same distance when the sheriff catches the robber. Thus,

$$120(x + 2) = 200x$$
$$80x = 240$$
$$x = 3 \text{ hours.}$$

The sheriff will catch the robber in three hours.

Check: The bank robber has traveled for five hours at 120 km/hr for a distance of

$$5(120) = 600 \text{ km.}$$

The sheriff has traveled for three hours at 200 km/hr for a distance of

$$3(200) = 600 \text{ km.}$$

Another type of distance problem involves two objects leaving from the same point and traveling in opposite directions.

Example 5   Pamela and Sue start at the same point and walk in opposite directions. The rate at which they are moving away from one another is 11 mph. At the end of three hours Pamela stops and Sue continues to walk for another hour. At the end of that time Pamela has walked twice as far as Sue. How far apart are they?

Use a table such as this one when two or more moving objects are involved. It allows for easy comparisons.

Notice that since the sheriff left two hours later, the robber will have been on the run for $x + 2$ hours when he is caught.

When the sheriff catches the robber, they will each have traveled the same distance.

The check shows the robber and sheriff have both traveled the same distance.

The problem tells us the time and lets us represent the distance. To represent the value for r we divide the distance by the time.

$$\left(d = rt \text{ or } \frac{d}{t} = r\right)$$

Solution   We use the following table:

	d	r	t
Pamela	$2x$	$\frac{2}{3}x$	3
Sue	x	$\frac{x}{4}$	4

The sum of their rates is 11 mph.

$$\frac{2}{3}x + \frac{x}{4} = 11$$

Solving, we obtain

$$x = 12$$
$$2x = 24.$$

Thus, they are 36 miles apart.

Check: Pamela's rate is $\frac{2}{3}(12) = 8$ mph.

Again, check to make sure the answers agree with the original problem.

Sue's rate is $\frac{12}{4} = 3$ mph. Their total rate is 11 mph. Also, Pamela's distance (24) is twice Sue's distance (12).

Still another type of distance problem involves two objects that leave from two different points and travel toward each other.

Example 6   Juan and Steven started 36 miles apart and walked toward each other, meeting in three hours. If Juan walked two miles per hour faster than Steven, find the rate of each.

If we let x represent the rate at which Steven walked, then Juan's rate would be represented by $x + 2$.

We represent the distance as the product of the rate and time ($d = rt$).

Solution   First set up the following table:

	d	r	t
Juan	$3(x + 2)$	$x + 2$	3
Steven	$3x$	x	3

The total distance they have traveled is 36 miles. Thus,

$$3(x + 2) + 3x = 36.$$

Solve this equation in the margin.

Solving, we obtain

$$x = 5$$
$$x + 2 = 7.$$

The answers are

$$\text{Juan's rate} = 7 \text{ mph}$$
$$\text{Steven's rate} = 5 \text{ mph}.$$

Check: Juan's rate (7) is two miles per hour faster than Steven's (5). Also

$$3(7) + 3(5) = 21 + 15 = 36.$$

Within the class of problems using $d = rt$ is a subclass of problems concerned with *parallel and opposing forces*.

Parallel forces travel in the same direction and opposing forces travel in opposite directions.

Example 7   A plane, whose speed in still air is 550 mph, flies against a headwind of 50 mph. How long will it take to travel 1,500 miles?

Solution   We use the formula $d = rt$. The distance is 1,500 miles and the rate of the plane against the wind will be its still air speed (550 mph) reduced by the headwind (50 mph).

$$d = rt$$
$$1,500 = (550 - 50)t$$
$$1,500 = 500t$$
$$t = 3$$

Thus, it will take three hours to travel the distance.

Check: $3(550 - 50) = 3(500) = 1,500$

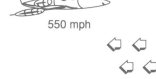

550 mph

50 mph

The normal speed of the plane will be reduced by the speed of the wind.

Example 8   Mike can row his boat from the hunting lodge upstream to the park in five hours. He can row back from the park to the lodge downstream in three hours. If Mike can row x kilometers per hour in still water, and if the stream is flowing at the rate of two kilometers per hour, how far is it from the lodge to the park?

Solution   In working the problem we assume that the rate of the stream will increase or decrease the rate of the boat by two kilometers per hour. Since x represents Mike's speed in still water, we obtain

	d	r	t
Upstream	$5(x - 2)$	$x - 2$	5
Downstream	$3(x + 2)$	$x + 2$	3

Setting the distance upstream equal to the distance downstream, we obtain

$$5(x - 2) = 3(x + 2)$$
$$x = 8.$$

Upstream in 5 hours

Downstream in 3 hours

When Mike is rowing upstream, he is opposing the direction of the current. When he rows downstream, he is parallel to the stream's current.

Notice that x is *not* the solution to the problem but is Mike's rate of rowing in still water. However, the question asked is "What is the distance from the lodge to the park?" To answer this use either the distance upstream or downstream since they are the same. Using the upstream distance, we have

See if you get the same answer using the distance downstream.

$$5(x - 2) = 5(8 - 2)$$
$$= 30 \text{ km}.$$

▼ **EXERCISE 3–4–1**
_____

Solve.

1. A train travels at the rate of 88 mph for three hours. How far has it traveled?

2. A car traveling at the rate of 1,680 ft/min crosses the Golden Gate Bridge in San Francisco in $2\frac{1}{2}$ minutes. How long is the bridge?

3. If a person walks at the rate of 5 mph, how long will it take to walk ten miles?

4. How long will it take for a plane to fly from Phoenix to Atlanta, a distance of 1,590 miles, at an average speed of 530 mph?

5. A cyclist traveled 57 miles in three hours. What was the average speed?

6. A. J. Foyt won the Indianapolis 500 by driving 500 miles in $3\frac{1}{8}$ hours. Find his average speed.

7. Sheila Young, of the United States, skated 500 meters in 42.76 seconds. What was her average rate of speed in meters per second? (Give answer to nearest hundredth.)

8. How long did it take Henry Rono, of Kenya, to complete the 3,000-meter run if his average speed was 400 meters per minute?

9. Radio waves travel at the speed of light (192,000 miles per second). If it takes 1.24 seconds for a radio wave to travel from the Moon to Earth, how far is the Moon from Earth?

10. How many days would it take a spaceship traveling at the speed of 20,000 mph to make the trip from Earth to Mars, a distance of 48,480,000 miles?

11. A car leaves a certain point and travels at the rate of 60 kilometers per hour. Another car leaves from the same point two hours later and travels the same route at 80 kilometers per hour. How long will it take the second car to catch the first car?

12. A car leaves a point and travels at the rate of 75 kilometers per hour. Another car leaves one hour later from the same point and follows the same route catching up with the first car in five hours. What was the speed of the second car?

13. Two cyclists leave the same point at the same time and travel in opposite directions. One cyclist travels at the rate of 20 mph and the other at 25 mph. How long will it take for them to be 90 miles apart?

14. Ellen leaves a certain point and walks at the rate of 6 mph. Maria leaves the same point one hour later and walks in the opposite direction at 8 mph. How long after Maria leaves will it take them to be 48 miles apart?

15. Frank and Mike live 22 miles apart. If each leaves his home at the same time and walks toward the other, Frank at a rate of 6 mph and Mike at 5 mph, how long will it take for them to meet?

16. Two boats, on opposite shores of an eleven-mile-wide lake, leave at the same time and travel toward each other meeting in fifteen minutes. If one boat travels four miles per hour faster than the other, find the rate of each.

17. A car leaves a certain point and travels east at 70 kilometers per hour. Three hours later a second car leaves the same point and travels west at 85 kilometers per hour. How long will the second car have to travel for the two cars to be 520 kilometers apart?

18. A plane leaves Dallas heading for Philadelphia at a rate of 420 mph. Another plane leaves Philadelphia for Dallas at a rate of 440 mph. If the air distance between the two cities is 1,290 miles, how long will it take for the two planes to pass each other?

19. Race-car driver Al Unser is maintaining an average speed of 104 mph, while A. J. Foyt is two miles behind him maintaining an average speed of 108 mph. How long will it take Foyt to catch Unser?

20. A spacecraft leaves the Moon for Earth (a distance of 238,250 miles) traveling at 6,500 mph. Three hours later a spacecraft leaves Earth for the Moon traveling at 6,000 mph. In how many hours after the moonbound spacecraft departs will they pass each other?

21. A train leaves a station and travels at the rate of 100 kilometers per hour. A second train leaves the same station two hours later and travels the same route at 120 kilometers per hour. How long after the second train leaves will the two trains be 60 kilometers apart?

22. In question 21 how long would it take for the second train to catch up with the first train?

23. A plane traveling at the rate of 654 kilometers per hour leaves New York bound for Houston. Thirty minutes later another plane traveling at the rate of 763 km/hr leaves New York bound for Houston. If both planes arrive at Houston at the same time, what is the distance between New York and Houston?

24. A car leaves a starting point and travels at the rate of 82 kilometers per hour. One hour later a second car leaves from the same point and travels in the same direction at 88 kilometers per hour. How soon after the second car leaves will the two cars be 58 km apart?

25. A man starts at a certain point and walks due east. One hour later another man starts at the same point and rides a bicycle due west, traveling nine miles per hour faster than the walker. After the cyclist has ridden for three hours they are sixty-nine miles apart. What was the average rate of each?

26. A jogger running at a rate of eight kilometers per hour starts from a given point. Fifteen minutes later a cyclist starts from the same point and heads toward the jogger at fourteen kilometers per hour. How long does it take the cyclist to catch the jogger?

27. Joan and Sally live five miles apart. They decide to meet for lunch in one-half hour at a restaurant that is located between them. They leave their homes at the same time and Joan walks at a rate two miles per hour faster than Sally. How fast does each walk if they both arrive at the restaurant in exactly the one-half hour?

28. A car enters the Lincoln Tunnel and exits the other end in 3 minutes. Another car traveling 548 ft/min faster than the first car travels through the tunnel in $2\frac{1}{2}$ minutes. How long is the Lincoln Tunnel?

29. Two towns, A and B, are 486 kilometers apart. A car leaves each town at the same time and travels toward the opposite town. The cars meet in three hours. If the car from town B traveled at eight kilometers per hour faster than the car from town A, what was the speed of each car?

30. A cyclist leaves from a starting point and travels at the rate of 12 kilometers per hour. One-half hour later a second cyclist starts from the same point and travels in the opposite direction at the rate of 15 kilometers per hour. How long has the second cyclist been traveling when the two cyclists are 87 kilometers apart?

31. A stream is flowing at the rate of four kilometers per hour. A motorboat that can travel at thirty-two kilometers per hour in still water runs downstream. How long will it take to travel ninety-nine kilometers?

32. A boat leaves a point 58 kilometers downstream from a town. If the rate of the current is 10 km/hr and the boat reaches the town in two hours, what is the rate of the boat in still water?

33. A plane travels a certain distance with a 30-kilometer-per-hour tailwind in three hours and returns against the wind in three hours, thirty minutes. Find the speed of the plane in still air.

34. A canoe was paddled upstream for six hours. The return downstream took two hours. If the rate of the current was eight miles per hour, find the speed of the canoe in still water.

35. A plane flew for $3\frac{1}{2}$ hours with a 30 mph tailwind. It took 5 hours for the return trip. Find the speed of the plane in still air.

36. A plane travels from New York to Rome with a 140-kilometer-per-hour tailwind in six hours. It returns against the same wind in eight hours. What is the distance between New York and Rome?

37. A motorboat travels upstream in five hours and makes the return trip downstream in three hours. If the rate of the stream's current is ten kilometers per hour, find the rate of the boat in still water.

38. The ground speed of a plane is 483 kilometers per hour against the wind and 805 kilometers per hour with the wind. If an upwind flight takes two hours longer than the downwind flight over the same distance, how long does the downwind flight take?

39. A motorboat traveling at the rate of thirty kilometers per hour in still water takes four hours to go upstream and three hours to return. Find the rate of the current.

40. A plane flying at a rate of 640 kilometers per hour in still air makes the outgoing trip against the wind in three hours and the return trip in two hours and thirty-six minutes. Find the speed of the wind.

3–5 MIXTURE PROBLEMS

OBJECTIVES

Upon completing this section you should be able to:
1. Identify mixture problems.
2. Construct an outline which allows for easy comparison of items and the formation of an equation.
3. Find the solution to mixture problems.

The final type of problem we will discuss in this chapter is the mixture problem. Mixture problems come in various settings. They may involve mixing candy, coffee, coins, liquids, and so on. They all have a central concept however, which helps in solving them.

Example 1   A merchant has one brand of coffee that sells for $1.80 per pound and a second brand that sells for $2.20 per pound. The merchant wishes to make a blend of the two coffees. How many pounds of each price coffee must be used to make 20 pounds of a mixture that would sell for $1.90 per pound?

Solution   We set up the following table from the information given in the problem.

Type	Pounds	Price	Total Value
First Brand	x	1.80	1.80x
Second Brand	20 − x	2.20	2.20(20 − x)
Mixture	20	1.90	1.90(20)

If x represents the number of pounds of the first brand, then 20 − x must represent the number of pounds of the second brand.

Looking at the "total value" column of the table, notice that the sum of the total values of the two brands must equal the total value of the mixture. This fact gives us the equation

$$1.80x + 2.20(20 - x) = 1.90(20).$$

Solve this equation in the margin.

Solving, we obtain

$$x = 15$$
$$20 - x = 5.$$

Thus, the merchant must use 15 pounds of the first brand and 5 pounds of the second brand.

Check: 1.80(15) + 2.20(5) = 38.00 and 38.00 ÷ 20 = 1.90

Always check your answers against the original problem.

Example 2   A total of 100 coins in nickels and quarters has a value of $14.40. How many of each type of coin are there?

Solution   From the given information we set up the following table:

Type	Number	Value	Total Value
Nickels	x	.05	.05x
Quarters	100 − x	.25	.25(100 − x)
Mixture	100		14.40

If there are x nickels and each nickel is worth $.05, then the total value of the nickels is .05x.

Again we see that the sum of the total value of nickels and the total value of quarters is equal to the total value of the mixture ($14.40). We therefore write the equation

$$.05x + .25(100 - x) = 14.40.$$

Solve this equation in the margin.

Solving, we obtain

$$x = 53$$
$$100 - x = 47.$$

Thus, there are 53 nickels and 47 quarters.

Check: .05(53) + .25(47) = 2.65 + 11.75 = 14.40

Example 3   A merchant has three types of hard candy. The first sells for $1.10 per kilogram, the second for $1.50 per kilogram, and the third for $1.60 per kilogram. How many kilograms of each candy would be used to make 100 kilograms that would sell for $1.36 per kilogram, if twice as much of the second type must be used as the third?

Solution   We may outline as follows:

Type	Kg	Price per Kg	Total Value
First	$100 - 3x$	1.10	$1.10(100 - 3x)$
Second	$2x$	1.50	$1.50(2x)$
Third	x	1.60	$1.60(x)$
Mixture	100	1.36	136.00

Notice that the equation must come from the value column. Any equation must have like quantities on each side. You can *never* equate pounds with value, number of coins with the dollar value of the coins, and so forth.

The equation is

$$1.10(100 - 3x) + 1.50(2x) + 1.60(x) = 136.$$

Solving, this gives $x = 20.$
Also $2x = 40$
and $100 - 3x = 40.$

Thus, the merchant would use

40 kg of the first type
40 kg of the second type
20 kg of the third type.

Check: 40 kg at $1.10 = $ 44
                  40 kg at $1.50 = $ 60
                  20 kg at $1.60 = $ 32
                  100 kg at $1.36 = $136

Example 4   How much pure alcohol must be added to nine quarts of water to obtain a mixture of 40% alcohol?

Solution   We set up the table.

	Quarts	% Alcohol	Amount of Alcohol
Alcohol	x	100%	$1x$
Mixture	$x + 9$	40%	$.40(x + 9)$

Notice that the amount of alcohol in the mixture will be equal to the amount added. We can therefore write the equation

$$x = .40(x + 9).$$

Again, in this type of problem the use of a table will allow easy comparisons.

We can only compare like quantities.

Use this margin to solve this equation. See if your answers agree.

Again the check provides the opportunity to see if the answers are correct.

We could also work this problem equating water instead of alcohol. We would start with 9 quarts that are 100% water and end up with $x + 9$ quarts that are 60% water. Solve:

$$1.00(9) = .60(x + 9)$$

Solving, we obtain

$$x = 6.$$

Therefore, six quarts of alcohol must be added.

Check: If we add 6 quarts to 9 quarts, there will be a total of 15 quarts in the mixture, of which 6 quarts, or 40%, is alcohol.

Example 5 The capacity of a car's radiator is nine liters. The mixture of antifreeze and water is 30% antifreeze. How much of the mixture in the radiator must be drawn off and replaced with pure antifreeze to raise the percentage of the mixture to 65% antifreeze?

In this problem the number of liters of solution remains constant.

Solution In outlining this problem again keep in mind that the equation must equate like quantities. We can choose to make an equation equating antifreeze with antifreeze, or water with water, but *not* water with antifreeze. If we choose to equate antifreeze with antifreeze, we proceed as follows. If x represents the number of liters of the mixture to be drawn off, the outline would be:

Before we set up a table outline, we must decide whether to talk about antifreeze or water.

	Total Liters in Radiator	% Antifreeze	Liters of Antifreeze
Originally	9	30%	.30(9)
Drawn Off	x	30%	.30(x)
Added Back	x	100%	1.00(x)
Final	9	65%	.65(9)

Do this problem again, equating water with water. Set up a similar table only this time label the heading water instead of antifreeze. Obtain the equation and solve it. Your answer should be the same.

Note that the original amount of antifreeze, less the amount of antifreeze drained off, plus the amount of antifreeze added back will equal the final amount of antifreeze. Thus,

$$.30(9) - .30(x) + 1.00(x) = .65(9)$$
$$2.7 + .7x = 5.85$$
$$.7x = 3.15$$
$$x = 4.5 \text{ liters.}$$

In this case solving for x solves the problem.

Check: $.30(9) - .30(4.5) + 1.00(4.5) = 2.7 - 1.35 + 4.5$
$$= 5.85$$

▼ **EXERCISE 3–5–1**

Solve.

1. A candy dealer wishes to mix caramels that sell for $3.50 per pound with creams that sell for $2.75 per pound to obtain thirty pounds of a mixture that will sell for $3.00 per pound. How many pounds of each must be used?

2. A merchant wishes to mix a brand of coffee that sells for $2.50 per pound with a second brand that sells for $2.00 per pound to obtain a 100-pound mixture that will sell for $2.20 per pound. How much of each brand should be used?

3. If peanuts sell for $2.50 per pound and cashews sell for $4.50 per pound, how many pounds of each must be used to produce twenty pounds of mixed nuts to sell for $3.25 per pound?

4. A merchant has two types of candy. The first type sells for $2.00 per kilogram and the second for $2.30 per kilogram. How much of each type of candy must be used to make a 75-kilogram mixture that will sell for $2.10 per kilogram?

5. A total of 75 coins in nickels and quarters has a value of $15.15. How many coins of each type are there?

6. How many dimes and nickels are there in a collection of 125 of these coins if the total value is $9.50?

7. A total of 35 coins in dimes and quarters has a value of $5.60. How many of each type of coin are there?

8. A collection of coins has twice as many quarters as dimes. If the total value of the two types of coins is $9.60, find the number of each type.

9. A fruit market wishes to prepare 35 bushels of a mixture of oranges and grapefruits to sell for $6.50 per bushel. If oranges sell for $6.00 per bushel and grapefruits sell for $7.75 per bushel, how many bushels of each must be used?

10. A nursery wishes to prepare 100 bags of a "weed and feed" mixture by mixing fertilizer and weed killer. If the mixture is to sell for $7.80 per bag, how many bags of fertilizer that sells for $7.00 per bag and weed killer that sells for $9.00 per bag must be used?

11. A candy store owner sells pecans for $6.00 per kilogram and cashews for $7.50 per kilogram. A 50-kilogram mixture of the two kinds of nuts is to be made that will sell for $6.60 per kilogram. How much of each kind should be used?

12. A collection of quarters and half-dollars has a total value of $45.00. If there are 130 coins in all, how many of each type are there?

13. A collection of coins has three times as many nickels as quarters. If the total value of the two types is $8.40, find the number of each type.

14. A pile of one-dollar bills and five-dollar bills has a total value of $452. If there are 100 bills in all, how many of each type are there?

15. How many liters of pure alcohol must be added to ten liters of water to obtain a mixture of 60% alcohol?

16. How many liters of pure alcohol must be added to ten liters of 50% alcohol to obtain a mixture of 60% alcohol?

17. A college sold tickets for a play. The tickets sold at $1.00 per child and $2.50 per adult. If 210 tickets were sold for a total of $450, how many of each type of ticket were sold?

18. A meat market mixes ground chuck that sells for $2.00 per pound with ground round that sells for $2.50 per pound to produce fifty pounds of a mixture that sells for $2.20 per pound. How much of each type is used?

19. A merchant mixes three brands of coffee that sell for $1.90, $2.30, and $2.60 per pound to produce fifty pounds of a blend that will sell for $2.19 per pound. If twice as much of the $1.90 brand is used as the $2.60 brand, how much of each brand is used?

20. A merchant produces a 55-pound mixture of nuts using peanuts that sell for $2.50 per pound, almonds that sell for $5.00 per pound, and cashews that sell for $4.50 per pound. The mixture contains three times as many peanuts as cashews and sells for $3.50 per pound. How many pounds of each type of nut are used?

21. A collection of 53 coins contains nickels, dimes, and quarters. The total value is $7.50. If there are seven more dimes than quarters, find the number of each type of coin.

22. A total of $37 in coins contains dimes, quarters, and half-dollars. If there are 97 coins in all and there are four times as many half-dollars as dimes, find the number of each type of coin.

23. Mary has $4.25 in nickels, dimes, and quarters. She has twice as many dimes as quarters. If the total number of coins is 37, find the number of each kind of coin.

24. Frank has a total of $1,295 in fives, tens, and twenties. If he has a total of 124 bills and there are six more tens than fives, how many of each bill does he have?

25. A dairy wishes to produce 225 gallons of a "low-fat" milk, which is 2% butterfat, by adding skim milk (no butterfat) to milk that is 3.6% butterfat. How much of each type of milk must they use?

26. A nurse has twelve milliliters of a 16% solution and wishes to dilute it to a 6% solution. How much distilled water must be added to obtain the desired solution?

27. How much water should be added to five liters of a 25% salt solution to obtain a 15% salt solution?

28. A 10% salt solution is added to a 25% salt solution to obtain ten liters of a 15% salt solution. How much of each is to be used?

29. An airline sold 176 tickets on a flight. A first-class ticket costs $250, a coach fare $200, and a super saver ticket $120. If they sold eight more super saver tickets than first-class tickets and the total revenue for the fares was $33,840, how many tickets of each class did they sell?

30. A tank contains five liters of a 30% acid solution. How much must be drained off and replaced with a 50% acid solution to obtain five liters of 38% acid solution?

Key Words

Section 3–1

- An **outline** expresses all unknowns in a problem in terms of one unknown.

Section 3–3

- **Number relation problems** involve comparing two or more numbers.

Section 3–4

- **Distance-rate-time problems** involve the distance formula for a constant rate given by $d = rt$.

Section 3–5

- **Mixture problems** involve combining quantities of various concentrations.

Procedures

Section 3–1

- Translate from words to algebraic expressions to properly outline a word problem.

Section 3–2

- Outline a problem completely and accurately before attempting to solve it.

- To solve a word problem follow these steps:
  Step 1 Write an expression for the unknown.
  Step 2 Write an equation that relates the unknowns to each other.
  Step 3 Solve the equation.
  Step 4 Make sure you have answered the question.
  Step 5 Check your answers to make sure they agree with the original problem.

A   N   S   W   E   R   S

Write an algebraic expression for problems 1–5.

1. _____

2. _____

3. _____

4. _____

5. _____

6. _____

7. _____

8. _____

9. _____

10. _____

1. Twice a number, decreased by seven

2. Half a number, increased by three

3. The sum of a given number and 5

4. The value in cents of x quarters

5. The number of meters in y centimeters

6. A certain number is seven more than three times another number. Write expressions for the numbers.

7. The sum of two numbers is 84. Write expressions for the numbers.

8. Write an expression for the interest when a certain amount is invested for a year at 16%.

9. The sum of three numbers is 181. The first number is twice the third. Write expressions for the three numbers.

10. Write an expression for the total amount received at the end of one year if x dollars is invested at 17.3% interest.

11. If x represents one number and $x + 5$ represents another number, write a mathematical statement that indicates their sum is 23.

12. If one number is represented by x and a second number is represented by $3x - 1$, write a mathematical statement that indicates their sum is 19.

11. _____

12. _____

13. Write an equation to show that a number represented by $3x$ less a number represented by $x - 7$ is equal to 13.

14. Write an equation to show that the sum of two numbers, represented by x and $x - 6$, less a third number, represented by $3x$, is equal to -11.

13. _____

14. _____

15. Express algebraically the fact that a certain number added to 9% of the number is 1.26.

16. One item costs $1.25 more than another item. Write an equation to show that the sum of the costs of the two items is $5.00.

15. _____

16. _____

17. A certain number, less a second number that is five more than three times the first number, is equal to -7. Find the numbers.

18. A certain number less 35% of that number is 312. Find the number.

17. _____

18. _____

19. How many kilograms of water must be evaporated from 90 kilograms of salt water that contains 5% salt to obtain a solution that is 7% salt? (Answer correct to nearest tenth of a kilogram.)

20. A certain amount of money is invested for one year at 8% interest. The total principal and interest at the end of the year is $1,620. Find the original amount invested.

19. _____

20. _____

21. _____

22. _____

23. _____

24. _____

25. _____

26. _____

21. The length of one side of a triangle is twice that of the second and four less than that of the third. If the perimeter is 64, find the three sides.

22. A meter stick is cut into two pieces so that one piece is 6 centimeters longer than the other. Find the lengths of the two pieces.

23. A man has twice as many dimes as nickels and two more quarters than dimes. The total value of the coins is $3.50. How many of each kind of coin does he have?

24. Roger is seven years older than Dave. The sum of their ages is 63. Find their ages.

25. A grocer has three grades of hamburger: Grade A at $3.80 per kilogram, Grade B at $3.50 per kilogram, and Grade C at $3.30 per kilogram. Using equal amounts of Grades A and C, the grocer wishes to mix the three grades to obtain 50 kilograms of hamburger that will sell for $3.53 per kilogram. How many kilograms of each grade must be used?

26. Ellen is four years older than Kathy, and Barbara is two years younger than twice Kathy's age. How old is each of the women if the sum of their ages is 58?

27. A boat travels upstream and back in 12 hours. If the speed of the boat in still water is 25 kilometers per hour and the current in the river is 10 kilometers per hour, how far upstream did the boat travel?

28. At the start of the semester there were twice as many students enrolled in psychology as in algebra, and three fewer students in history than in psychology. After the first week of classes, the algebra class had doubled in enrollment, four students had dropped psychology, and five had added history. If the total enrollment after the first week of class was 148, how many students were enrolled in each class at the start of the semester?

A N S W E R S

27. _____

28. _____

29. Two towns, A and B, are 435 kilometers apart. A car leaves town B and travels toward town A at a speed of 65 kilometers per hour. At the same time a car leaves town A and travels toward town B. At what speed must this car travel to meet the other car in three hours?

30. A candy dealer has three types of candy selling at $4.50 a box, $6.95 a box, and $8.00 a box. The dealer has four more $6.95 boxes than $8.00 boxes, and the $4.50 boxes total two more than twice the number of $8.00 boxes. The dealer wishes to mix the candy in the same total number of boxes and sell the new boxes for $6.00 each. How many boxes of each candy does the dealer have? (Assume his total revenue remains the same.)

29. _____

30. _____

A N S W E R S

31. _____

32. _____

33. _____

34. _____

35. _____

36. _____

37. _____

38. _____

SCORE: _____

31. A 40-meter length of rope is cut into three pieces. The second piece is three times as long as the first, and the third piece is five meters shorter than twice the length of the second. Find the length of each piece.

32. A plane with a still air speed of 680 kilometers per hour makes an outgoing trip against a headwind in five hours and the return trip in three hours and thirty minutes. Find the speed of the headwind.

33. A girl has 107 coins consisting of nickels, dimes, and quarters. She has seven less than twice as many quarters as nickels. If the total value of the coins is $17.15, how many of each type of coin does she have?

34. The length of a rectangle is four more than twice the width. If the width is tripled and the length is doubled, the perimeter of the new rectangle is twelve more than twice the perimeter of the original. Find the dimensions of the original rectangle.

35. A tank contains 20 liters of a 10% acid solution. How much pure acid should be added to the solution to obtain a 15% acid solution? (Answer correct to the nearest tenth of a liter.)

36. The sum of three numbers is 123. The second number is five less than twice the first. The third number is seven more than the sum of the first two numbers. Find the numbers.

37. A boat leaves a dock and travels at the rate of 36 kilometers per hour. Twenty minutes later a second boat leaves the same dock and travels the same route at 50 kilometers per hour. How long will it take the second boat to be 5 kilometers behind the first boat?

38. A full two-liter bottle contains a 20% alcohol solution. How much of the solution must be poured out and replaced with pure alcohol to obtain two liters of a 30% alcohol solution?

1. Write an algebraic expression for the following:

   a. Three times a number x, increased by six

   b. Half a number y, decreased by seven

   c. 9.4% of a number x

   d. The number of weeks in d days

1a. _____

1b. _____

1c. _____

1d. _____

2. _____

3. _____

4. _____

5. _____

2. Sally has three more dimes than nickels and twice as many quarters as dimes. If she has a total of $5.05, find how many she has of each coin.

3. A piece of rope that is 63 meters long is cut into three pieces such that the second piece is twice as long as the first and the third piece is three meters longer than the second. Find the length of each piece.

4. A man can row at the rate of 12 kilometers per hour in still water. He rows upstream for two hours and returns in one hour and twelve minutes. Find the rate of the current.

5. How many liters of water must be added to 10 liters of a 12% alcohol solution to obtain an 8% alcohol solution?

SCORE: _____

NAME:

CLASS / SECTION:

DATE:

A   N   S   W   E   R   S

1. _____

2. _____

3. _____

4. _____

5. _____

6. _____

7. _____

8. _____

9. _____

10. _____

11. _____

12. _____

1. Find the negative of -1.5.

2. Evaluate: $-|-20|$

3. Combine: $(-39) + (+53)$

4. Find the product:
$(-5)(-2)(-1)(-3)$

5. Simplify: $(-2x^2y^2)(4xy^3)(-5x^2y)$

6. Remove grouping symbols and simplify:
$5a + 2\{3a - 4[2a - (3a - 1)]\}$

7. Evaluate $3a^2 - 4ab - 2b^2$ if $a = 3, b = -2$.

8. Classify $4(x - 3) = 10x - 2(3x + 6)$ as a conditional equation or an identity.

9. Solve for x: $3x = -21$

10. Solve for x: $3x - 5 = 2x - 1$

11. On a certain math test the ratio of passing grades to failing grades was 7 to 1. If 32 students took the test, how many passed?

12. Solve for c:
$2(x - c) - 5y = 4x - 5c$

A N S W E R S

13. Solve for x and graph the solution on the number line:

$$2(x - 3) + 4x \leq \frac{2}{3}(x + 1) + 2x + 3$$

13. _____

14. _____

14. Solve for x and graph the solution on the number line:

$$|3x - 5| < 7$$

15. _____

16. _____

15. Write an algebraic expression for the number of cents in x dimes.

16. Write an equation to show that a number represented by $8x - 3$ less a number represented by $5x$ is equal to 6.

17. _____

17. The length of a rectangle is 2 feet longer than twice the width. Find the length and width if the perimeter of the rectangle is 22 feet.

18. The sum of three consecutive integers is 39. Find the integers.

18. _____

19. _____

19. A motorboat travels up a river against a 5 mph current for 2 hours and returns in $1\frac{1}{2}$ hours. Find the rate of the motorboat in still water.

20. How many liters of pure alcohol must be added to fifteen liters of water to obtain a solution of 40% alcohol?

20. _____

SCORE: _____

C H A P T E R

4

PRETEST

Before beginning this chapter, answer as many of the following questions as you can. When you have finished the chapter, take the practice test at the end of the chapter and compare the scores of the two tests to see how much you have learned.

A N S W E R S

1. Multiply:

   a. $(2x - 1)(x + 7)$

   $x(2x-1) + 7(2x-1$
   $2x^3 - x + 14x - 7$
   $2x^3 + 13x - 7$

   b. $(2x - 3)(x^2 - 3x + 4)$

   $x^2(2x-3) - 3x(2x-3) + 4(2x-3)$
   $2x^3 - 3x^2 - 6x^2 + 9x + 8x - 12$
   $2x^3 - 9x^2 + 17x - 12$

2. Divide:

   a. $\dfrac{24ab^3 + 8a^2b^4 - 48a^3b^5}{-8ab^3}$

   $-3 + ab + 6A^2b^2$

   b. $(x^4 - x^2 + 4) \div (x + 3)$

Factor completely.

3. $35x^2 - 28x$

4. $6x^3y^3 - 8x^3y^2 + 10x^2y^3$

5. $a(x - y) - (x - y)$

6. $x^2 - 121$

7. $25x^2 - 20x + 4$

8. $ax - 3b + 3a - bx$

9. $x^2 + 20x + 100$

10. $3x^2 + 17x + 20$

1a. _____

1b. _____

2a. _____

2b. _____

3. _____

4. _____

5. _____

6. _____

7. _____

8. _____

9. _____

10. _____

219

11. $6x^2 + 7x - 5$

12. $x^3 + 8y^3$

11. _____

12. _____

13. $36x^2 - 121$

14. $x^2 - 2x - 63$

13. _____

14. _____

15. $a^2x + 12 - 3a^2 - 4x$

16. $6x^2 + 27x - 105$

15. _____

16. _____

17. $17x^3 - 17x$

18. $6x^3 - 27x^2 - 15x$

17. _____

19. $8x^2 + 14x - 15$

20. $4x^2 - y^2 - 4x + 1$

18. _____

19. _____

20. _____

SCORE: _____

4

Polynomials: Products and Factoring

Factoring is a process of changing an expression from a sum or difference of terms to a product of factors.

Factoring is a basic technique used for simplifying algebraic expressions and solving certain types of equations. Very little can be accomplished beyond this point in algebra without a knowledge of and a proficiency in factoring. Factoring cannot be separated from multiplication since both are based on the distributive law.

In this chapter we will study multiplication and division of polynomials and useful types of factoring, including removing the greatest common factor, special types of factoring, factoring the general trinomial, and factoring by grouping.

The techniques of factoring are mastered only by practice. The student should therefore work all problems in this chapter.

◥ 4–1 PRODUCTS OF POLYNOMIALS

OBJECTIVES

Upon completing this section you should be able to:

1. Recall how to add and subtract two polynomials.
2. Multiply any two polynomials.

Recall how to remove parentheses.

You may want to review section 1–8.

In chapter 1 we added and subtracted polynomials by combining like terms. We also multiplied polynomials by monomials.

Example 1   Add: $(5x^2 + 2x - 3) + (2x^2 + 4x + 5)$

  Solution
$$(5x^2 + 2x - 3) + (2x^2 + 4x + 5)$$
$$= 5x^2 + 2x - 3 + 2x^2 + 4x + 5$$
$$= 7x^2 + 6x + 2$$

Example 2   Subtract: $(5x^2 + 2x - 3) - (2x^2 + 4x + 5)$

  Solution
$$(5x^2 + 2x - 3) - (2x^2 + 4x + 5)$$
$$= 5x^2 + 2x - 3 - 2x^2 - 4x - 5$$
$$= 3x^2 - 2x - 8$$

Example 3   Multiply: $3x^3(2x^2 + 3x - 1)$

  Solution
$$3x^3(2x^2 + 3x - 1) = 6x^5 + 9x^4 - 3x^3$$

Example 4   Multiply: $-4x(3x - 2)$

  Solution
$$-4x(3x - 2) = -12x^2 + 8x$$

◢ EXERCISE 4–1–1

As a review of the techniques studied in chapter 1, perform the indicated operations in the following problems.

1. $(3x^2 + 2x - 4) + (6x^2 - 7x + 10)$

$9x^2 - 5x + 6$

2. $(8x^2 - 4x + 7) + (-5x^2 + 2x - 6)$

$3x^2 - 2x + 1$

3. $(2a^2 - 3a + 2) + (a^2 + 3a - 1)$

$a^2 + 1$

4. $(3a^2 - 7a + 5) + (13a^2 - 21a - 6)$

$16a^2 - 28a - 1$

5. $(9x^2 - 2x + 3) - (8x^2 + 3x - 4)$

$9x^2 - 2x + 3 + 8x^2 - 3x + 4$
$x2 - 5x + 7$

6. $(16x^2 + 5x - 1) - (12x^2 + 8x + 3)$

$4x^2 + 13x - 4$

7. $(12a^3 + 5a^2 - a + 9) - (10a^3 + 6a^2 - a - 4)$

$12a^3 + 5a^2 - a + 9 + (-10a^3) - 6a^2 + a + 4$
$2a^3 - a^2 + 13$

8. $(8a^3 + 4a^2 + 5a - 1) - (4a^3 + 8a^2 - 5a - 1)$

$8a^3 + 4a^2 + 5a - 1 - 4a^3 - 8a^2 + 5a + 1$
$4a^3 - 4a^2 + 10a$

9. $5(x^2 - x + 2)$

$5x^2 - 5x + 10$

10. $6(x^2 + 2x - 3)$

$6x^2 + 12x - 18$

11. $2x(3x - 1)$

$6x^2 - 2x$

12. $3x(4x + 5)$

$12x^2 + 15x$

13. $2x(x^2 - 2x + 1)$

$2x^3 - 4x^2 + 2x$

14. $5x(2x^2 + 3x - 2)$

$10x^3 + 15x^2 - 10x$

15. $3x^2(2x^2 + x - 5)$

$6x^4 + 3x^3 - 15x^2$

16. $2x^2(5x^2 - 4x + 3)$

$10x^4 - 8x^3 + 6x^2$

17. $-5a(2a^2 - a + 7)$

$-10a^3 + 5a^2 - 35a$

18. $-3a(3a^2 + 9a - 2)$

$-9a^3 - 27a^2 + 6a$

19. $4a^3(2a^3 - 4a^2 + 3a - 5)$

$8a^6 - 16a^5 + 12a^4 - 20a^3$

20. $2a^3(4a^3 + 6a^2 - a + 3)$

$8a^6 + 12a^5 - 2a^4 + 6a^3$

We now turn to the multiplication of polynomials by polynomials other than monomials.

$(3x + 1)$ is a binomial.

$(2x^2 + 3x - 5)$ is a trinomial.

Example 5   Multiply: $(3x + 1)(2x^2 + 3x - 5)$

Solution   To multiply these two polynomials we resort to two ideas previously studied—the meaning of grouping symbols and the distributive law. First we will note that since $3x + 1$ is in parentheses, it can be regarded as a single factor. We now use the distributive law.

Multiply each term of the trinomial by the binomial.

$$(3x + 1)(2x^2 + 3x - 5) = 2x^2(3x + 1) + 3x(3x + 1) - 5(3x + 1)$$

Note that we have indicated the product of the factor $(3x + 1)$ and each term of $(2x^2 + 3x - 5)$. We now use the distributive law again to remove the resulting parentheses.

Now combine like terms.

$$2x^2(3x + 1) + 3x(3x + 1) - 5(3x + 1)$$
$$= 6x^3 + 2x^2 + 9x^2 + 3x - 15x - 5$$
$$= 6x^3 + 11x^2 - 12x - 5$$

Example 6   Multiply: $(x + y + 3)(x - 2y + 5)$

The same technique is used for multiplying any two polynomials.

Solution   Here we are asked to find the product of two trinomials. As before, we use the distributive law.

$$(x + y + 3)(x - 2y + 5)$$
$$= x(x + y + 3) - 2y(x + y + 3) + 5(x + y + 3)$$

Make sure to combine like terms.

$$= x^2 + xy + 3x - 2xy - 2y^2 - 6y + 5x + 5y + 15$$
$$= x^2 - xy + 8x - 2y^2 - y + 15$$

▼ EXERCISE 4–1–2

Find the products.

$x^3 - 4x^2 + 3x - x^2 + 4x - 3$

1. $(x + 5)(x^2 + 3x + 1)$

$x^3 + 3x^2 + x + 5x^2 + 15x + 5$
$x^3 + 8x^2 + 16x + 5$

2. $(a + 3)(a^2 - 2a + 4)$

$a^3 - 2a^2 + 4a + 3a^2 - 6a + 12$
$a^3 - a^2 - 2a + 12$

3. $(x - 1)(x^2 - 4x + 3)$

$x^3 - 5x^2 + 7x - 3$

4. $(a - 4)(2a^2 + 3a - 11)$

5. $(2x - 1)(x^2 - 4x + 3)$

$2x^3 - 8x^2 + 6x - x^2 - 4x - 3$
$2x^3 - 7x^2 + 2x - 3$

6. $(4a - 5)(a^2 + 2a - 1)$

7. $(5a + b)(3a^2 + ab - 4b^2)$

$15a^3 + 5a^2b - 20ab^2 + 3a^2b + ab^2 - 4b^3$
$15a^3 + 8a^2b - 19ab^2 - 4b^3$

8. $(2x + y)(4x^2 - 2xy + y^2)$

9. $(2x - 3y + z)(4x + 3y)$

$8x^2 + 6xy - 12xy + 9y^2 + 4xz + 3yz$
$8x^2 - 6xy - 9y^2 + 4xz + 3yz$

10. $(8a + b - 4c)(3a - 5c)$

11. $(x + 3)(x^2 - 3x + 9)$

$x^2 - 3x^2 + 9x + 3x^2 - 9x + 27$

$x^3 + 27$

12. $(2x - 5)(3x^2 + x - 2)$

13. $(2x^2 - x + 1)(x - 6)$

$2x^3 - 12x^2 - x^2 + 6x + x - 6$

$2x^3 - 13x^2 + 7x - 6$

14. $(3x + 7)(2x^2 - 5x + 8)$

15. $(x - 1)(x^2 + x + 1)$

$x^3 + x^2 + x - x^2 - x - 1$

$x^3 - 1$

16. $(x - 5)(x^2 + 5x + 25)$

17. $(x + 1)(x^3 + 4x^2 - x + 3)$

$x^4 + 4x^3 - x^2 + 3x + x^3 + 4x^2 - x + 3$

$x^4 + 5x^3 + 3x^2 + 2x + 3$

18. $(a - 3)(a^3 - 2a^2 + 5a - 9)$

19. $(x - 3)(x^2 + 3x + 9)$

$x^3 + 3x^2 + 9x - 3x^2 - 9x - 27$

$x^3 - 27$

20. $(2a - b)(4a^2 + 2ab + b^2)$

21. $(a + b + c)(a + b + c)$

$a^2 + ab + ac + ab + b^2 + bc + ac$

$+ bc + c^2$

$a^2 + 2ab + 2ac + b^2 + 2bc + c^2$

22. $(x + y - z)(x + y - z)$

23. $(x + y - 3)(x - y + 4)$

$x^2 - xy + 4x + xy - y^2 + 4y - 3x + 3y - 12$

$x^2 - y^2 + x + 7y - 12$

24. $(x + y + 4)(x - y - 2)$

25. $(a + 2b - 1)(2a - b - 5)$

$2a^2 - ab - 5a + 4ab - 2b^2 - 10b - 2a + b + 5$

$2a^2 - 12b - 7A + 4ab - 2b^2 + b + 5$

26. $(3a - b - c)(2a + 4b - c)$

27. $(5x - 3y + z)(x + y - 3z)$

$5x^2 + 5xy - 15xz - 3xy - 3y^2 + 9yz + xz + yz - 3z^2$

$5x^2 + 2xy - 14xz + 3y^2 + 10yz - 3z^2$

28. $(2a + b - 4c)(a - 5b + c)$

29. $(x^2 + xy + 1)(x - xy + 5)$

$x^3 - x^2y + 5x^2 + x^2y - x^2y^2 + 5xy + x - xy + 5$

$x^3 - x^3y + 5x^2 + 4xy + x^2y - x^2y^2 + x + 5$

30. $(a^2 - 2ab - 3)(a + ab - 4)$

 **4-2 DIVISION OF POLYNOMIALS**

We must remember that coefficients and exponents are controlled by different laws because they have different definitions. In division of monomials the coefficients are divided while the exponents are subtracted according to the division law of exponents.

Example 1 $\dfrac{8x^8}{4x^4} = \left(\dfrac{8}{4}\right)(x^{(8-4)}) = 2x^4$

If no division is possible or if only reducing a fraction is possible with the coefficients, this does not affect the use of the law of exponents for division.

Reduce this type of fraction in two steps:

1. Reduce the coefficients.
2. Use the third law of exponents.

Example 2 $\dfrac{3x^5}{5x^2} = \dfrac{3x^3}{5}$

Example 3 $\dfrac{6xy^2}{-15x^3y} = \dfrac{2y}{-5x^2} = -\dfrac{2y}{5x^2}$

EXERCISE 4-2-1

Simplify.

1. $\dfrac{4x^5}{2x^3}$ $2x^2$

2. $\dfrac{6a^8}{2a^6}$

3. $\dfrac{10x^9}{5x^3}$ $2x^6$

4. $\dfrac{8a^{10}}{2a^5}$

5. $\dfrac{4b^7}{12b^2}$ $\dfrac{b^5}{3}$

6. $\dfrac{18x^4}{6x^8}$

7. $\dfrac{9a^2}{15a^{10}}$ $\dfrac{3}{5a^8}$

8. $\dfrac{18w^{16}}{32w^4}$

9. $\dfrac{10x^3y^2}{2x^2y}$ $5xy$

10. $\dfrac{50a^3b^4}{15a^6b^8}$

11. $\dfrac{-8x^3}{2x^5}$ $\dfrac{-4}{x^2}$

12. $\dfrac{10a^2}{-5a^3}$

13. $\dfrac{-4a^3b}{-12ab^4}$ $\dfrac{a^2}{3b^3}$

14. $\dfrac{-3w^2z^3}{15wz^9}$

15. $\dfrac{18x^3y^5}{24xy^7}$ $\dfrac{3x^2}{4y^2}$

16. $\dfrac{34a^3b^4}{17a^3b}$

17. $\dfrac{-a^3}{-a^4}$ $\dfrac{1}{a}$

18. $\dfrac{-14x^5}{2x^3}$

19. $\dfrac{x^{14}}{4x^7}$ $\dfrac{x^7}{4}$

20. $\dfrac{-15x^4}{-3xy}$

21. $\dfrac{21x^2y^7}{-6x^5y^3}$ $\dfrac{7y^4}{-2x^3}$

22. $\dfrac{12x^5y^6}{-4x^2y}$

23. $\dfrac{-36x^3y^5}{9x^6y}$ $\dfrac{-4y^4}{x^3}$

24. $\dfrac{-3x^4y^3}{27x^7y^5}$

25. $\dfrac{-x^2y^2}{-x^2y^2}$ 1

26. $\dfrac{12ab}{-24a^3b^2}$

27. $\dfrac{-15xy}{21yz}$ $\dfrac{-5x}{7z}$

28. $\dfrac{7a^2bc}{84ab^2c}$

29. $\dfrac{27w^3x^2y^4}{-36w^2y^4}$ $\dfrac{-3wx^2}{4}$

30. $\dfrac{-132a^3bc^4}{77a^5c}$

To divide a polynomial by a monomial involves one very important fact in addition to things we already have used. That fact is this: When there are several terms in the numerator of a fraction, then *each* term must be divided by the denominator.

Example 4 $\dfrac{3xy + 9x^2}{3x}$ means that both $3xy$ and $9x^2$ are divided by $3x$.

$$\dfrac{3xy + 9x^2}{3x} = \dfrac{3xy}{3x} + \dfrac{9x^2}{3x} = y + 3x$$

Example 5 $\dfrac{7x^2y - 8x^3y^2}{xy} = \dfrac{7x^2y}{xy} - \dfrac{8x^3y^2}{xy} = 7x - 8x^2y$

We could rewrite $\dfrac{3xy + 9x^2}{3x}$ as

$\dfrac{1}{3x}(3xy + 9x^2)$

$= \left(\dfrac{1}{3x}\right)(3xy) + \left(\dfrac{1}{3x}\right)(9x^2)$

$= \dfrac{3xy}{3x} + \dfrac{9x^2}{3x}.$

Thus we are actually using the distributive property in this process.

Example 6 $\dfrac{2A - 4B}{-2} = \dfrac{2A}{-2} - \dfrac{4B}{-2} = -A + 2B \text{ or } 2B - A.$

▼ **EXERCISE 4-2-2**

Simplify.

1. $\dfrac{3a + 15b}{3}$ $a+5b$

2. $\dfrac{4x + 12x^2}{4}$

3. $\dfrac{4x^2 + 10y^2}{2}$ $2x^2+5y^2$

4. $\dfrac{15a + 25b}{5}$

5. $\dfrac{a^2b + ab^2}{a}$ $ab+b^2$

6. $\dfrac{3x + 5x^2}{x}$

7. $\dfrac{3a^2 + 9ab}{3a}$ $a+3b$

8. $\dfrac{7x + 14xy}{7x}$

9. $\dfrac{14x - 21x^3}{7x}$ $2-3x^2$

10. $\dfrac{20a - 50a^2}{10a}$

11. $\dfrac{ab + a^2b^2}{a}$ $b+ab^2$

12. $\dfrac{x^2y + xy^2}{x}$

13. $\dfrac{25x^2 - 30xy}{5x}$ $5x-6y$

14. $\dfrac{24a - 18a^2b}{6a}$

15. $\dfrac{x^2y + xy^2}{xy}$ $x+y$

16. $\dfrac{ab^3 - a^3b}{ab}$

17. $\dfrac{x^3y^2 + x^2y^3}{x^2y^2}$ $x+y$

18. $\dfrac{a^3b^5 + a^4b^4}{a^3b^3}$

19. $\dfrac{3xy - 18x^2y^3}{3xy}$ $1-6xy^2$

20. $\dfrac{16a^2b - 8ab}{8ab}$

21. $\dfrac{21a^2b^3 - 49a^3b^5}{7a^2b^2}$

$3b-7ab^3$

22. $\dfrac{36x^2y^4 - 12x^3y^3}{6x^2y^3}$

23. $\dfrac{8x^2yz + 20x^3y^2z^4}{4x^2z}$

$2y + 5xy^2z^3$

24. $\dfrac{22a^2b^2c + 55a^3bc^3}{11a^2b}$

25. $\dfrac{32x^2y - 48xy^2 + 8x}{4x}$

$8xy - 12y^2 + 2$

26. $\dfrac{18a^3b - 36a^4 + 9a^3}{9a^2}$

27. $\dfrac{3x + 6y}{-3}$ $-x - 2y$

28. $\dfrac{15a + 5b}{-5}$

29. $\dfrac{28a^3 - 14b^3 - 7}{-7}$

$-4a^3 + 2b^3 + 1$

30. $\dfrac{46x - 23y + 69}{-23}$

31. $\dfrac{14x^2 - 20x + 4xy}{-2x}$

$-7x + 10 - 2y$

32. $\dfrac{8a^3 - 12a^4 - 4a^2}{-4a^2}$

33. $\dfrac{16x^2y + 12xy^2}{-4xy}$

$-4x - 3y$

34. $\dfrac{22ab^2 + 8a^2b - 2ab}{-2ab}$

35. $\dfrac{49x^2y^2 - 63x^2y^3}{-7x^2y^2}$

$-7 + 9y$

36. $\dfrac{9x^3y^5 - 33x^4y^4}{-3x^3y^4}$

37. $\dfrac{9a^2b - 72a^3b^2 - 18a^2b^2}{-9a^2b}$

$-1 + 8ab + 2b$

38. $\dfrac{14xy^2 - 2x^2y^3z + 6x^2y^2}{-2xy^2}$

39. $\dfrac{-21xy^3 - 15x^2y + 3xy}{-3xy}$

$7y^2 + 5x - 1$

40. $\dfrac{-11a^2b^2 - 121a^3b - 22a^2b^3}{-11a^2b}$

The process for dividing a polynomial by another polynomial will be a valuable tool in later topics. Here we will develop the technique and discuss the reasons why it works in the future.

This technique is called the **long division algorithm.** An *algorithm* is simply a method that must be precisely followed. Therefore, we will present it in a step-by-step format and by example.

Recall the three expressions in division:

$$\text{divisor } \overline{)\,\text{dividend}}^{\text{quotient}}$$

Example 7   Find $(x^2 + 5x - 14) \div (x + 7)$.

If we were asked to arrange the expression
$$3x - 4x^2 + x^5 - x^4 + 8$$
in descending powers, we would write
$$x^5 - x^4 + 0x^3 - 4x^2 + 3x + 8.$$

The coefficient zero gives $0x^3 = 0$. That is the reason the x^3 term was missing or not written in the original expression.

Step 1   Arrange both the divisor and dividend in descending powers of the variable (this means highest exponent first, next highest second, and so on) and supply a zero coefficient for any missing terms. (In this example the arrangement need not be changed and there are no missing terms.) Then arrange the divisor and dividend in the following manner:

$$x + 7\overline{)x^2 + 5x - 14}$$

Step 2   To obtain the first term of the quotient, divide the first term of the dividend by the first term of the divisor, in this case $x^2 \div x = x$. We record this as follows:

$$\overset{\displaystyle x}{\textcircled{x} + 7\overline{)\textcircled{x^2} + 5x - 14}}$$

Make sure you write the quotient directly over the quantity you are dividing into. In this case x divides into x^2 x times. Thus, the x is written directly over the quantity x^2.

Step 3   Multiply the entire divisor by the term obtained in step 2. Subtract the result from the dividend as follows:

$$\begin{array}{r} \textcircled{x} \\ x + 7\overline{)x^2 + 5x - 14} \\ -(x^2 + 7x) \\ \hline -2x - 14 \end{array}$$

The first term of the remainder $(-2x - 14)$ is $-2x$.

Step 4   Divide the first term of the remainder by the first term of the divisor to obtain the next term of the quotient. Then multiply the entire divisor by the resulting term and subtract again as follows:

$$\begin{array}{r} x - 2 \\ \textcircled{x} + 7\overline{)x^2 + 5x - 14} \\ -(x^2 + 7x) \\ \hline \boxed{-2x} - 14 \\ -(-2x - 14) \\ \hline 0 \end{array}$$

Multiply $(x + 7)$ by -2.

$$\begin{array}{r} 4 \\ 3\overline{)14} \\ 12 \\ \hline 2 \text{ remainder} \end{array}$$

Thus $(4) \times (3) + 2 = 14$.

quotient   divisor   remainder
                                    dividend

This process is repeated until either the remainder is zero (as in this example) or the power of the first term of the remainder is less than the power of the first term of the divisor.

As in arithmetic, division is checked by multiplication. We must remember that (quotient) $\times$ (divisor) + (remainder) = (dividend).

To check this example we multiply $(x + 7)$ and $(x - 2)$ to obtain $x^2 + 5x - 14$.

Example 8   Find $(2x^3 + 11x^2 + 8x - 15) \div (x + 3)$.

$$
\begin{array}{r}
2x^2 + 5x - 7 \\
x + 3 \overline{\smash{)}\, 2x^3 + 11x^2 + 8x - 15} \\
\underline{-(2x^3 + 6x^2)} \\
5x^2 + 8x - 15 \\
\underline{-(5x^2 + 15x)} \\
-7x - 15 \\
\underline{-(-7x - 21)} \\
+6
\end{array}
$$

The answer is $2x^2 + 5x - 7$, remainder $+6$.

Check: $(2x^2 + 5x - 7)(x + 3) + 6$ (quotient) $\times$ (divisor)
$2x^3 + 5x^2 - 7x + 6x^2 + 15x - 21 + 6$ $+$ (remainder) $=$ (dividend)
$2x^3 + 11x^2 + 8x - 15$

Since this is the dividend, the answer is correct.

Example 9   Find $(x^2 - 9) \div (x + 3)$.

Note here that the dividend $x^2 - 9$ has a missing term. We must supply a zero coefficient, so the expression will be $x^2 + 0x - 9$.

> **WARNING**
>
> A common error is to forget to write the missing term with a zero coefficient.

$$
\begin{array}{r}
x - 3 \\
x + 3 \overline{\smash{)}\, x^2 + 0x - 9} \\
\underline{-(x^2 + 3x)} \\
-3x - 9 \\
\underline{-(-3x - 9)} \\
0
\end{array}
$$

The answer is $x - 3$. Checking, we find $(x + 3)(x - 3)$
$= x^2 - 9$.

▼   **EXERCISE 4–2–3**

Divide.

1. $(x^2 + 3x + 2) \div (x + 1)$ $x + 2$

2. $(x^2 + 8x + 15) \div (x + 3)$

3. $(x^2 + 2x - 8) \div (x + 4)$ $x - 2$

4. $(x^2 - 2x - 3) \div (x - 3)$

5. $(x^2 - 5x + 6) \div (x - 3)$ $x - 2$

6. $(x^2 - 13x + 36) \div (x - 9)$

7. $(x^2 - 12x - 40) \div (x + 3)$ $x - 5$ $R5$

8. $(x^2 + 4x - 1) \div (x - 8)$

9. $(x^2 + 3x - 10) \div (x + 5)$ $x - 2$

10. $(x^2 - 25) \div (x + 5)$

11. $(x^2 - 17x + 66) \div (x - 6)$ $x - 11$

12. $(x^2 - 18x + 65) \div (x - 5)$

13. $(x^2 - 16) \div (x - 4)$ $x + 4$

14. $(x^2 - 36) \div (x - 6)$

15. $(x^3 - x^2 - 2x + 8) \div (x + 2)$
$x^2 - 3x + 4$

16. $(x^3 + 2x^2 - x - 2) \div (x + 1)$

17. $(2x^3 - 2x^2 - 5x + 6) \div (x - 3)$
$2x^2 + 4x + 7$ $R 27$

18. $(x^3 + 2x^2 - 30x + 2) \div (x - 5)$

19. $(x^3 + 3x^2 + 500) \div (x + 9)$
$x^2 - 6x + 54$ $R 14$

20. $(x^3 + 2x^2 - 32) \div (x + 4)$

21. $(x^3 + x^2 + 10x - 1,000) \div (x - 10)$

22. $(x^3 - 5x^2 - 60x - 68) \div (x - 11)$

23. $(3x^3 + 11x^2 + 9x - 10) \div (3x + 2)$

24. $(4x^3 + 8x^2 + x - 1) \div (2x + 1)$

25. $(x^3 - 50x + 1) \div (x - 7)$

26. $(x^3 + 3x^2 + 110) \div (x + 6)$

27. $(x^4 - x^2 - 72) \div (x + 3)$

28. $(x^4 + 5x^2 - 30) \div (x + 2)$

29. $(x^4 - 1) \div (x + 1)$

30. $(x^4 - 2x) \div (x + 2)$

◥◣ 4–3 REMOVING THE GREATEST COMMON FACTOR

The process of factoring is essential to the simplification of many algebraic expressions and is a useful tool in solving higher degree equations. In fact, the process of factoring is so important that very little of algebra beyond this point can be accomplished without understanding it.

In earlier chapters the distinction between **terms** and **factors** has been stressed. You should remember that terms are added or subtracted and factors are multiplied. Three important definitions follow.

Terms occur in an indicated sum or difference. **Factors** occur in an indicated product.

An expression is in **factored form** only if the *entire* expression is an indicated product.

O B J E C T I V E S

▼

Upon completing this section you should be able to:
1. Determine which factors are common to all terms in an expression.
2. Factor common factors.

$3(x + y)$ indicates the product of 3 and the quantity $(x + y)$.

	In Factored Form	**Not in Factored Form**
Example 1	$2x(x + y)$	$2x + 3y + z$
Example 2	$(x + y)(3x - 2y)$	$2(x + y) + z$
Example 3	$(x + 4)(x^2 + 3x - 1)$	$(x + y)(2x - y) + 5$

Note in the expression $2(x + y) + z$ that part of it is factored but *not* the entire expression.

Note in these examples that we must always regard the entire expression. Factors can be made up of terms and terms can contain factors, but **factored form** must conform to the definition above.

Factoring is a process of changing an expression from a sum or difference of terms to a product of factors.

Note that in this definition it is implied that the value of the expression is not changed—only its form.

In the previous chapter we multiplied an expression such as $5(2x + 1)$ to obtain $10x + 5$. In general, factoring will "undo" multiplication. Each term of $10x + 5$ has 5 as a factor, and $10x + 5 = 5(2x + 1)$.

To factor an expression by removing common factors, proceed as in example 4.

Example 4   Factor $3x^2 + 6xy + 9xy^2$.

First list the factors of each term.

3x is the greatest common factor of all three terms.

$3x^2$ has factors 1, 3, x, x^2, ③x, and $3x^2$.

$6xy$ has factors 1, 2, 3, 6, x, $2x$, ③x, $6x$, y, and so on.

$9xy^2$ has factors 1, 3, 9, x, ③x, $9x$, xy, xy^2, and so on.

Next look for factors that are common to all terms, and search out the greatest of these. This is the **greatest common factor.** In this case, the greatest common factor is $3x$.

Proceed by placing $3x$ before a set of parentheses.

$$3x(\qquad)$$

The terms within the parentheses are found by *dividing* each term of the original expression by $3x$.

$$3x^2 + 6xy + 9xy^2 = 3x(x + 2y + 3y^2)$$

Note that this is the distributive property. It is the reverse of the process that we have been using until now.

The original expression is now changed to factored form. To check the factoring keep in mind that factoring changes the *form* but not the value of an expression. If the answer is correct, it must be true that $3x(x + 2y + 3y^2) = 3x^2 + 6xy + 9xy^2$. Multiply to see that this is true. A second check is also necessary for factoring—we must be sure that the expression has been **completely factored.** In other words, "Did we remove all common factors? Can we factor further?"

If we had only removed the factor "3" from $3x^2 + 6xy + 9xy^2$, the answer would be

This expression is factored but not completely factored.

$$3(x^2 + 2xy + 3xy^2).$$

Multiplying to check, we find the answer is actually equal to the original expression. However, the factor x is still present in all terms. Hence the expression is not completely factored.

For factoring to be correct the solution must meet two criteria.

1. It must be possible to multiply the factored expression and get the original expression.
2. The expression must be *completely* factored.

Example 5   Factor $12x^3 + 6x^2 + 18x$.

Say to yourself, "What is the largest common factor of 12, 6, and 18?"

At this point it should not be necessary to list the factors of each term. You should be able to mentally determine the greatest common factor. A good procedure to follow is to think of the elements individually. In other words, don't attempt to obtain all common factors at once but get first the number, then each letter involved. For instance, 6 is a factor of 12, 6, and 18, and x is a factor of each term. Hence $12x^3 + 6x^2 + 18x = 6x(2x^2 + x + 3)$. Multiplying, we get the original and can see that the terms within the parentheses have no other common factor, so we know the solution is correct.

Then, "What is the largest common factor of x^3, x^2, and x?"

Remember, this is a check to make sure we have factored correctly.

Example 6 Factor $a^2b^2c + 2ab^2c^2 - 3ab^3c$.

Solution We note that a, b^2, and c are common factors.

$$\text{Hence } a^2b^2c + 2ab^2c^2 - 3ab^3c = ab^2c(a + 2c - 3b).$$

Check: $ab^2c(a + 2c - 3b) = a^2b^2c + 2ab^2c^2 - 3ab^3c$

Again, multiply out as a check.

Example 7 Factor $2x - 4y + 2$.

Solution The only common factor is 2.

$$2x - 4y + 2 = 2(x - 2y + 1)$$

Example 8 Factor $5x^3y + 10x^2y^2 + 5xy^2$.

Solution $$5x^3y + 10x^2y^2 + 5xy^2 = 5xy(x^2 + 2xy + y)$$

Again, find the greatest common factor of the numbers and each letter separately.

If an expression cannot be factored it is said to be **prime.**

Example 9 Factor $3x^2y + 5x + 2y^2$.

Solution Since there is no common factor (except 1), this expression is *prime*.

Remember that 1 is always a factor of any expression.

Sometimes a common factor may be made up of terms.

Example 10 Factor $2x(a + b) - 4(a + b)$.

Solution Here the common factor is $(a + b)$.

$$2x(a + b) - 4(a + b) = (a + b)(2x - 4)$$

Divide each term by $(a + b)$.

Example 11 Factor $3x^2(x + y) + 5x(x + y)$.

Solution Note that we have two common factors, x and $(x + y)$.

$$3x^2(x + y) + 5x(x + y) = x(x + y)(3x + 5)$$

Divide each term by $x(x + y)$.

Example 12 Factor $3(a + b)^2 + (a + b)$.

Solution $(a + b)$ is the common factor.

$$3(a + b)^2 + (a + b) = (a + b)[3(a + b) + 1]$$

Again, don't forget to write the "1."

EXERCISE 4-3-1

Factor each of the following by removing all common factors.

1. $12a + 10b$

$2(6a+5b)$

2. $6x + 8y$

3. $12x - 20y$

$4(3x-5y)$

4. $11a - 33b$

5. $a^2 + 5a$

$a(a+5)$

6. $x^2 + 3x$

7. $x^2 - xy$

$x(x-y)$

8. $a^2 - a$

9. $2a^2 + 6a$

$2a(a+3)$

10. $5x^2 + 20x$

11. $15x^2 - 35x$

$5x(3x-7)$

12. $18a^2 - 6a$

13. $9x^2 - 3x$

$3x(3x-1)$

14. $4a^3 - 2a^2$

15. $6a^2 + 3a$

$3a(2a+1)$

16. $21x^3 + 7x^2$

17. $8x^2y + 12xy$

$4xy(2x+3)$

18. $5a^2b - 25ab^2$

19. $x^3y - x^2y^2$

$x^2y(x-y)$

20. $9a^4b^2 + 15a^2b^2$

21. $10a^2b - 2ab^2 + 6ab$

$2ab(5a-b+3)$

22. $12x^3y^2 + 20x^2y + 28x^2y^2$

23. $15x^4y - 20x^2y^2 + 5x^2y$

$5x^2y(3x^2-4y+1)$

24. $11a^3b^4 + 44ab^3 - 33a^2b^4$

25. $3x + 4y - 5z$

Prime

26. $3a - 27b + 9c$

27. $6x + 21y - 27z$

$3(2x+7y-9z)$

28. $4a^2 - 29b + 10ab$

29. $14a^2b - 35ab - 63a$

$7a(2ab-5b-9)$

30. $27x^3y^2 + 18x^2y^2 - 36x^2y^3$

31. $2x(x + 4) + 3(x + 4)$

$(2x+3)(x+4)$

32. $4a(a + b) + 3(a + b)$

33. $a(a - 1) + 5(a - 1)$

$(a+5)(a-1)$

34. $2x(x - 5) + 7(x - 5)$

35. $6x(y + 4) - 7(y + 4)$

$(6x-7)(4+4)$

36. $3a(b - c) - 4(b - c)$

37. $3x(x - 2) + (x - 2)$

$(3x+1)(x-2)$

38. $10x(y + 1) - (y + 1)$

39. $7a(a - 4)^2 - 15(a - 4)$

40. $14a(b + 1)^2 + 5(b + 1)$

41. $x(x + 3) + 3(x + 3)$

$(x+3)(x+3)$

$(x+3)^2$

42. $a(a + b) + b(a + b)$

43. $3x^2(x + 1) + 2x(x + 1)$

$(3x^2+2x)(x+1)$

44. $a^2(a - 1) + 3a(a - 1)$

a^3-a+3a^2-1

$(a-1)(a^2+3a)$

45. $a^2(a + b)^2 + 4a(a + b)$

46. $3a^2(a - b)^2 - 5a(a - b)$

47. $8a^2(a + 4) + 2a(a + 4)$

$(8a^2+2a)(a+4)$

$2a(4a+2)(a+4$

48. $15x^3(x - 3) - 12x^2(x - 3)$

49. $12x^3(x - 5)^2 - 18x^2(x - 5)$

50. $9a^4(a + 2)^3 + 21a^3(a + 2)^2$

◢◣ 4-4 OBTAINING A COMMON FACTOR BY GROUPING

In section 4–3 we saw that a common factor does not need to be a single term. For instance, in the expression $2y(x + 3) + 5(x + 3)$ we have a common factor of $(x + 3)$ that is made up of two terms. This observation is helpful in certain types of factoring.

OBJECTIVES ▼

Upon completing this section you should be able to:

1. Group terms that have a common factor.
2. Completely factor these expressions.

Example 1   Factor $3ax + 6y + a^2x + 2ay$.

Solution   First note that not all four terms in the expression have a common factor, but that some of them do. For instance, we can factor 3 from the first two terms, giving $3(ax + 2y)$. If we factor a from the remaining two terms, we get $a(ax + 2y)$. The expression is now $3(ax + 2y) + a(ax + 2y)$, and we have a common factor of $(ax + 2y)$ and can factor as $(ax + 2y)(3 + a)$. Multiplying $(ax + 2y)(3 + a)$, we get the original expression $3ax + 6y + a^2x + 2ay$ and see that the factoring is correct.

$$3ax + 6y + a^2x + 2ay = (ax + 2y)(3 + a)$$

The first two terms
$$3ax + 6y = 3(ax + 2y)$$
The remaining two terms
$$a^2x + 2ay = a(ax + 2y)$$
$3(ax + 2y) + a(ax + 2y)$ is now a factoring problem similar to the problems you did in exercise 4–3–1.

This is an example of **factoring by grouping** since we "grouped" the terms two at a time.

Example 2   Factor $ax - ay + 2x - 2y$.

Multiply $(x - y)(a + 2)$ and see if you get the original expression.

Solution $ax - ay + 2x - 2y = a(x - y) + 2(x - y)$
$$= (x - y)(a + 2)$$

Example 3   Factor $2ax + 3a + 4x + 6$.

Solution $2ax + 3a + 4x + 6 = a(2x + 3) + 2(2x + 3)$
$$= (2x + 3)(a + 2)$$

Again, multiply as a check.

Sometimes the terms must first be rearranged before factoring by grouping can be accomplished.

Example 4   Factor $3ax + 2y + 3ay + 2x$.

Remember, the commutative property allows us to rearrange these terms.

We could also rearrange the terms as

$3ax + 2x + 3ay + 2y$
$= x(3a + 2) + y(3a + 2)$
$= (3a + 2)(x + y)$.

Solution   The first two terms have no common factor, but the first and third terms do, so we will rearrange the terms to place the third term after the first. Always look ahead to see the order in which the terms could be arranged.

$$3ax + 2y + 3ay + 2x = 3ax + 3ay + 2x + 2y$$
$$= 3a(x + y) + 2(x + y)$$
$$= (x + y)(3a + 2)$$

In all cases it is important to be sure that the factors within parentheses are exactly alike. This may require factoring a negative number or letter.

Example 5   Factor $ax - ay - 2x + 2y$.

$ax - ay - 2x + 2y$

factor (a) factor (-2)

Solution   Note that when we factor a from the first two terms, we get $a(x - y)$. Looking at the last two terms, we see that factoring $+2$ would give $2(-x + y)$ but factoring "-2" gives $-2(x - y)$. We want the terms within parentheses to be $(x - y)$, so we proceed in this manner.

$$ax - ay - 2x + 2y = a(x - y) - 2(x - y)$$
$$= (x - y)(a - 2)$$

▼   **EXERCISE 4-4-1**

Completely factor.

1. $ax + ay + 3x + 3y$

$a(x+y) + 3(x+y)$
$(a+3)(x+y)$

2. $ax + ay + 4x + 4y$

3. $ab + ac - 2b - 2c$

$a(b+c) - 2(b+c)$
$(a-2) \cdot (b+c)$

4. $xz + yz - 3x - 3y$

5. $2ax + ay + 6x + 3y$

$a(2x+y) + 3(2x+y)$
$(a+3)(2x+y)$

6. $2ax + 3a + 4x + 6$

7. $ax - ay + 2x - 2y$

$a(x-y) + 2(x-y)$
$(a+2)(x-y)$

8. $ax + 2ay - 3x - 6y$

9. $5x + 10y + ax + 2ay$

$5(x+2y) + a(x+2y)$
$(5+a)(x+2y)$

10. $2ax + 2ay - 5x - 5y$

11. $ac - 5a - bc + 5b$

$a(c-5) - b(c-5)$
$(a-b)(c-5)$

12. $ax - x - ay + y$

13. $3ax - 2y - 3ay + 2x$

$3ax - 3ay - 2y + 2x$

$3a(x-y) + 2(x-y)$
$(3a+2)(x-y)$

14. $ax + y + ay + x$

15. $ax - 3y + 3x - ay$

16. $ax - 4b + 4a - bx$

17. $3ax - y + 3x - ay$

18. $2ax - 5b + 10a - bx$

19. $2ax - y + 2ay - x$

20. $xy + 3y + 2x + 6$

21. $6ax - y + 2ay - 3x$

$2a(3x+y) + 3x - y$

22. $cx - 4y + cy - 4x$

23. $4ax + 3c + 6a + 2cx$

24. $6ax - y - 3x + 2ay$

25. $a^2 - 3b - ab + 3a$

26. $2a^2 - 5c - ac + 10a$

27. $3x - 2y - 6 + xy$

28. $xy - 8 - 2x + 4y$

29. $5y - 2x - 10 + xy$

$y(5+x) - 2(5+x)$
$(y-2)(5+x)$

30. $3y - x - 3 + xy$

◥ 4-5 FACTORING TRINOMIALS

A large number of future problems will involve factoring trinomials as products of two binomials. In section 4-1, you learned how to multiply polynomials. We now wish to look at the special case of multiplying two binomials and develop a pattern for this type of multiplication.

Example 1   Find the product of $(2x + 3)(3x - 4)$.

Solution
$$(2x + 3)(3x - 4) = 2x(3x - 4) + 3(3x - 4)$$
$$= 6x^2 - 8x + 9x - 12$$
$$= 6x^2 + x - 12$$

OBJECTIVES

Upon completing this section you should be able to:

1. Mentally multiply two binomials.
2. Factor a trinomial having a first term coefficient of 1.
3. Find the factors of any factorable trinomial.

Since this type of multiplication is so common, it is helpful to be able to find the answer without going through so many steps. Let us look at a pattern for this.

From the example $(2x + 3)(3x - 4) = 6x^2 + x - 12$, note that the first term of the answer $(6x^2)$ came from the product of the two first terms of the factors, that is $(2x)(3x)$.

$$6x^2$$
$$(2x + 3)(3x - 4)$$

Also note that the third term (-12) came from the product of the second terms of the factors, that is $(+3)(-4)$.

$$-12$$
$$(2x + 3)(3x - 4)$$

We now have the following part of the pattern:

1st term   3rd term
$$(2x + 3)(3x - 4)$$

Now looking at the example again, we see that the middle term $(+x)$ came from a sum of two products $(2x)(-4)$ and $(3)(3x)$.

$$(2x + 3)(3x - 4)$$
$$+9x$$
$$-8x$$

This method of multiplying two binomials is sometimes called the FOIL method.

FOIL stands for First, Outer, Inner, Last.

It is a shortcut method for multiplying two binomials and its usefulness will be seen when we factor trinomials.

For any two binomials we now have these four products:

1. First term by first term
2. Outside terms
3. Inside terms
4. Last term by last term

These products are shown by this pattern.

$$(\quad)(\quad)$$

When the products of the outside terms and inside terms give like terms, they can be combined and the solution is a trinomial.

$$15x^2 \qquad 7$$

Example 2 $(5x + 1)(3x + 7) = 15x^2 + 38x + 7$

$$3x$$
$$35x$$

You should memorize this pattern.

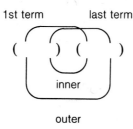

1st term   last term

inner

outer

Again, maybe memorizing the word FOIL will help.

Not only should this pattern be memorized, but the student should also learn to go from problem to answer without any written steps. This mental process of multiplying is necessary if proficiency in factoring is to be attained.

Example 3 $(4x + 5)(3x - 2) = 12x^2 + 7x - 10$

As you work the following exercises, attempt to arrive at a correct answer without writing anything except the answer. The more you practice this process, the better you will be at factoring.

▼ **EXERCISE 4–5–1**

Find the products.

1. $(a + 3)(a + 2)$
$a^2 + 5a + 6$

2. $(a + 1)(a + 5)$

3. $(x + 5)(x + 2)$
$x^2 + 7x + 10$

4. $(x - 1)(x - 4)$

5. $(a - 3)(a - 2)$
$a^2 - 5a + 6$

6. $(a - 4)(a - 2)$

7. $(a - 3)(a - 5)$
$a^2 - 8a + 15$

8. $(x + 2)(x - 3)$

9. $(x - 1)(x + 5)$
$x^2 + 4x - 5$

10. $(x + 7)(x - 2)$

11. $(a - 5)(a + 9)$
$a^2 + 4a + 45$

12. $(2x + 1)(x + 2)$

13. $(3x + 2)(2x + 1)$
$6x^2 + 3x + 4x + 2$
$6x^2 + 7x + 2$

14. $(a + 2)(3a - 5)$

15. $(2a - 1)(3a + 4)$
$6a^2 + 8a - 3a + 4$
$6a^2 + 5a - 4$

16. $(5x + 2)(x - 4)$

17. $(3x - 5)(2x - 3)$
$6x^2 - 9x - 10x + 15$
$6x^2 - 19x + 15$

18. $(2x + y)(3x + 5y)$

19. $(2a - b)(5a + 3)$
$10a^2 + 6a - 5ab - 3b$
$10a^2 + 6a - 5ab - 3b$

20. $(3x + 1)(x + 7y)$

21. $(a - 5)(4a + c)$
$4a^2 + ac - 20a + 5c$

22. $(x + 2)(x + 2)$

23. $(x + 5)(x + 5)$
$x^2 + 10x + 25$

24. $(x + 3)^2$

25. $(x - 7)^2$ **26.** $(x - 1)^2$ **27.** $(x + 3)(x - 3)$ **28.** $(x - 5)(x + 5)$

$(x-7)(x-7)$
$x^2 - 14x + 49$

$x^2 = 9$

29. $(2x + 7)(2x - 7)$ **30.** $(a + b)(a - b)$

$4x^2 - 14x + 14x - 49$

$4x^2 - 49$

Now that we have established the pattern of multiplying two binomials, we are ready to factor trinomials. We will first look at factoring only those trinomials with a first term coefficient of 1.

Example 4   Factor $x^2 + 11x + 24$.

Solution   Since this is a trinomial and has no common factor we will use the multiplication pattern to factor.

First write parentheses under the problem.

$x^2 + 11x + 24$

(   )(   )

We now wish to fill in the terms so that the pattern will give the original trinomial when we multiply. The first term is easy since we know that $(x)(x) = x^2$.

x^2

$(x\quad)(x\quad)$

We must now find numbers that multiply to give 24 and *at the same time* add to give the middle term. Notice that in each of the following we will have the correct first and last term.

$(x + 1)(x + 24) = x^2 + 25x + 24$
$(x + 2)(x + 12) = x^2 + 14x + 24$
$(x + 4)(x + 6) = x^2 + 10x + 24$
$(x + 3)(x + 8) = x^2 + 11x + 24$

Only the last product has a middle term of $11x$, and the correct solution is

$x^2 + 11x + 24 = (x + 3)(x + 8).$

This method of factoring is called *trial and error*—for obvious reasons.

We will actually be working in reverse the process developed in the last exercise set.

Remember, the product of the first two terms of the binomials gives the first term of the trinomial.

Some number facts from arithmetic might be helpful here.

1. *The product of two odd numbers is odd.*
2. *The product of two even numbers is even.*
3. *The product of an odd and an even number is even.*
4. *The sum of two odd numbers is even.*

Example 5 Factor $x^2 - 11x + 24$.

Solution Here the problem is only slightly different. We must find numbers that multiply to give (24) and *at the same time* add to give (-11). You should always keep the pattern in mind. The last term is obtained strictly by multiplying, but the middle term comes finally from a sum. Knowing that the product of two negative numbers is positive, but the sum of two negative numbers is negative, we obtain

$$x^2 - 11x + 24 = (x - 8)(x - 3).$$

Example 6 Factor $x^2 - 5x - 24$.

Solution We are here faced with a negative number for the third term, and this makes the task slightly more difficult. Since (-24) can only be the product of a positive number and a negative number, and since the middle term must come from the sum of these numbers, we must think in terms of a difference. We must find numbers whose product is 24 and that differ by 5. Furthermore, the larger number must be negative, because when we add a positive and negative number the answer will have the sign of the larger. Keeping all of this in mind, we obtain

$$x^2 - 5x - 24 = (x - 8)(x + 3).$$

The following points will help as you factor trinomials:

1. When the sign of the third term is positive, both signs in the factors must be alike—and they must be like the sign of the middle term.
2. When the sign of the last term is negative, the signs in the factors must be unlike—and the sign of the larger must be like the sign of the middle term.

5. The sum of two even numbers is even.
6. The sum of an odd and even number is odd.

Therefore, when we factor an expression such as $x^2 + 11x + 24$, we know that the product of the last two terms in the binomials must be 24, which is *even*, and their sum must be 11, which is *odd*.

Thus, only an odd and an even number will work. We need not even try combinations like 6 and 4 or 2 and 12, and so on.

The order of factors is insignificant.

$$(x - 8)(x + 3)$$
$$= (x + 3)(x - 8)$$

by the commutative law of multiplication.

▼ **EXERCISE 4–5–2**

Factor.

1. $x^2 + 4x + 3$ **2.** $x^2 + 7x + 10$ **3.** $x^2 + 5x + 6$ **4.** $x^2 + 6x + 8$

$(x+3)(x+1)$ $(x+3)(x+2)$

5. $x^2 + 10x + 9$ **6.** $x^2 + 9x + 20$ **7.** $x^2 + 8x + 16$ **8.** $x^2 + 10x + 16$

$(x+1)(x+9)$ $(x+4)(x+4)$

9. $x^2 + 14x + 24$ **10.** $x^2 + 10x + 24$ **11.** $x^2 - 4x + 3$ **12.** $x^2 - 6x + 5$

13. $x^2 - 3x + 2$ **14.** $x^2 - 8x + 7$ **15.** $x^2 - 10x + 21$ **16.** $x^2 - 8x + 12$

17. $x^2 - 8x + 15$ **18.** $x^2 - 12x + 35$ **19.** $x^2 - 8x + 16$ **20.** $x^2 - 6x + 9$

21. $x^2 - 2x - 3$ **22.** $x^2 - 3x - 10$ **23.** $x^2 - 2x - 8$ **24.** $x^2 + 5x - 6$

25. $x^2 + x - 6$ **26.** $x^2 + 2x - 35$ **27.** $x^2 - x - 20$ **28.** $x^2 - 7x - 18$

29. $x^2 + 4x - 21$ **30.** $x^2 - 9x - 22$ **31.** $x^2 + 41x + 40$ **32.** $x^2 + 25x + 24$

33. $x^2 + 29x - 30$ **34.** $x^2 + 9x - 22$ **35.** $x^2 + 2x - 48$ **36.** $x^2 + 4x - 21$

37. $x^2 + 7x - 18$ **38.** $x^2 + 11x + 18$ **39.** $x^2 - 8x + 12$ **40.** $x^2 + x - 12$

41. $x^2 + 34x - 35$ **42.** $x^2 + 13x + 36$ **43.** $x^2 + 9x - 36$ **44.** $x^2 - 12x + 35$

45. $x^2 - 11x + 24$ **46.** $x^2 + 22x + 21$ **47.** $x^2 + 13x - 48$ **48.** $x^2 - 13x + 40$

49. $x^2 + 6x - 40$ **50.** $x^2 - 13x + 30$ **51.** $x^2 + 15x + 50$ **52.** $x^2 + 55x + 54$

53. $x^2 + 49x - 50$ **54.** $x^2 + 10x - 56$ **55.** $x^2 + 22x + 72$ **56.** $x^2 + 21x - 72$

57. $x^2 - 5x - 24$ **58.** $x^2 + 18x - 40$ **59.** $x^2 + 2x - 15$ **60.** $x^2 + 13x + 22$

61. $x^2 + 14x + 48$ **62.** $x^2 + 20x - 96$ **63.** $x^2 + 31x + 84$ **64.** $x^2 - 28x + 96$

65. $x^2 + 83x - 84$ **66.** $x^2 + x - 30$ **67.** $x^2 - 3x - 40$ **68.** $x^2 + 16x + 15$

69. $x^2 + 21x - 22$ **70.** $x^2 - 20x + 96$ **71.** $x^2 + 8x - 84$ **72.** $x^2 + 131x + 130$

73. $x^2 - 12x - 28$ **74.** $x^2 - 39x + 140$ **75.** $x^2 + 31x + 130$ **76.** $x^2 - 8x + 15$

77. $x^2 + 3x - 130$ **78.** $x^2 + 16x + 48$ **79.** $x^2 - 50x + 96$ **80.** $x^2 + 23x - 140$

In the previous exercise the coefficient of each of the first terms was 1. When the coefficient of the first term is not 1, the problem of factoring is much more complicated because the number of possibilities is greatly increased.

Having done the previous exercise set, you are now ready to try some more challenging trinomials.

Example 7   Factor $6x^2 + 17x + 12$.

Solution   Factors of $6x^2$ are x, $2x$, $3x$, and $6x$. Factors of 12 are 1, 2, 3, 4, 6, and 12.

Notice that there are twelve ways to obtain the first and last terms, but only one has $17x$ as a middle term.

$(6x + 1)(x + 12)$	Middle term $73x$
$(6x + 2)(x + 6)$	Middle term $38x$
$(6x + 3)(x + 4)$	Middle term $27x$
$(6x + 12)(x + 1)$	Middle term $18x$
$(6x + 6)(x + 2)$	Middle term $18x$
$(6x + 4)(x + 3)$	Middle term $22x$

You could, of course, try each of these mentally instead of writing them out.

$(2x + 2)(3x + 6)$ Middle term $18x$

$(2x + 6)(3x + 2)$ Middle term $22x$

$(2x + 1)(3x + 12)$ Middle term $27x$

$(2x + 12)(3x + 1)$ Middle term $38x$

$(2x + 4)(3x + 3)$ Middle term $18x$

$(2x + 3)(3x + 4)$ Middle term $17x$

There is only one way to obtain all three terms:

$$(2x + 3)(3x + 4) = 6x^2 + 17x + 12.$$

In this example one out of twelve possibilities is correct. Thus *trial and error* can be very time-consuming.

Even though the method used is one of guessing, it should be "educated guessing" in which we apply all of our knowledge about numbers and exercise a great deal of mental arithmetic. In the preceding example we would immediately dismiss many of the combinations. Since we are searching for $17x$ as a middle term, we would not attempt those possibilities that multiply 6 by 6, or 3 by 12, or 6 by 12, and so on, as those products will be larger than 17. Also, since 17 is odd, we know it is the sum of an even number and an odd number. All of these things help reduce the number of possibilities to try.

First find numbers that give the correct first and last terms of the trinomial. Then add the outer and inner product to check for the proper middle term.

Example 8   Factor $6x^2 - 23x + 15$.

Solution   First we should analyze the problem.

1. The last term is positive, so two like signs.
2. The middle term is negative, so both signs will be negative.
3. The factors of $6x^2$ are x, $2x$, $3x$, $6x$. The factors of 15 are 1, 3, 5, 15.
4. Eliminate as too large the product of 15 with $2x$, $3x$, or $6x$. Try some reasonable combinations.

These would automatically give too large a middle term.

$(2x - 5)(3x - 3)$ Middle term $(-21x)$
                                         Incorrect

$(2x - 3)(3x - 5)$ Middle term $(-19x)$
                                         Incorrect

$(6x - 3)(x - 5)$ Middle term $(-33x)$
                                         Incorrect

$(6x - 5)(x - 3)$ Middle term $(-23x)$
                                         CORRECT

See how the number of possibilities is cut down.

so

$$6x^2 - 23x + 15 = (6x - 5)(x - 3).$$

Example 9   Factor $4x^2 - 5x - 6$.

Solution   Analyze:

1. The last term is negative, so unlike signs.
2. We must find products that differ by 5 with the larger number negative.
3. We eliminate a product of $4x$ and 6 as probably too large.
4. Try some combinations.

$(2x - 3)(2x + 2)$	Incorrect
$(4x - 2)(x + 3)$	Incorrect
$(4x - 3)(x + 2)$	Here the middle term is $+5x$, which is the right number but the wrong sign. Be careful not to accept this as the solution, but switch signs so the larger product agrees in sign with the middle term.

Remember, mentally try the various possible combinations that are reasonable. This is the process of "trial and error" factoring. You will become more skilled at this process through practice.

$(4x + 3)(x - 2)$ gives the correct sign for the middle term.

$$4x^2 - 5x - 6 = (4x + 3)(x - 2)$$

By the time you finish the following exercise set you should feel much more comfortable about factoring a trinomial.

▼ **EXERCISE 4–5–3**

Factor.

1. $2x^2 + 5x + 2$
   $(x + 2)(2x + 1)$

2. $2x^2 + 11x + 5$
   $(2x + 1)(x + 5)$

3. $2x^2 + 5x + 3$
   $(x + 1)(2x + 3)$

4. $3x^2 + 8x + 4$

5. $2x^2 - 7x + 3$

6. $2x^2 + 7x + 3$

7. $6x^2 + 7x + 2$

8. $10x^2 + 19x + 6$

9. $2x^2 + 3x - 20$

10. $6x^2 - 11x - 10$

11. $3x^2 - x - 10$

12. $15x^2 + 23x + 4$

13. $8x^2 - 2x - 3$

14. $8x^2 + 22x + 5$

15. $6x^2 + 31x + 5$

16. $3x^2 - 13x - 10$

17. $5x^2 + 8x + 3$

18. $12x^2 - 11x + 2$

19. $16x^2 - 6x - 1$

20. $9x^2 + 9x + 2$

21. $4x^2 - x - 18$ **22.** $3x^2 - 4x - 7$ **23.** $4x^2 + 4x + 1$ **24.** $3x^2 + x - 14$

25. $6x^2 - 49x + 30$ **26.** $10x^2 + 21x + 9$ **27.** $9x^2 - 12x + 4$ **28.** $10x^2 + 21x - 10$

29. $6x^2 - x - 1$ **30.** $8x^2 - 14x - 15$ **31.** $6x^2 + 13x + 6$ **32.** $4x^2 - 11x + 6$

33. $15x^2 - 19x - 10$ **34.** $6x^2 - 5x - 6$ **35.** $12x^2 - 16x + 5$ **36.** $16x^2 + 24x + 9$

37. $6x^2 + x - 15$ **38.** $8x^2 - 2x - 15$ **39.** $15x^2 + 2x - 24$ **40.** $32x^2 + 4x - 3$

◤ 4–6 OPTIONAL SHORTCUTS TO TRIAL AND ERROR FACTORING

OBJECTIVES

Upon completing this section you should be able to:

1. Find the key number of a trinomial.
2. Use the key number to factor a trinomial.

The product of these two numbers is the "key number."

In this section we wish to discuss some shortcuts to trial and error factoring. These are optional for two reasons. First, some might prefer to skip these techniques and simply use the trial and error method; second, these shortcuts are not always practical for large numbers. However, they will increase speed and accuracy for those who master them.

The first step in these shortcuts is finding the **key number.** After you have found the key number it can be used in more than one way.

> In a trinomial to be factored the **key number** is the product of the coefficients of the first and third terms.

Example 1 In $6x^2 - 23x + 12$ the key number is $(6)(12) = 72$.

Example 2 In $5x^2 + 13x - 6$ the key number is $(5)(-6) = -30$.

▼ **EXERCISE 4-6-1**

Find the key number in each of the trinomials.

1. $2x^2 + 7x + 3$ **2.** $6x^2 + 7x + 2$ **3.** $3x^2 - x - 10$ **4.** $8x^2 - 2x - 3$

5. $2x^2 + 3x - 20$ **6.** $3x^2 - 13x - 10$ **7.** $10x^2 + 19x + 6$ **8.** $15x^2 + 23x + 4$

9. $16x^2 - 6x - 1$ **10.** $4x^2 - x - 18$ **11.** $8x^2 + 22x + 5$ **12.** $6x^2 + 31x + 5$

13. $3x^2 - 4x - 7$ **14.** $3x^2 + x - 14$ **15.** $6x^2 - 49x + 30$ **16.** $10x^2 + 21x - 10$

17. $4x^2 - 11x + 6$ **18.** $9x^2 - 12x + 4$ **19.** $15x^2 - 19x - 10$ **20.** $8x^2 - 14x - 15$

The first use of the key number is shown in example 3.

Example 3 Use the key number to factor $4x^2 + 3x - 10$.

Solution **Step 1** Find the key number. In this example $(4)(-10)$
$= -40$.

Step 2 Find factors of the key number (-40) that will add to
give the coefficient of the middle term $(+3)$. In this case
$(+8)(-5) = -40$ and $(+8) + (-5) = +3$.

Step 3 The factors $(+8)$ and (-5) will be the cross products in
the multiplication pattern.

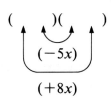

The possible pairs of factors of
(-40) are

$(1)(-40)$
$(-1)(40)$
$(2)(-20)$
$(-2)(20)$
$(4)(-10)$
$(-4)(10)$
$(8)(-5)$
$(-8)(5)$. This is the
only pair
whose sum
is 3.

Step 4 Using only the outside cross product, find factors of the
first and third terms that will multiply to give the
product. In this example we must find factors of $4x^2$ and
-10 that will multiply to give $+8x$. These are $4x$ from
$4x^2$ and $(+2)$ from (-10).

Place these factors in the first and last positions in the pattern

$$(4x \quad)(\quad + 2)$$

$$8x$$

There is only one way it can be done correctly.

Step 5 Forget the key number at this point and look back at the original problem. Since the first and last positions are correctly filled, it is now only necessary to fill the other two positions.

$$4x^2 + 3x - 10$$

$$4x^2$$

$$(4x \quad)(\quad + 2)$$

We know the product of the two first terms must give $4x^2$ and $4x$ is already in place. There is no choice other than x.

$$4x^2$$

$$(4x \quad)(x + 2)$$

Again, this can be done in only one way.

We know that the product of the two second terms must be (-10) and $(+2)$ is already in place. We have no choice other than (-5).

$$-10$$

$$(4x - 5)(x + 2)$$

We now have the factors of $4x^2 + 3x - 10$ as $(4x - 5)(x + 2)$.

Note that in step 4 we could have started with the inside product instead of the outside product. We would have obtained the same factors. The most important thing is to have a systematic process for factoring.

Example 4   Factor $3x^2 - 10x - 8$.

Solution   Key number $= (3)(-8) = -24$

Factors of -24 that add to give (-10) are $(-12)(+2)$.

Using $(-12x)$ as the product of the outside terms, we find $(3x)(-4) = -12x$.

$$(3x \qquad)(\qquad -4)$$

Remember, if a trinomial is factorable, there is only one possible set of factors.

Looking back to the original problem $3x^2 - 10x - 8$, we find the other terms to be x and $+2$.

Therefore, $3x^2 - 10x - 8 = (3x + 2)(x - 4)$.

If no factors of the key number can be found whose sum is the coefficient of the middle terms, then the trinomial is *prime* and does not factor.

▼ **EXERCISE 4-6-2**

Use the method discussed above to factor the following.

1. $2x^2 + 7x + 6$ **2.** $2x^2 + 7x + 3$ **3.** $2x^2 + 13x + 20$ **4.** $2x^2 + 13x + 15$

5. $3x^2 + 14x + 8$ **6.** $3x^2 + 11x + 6$ **7.** $3x^2 + 4x - 4$ **8.** $2x^2 + 11x - 6$

9. $2x^2 - 9x - 5$ **10.** $3x^2 - 4x - 15$ **11.** $5x^2 + 19x - 4$ **12.** $2x^2 - 11x + 15$

13. $3x^2 - 17x + 10$ **14.** $4x^2 + 13x + 3$ **15.** $4x^2 - 11x + 6$ **16.** $4x^2 - 12x + 5$

17. $6x^2 + 7x + 2$ **18.** $6x^2 + 11x + 3$ **19.** $6x^2 - x - 2$ **20.** $5x^2 + 24x + 16$

21. $4x^2 - 33x - 27$ **22.** $6x^2 + 23x - 4$ **23.** $8x^2 + 18x - 5$ **24.** $6x^2 - 5x - 4$

25. $6x^2 - 41x + 30$ **26.** $8x^2 - 35x + 12$ **27.** $8x^2 + 10x - 7$ **28.** $10x^2 - 77x - 24$

29. $18x^2 - 3x - 10$ **30.** $50x^2 + 35x - 9$

A second use for the key number as a shortcut involves factoring by grouping.
It works as in example 5.

Example 5   Factor $4x^2 + 3x - 10$.

Solution   **Step 1** Find the key number $(4)(-10) = -40$.

Step 2 Find factors of (-40) that will add to give the coefficient of the middle term $(+3)$.

Steps 1 and 2 in this method are the same as in the previous method.

$4x^2 + \underset{\displaystyle 3x}{\overbrace{}} - 10$

$4x^2 \underbrace{(+ 8x)}\underbrace{(- 5x)} - 10$

This now becomes a regular
factoring by grouping problem.

Step 3 Rewrite the original problem by breaking the middle
term into the two parts found in step 2. $8x - 5x = 3x$,
so we may write

$$4x^2 + 3x - 10 = 4x^2 + 8x - 5x - 10.$$

Step 4 Factor this problem from step 3 by the grouping method
studied in section 4–4.

$$4x^2 + 8x - 5x - 10 = 4x(x + 2) - 5(x + 2)$$
$$= (x + 2)(4x - 5)$$

Again, there is only one possible
pair of factors that can be
obtained from a given trinomial.

Hence

$$4x^2 + 3x - 10 = (x + 2)(4x - 5).$$

Example 6   Factor $3x^2 - 10x - 8$.

Solution   Key number = (-24).

Factors of key number that add to give -10 are -12 and 2.
Hence

$$3x^2 - 10x - 8 = 3x^2 - 12x + 2x - 8$$
$$= 3x(x - 4) + 2(x - 4)$$
$$= (x - 4)(3x + 2).$$

Remember, if step 2 is impossible,
the trinomial is prime and cannot
be factored.

▼ **EXERCISE 4–6–3**
_____

Use the method discussed to factor the following.

1. $x^2 + 9x + 20$ **2.** $x^2 + 6x + 8$ **3.** $x^2 - x - 6$ **4.** $x^2 + x - 12$

5. $x^2 - 8x + 12$ **6.** $x^2 - 7x + 12$ **7.** $2x^2 + 7x + 3$ **8.** $2x^2 + 11x + 15$

9. $3x^2 + 5x + 2$ **10.** $2x^2 + 11x + 12$ **11.** $2x^2 - 5x - 12$ **12.** $3x^2 - 8x - 3$

13. $5x^2 + 9x - 2$ **14.** $4x^2 + x - 3$ **15.** $2x^2 - 11x + 15$ **16.** $3x^2 - 7x + 2$

17. $3x^2 - 14x + 8$ **18.** $2x^2 - 15x + 27$ **19.** $3x^2 - 28x + 32$ **20.** $2x^2 - 15x + 18$

21. $4x^2 + 9x - 9$ **22.** $6x^2 + 23x - 4$ **23.** $x^2 + 8x + 16$ **24.** $x^2 - 12x + 36$

25. $4x^2 + 8x + 3$ **26.** $4x^2 - 8x + 3$ **27.** $6x^2 - x - 12$ **28.** $8x^2 - 10x - 3$

29. $16x^2 - 8x - 3$ **30.** $20x^2 - 56x + 15$

◣ 4-7 SPECIAL CASES IN FACTORING

In this section we wish to examine some special cases of factoring that occur often in problems. If these special cases are recognized, the factoring is then greatly simplified.

The first special case we will discuss is the **difference of two perfect squares.**

Recall that in multiplying two binomials by the pattern, the middle term comes from the sum of two products.

middle term

From our experience with numbers we know that the sum of two numbers is zero only if the two numbers are negatives of each other.

Example 1 $(3x - 4)(3x + 4) = 9x^2 - 16$

Example 2 $(2x + 1)(2x - 1) = 4x^2 - 1$

Example 3 $(5x - 6)(5x + 6) = 25x^2 - 36$

In each example the middle term is zero. Note that if two binomials multiply to give a binomial (middle term missing), they must be in the form of $(a - b)$ $(a + b)$.

$(a - b)(a + b) = a^2 - b^2$

Reading this rule from right to left tells us that if we have a problem to factor and if it is in the form of $a^2 - b^2$, the factors will be $(a - b)(a + b)$.

OBJECTIVES

Upon completing this section you should be able to:

1. Identify and factor the differences of two perfect squares.
2. Identify and factor a perfect square trinomial.

When the sum of two numbers is zero, one of the numbers is said to be the *additive inverse* of the other.

For example:

$$(+3) + (-3) = 0$$

so $+3$ is the additive inverse of -3, also -3 is the additive inverse of $+3$.

The rule may be written as
$$a^2 - b^2 = (a - b)(a + b).$$

This is the form you will find most helpful in factoring.

Example 4   Factor $25x^2 - 16$.

Solution   Here both terms are perfect squares and they are separated by a negative sign.

Where $a = 5x$ and $b = 4$

The form of this problem is $(5x)^2 - (4)^2$ (the form of $a^2 - b^2$). So

$$25x^2 - 16 = (5x + 4)(5x - 4).$$

The sum of two squares is not factorable.

$a^2 + b^2 \neq (a + b)^2$. Why?

Special cases do make factoring easier, but be certain to recognize that a special case is just that—very special. In this case *both* terms must be perfect squares *and* the sign must be negative, hence "the difference of two perfect squares."

Example 5 $x^2 - 7$ Not the special case. Why?

Example 6 $4x^2 + 9$ Not the special case. Why?

You must also be careful to recognize perfect squares. Remember that perfect square numbers are numbers that have square roots that are integers. Also, perfect square exponents are even.

Example 7 $4x^6 - 9 = (2x^3 - 3)(2x^3 + 3)$

Example 8 $x^4 - 25 = (x^2 - 5)(x^2 + 5)$

Example 9 $(a + b)^2 - c^2 = [(a + b) - c][(a + b) + c]$

Example 10 $(x + y)^2 - (p + k)^2$
$= [(x + y) - (p + k)][(x + y) + (p + k)]$

> **WARNING**
>
> Students often overlook the fact that 1 *is* a perfect square. Thus an expression such as $x^2 - 1$ is the difference of two perfect squares and can be factored by this method.

▼ **EXERCISE 4-7-1**

Factor.

1. $x^2 - 4$

$(x+2)(x-2)$

2. $a^2 - 25$

3. $x^2 - 1$

$(x+1)(x-1)$

4. $b^2 - 49$

5. $a^2 - 81$

$(a+9)(a-9)$

6. $x^2 - 9$

7. $x^2 - 100$

$(x+10)(x-10)$

8. $a^2 - 36$

9. $y^2 - 121$

$(y-11)(y+11)$

10. $x^2 - 169$

11. $4x^2 - 9$

$(2x+3)(2x-3)$

12. $9x^2 - 25$

13. $16x^2 - 1$

$(4x - 1)(4x + 1)$

14. $9a^2 - 49$

15. $4x^2 - y^2$

$(2x - y)(2x + y)$

16. $a^2 - 16b^2$

17. $4x^2 - 25y^2$

18. $25x^2 - 144$

19. $16a^2 - 121b^2$

20. $169a^2 - 225b^2$

21. $(x + y)^2 - 64$

22. $(a + b)^2 - 100$

23. $x^4 - 9$

$(x^2 - 3)(x^2 + 3)$

24. $a^4 - b^2$

25. $x^6 - 36$

26. $4x^8 - 9$

27. $25a^4 - 49$

28. $16x^2 - 25y^4$

29. $(x + y)^4 - z^2$

30. $(a - b)^6 - c^4$

Another special case in factoring is the perfect square trinomial. Observe that squaring a binomial gives rise to this case.

$$(a + b)^2 = (a + b)(a + b) = a^2 + 2ab + b^2$$

For factoring purposes it is more helpful to write the statement as

$$a^2 + 2ab + b^2 = (a + b)^2.$$

We recognize this case by noting the special features. Three things are evident.

1. The first term is a perfect square.
2. The third term is a perfect square.
3. The middle term is twice the product of the square root of the first and third terms.

perfect squares

a^2 + $2ab$ + b^2

twice the product of the square root of the first and third terms

Example 11 Factor $25x^2 + 20x + 4$.

Solution **1.** $25x^2$ is a perfect square—principal square root $= 5x$.
2. 4 is a perfect square—principal square root $= 2$.
3. $20x$ is twice the product of the square roots of $25x^2$ and
4. $20x = 2(5x)(2)$.

> *To factor a perfect square trinomial* form a binomial with the square root of the first term, the square root of the last term and the sign of the middle term, and indicate the square of this binomial.

$25x^2$ + $20x$ + 4

square root   square root

$(5x$ + $2)^2$

Always square the binomial as a check to make sure the middle term is correct.

Thus, $25x^2 + 20x + 4 = (5x + 2)^2$.

Example 12 $25x^2 - 20x + 4 = (5x - 2)^2$

Example 13 $9x^2 - 6x + 1 = (3x - 1)^2$

Example 14 $x^2 + 8x + 16 = (x + 4)^2$

$15x \neq 2(2x)(3)$.

Example 15 $4x^2 + 15x + 9$ Not the special case of a perfect square trinomial.

▼ **EXERCISE 4-7-2**

Square the following binomials.

1. $(x + y)^2$

$(x+y)(x+y)$
$x^2 + 2xy + y^2$

2. $(x + 4)^2$

3. $(x + 6)^2$

$(x+6)(x+6)$
$x^2 + 12x + 36$

4. $(a + 1)^2$

5. $(a - 2)^2$

$(a-2)(a-2)$
$a^2 - 4a + 4$

6. $(x - 5)^2$

7. $(2x + 7)^2$

$(2x+7)(2x+7)$
$4x^2 + 14x + 14x + 49$
$4x^2 + 28x + 49$

8. $(2x - 1)^2$

9. $(3a + 1)^2$

$(3a+1)(3a+1)$
$9a^2 + 3a + 3a + 1$
$9a^2 + 6a + 1$

10. $(4x - 3)^2$

Supply the missing middle terms so the following expressions are perfect square trinomials.

11. $x^2 + \underline{\ 4x\ } + 4$ $(x+2)^2$

12. $x^2 + \underline{\hspace{2cm}} + 9$

13. $4x^2 + \underline{\ 20x\ } + 25$ $(2x+5)^2$

14. $9x^2 - \underline{\hspace{2cm}} + 4$

15. $x^2 + \underline{\ 2x\ } + 1$ $(x+1)^2$

16. $25a^2 + \underline{\hspace{2cm}} + 1$

17. $4x^2 - \underline{\ 4xy\ } + y^2$ $(2x-y)^2$

18. $16x^2 + \underline{\hspace{2cm}} + 4y^2$

19. $x^2 - \underline{\ 14x\ } + 49y^2$

20. $36x^4 + \underline{\hspace{2cm}} + 25y^2$

Factor.

21. $x^2 + 6x + 9$

$(x+3)^2$

22. $x^2 + 10x + 25$

23. $x^2 - 14x + 49$

$(x-7)^2$

24. $a^2 - 2a + 1$

25. $4x^2 + 12x + 9$

26. $4a^2 - 36ab + 81b^2$

27. $9x^2 + 30x + 25$

28. $49a^2 + 14a + 1$

29. $36x^2 - 60xy + 25y^2$

30. $25x^2 + 90x + 81$

Another special case of factoring occurs when we have a sum or difference of two cubes.

In exercise 4–1–2 number 41 you were asked to find the product $(x + 3)(x^2 - 3x + 9)$. The answer is $x^3 + 27$, which is the sum of two cubes. Thus, we can write $x^3 + 27 = (x + 3)(x^2 - 3x + 9)$. This leads us to the following rule for the **sum or difference of two cubes.**

$x^3 + 27 = (x)^3 + (3)^3$

sum of two cubes

$$a^3 + b^3 = (a + b)(a^2 - ab + b^2)$$
$$a^3 - b^3 = (a - b)(a^2 + ab + b^2)$$

Multiply the right side of each of these to see if you get the left side.

Example 16 Factor $x^3 + 125$.

> *Solution* $x^3 + 125 = (x)^3 + (5)^3$
>
> $= (x + 5)(x^2 - 5x + 25)$

Use the rule for $a^3 + b^3$.

Example 17 Factor $8x^3 - 1$.

> *Solution* $8x^3 - 1 = (2x)^3 - (1)^3$
>
> $= (2x - 1)(4x^2 + 2x + 1)$

Use the rule for $a^3 - b^3$.

▼ EXERCISE 4-7-3

Factor.

1. $x^3 + y^3$

2. $x^3 + 8$

3. $x^3 - y^3$

4. $x^3 - 8$

5. $x^3 - 1$

6. $a^3 - 1$

7. $27a^3 + 1$

8. $8x^3 + 1$

9. $8x^3 - y^3$

10. $27a^3 - b^3$

11. $27a^3 + 8b^3$

12. $8x^3 + 125y^3$

13. $125a^3 - 27b^3$

14. $64x^3 - 27y^3$

15. $x^3 - y^6$

16. $8a^3 - 125$

17. $8a^6 + b^3$

18. $64x^6 - 1$

19. $a^9 - 1$

20. $x^9 + 1$

◥◣ 4-8 COMPLETE FACTORIZATION

OBJECTIVES
▼

Upon completing this section you should be able to factor a trinomial using the following two steps:

1. First look for common factors.
2. Factor the remaining trinomial by applying the methods of this chapter.

We have now studied many of the methods of factoring found in algebra. However, you must be aware that a single problem can require more than one of these methods. Remember that there are two checks for correct factoring.

1. Will the factors multiply to give the original problem?
2. Are all factors prime?

Example 1   Factor $2x^2 + 10x + 12$.

Solution   We notice that (2) is a common factor.

$$2x^2 + 10x + 12 = 2(x^2 + 5x + 6)$$

Now if we leave the problem at this point, the first check works but how about the second?

We see that the factor $(x^2 + 5x + 6)$ can now be factored into $(x + 3)(x + 2)$. The complete factorization then is $2x^2 + 10x + 12 = 2(x + 3)(x + 2)$.

Suppose in this example we had looked at the problem $2x^2 + 10x + 12$ and simply used the trial and error method. We could get $2x^2 + 10x + 12 = (2x + 4)(x + 3)$.

Once again the first check works because the product gives the original. However, the factor $(2x + 4)$ has a common factor of (2). $2x + 4 = 2(x + 2)$.

Thus $2x^2 + 10x + 12 = 2(x + 2)(x + 3)$. Nothing short of this complete factorization is correct.

> **WARNING**
>
> Once a common factor has been found, you must check to see if the resulting trinomial is factorable.

If a trinomial has any common factors, it is usually easier if they are factored first.

Example 2   Factor $12x^2 - 27$.

Solution   We note a common factor of (3).

$$12x^2 - 27 = 3(4x^2 - 9)$$

At this point we recognize $4x^2 - 9$ as the difference of two perfect squares.

$$3(4x^2 - 9) = 3(2x - 3)(2x + 3)$$

So $12x^2 - 27 = 3(2x - 3)(2x + 3)$.

A good procedure to follow in factoring is to always remove the greatest common factor first and then factor what remains, if possible.

Example 3   Factor $16x^3y - 2y$.

Solution   We note a common factor of $2y$.

$$16x^3y - 2y = 2y(8x^3 - 1)$$

Here we have a common literal factor as well as a numerical factor.

We should recognize that $8x^3 - 1$ is the difference of two cubes. Thus

$$2y(8x^3 - 1) = 2y(2x - 1)(4x^2 + 2x + 1).$$

Example 4   Factor $x^4y^5 + 8xy^2$.

Solution $$\begin{aligned} x^4y^5 + xy^2 &= xy^2(x^3y^3 + 8) \\ &= xy^2(xy + 2)(x^2y^2 - 2xy + 4) \end{aligned}$$

xy^2 is the common factor.

▼ **EXERCISE 4-8-1**

Factor completely. If the expression is prime, so state.

1. $4x^2 + 14x + 6$

$2(2x^2 + 7x + 3)$

$2(2x + 1)(x + 3)$

2. $50x^2 + 95x + 30$

3. $6x^2 + 5x + 1$

4. $4x^2 + 4x - 3$

5. $3x^2 + 6x - 144$

6. $2x^2 + 14x - 36$

7. $3x^2 - 8x - 3$

8. $3x^2 + 13x - 10$

9. $3x^2 - 12$

10. $2x^2 - 32$

11. $4x^2 + 24x + 36$

12. $3x^2 + 12x + 12$

13. $x^2 + x + 1$

14. $16x^2 - 4x - 6$

15. $3x^3 + 10x^2 - 8x$

16. $2x^2 + 7x + 6$

17. $42x^2 - 7x - 7$

18. $x^2 - x - 30$

19. $5x^2 + 5x - 60$

20. $36x^2 - 33x + 6$

21. $x^3 - 8x^2 + 12x$

22. $2x^2 + 5x + 2$

23. $20x^2 - 5$

24. $18x^2 - 8$

25. $3x^2 + 2x + 5$

26. $x^2 - 5x - 14$

27. $14x^2 + 11x + 2$

28. $2x^3 - 2x^2 - 4x$

29. $8x^2 - 8x + 2$

30. $36x^2 - 48x + 16$

31. $3x^3 - 39x^2 + 120x$

32. $2x^2 + 7x - 4$

33. $x^2 + 5x - 14$

34. $15x^2 + 5x - 70$

35. $6x^2 - 3x + 2$

36. $3x^2y + 30xy - 168y$

37. $4x^2 - 4x - 3$

38. $10x^2 - x - 2$

39. $x^2 + 2x - 35$

40. $4x^3 + 8x^2 - 60x$

41. $27x^3 - 75x$

42. $3x^2 - 11x - 4$

43. $2x^2 + 4x + 6$

44. $2x^2 - 9x - 5$

45. $x^2 + x - 30$

46. $16x^3 + 48x^2 + 36x$

47. $3x^2 + 93x + 252$

48. $6x^2 + 15x - 24$

49. $7x^2 + 3x - 2$

50. $2x^4 - 7x^2 - 4$

51. $2a^4 - 128a$

52. $x^5 + 125x^2$

53. $81x^3y + 3y$

54. $16a^3b^2 + 54b^2$

55. $a^5b^4 - a^2b$

56. $x^4y^5 - 8xy^2$

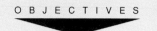 **4–9 OTHER TYPES OF FACTORING BY GROUPING**

O B J E C T I V E S

Upon completing this
section you should be able
to factor certain
expressions by locating
perfect square trinomials.

Many algebraic expressions do not seem to fall into any of the categories we
have discussed and at first seem not to be factorable. Sometimes, however, by
rearranging the terms and/or regrouping, they can be factored.

In section 4–4 we saw that by grouping terms in pairs we sometimes could factor
a common binomial and thereby factor an expression having four terms. There
are other ways of using this grouping idea to factor expressions.

This expression has four terms but
cannot be factored by the method
discussed in section 4–4.

Example 1   Factor $x^2 - y^2 + 6x + 9$.

Solution   As we look at this problem our first impression might be to
factor $x^2 - y^2$ and remove 3 from $6x + 9$, but this will not
produce any common factor. We might assume the expression
not to be factorable unless we look at a three and one grouping
instead of a two and two grouping. We should note that $x^2 + 6x$
$+ 9$ is a perfect square trinomial. So we rearrange the terms,
obtaining

This is the difference of two
squares.

$$x^2 + 6x + 9 - y^2 = (x + 3)^2 - y^2$$
$$= [(x + 3) - y][(x + 3) + y]$$
$$= (x + 3 - y)(x + 3 + y).$$

Example 2   Factor $25 - 4a^2 + 4ab - b^2$.

Note a perfect square trinomial

Solution $25 - 4a^2 + 4ab - b^2 = 25 - (4a^2 - 4ab + b^2)$
$$= 25 - (2a - b)^2$$
$$= [5 - (2a - b)][5 + (2a - b)]$$
$$= (5 - 2a + b)(5 + 2a - b)$$

This is simplified.

Note that in the previous two examples we had a perfect square trinomial and
then the difference of two squares.

There can be two perfect square trinomials.

Example 3   Factor $x^2 - y^2 - 2x - 6y - 8$.

Solution   If we note that $x^2 - 2x + 1$ would be a perfect square
trinomial, and that $y^2 + 6y + 9$ would be a perfect square
trinomial, we can write -8 as $-9 + 1$ and obtain

$-8 = -9 + 1$

$$x^2 - y^2 - 2x - 6y - 8 = x^2 - 2x + 1 - y^2 - 6y - 9$$
$$= (x^2 - 2x + 1) - (y^2 + 6y + 9)$$
$$= (x - 1)^2 - (y + 3)^2$$
$$= [(x - 1) - (y + 3)][(x - 1) + (y + 3)]$$
$$= (x - y - 4)(x + y + 2).$$

Again, the difference of two
squares.

Of course these are all very special situations, but we cannot look at an expression and simply assume it will not factor just because it does not neatly fall into
some simple type of factoring.

Example 4   Factor $x^4 + 4$.

Solution   Our first impression might be that this is not factorable since it is not the difference of two squares, but rather the sum of two squares. However, if we supply a middle term of $4x^2$ to make a perfect square trinomial, we obtain

$$x^4 + 4 = x^4 + 4x^2 + 4 - 4x^2$$
$$= (x + 2)^2 - 4x^2$$
$$= (x^2 + 2 - 2x)(x^2 + 2 + 2x).$$

Such problems require much "looking ahead" and perhaps some "dead ends," but don't say "not factorable" until all avenues have been exhausted.

At first glance this expression does not look like it factors.

Of course, if we add $4x^2$, we must also subtract $4x^2$.

▼ **EXERCISE 4–9–1**

Factor.

1. $x^2 - y^2 + 4x + 4$

2. $a^2 - b^2 + 10a + 25$

3. $x^2 + y^2 - 25 + 2xy$

4. $a^2 + b^2 - 16 + 2ab$

5. $4a^2 + b^2 - 4ab - 9$

6. $9x^2 + y^2 - 6xy - 4$

7. $x^2 - y^2 - 4y - 4$

8. $a^2 - b^2 - 8b - 16$

9. $36 - a^2 - 4ab - 4b^2$

10. $1 - x^2 - 6xy - 9y^2$

11. $25 + 6xy - 9x^2 - y^2$

12. $16 + 4xy - 4x^2 - y^2$

13. $x^2 - y^2 + 2x - 4y - 3$

14. $a^2 - b^2 + 6a - 4b + 5$

15. $a^2 - b^2 - 10a + 6b + 16$

16. $x^2 - y^2 - 4x + 8y - 12$

17. $4x^4 + 1$

18. $64a^4 + 1$

19. $x^4 + x^2 + 1$

20. $x^4 + 2x^2 + 9$

Key Words

Section 4–1

- The **distributive law** is used when multiplying polynomials.

Section 4–2

- The **long division algorithm** is used when dividing polynomials.

Section 4–3

- An expression is in **factored form** only if the entire expression is an indicated product.

- **Factoring** is a process that changes a sum or difference of terms to a product of factors.

- A **prime** expression cannot be factored.

- The **greatest common factor** is the greatest factor common to all terms.

- An expression is **completely factored** when no further factoring is possible.

Section 4–4

- The possibility of **factoring by grouping** exists when an expression contains four or more terms.

Section 4–5

- The **FOIL** method can be used to multiply two binomials.

Section 4–6

- The **key number** is the product of the coefficients of first and third terms of a trinomial.

Section 4–7

- Special cases in factoring include the **difference of two squares, perfect square trinomials,** and the **sum** or **difference of two cubes.**

Procedures

Section 4–1

- To multiply a polynomial by another polynomial multiply each term of one polynomial by each term of the other and combine like terms.

Section 4–2

- To divide a monomial by a monomial divide the numerical coefficients and use the third law of exponents for the literal numbers.

- To divide a polynomial by a monomial divide each term of the polynomial by the monomial.

- To divide a polynomial by another polynomial use the long division method.

Section 4–3

- To remove common factors find the greatest common factor and divide each term by it.

Section 4–5

- Trinomials can be factored by using the trial and error method. This uses the pattern for multiplication to find factors that will give the original trinomial.

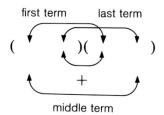

first term   last term

middle term

Section 4–6

- Use the key number as an aid in determining factors whose sum is the coefficient of the middle term of a trinomial.

Section 4–7

- To factor the difference of two squares use the rule

$$a^2 - b^2 = (a - b)(a + b).$$

- To factor a perfect square trinomial form a binomial with the square root of the first term, the square root of the last term and the sign of the middle term, and indicate the square of this binomial.

- To factor the sum of two cubes, use the rule

$$a^3 + b^3 = (a + b)(a^2 - ab + b^2).$$

- To factor the difference of two cubes use the rule

$$a^3 - b^3 = (a - b)(a^2 + ab + b^2).$$

Perform the indicated operation and simplify.

A N S W E R S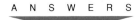

1. $(x + 9)^2$　　　　　　　　**2.** $(x + 8)(x + 2)$

1. _____

3. $(2x - 5)(3x - 1)$　　　　**4.** $(3x - 2)^2$

2. _____

3. _____

5. $(x + 2)(x^2 - 5x + 4)$　　**6.** $(2x + 1)(3x^2 - 5x + 2)$

4. _____

5. _____

7. $(a + 2b + 1)(a - b + 2)$　　**8.** $(x - y + z)(2x - 3y - z)$

6. _____

7. _____

9. $\dfrac{12x^2}{6x^6}$　　　　　　**10.** $\dfrac{24a^3}{8a^9}$

8. _____

9. _____

10. _____

11. $\dfrac{10x^4y^3 - 15x^2y^5 + 25x^3y^4}{5x^2y^3}$　　**12.** $\dfrac{12a^2b^4 - 8a^4b^3 + 16a^3b^5}{-4a^2b}$

11. _____

12. _____

13. $(x^2 + 5x - 24) \div (x - 3)$　　**14.** $(x^2 + 11x + 28) \div (x + 4)$

13. _____

14. _____

15. $(x^2 - 13x + 30) \div (x - 5)$　　**16.** $(x^2 - 20x + 100) \div (x - 9)$

15. _____

16. _____

17. $(2x^3 + x^2 - x + 5)$
$\div (2x + 3)$　　　　　　**18.** $(3x^3 + 7x^2 - x + 5)$
$\div (3x + 1)$

17. _____

18. _____

19. $(x^3 + 1) \div (x + 3)$　　　**20.** $(x^4 - 3x^2 + 3) \div (x - 2)$

19. _____

20. _____

A N S W E R S

Factor completely or identify if prime.

21. $9x + 6y$

22. $21a + 35b$

21. _____

22. _____

23. $a^2 - a$

24. $4x^2 - 28x$

23. _____

24. _____

25. $6x^3y + 10x^2y^2$

26. $8a^3b^2 - 12ab^4$

25. _____

26. _____

27. $6a^2b + 9ab^2 - 6ab$

28. $9x^3y^2 - 6x^4y + 15x^2y^2$

27. _____

28. _____

29. $x(a + b) - 2(a + b)$

30. $3x(x + 2) - 4(x + 2)$

29. _____

30. _____

31. $cx + dx + cy + dy$

32. $ax - ay + 2x - 2y$

31. _____

32. _____

33. $ax - 3y - 3x + ay$

34. $ax + 20 - 5a - 4x$

33. _____

34. _____

35. $a^2x - 8 + 2a^2 - 4x$

36. $2a^2x - 3 - 2x + 3a^2$

35. _____

36. _____

37. $x^2 + 4x + 3$

38. $a^2 - 9a + 14$

37. _____

38. _____

39. $x^2 - 3x - 10$

40. $x^2 + 3x + 5$

39. _____

40. _____

41. $2ab - 4a + 3b - 6$

42. $6x^2 - x - 15$

41. _____

42. _____

43. $x^2 + 6x + 9$

44. $x^2 - 10x + 25$

45. $25x^2 - 49$

46. $81a^2 - 121$

47. $x^2 + 8x + 16$

48. $9x^2 - 30x + 25$

49. $9x^2 - 12x + 4$

50. $4x^2 - 64$

51. $a^3 + 27$

52. $a^3b^3 + 1$

53. $4x^2 - 15x + 12$

54. $x^2 + 5x + 6$

55. $9x^2 - y^2 + 6x + 1$

56. $4x^2 - y^2 + 12x + 9$

57. $x^2 + 5x - 14$

58. $x^2 - 4x + 3$

59. $5x^2 - 45$

60. $x^2 + 2x - 15$

61. $x^2 + x - 72$

62. $2x^2 + 5x + 2$

63. $a^2 + 2b - 2a - ab$

64. $2x^2 - x - 3$

43. _____

44. _____

45. _____

46. _____

47. _____

48. _____

49. _____

50. _____

51. _____

52. _____

53. _____

54. _____

55. _____

56. _____

57. _____

58. _____

59. _____

60. _____

61. _____

62. _____

63. _____

64. _____

A N S W E R S

65. _____

66. _____

67. _____

68. _____

69. _____

70. _____

71. _____

72. _____

73. _____

74. _____

75. _____

76. _____

77. _____

78. _____

79. _____

80. _____

81. _____

82. _____

SCORE: _____

65. $4a^2 - b^2 - 4b - 4$

66. $9a^2 - b^2 - 10b - 25$

67. $3x^2 + 13x - 10$

68. $6x^2 - 5x - 4$

69. $9x^2 - 64$

70. $12x^3 - 75x$

71. $3x^3 - 24$

72. $3x^4 - 192x$

73. $6x^2 - 7x + 2$

74. $4x^2 + 14x + 6$

75. $x^2 - y^2 - 6x + 4y + 5$

76. $x^2 - y^2 - 8x - 4y + 12$

77. $6x^3 + 8x^2 - 8x$

78. $18x^2 + 3x - 6$

79. $5x^2 - 2x + 3$

80. $3x^2 + 9x + 15$

81. $x^4 - 8x^2 + 4$

82. $x^4 - 7x^2 + 9$

4 **TEST**

1. Multiply:

   a. $(x + 5)(x - 8)$
               b. $(2a + b)(a^2 - ab + b)$

A N S W E R S

1a. _____

2. Divide:

   a. $\dfrac{12x^2y - 3x^3y^4 - 15x^4y^2}{-3x^2y}$
        b. $(x^4 + 2x^3 + 32) \div (x - 2)$

1b. _____

2a. _____

2b. _____

Factor completely.

3. _____

3. $22a^2 + 33a$
             4. $10x^3y - 6x^2y^2 + 2xy^3$

4. _____

5. $2x(a - 1) - 3(a - 1)$
      6. $x^2 - 81$

5. _____

6. _____

7. $x^2 - 22x + 121$
       8. $ax + bx + 3a + 3b$

7. _____

9. $x^2 - 14x + 49$
       10. $4x^2 - 19x + 12$

8. _____

9. _____

11. $2x^2 + 13x + 15$
      12. $27a^3 - 8b^3$

10. _____

11. _____

13. $25x^2 - 144$
       14. $x^2 - 6x - 27$

12. _____

13. _____

14. _____

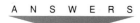

A N S W E R S

15. _____

16. _____

17. _____

18. _____

19. _____

20. _____

SCORE: _____

15. $ax^2 - 27 + 3x^2 - 9a$

16. $6x^2 - 10x - 4$

17. $32x^3 - 72x$

18. $6x^3 + 3x^2 - 18x$

19. $6x^2 - x - 15$

20. $x^2 - y^2 - 4x - 10y - 21$

C   H   A   P   T   E   R

5

PRETEST

Before beginning this chapter, answer as many of the following questions as you can. When you have finished the chapter, take the practice test at the end of the chapter and compare the scores of the two tests to see how much you have learned.

A   N   S   W   E   R   S

1. Simplify: $\dfrac{x^2 - 9}{2x^3 - x^2 - 15x}$ $= \dfrac{(x+3)(x-3)}{x-}$

1. _____

2. Multiply:

$$\frac{x^2 - 8x + 16}{2x^2 + 11x + 5} \cdot \frac{4x^2 - 1}{x^2 - x - 12} \cdot \frac{x^2 + 8x + 15}{2x^2 - 9x + 4}$$

2. _____

3. Divide: $\dfrac{3x^2 - 4x - 4}{x^2 + 3x - 10} \div \dfrac{3x^2 + 11x + 6}{x^2 - x - 12}$

3. _____

4. a. Find the LCD for the following fractions:

$$\frac{1}{x^2 + 10x + 24}, \frac{x - 1}{3x^2 + 10x - 8}$$

b. Find the missing numerator:

$$\frac{x - 3}{x + 3} = \frac{?}{2x^2 - x - 21}$$

4a. _____

4b. _____

271

5. Add:

$$\frac{5}{x^2 + 2x - 8} + \frac{x + 3}{x^2 - 3x + 2}$$

6. Subtract:

$$\frac{x + 5}{x^2 + 2x - 3} - \frac{1}{x^2 + 4x + 3}$$

5. _____

6. _____

7. Simplify:

$$\frac{\dfrac{1}{x^2 - 2x - 15} - \dfrac{1}{x + 3}}{1 - \dfrac{1}{x - 5}}$$

8. Solve:

$$\frac{x}{x + 2} + \frac{3}{x - 2} = 1 - \frac{1}{x^2 - 4}$$

7. _____

8. _____

SCORE: _____

Algebraic Fractions

A camera utilizes the formula from optics $\frac{1}{f} = \frac{1}{d_o} + \frac{1}{d_i}$, where f represents the focal length of the lens, d_o represents the distance of the object from the lens, and d_i represents the distance of the image from the lens. If a lens has a focal length of 12 centimeters, how far from the lens will the image appear when the object is 36 centimeters from the lens?

Working with fractions in algebraic expressions and equations is a necessary skill. Your knowledge of the arithmetic of fractions will form a foundation for work in algebraic fractions. The methods used in reducing, multiplying, dividing, adding, and subtracting fractions and simplifying complex fractions are identical with the methods from arithmetic. Equations with algebraic fractions are also covered in this chapter.

To work the problems in this chapter you must know the processes of factoring from the previous chapter. Mastery of one topic is almost always necessary for understanding another topic in algebra.

5-1 SIMPLIFYING FRACTIONS

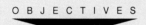

OBJECTIVES

Upon completing this section you should be able to:
1. Factor the numerator and denominator of a fraction.
2. Simplify algebraic fractions.

An **algebraic fraction** is the indicated ratio of two algebraic expressions.

In your study of arithmetic you were instructed that fractional answers were always to be left in *reduced*, or simplified form. For the fraction $\frac{8}{12}$ you "reduced" it to $\frac{2}{3}$ by dividing the numerator and the denominator by 4. The fraction $\frac{2}{3}$ cannot be reduced because no number (other than 1) will divide both numerator and denominator. In simplifying fractions this way you were using the following definition.

A fraction such as $\frac{2}{3}$ is in simplified form since the numerator 2 and the denominator 3 have no common factor other than one.

The fraction $\frac{4}{6}$ however is not in simplified form since numerator and denominator have a common factor of 2.

A fraction is in **simplified** (or reduced) **form** if the numerator and denominator contain no common factor (other than 1).

To obtain the simplified form of a fraction apply the following rule.

To simplify a fraction factor the numerator and denominator completely and then divide both numerator and denominator by all common factors.

Example 1   Simplify: $\dfrac{28}{42}$

Solution   First factor the numerator and denominator completely.

$$\frac{28}{42} = \frac{(2)(2)(7)}{(2)(3)(7)}$$

Next divide by the common factors, giving

$$\frac{\overset{1}{\cancel{(2)}}(2)\overset{1}{\cancel{(7)}}}{\underset{1}{\cancel{(2)}}(3)\underset{1}{\cancel{(7)}}} = \frac{2}{3} \text{ , which is the simplified form of } \frac{28}{42} \text{ .}$$

Remember, a factor divided by itself is equal to 1.

Example 2   Simplify: $\dfrac{2x + 4}{x^2 + 2x}$

Solution   First factor, giving $\dfrac{2(x + 2)}{x(x + 2)}$.

Now divide by the common factor $(x + 2)$ in both numerator and denominator to get

$$\frac{2x + 4}{x^2 + 2x} = \frac{2\overset{1}{\cancel{(x + 2)}}}{x\underset{1}{\cancel{(x + 2)}}} = \frac{2}{x} \text{ .}$$

> **WARNING**
>
> We can only divide common factors, never common terms.

Example 3   Simplify: $\dfrac{x^2 - 4}{x^2 + 8 + 12}$

Solution $\dfrac{x^2 - 4}{x^2 + 8x + 12} = \dfrac{(x - 2)\overset{1}{\cancel{(x + 2)}}}{\underset{1}{\cancel{(x + 2)}}(x + 6)} = \dfrac{x - 2}{x + 6}$

In an expression such as $\dfrac{a + 3}{b + 3}$ some students are tempted to divide the 3s. Note that this is *incorrect* since they are *terms* and not factors.

Example 4   Simplify: $\dfrac{2x + 7}{x^2 + 5}$

Solution   Since $2x + 7$ and $x^2 + 5$ have no common factors (other than 1), this fraction is already in simplified form.

Example 5   Simplify: $\dfrac{x^2 + 5x + 6}{x^2 - 7x + 10}$

Solution $\dfrac{x^2 + 5x + 6}{x^2 - 7x + 10} = \dfrac{(x + 3)(x + 2)}{(x - 5)(x - 2)}$

Note that $(x + 2)$ and $(x - 2)$ are not exactly alike and therefore you should not try to divide them.

Notice that even though we were able to factor the numerator and denominator, we still cannot divide since no factors are common to both. The given fraction is already in simplified form.

The fact that a given fraction might require any of the methods of factoring you have studied emphasizes again the importance of being proficient in factoring.

Example 6 Simplify: $\dfrac{3x^2 + x - 10}{ax + 6 + 2a + 3x}$

Solution Here you may use "trial and error" for the numerator and "grouping" for the denominator.

$$\frac{3x^2 + x - 10}{ax + 6 + 2a + 3x} = \frac{(3x - 5)(x + 2)}{(a + 3)(x + 2)} = \frac{3x - 5}{a + 3}$$

Here $(x + 2)$ is a common factor, so both numerator and denominator may be divided.

Example 7 Simplify: $\dfrac{2x + 5}{4x^2 - 25}$

Solution $\dfrac{2x + 5}{4x^2 - 25} = \dfrac{(2x + 5)}{(2x + 5)(2x - 5)} = \dfrac{1}{2x - 5}$

Note that the numerator $2x + 5$ can be written as $(2x + 5) \cdot 1$. Thus when the factor $(2x + 5)$ is divided, the factor 1 remains.

Example 8 Simplify: $\dfrac{2x - 3}{3 - 2x}$

Solution This type of problem requires special attention because it is a common cause for error. At first glance the factors might be mistakenly considered as common, or the fraction might be mistakenly considered as already simplified. Note that the factors cannot be divided since the signs keep them from being identical. If, however, negative 1 is factored from one of the factors, then there are like factors and division can be accomplished.

Any factors in the form of $a - b$ and $b - a$ are negatives of each other, thus $2x - 3$ and $3 - 2x$ are negatives of each other.

$$2x - 3 = -(3 - 2x)$$

$$\frac{2x - 3}{3 - 2x} = \frac{-1(3 - 2x)}{(3 - 2x)} = -1$$

Note that some answers may be left in many different forms. For example, the expression $\dfrac{b - a}{x - y}$ could be written as $\dfrac{-(a - b)}{x - y}$ or $-\dfrac{a - b}{x - y}$ or $\dfrac{a - b}{-(x - y)}$ or $\dfrac{a - b}{y - x}$.

All of these are equivalent forms of the same expression. The preferred form would be one that uses the least number of written signs.

Always check your answer to see if it is equivalent to the form given in the answer section.

Example 9 Simplify: $\dfrac{x^2 - 6x + 8}{8 + 2x - x^2}$

Solution

$$\frac{x^2 - 6x + 8}{8 + 2x - x^2} = \frac{(x - 4)(x - 2)}{(4 - x)(2 + x)}$$

$$= \frac{-1(4 - x)(x - 2)}{(4 - x)(2 + x)}$$

$$= \frac{-1(x - 2)}{(2 + x)}$$

$$= \frac{2 - x}{2 + x}$$

▼ **EXERCISE 5-1-1**

Simplify when possible.

1. $\dfrac{10}{15}$ $\dfrac{2 \cdot 5}{3 \cdot 5}$ $\dfrac{2}{3}$

2. $\dfrac{12}{20}$

3. $\dfrac{18}{24}$ $\dfrac{6 \cdot 3}{6 \cdot 4} = \dfrac{3}{4}$

4. $\dfrac{6}{18}$ $\dfrac{2 \cdot 3}{3 \cdot 6}$ $\dfrac{2}{6}$

5. $\dfrac{16}{40}$ $\dfrac{4 \cdot 4}{4 \cdot 10} = \dfrac{2 \cdot 2}{2 \cdot 5} = \dfrac{2}{5}$

6. $\dfrac{42}{48}$

7. $\dfrac{3x}{5x}$ $= \dfrac{3}{5}$

8. $\dfrac{2x}{4x}$ $= \dfrac{1}{2}$

9. $\dfrac{x}{x^2}$ $\dfrac{1}{x}$

10. $\dfrac{x}{x^3}$ $= \dfrac{1}{x^2}$

11. $\dfrac{14xy}{21xz}$ $\dfrac{2 \cdot 7 \cdot x \cdot y}{3 \cdot 7 \cdot x \cdot z} = \dfrac{2y}{3z}$

12. $\dfrac{12xy^2}{16x^2y}$ $\dfrac{3y}{4x}$

13. $\dfrac{3x - 9}{4x - 12}$ $\dfrac{3(x-3)}{4(x-3)} = \dfrac{3}{4}$

14. $\dfrac{2a + 4}{5a + 10}$ $\dfrac{2(a+2)}{5(a+2)} = \dfrac{2}{5}$

15. $\dfrac{x^2 + 3x}{x^2 - 2x}$ $\dfrac{x(x+3)}{x(x-2)} = \dfrac{x+3}{x-2}$

16. $\dfrac{x^2 + 4x}{x^2 + x}$

17. $\dfrac{ab - 3a}{b^2 - 3b}$ $\dfrac{a(b-3)}{b(b-3)} = \dfrac{a}{b}$

18. $\dfrac{a^2 + 4a}{ab + 4b}$

19. $\dfrac{2a^2 - a}{a^2 + 4a}$

20. $\dfrac{x^2 + 3x}{3x^2 - 2x}$

21. $\dfrac{x - 2}{2 - x}$

22. $\dfrac{5 - x}{x - 5}$

23. $\dfrac{2x - 8}{12 - 3x}$

24. $\dfrac{9 - 3x}{4x - 12}$

25. $\dfrac{3a - 2a^2}{8a - 12}$

26. $\dfrac{2a^2 - 8a}{20a - 5a^2}$

27. $\dfrac{2x - 6}{3x + 9}$

28. $\dfrac{4x - 2}{8x + 4}$

29. $\dfrac{x^2}{3x^2 - 2x}$

30. $\dfrac{x}{x^2 + 3x}$

31. $\dfrac{2x^2 + 5x - 12}{ax + 4a - 2x - 8}$

32. $\dfrac{3x^2 + 5x - 2}{3ax + 6x - a - 2}$

33. $\dfrac{x^2 - 1}{x^2 + x - 2}$

34. $\dfrac{x + 3}{x^2 - 9}$

35. $\dfrac{x^3 - 27}{x^2 + 3x + 9}$

36. $\dfrac{x^3 + 8}{x^2 - 4}$

37. $\dfrac{x^2 - x}{x^3 - 2x^2 - 3x}$

38. $\dfrac{x^2 + 10x + 16}{x^2 - 4}$

39. $\dfrac{12 + x - x^2}{x^2 + 3x}$

40. $\dfrac{x^2 - x - 2}{x^2 + x - 2}$

41. $\dfrac{x^2 + 5x}{x^2 + 8x + 15}$

42. $\dfrac{x^2 - 2x - 8}{x + 2}$

43. $\dfrac{6 - 3x - 2a + ax}{a^2 + 4a - 21}$

44. $\dfrac{16 - 8a - 2x + ax}{x^2 - 5x - 24}$

45. $\dfrac{4x^2 - 1}{8x^3 + 1}$

46. $\dfrac{2x - 3x^2}{27x^3 - 8}$

47. $\dfrac{3x^2 - 2x}{3x^2 + x - 2}$

48. $\dfrac{3 + 5x - 2x^2}{2x^2 + 11x + 5}$

49. $\dfrac{3a^2 + 17a + 10}{2a^2 + 7a - 15}$

50. $\dfrac{6x^2 - x - 2}{6x^2 + 5x - 6}$

5-2 MULTIPLICATION OF ALGEBRAIC FRACTIONS

$\dfrac{a}{b} \cdot \dfrac{c}{d} = \dfrac{ac}{bd}$ is the definition of the product of two fractions. In words this says, "multiply numerator by numerator and denominator by denominator." You have used this rule many times in arithmetic as you multiplied fractions.

However, remember that all fractional answers must be in simplified form. We could follow the above definition and then simplify the answer as in the previous section. But with algebraic fractions this can lead to very difficult expressions. The following rule allows us to simplify as we multiply, so the answer will then be in simplified form.

> *When multiplying algebraic fractions,* factor all numerators and denominators completely, then divide by all factors common to a numerator and denominator before multiplying.

O B J E C T I V E S

Upon completing this section you should be able to:

1. Factor numerators and denominators of all fractions being multiplied.
2. Identify and divide by all common factors.
3. Write the product in simplest form.

Again, remember common factors must be *exactly* alike.

The product of the remaining factors of the numerator will be the numerator of the answer and the product of the remaining factors of the denominator will be the denominator of the answer.

Example 1   Multiply: $\dfrac{x^2 - 9}{2x^2 + 5x + 2} \cdot \dfrac{x^2 + 5x + 6}{x^2 + 6x + 9}$

Solution

$$\dfrac{x^2 - 9}{2x^2 + 5x + 2} \cdot \dfrac{x^2 + 5x + 6}{x^2 + 6x + 9} = \dfrac{(x - 3)\cancel{(x + 3)}}{(2x + 1)\cancel{(x + 2)}} \cdot \dfrac{\cancel{(x + 2)}\cancel{(x + 3)}}{\cancel{(x + 3)}\cancel{(x + 3)}}$$

$$= \dfrac{x - 3}{2x + 1}$$

We will use the dot · to indicate multiplication since using × might be confused with the variable x.

Example 2   Multiply: $\dfrac{2x^2 - 5x - 12}{x^2 + 6x + 8} \cdot \dfrac{x^2 - 1}{2x^2 + 5x + 3}$

Solution

$$\dfrac{2x^2 - 5x - 12}{x^2 + 6x + 8} \cdot \dfrac{x^2 - 1}{2x^2 + 5x + 3}$$

$$= \dfrac{\cancel{(2x + 3)}(x - 4)}{(x + 4)(x + 2)} \cdot \dfrac{(x - 1)\cancel{(x + 1)}}{\cancel{(2x + 3)}\cancel{(x + 1)}}$$

$$= \dfrac{(x - 4)(x - 1)}{(x + 4)(x + 2)}$$

The answer could be given as $\dfrac{x^2 - 5x + 4}{x^2 + 6x + 8}$, but the factored form is generally used.

Example 3 Multiply: $\dfrac{x^2 - 6x + 8}{x^2 + 5x - 14} \cdot \dfrac{x^2 - x - 6}{8 + 2x - x^2}$

Note that $(x + 2)$ and $(2 + x)$ are the same but $(x - 4)$ and $(4 - x)$ are negatives of each other.

Solution $= \dfrac{(x - 4)(x - 2)}{(x - 2)(x + 7)} \cdot \dfrac{(x + 2)(x - 3)}{(4 - x)(2 + x)}$

$= \dfrac{-1(4 - x)(x - 2)}{(x - 2)(x + 7)} \cdot \dfrac{(x + 2)(x - 3)}{(4 - x)(2 + x)}$

Again, there are many possible forms for the final answers.

The form given here is preferred because it contains the fewest number of signs.

$= \dfrac{-1(x - 3)}{x + 7}$

$= \dfrac{3 - x}{x + 7}$

Example 4 Multiply: $\dfrac{2x^2 - x - 3}{3x^2 + 14x + 8} \cdot \dfrac{x^2 - 16}{2x^2 - 6x - 8} \cdot \dfrac{6x^2 + x - 2}{4x^2 - 8x + 3}$

There are a lot of factors in this problem. *Be careful!*

Solution $= \dfrac{(2x - 3)(x + 1)}{(3x + 2)(x + 4)} \cdot \dfrac{(x + 4)(x - 4)}{2(x - 4)(x + 1)} \cdot \dfrac{(2x - 1)(3x + 2)}{(2x - 3)(2x - 1)}$

$= \dfrac{1}{2}$

▼ **EXERCISE 5–2–1** ⟨ok⟩

Multiply.

1. $\dfrac{2}{3} \cdot \dfrac{9}{10} = \dfrac{3}{5}$

2. $\dfrac{3}{5} \cdot \dfrac{10}{15} \cdot \dfrac{2}{5}$

3. $\dfrac{2}{7} \cdot \dfrac{7}{12}$ $\dfrac{7}{42} = \dfrac{1}{6}$

4. $\dfrac{6}{10} \cdot \dfrac{15}{24}$

5. $\dfrac{3xyz}{5} \cdot \dfrac{10yz}{12xy^2}$ $\dfrac{2yz^2}{4y} = \dfrac{z^2}{2}$

6. $\dfrac{4a^2c}{7b} \cdot \dfrac{21b^2}{8ac^2}$

7. $\dfrac{3x}{5x + 10} \cdot \dfrac{5x - 15}{x^2 - 3x}$ $\dfrac{3x}{x(x + 2)} = \dfrac{3}{(x + 2)}$

8. $\dfrac{2x - 8}{5x} \cdot \dfrac{x^2 + 2x}{4x - 16}$

9. $\dfrac{2x}{x+4} \cdot \dfrac{x^2+3x-4}{2x^2-2x} = \dfrac{x-1}{x-2}$

(handwritten: $(x+4)(x-1)$ over; $2x(x-2)$ below)

10. $\dfrac{4x}{x^2+2x-8} \cdot \dfrac{x-2}{2x}$

11. $\dfrac{x^2}{x^2+2x} \cdot \dfrac{x^2+5x+6}{x^2+3x}$

(handwritten: $(x+3)(x+2)$; $\dfrac{x^2}{x \cdot x} = 1$; $x(x+2)$ $x(x+3)$)

12. $\dfrac{x^2-3x}{x^2-6x+9} \cdot \dfrac{x^2-2x-3}{x^2+x}$

13. $\dfrac{5x+10}{x-5} \cdot \dfrac{2x-10}{5x-5} = \dfrac{x+2}{x-1}$

(handwritten: $5(x+2)$ $2(x-5)$; $5(x-1)$)

14. $\dfrac{2x+2}{x+3} \cdot \dfrac{x+3}{2x-6}$

15. $\dfrac{x^2-4}{x^2-4x+4} \cdot \dfrac{x^2+x-2}{x^2+4x+4} = \dfrac{x-1}{x-2}$

(handwritten: $(x-2)(x+2)$ $(x-1)(x+2)$; $(x-2)(x-2)$ $(x+2)(x+2)$)

16. $\dfrac{x^2-8x+16}{x^2-x-12} \cdot \dfrac{x^2-9}{x^2-3x-4}$

17. $\dfrac{3x+15}{x^2+6x+5} \cdot \dfrac{x^2+2x+1}{x^2-1} = \dfrac{3}{(x-1)}$

(handwritten: $3(x+5)$ $(x+1)(x+1)$; $(x+5)(x+1)$ $(x+1)(x-1)$)

18. $\dfrac{x^2-1}{5x-5} \cdot \dfrac{5}{x^2+5x+4}$

19. $\dfrac{x^2+4x+3}{x-5} \cdot \dfrac{x^2-3x-10}{x+3} = (x+1)(x+2)$

(handwritten: $(x+1)(x+3)$ $(x-5)(x+2)$; or x^2+3x+2)

20. $\dfrac{x+3}{x^2+3x-10} \cdot \dfrac{x^2-3x+2}{x^2+4x+3}$

21. $\dfrac{x^2+5x-14}{x-3} \cdot \dfrac{6-2x}{x^2+12x+35} = \dfrac{-2(x-2)}{x+5}$

(handwritten: $(x+7)(x-2)$ $-2(x-3)$; $(x+7)(x+5)$)

22. $\dfrac{2x-1}{3x^2-2x-8} \cdot \dfrac{3x+4}{2x^2+x-1}$

23. $\dfrac{x^2+6x+5}{2x^2-2x-12} \cdot \dfrac{4x^2-36}{x^2+8x+15} = \dfrac{2(x+1)}{(x+2)}$

(handwritten: $(x+5)(x+1)$; $4(x^2-9)$ $4(x-3)(x+3)$; $2(x^2-x-6)$ $(x+5)(x+3)$; $2(x+2)(x-3)$)

24. $\dfrac{2x^2-9x-5}{20+x-x^2} \cdot \dfrac{3x^2+14x+8}{2x^2-11x-6}$

25. $\dfrac{2x-6}{4x-3} \cdot \dfrac{x^2+4x+4}{x^2-x-6}$

(handwritten: $\dfrac{2(x-3)}{4x-3} \cdot \dfrac{(x+2)(x+2)}{(x-3)(x+2)} = \dfrac{2(x+2)}{4x-3}$)

26. $\dfrac{3x^2+10x+3}{x^2+10x+21} \cdot \dfrac{2x^2+9x-35}{6x^2-13x-15}$

27. $\dfrac{x+3}{x^2-4x-5} \cdot \dfrac{x^2-2x-3}{x^2+5x+6} \cdot \dfrac{x^2+x-30}{x^2+x-12}$ $= \dfrac{x+6}{(x+2)(x+4)}$

28. $\dfrac{2x^2+9x-5}{x^2+10x+21} \cdot \dfrac{3x^2+11x+6}{2x^2-13x+6} \cdot \dfrac{x^2+3x-28}{3x^2-10x-8}$

29. $\dfrac{2x^2-7x-15}{5x^2-24x-5} \cdot \dfrac{20x^2+14x+2}{2x^2+11x+12} \cdot \dfrac{3x^2+x-2}{10x^2+35x+15}$

30. $\dfrac{3x^2+17x+10}{x^2+10x+25} \cdot \dfrac{x^2-25}{3x^2+11x+6} \cdot \dfrac{5x^2+16x+3}{5x^2-24x-5}$

5-3 DIVISION OF ALGEBRAIC FRACTIONS

OBJECTIVES

Upon completing this section you should be able to:

1. Change a division problem to a related multiplication problem.
2. Divide algebraic fractions.

Division of fractions is defined in terms of multiplication.

> To **divide** by a fraction multiply by the inverse of the divisor.

Thus, division by a fraction is performed by inverting the divisor and then multiplying. In symbols this is stated

$$\frac{a}{b} \div \frac{c}{d} = \frac{a}{b} \cdot \frac{d}{c}$$

We have often used this definition in arithmetic to divide fractions.

WARNING

The divisor follows the ÷ sign. Do not invert the wrong fraction.

Example 1 $\dfrac{3}{5} \div \dfrac{2}{3} = \dfrac{3}{5} \cdot \dfrac{3}{2} = \dfrac{9}{10}$

We use this same definition to establish a rule for dividing algebraic fractions.

> *To divide one algebraic expression by another* invert the divisor and change the operation to multiplication.

Example 2 $\dfrac{2x^2 + 7x + 5}{3x^2 - x - 14} \div \boxed{\dfrac{2x^2 - x - 15}{x^2 - x - 6}}$ ← Divisor

Solution $= \dfrac{2x^2 + 7x + 5}{3x^2 - x - 14} \cdot \dfrac{x^2 - x - 6}{2x^2 - x - 15}$

Now proceed as in multiplication.

Once the problem is changed from division to multiplication, it is completed as in the previous section.

$$= \dfrac{\cancel{(2x+5)}(x+1)}{(3x-7)\cancel{(x+2)}} \cdot \dfrac{\cancel{(x+2)}\cancel{(x-3)}}{\cancel{(x-3)}\cancel{(2x+5)}}$$

$$= \dfrac{x+1}{3x-7}$$

Example 3 $\dfrac{6x^2 - 5x - 4}{3x^2 - 10x + 8} \div (2x^2 + 5x + 2)$

If no denominator is shown, it is understood to be 1.

Solution In this case we must recognize that the divisor, $2x^2 + 5x + 2$, can be written as the fraction $\dfrac{2x^2 + 5x + 2}{1}$. When we invert and multiply, we have

$$= \dfrac{6x^2 - 5x - 4}{3x^2 - 10x + 8} \cdot \dfrac{1}{2x^2 + 5x + 2}$$

$$= \dfrac{\cancel{(2x+1)}\cancel{(3x-4)}}{\cancel{(3x-4)}(x-2)} \cdot \dfrac{1}{\cancel{(2x+1)}(x+2)}$$

$$= \dfrac{1}{(x-2)(x+2)} \text{ or } \dfrac{1}{x^2 - 4}.$$

▼ **EXERCISE 5–3–1**

Divide.

1. $\dfrac{2}{3} \div \dfrac{8}{3}$

$6 \div 24 = \frac{1}{4}$

2. $3 \div \dfrac{9}{11}$

3. $\dfrac{6}{5} \div \dfrac{9}{10}$

4. $\dfrac{12}{14} \div 6$

5. $\dfrac{3x}{7y} \div \dfrac{6x}{14y}$

6. $\dfrac{4ab}{3c} \div \dfrac{2a}{9c}$

7. $\dfrac{2a^2}{15bc} \div \dfrac{4a}{5b^2c}$

8. $\dfrac{6xz}{11y^2} \div \dfrac{12x}{44yz}$

9. $\dfrac{3x + 12}{x + 4} \div \dfrac{3x - 3}{x + 1}$

10. $\dfrac{5x - 10}{3x - 6} \div \dfrac{5x + 10}{6x + 12}$

11. $\dfrac{x^2 - 4}{4x + 8} \div \dfrac{x^2 - 4x + 4}{4 - 2x}$

12. $\dfrac{x^2 + 6x + 9}{x^2 + x - 6} \div \dfrac{x + 2}{x^2 - 4}$

13. $3x + 2 \div \dfrac{3x^2 - x - 2}{x^2 + x - 2}$

14. $\dfrac{4x^2 + 4x + 1}{2x - 1} \div (2x + 1)$

15. $\dfrac{x^2 - 2x - 3}{x^2 - 4x - 5} \div \dfrac{x - 3}{x - 5}$

16. $\dfrac{x^2 + 3x - 10}{x^2 + x - 6} \div \dfrac{x + 7}{x + 3}$

17. $\dfrac{x^2 + 3x - 4}{x^2 - 1} \div \dfrac{x^2 + 6x + 8}{x + 1}$

18. $\dfrac{x^2 - 25}{x^2 - 3x - 10} \div \dfrac{x^2 - x - 30}{x^2 + 4x + 4}$

19. $\dfrac{2x^2 + 11x + 12}{4x^2 + 16x + 15} \div \dfrac{2x^2 + 3x - 20}{8x^3 + 125}$

20. $\dfrac{3x^2 - 13x + 4}{x^3 - 64} \div \dfrac{3x^2 + 2x - 1}{3x^2 + 12x + 48}$

21. $\dfrac{x^2 - 2x - 63}{x^2 - 81} \div \dfrac{2x^2 - 5x - 12}{2x + 3}$

22. $\dfrac{3x^2 + 11x + 10}{x^2 - 4} \div \dfrac{2x^2 + 7x - 4}{x^2 + 2x - 8}$

23. $\dfrac{3x^2 + 9x - 54}{2x^2 - 2x - 12} \div \dfrac{3x^2 + 21x + 18}{4x^2 - 12x - 40}$

24. $\dfrac{6 + x - x^2}{2x^2 + 3x - 2} \div \dfrac{x^2 + 4x - 21}{3x^2 + 25x + 28}$

25. $\dfrac{4x^2 + 12x + 9}{3x^2 + 2x - 1} \div \dfrac{4x^2 - 9}{3x^2 + 14x - 5}$

26. $\dfrac{x^2 - 7x - 18}{2x^2 + 9x + 10} \div (2x + 5)$

27. $\dfrac{14x^2 + 23x + 3}{2x^2 + x - 3} \div \dfrac{7x^2 + 15x + 2}{2x^2 - 3x + 1}$

(handwritten above and below:) $(7x+1)(2x+3)$ $(7x+1)(x+2)$ $= \dfrac{2x-1}{x+2}$

(handwritten below:) $(2x+3)(x-1)$ $(2x-1)(x-1)$

28. $\dfrac{3x^2 - 17x + 10}{9x^2 - 12x + 4} \div \dfrac{20 + x - x^2}{6x^2 - 7x + 2}$

29. $\dfrac{25x^2 - 1}{10x^2 + 17x + 3} \times (1 - 5x)$

(handwritten:) $(5x+1)(5x-1)$; $\dfrac{1}{(2x+3)}$; $= \dfrac{-1}{(2x+3)}$; $(5x+1)(2x+3)$; $(1-5x)$; $-1(5x-1)$

30. $\dfrac{3x^2 + 22x - 16}{2x^2 + 19x + 24} \div \dfrac{3x^2 + 13x - 10}{2x^2 + 13x + 15}$

▶ 5–4 FINDING THE LEAST COMMON DENOMINATOR

O B J E C T I V E S

Upon completing this section you should be able to:

1. Factor a denominator of a fraction completely.
2. Find the least common denominator of two or more fractions.

The least common denominator of a set of fractions is sometimes referred to as the *least common multiple* of the denominators.

The rule for addition and subtraction of fractions requires that the fractions to be combined must have the same denominator. As preparation for performing these operations we will now investigate the method of finding the least common denominator for any group of fractions.

> A **common denominator** for two or more fractions is an expression that contains all factors of the denominator of each fraction. A **least common denominator** contains the minimum number of factors to be a common denominator.

Mental arithmetic will allow you to find the least common denominator for small numbers. If asked to add $\dfrac{1}{4}$ and $\dfrac{5}{6}$, it is easy to arrive at a least common denominator of 12. If asked how we arrived at 12, we just know that 12 is the least number divisible by both 4 and 6. However, a more involved method is necessary if the numbers are larger or if the fractions are algebraic fractions.

We could obtain a common denominator of these fractions by finding the product

$12 \times 14 \times 15 \times 18 = 45,360.$

While this number is a common denominator, it is not the *least* common denominator.

Example 1   Find the least common denominator for $\dfrac{1}{12}, \dfrac{1}{14}, \dfrac{1}{15},$ and $\dfrac{1}{18}.$

Solution   This problem would require a considerable amount of guesswork, or testing possibilities, if we had no general method.

Let's consider the definition. From it we know that a common denominator for these numbers must contain all factors of each. In other words, we are looking for the smallest number divisible by 12, 14, 15, and 18.

First completely factor each number.

$$12 = (2)(2)(3)$$
$$14 = (2)(7)$$
$$15 = (3)(5)$$
$$18 = (2)(3)(3)$$

The number we are looking for must contain $(2)(2)(3)$ in order to be divisible by 12. It must contain $(2)(7)$ in order to be divisible by 14, and so on. Proceed as follows:

Write the factors of the first number, 12.

$$(2)(2)(3)$$

Now look at the factors of the next number, 14, and see that we need $(2)(7)$. But since we already have a 2, we only need the factor (7). This gives

$$(2)(2)(3)(7).$$

This number is now divisible by 12 and by 14. The factors of the next number, 15, are (3) and (5). Since we already have a 3, we need only the factor 5, giving

$$(2)(2)(3)(7)(5).$$

This number is now divisible by 12, 14, and 15. The factors of the next number, 18, are $(2)(3)(3)$. We already have 2 and one 3. Hence, we need another 3.

$$(2)(2)(3)(7)(5)(3) = 1,260$$

This number, 1,260, is a common denominator of 12, 14, 15, and 18 because it contains all factors of each and is therefore divisible by each. It is the *least* common denominator because it contains only those factors necessary to make it divisible by 12, 14, 15, and 18.

Note that 1,260 is considerably smaller than the number obtained by simply finding the product of all the denominators.

The preceding discussion gives rise to a rule for obtaining a least common denominator for any number of fractions, whether they be numbers or algebraic expressions.

To find the least common denominator for two or more fractions

1. Factor each denominator completely.
2. Write the denominator of the first fraction in factored form as the proposed common denominator.
3. By inspection determine which factors of the second denominator are not already in the proposed common denominator and include them.
4. Repeat step three for each fraction.

Once mastered, this step-by-step procedure will greatly simplify your work.

Note that in finding the least common denominator we pay no attention to the numerator.

Example 2   Find the least common denominator for $\dfrac{x + 2}{x + 3}$ and $\dfrac{x + 1}{x^2 + 7x + 12}$.

Solution   First denominator: $(x + 3)$

This is just the denominator of the first fraction.

Second denominator: $(x + 3)(x + 4)$

Proposed common denominator: $(x + 3)$

By inspection of the second denominator we need the additional factor $(x + 4)$. The least common denominator is $(x + 3)(x + 4)$.

Again, the numerators have no effect on what the least common denominator will be.

Example 3   Find the least common denominator for $\dfrac{2x - 3}{6x^2 - 5x - 4}$ and $\dfrac{x + 1}{3x^2 - 10x + 8}$.

Solution   First denominator: $(3x - 4)(2x + 1)$

Second denominator: $(3x - 4)(x - 2)$

Proposed common denominator: $(3x - 4)(2x + 1)$

By inspection of the second denominator we need the additional factor $(x - 2)$. The least common denominator is $(3x - 4)(2x + 1)(x - 2)$.

Sometimes the least common denominator is abbreviated as LCD.

Example 4   Find the least common denominator for $\dfrac{5}{6x^2}$ and $\dfrac{x + 1}{2x^2 + 4}$.

Solution   First denominator: $(2)(3)(x)(x)$

Note that x^2 is a factor in the denominator of the first fraction but *not* in the second fraction.

Second denominator: $2(x^2 + 2)$

Proposed common denominator: $(2)(3)(x)(x)$

By inspection of the second denominator we need an additional factor of $(x^2 + 2)$. The least common denominator is

$$(2)(3)(x)(x)(x^2 + 2) \text{ or } 6x^2 (x^2 + 2).$$

Example 5   Find the least common denominator for $\dfrac{2x}{x^2 - 4}$ and $\dfrac{3}{x^2 + 4x + 4}$.

Solution   First denominator: $(x + 2)(x - 2)$

Second denominator: $(x + 2)(x + 2)$

Proposed common denominator: $(x - 2)(x + 2)$

Note that the second denominator requires *two* $(x + 2)$ factors.

By inspection we see that part of the second denominator is included but another $(x + 2)$ factor is needed. The least common denominator is $(x + 2)(x + 2)(x - 2)$.

Example 6   Find the least common denominator for $\dfrac{x+4}{3x+6}, \dfrac{7}{12}$, and

$\dfrac{x-1}{2x^2+10x+12}$.

Solution   First denominator: $3(x+2)$

Second denominator: $2(2)(3)$

Third denominator: $2(x+3)(x+2)$

Proposed common denominator: $3(x+2)$

By inspection of the second denominator we see we need to include the factors (2) and (2). We now have $2(2)(3)(x+2)$. By inspection of the third denominator we see we need the factor $(x+3)$. The least common denominator is $2(2)(3)(x+2)(x+3)$ or $12(x+2)(x+3)$.

Here we have three denominators, but the procedure is the same.

We usually multiply the numerals.

Example 7   Find the least common denominator for $\dfrac{2x+1}{x^2+3x+2}, \dfrac{x-1}{x+4}$,

and $\dfrac{3}{x^2+5x+4}$.

Solution   First denominator: $(x+1)(x+2)$

Second denominator: $(x+4)$

Third denominator: $(x+1)(x+4)$

Proposed common denominator: $(x+1)(x+2)$

Inspecting the second denominator we see that $(x+4)$ must be a factor, so we now have $(x+1)(x+2)(x+4)$. Inspecting the third denominator we see that $(x+1)$ and $(x+4)$ have already been included. Therefore the least common denominator is $(x+1)(x+2)(x+4)$.

It will help avoid errors if you inspect only one denominator at a time.

▼ **EXERCISE 5-4-1** $1-20$

Find the least common denominator.

1. $\dfrac{1}{3}, \dfrac{2}{5}$

2. $\dfrac{1}{2}, \dfrac{3}{7}$

3. $\dfrac{3}{4}, \dfrac{5}{6}$

4. $\dfrac{1}{6}, \dfrac{5}{9}$

5. $\dfrac{3}{8}, \dfrac{11}{12}$ $\dfrac{9}{24}$ $\dfrac{22}{24}$

6. $\dfrac{5}{18}, \dfrac{7}{30}$

7. $\dfrac{1}{6}, \dfrac{4}{15}, \dfrac{3}{10}$ $\dfrac{5}{30}$ $\dfrac{8}{30}$ $\dfrac{9}{30}$

8. $\dfrac{2}{21}, \dfrac{3}{14}, \dfrac{1}{28}$

9. $\dfrac{5}{12}, \dfrac{3}{16}, \dfrac{1}{20}, \dfrac{2}{15}$

10. $\dfrac{1}{14}, \dfrac{6}{35}, \dfrac{7}{30}, \dfrac{2}{21}$

11. $\dfrac{1}{x}, \dfrac{1}{y}, \dfrac{3}{z}$

12. $\dfrac{1}{a}, \dfrac{2}{a^2 b}, \dfrac{1}{b^2}$

13. $\dfrac{1}{2x}, \dfrac{5}{4x^2 y}, \dfrac{1}{x^3 y^2}$ $=$ $4x^3 y^2$

$\dfrac{2x^2 y^2}{4x^3 y^2}$ $\dfrac{5x}{4x^3 y^2}$ $\dfrac{4}{4x^3 y^2}$

14. $\dfrac{1}{3a^2}, \dfrac{1}{12ab^2}, \dfrac{1}{b^2}$

15. $\dfrac{1}{x}, \dfrac{1}{x+2}$ $=$ $\dfrac{(x+2)}{x(x+2)}$

$\dfrac{x}{x(x+2)}$

16. $\dfrac{x}{x+1}, \dfrac{1}{x}$

17. $\dfrac{1}{x+1}, \dfrac{1}{x+3}$

18. $\dfrac{1}{x-2}, \dfrac{1}{x+2}$

19. $\dfrac{1}{6x^2}, \dfrac{x+1}{3x^2-6}$

20. $\dfrac{3}{10x}, \dfrac{x+3}{4x^2-8x}$

21. $\dfrac{a}{a-3}, \dfrac{1}{a^2-9}$

$LCD =$

$(a-3)(a+3) \qquad (a+3)(a-3)$

22. $\dfrac{x}{x^2-49}, \dfrac{1}{x+7}$

23. $\dfrac{1}{x+3}, \dfrac{1}{x^2-x-12}$

24. $\dfrac{x}{x^2+8x+12}, \dfrac{1}{x+2}$

25. $\dfrac{1}{x-3}, \dfrac{x}{x^2-6x+9}$

26. $\dfrac{x}{x+3}, \dfrac{1}{x^2+6x+9}$

27. $\dfrac{1}{a-2}, \dfrac{2a}{a^2-3a-10}$

28. $\dfrac{1}{x^2-x-12}, \dfrac{x}{x+4}$

29. $\dfrac{x-1}{x+1}, \dfrac{3x}{x^2+7x+6}$

30. $\dfrac{x+1}{x^2+7x+12}, \dfrac{x+2}{x+3}$

31. $\dfrac{2x}{x^2-6x+5}, \dfrac{x+4}{x^2-2x+1}$

32. $\dfrac{x-7}{x^2+6x+9}, \dfrac{5x^2}{2x^2+5x-3}$

33. $\dfrac{x-1}{x^2+5x+6}, \dfrac{x}{x^2-x-12}$

34. $\dfrac{2x+1}{x^2+3x-10}, \dfrac{x+6}{x^2-x-2}$

35. $\dfrac{3x+1}{2x^2+11x+12}, \dfrac{x-6}{2x^2+5x-12}$

36. $\dfrac{2x-1}{2x^2-x-15}, \dfrac{x+4}{4x^2-25}$

37. $\dfrac{x+1}{5x+5}, \dfrac{3}{20}, \dfrac{2x-1}{2x^2-6x-8}$

38. $\dfrac{4x}{3x^2-12x-15}, \dfrac{x-2}{10x+5}, \dfrac{5}{12}$

39. $\dfrac{3x^2}{x^2+x-6}, \dfrac{x}{x-5}, \dfrac{x+2}{x^2-7x+10}$

40. $\dfrac{4x-1}{2x^2+7x-4}, \dfrac{2x}{2x-1}, \dfrac{x-3}{x^2-3x-28}$

41. $\dfrac{x-7}{x^2-3x-18}, \dfrac{2x+3}{x^2+8x+12}, \dfrac{x+4}{x^2+9x+18}$

42. $\dfrac{x+1}{x^2-x-6}, \dfrac{x}{x^2-6x+9}, \dfrac{2x-1}{x^2-2x-8}$

5-5 EQUIVALENT FRACTIONS

In the previous section we learned how to find a common denominator for two or more fractions. As a further step toward adding fractions, we now give our attention to changing an algebraic fraction to an equivalent fraction with a different denominator.

$\dfrac{x}{y}=\dfrac{ax}{ay}$, when $a \neq 0$, is called the **fundamental principle of fractions.** When we analyze this statement, we see two equivalent fractions and note that the *numerator* and *denominator* of $\dfrac{x}{y}$ have both been multiplied by the *same* nonzero number, a.

> *To change a fraction to an equivalent fraction* multiply numerator and denominator by the same nonzero expression.

Why must the expression be nonzero?

Example 1 Change $\dfrac{2}{3}$ to a fraction with a denominator of 15.

Solution We have $\dfrac{2}{3}=\dfrac{?}{15}$. To solve this problem we only need to know what number the original denominator, 3, has been multiplied by to get 15, and then multiply the numerator by this *same* number. Since $15=(5)(3)$, we have $\dfrac{2}{3}=\dfrac{?}{(5)(3)}$ and see that the 3 has been multiplied by 5. So the numerator, 2, must also be multiplied by 5, giving $\dfrac{2}{3}=\dfrac{(5)(2)}{(5)(3)}=\dfrac{10}{15}$.

You could think of this process as the reverse of reducing fractions.

Example 2 Change $\dfrac{x+1}{2x+3}$ to a fraction having a denominator of $(2x+3)(x-4)$. This also may be stated as $\dfrac{x+1}{2x+3}=\dfrac{?}{(2x+3)(x-4)}$.

Solution Since the new denominator is in factored form, by inspection we see that the original denominator $(2x+3)$ has been multiplied by the factor $(x-4)$. Therefore, the original numerator $(x+1)$ must also be multiplied by the factor $(x-4)$, giving $\dfrac{x+1}{2x+3}=\dfrac{(x+1)(x-4)}{(2x+3)(x-4)}=\dfrac{x^2-3x-4}{(2x+3)(x-4)}$.

Note that in the final form of the fraction we have multiplied the factors in the numerator but have left the denominator in factored form. This is the preferred way of writing the answer.

Example 3   Change $\dfrac{2x+1}{x-3}$ to a fraction with denominator

$(2)(x-3)(x+1)$. That is $\dfrac{2x+1}{x-3} = \dfrac{?}{2(x-3)(x+1)}$.

Solution   Since the original denominator $(x-3)$ has been multiplied by (2) and $(x+1)$, the original numerator $(2x+1)$ must also be multiplied by (2) and $(x+1)$.

Again, note the form of the answer.

$$\frac{2x+1}{x-3} = \frac{2(2x+1)(x+1)}{2(x-3)(x+1)} = \frac{4x^2+6x+2}{2(x-3)(x+1)}$$

Example 4   Change both fractions $\dfrac{x-3}{x+2}$ and $\dfrac{x+1}{x-4}$ to fractions with a

denominator of $(x+2)(x-4)$ or $\dfrac{x-3}{x+2} = \dfrac{?}{(x+2)(x-4)}$ and

$\dfrac{x+1}{x-4} = \dfrac{?}{(x+2)(x-4)}$.

Solution   For the first fraction we have $\dfrac{x-3}{x+2} = \dfrac{(x-3)(x-4)}{(x+2)(x-4)}$

$$= \frac{x^2-7x+12}{(x+2)(x-4)}.$$

For the second fraction we have $\dfrac{x+1}{x-4} = \dfrac{(x+1)(x+2)}{(x+2)(x-4)}$

$$= \frac{x^2+3x+2}{(x+2)(x-4)}.$$

▼ **EXERCISE 5–5–1**

Find the missing numerator in each fraction.

1. $\dfrac{1}{2} = \dfrac{?}{6}$ $\dfrac{3}{6}$

2. $\dfrac{1}{3} = \dfrac{?}{15}$

3. $\dfrac{2}{5} = \dfrac{?}{30}$ $\dfrac{12}{30}$

4. $\dfrac{3}{4} = \dfrac{?}{36}$

5. $\dfrac{5}{7} = \dfrac{?}{147}$ $\dfrac{105}{147}$

6. $\dfrac{4}{9} = \dfrac{?}{126}$

7. $\dfrac{1}{x+2} = \dfrac{?}{(x+2)(x-3)}$ *(handwritten: X-3)*

8. $\dfrac{1}{x-3} = \dfrac{?}{(x-3)(x-5)}$

9. $\dfrac{x+1}{2x+1} = \dfrac{?}{(2x+1)(x-2)}$ *(handwritten: X²-X-2)*

(handwritten: $(x-2)(x+1)$ x²-x-2)

10. $\dfrac{x-2}{3x+4} = \dfrac{?}{(3x+4)(x+5)}$

11. $\dfrac{2x+3}{x-4} = \dfrac{?}{4x-16}$ *(handwritten: 8X+12)*

(handwritten: 4(x-4))

12. $\dfrac{3x-2}{2x+3} = \dfrac{?}{6x+9}$

13. $\dfrac{x-3}{x+5} = \dfrac{?}{x^2+9x+20}$

(*Hint:* Factor the second denominator.)

(handwritten: (x+5)(x+4))

14. $\dfrac{x+8}{x-7} = \dfrac{?}{x^2-11x+28}$

15. $\dfrac{2x-1}{3x+4} = \dfrac{?}{3x^2-14x-24}$

(handwritten: (3x-4)(x-6))

16. $\dfrac{4x+3}{2x+5} = \dfrac{?}{6x^2+11x-10}$

17. $\dfrac{2x-5}{3x+4} = \dfrac{?}{9x^2+24x+16}$

18. $\dfrac{3x-7}{2x+5} = \dfrac{?}{4x^2-25}$

19. $\dfrac{x+4}{2x-3} = \dfrac{?}{2(2x-3)(x+3)}$

20. $\dfrac{x-6}{3x+1} = \dfrac{?}{4(3x+1)(x-2)}$

21. $\dfrac{2x+3}{x-5} = \dfrac{?}{6x^2-9x-105}$

22. $\dfrac{3x-4}{2x+3} = \dfrac{?}{20x^2+20x-15}$

23. $\dfrac{x-5}{x^2+2x-15} = \dfrac{?}{(x+5)(x-3)(x+2)}$

24. $\dfrac{2x+1}{x^2-3x-28} = \dfrac{?}{(x-7)(x+3)(x+4)}$

25. $\dfrac{x + 7}{x^2 - 6x + 5} = \dfrac{?}{2(x + 7)(x - 5)(x - 1)}$

26. $\dfrac{3x + 1}{2x^2 - 5x - 3} = \dfrac{?}{(2x + 1)(x - 3)(x + 2)}$

27. $\dfrac{3x + 2}{x^2 + 4x - 12} = \dfrac{?}{(2x - 3)(x + 6)(x - 2)}$

28. $\dfrac{2x - 1}{3x^2 - 16x + 5} = \dfrac{?}{3(3x - 1)(2x + 1)(x - 5)}$

29. Change both fractions (a) $\dfrac{x + 1}{x - 3}$ and

   (b) $\dfrac{x - 2}{x + 4}$ to fractions with a denominator of $(x - 3)(x + 4)$.

30. Change both fractions (a) $\dfrac{x + 3}{2x - 3}$ and

   (b) $\dfrac{x}{x - 8}$ to fractions with a denominator of $(2x - 3)(x - 8)$.

31. Change both fractions (a) $\dfrac{x - 1}{2x - 8}$ and

   (b) $\dfrac{2x + 1}{2x + 3}$ to fractions with a denominator of $2(x - 4)(2x + 3)$.

32. Change both fractions (a) $\dfrac{x + 3}{6x + 3}$ and

   (b) $\dfrac{6x}{4x - 2}$ to fractions with a denominator of $6(2x + 1)(2x - 1)$.

33. Change both fractions (a) $\dfrac{3x + 2}{x^2 + 2x - 3}$ and

   (b) $\dfrac{x - 2}{x^2 + 3x - 4}$ to fractions with a denominator of $(x - 1)(x + 3)(x + 4)$.

34. Change both fractions (a) $\dfrac{2x - 7}{x^2 + 6x + 8}$ and

   (b) $\dfrac{x + 9}{x^2 + 5x + 4}$ to fractions with a denominator of $(x + 1)(x + 2)(x + 4)$.

◣ 5-6 ADDITION OF ALGEBRAIC FRACTIONS

From our knowledge of arithmetic we know that only fractions with like denominators can be added. The same is true of algebraic fractions. With the skills learned in the two previous sections, we are now ready to combine algebraic fractions.

OBJECTIVES

Upon completing this section you should be able to:
1. Add fractions having the same denominator.
2. Find the least common denominator of two or more fractions.
3. Apply the rule for adding fractions.

> The sum of two or more fractions which have the same denominator is the sum of the numerators over their common denominator.

Example 1 $\dfrac{3}{8} + \dfrac{4}{8} = \dfrac{3+4}{8} = \dfrac{7}{8}$

Example 2 $\dfrac{x+1}{x+2} + \dfrac{5}{x+2} = \dfrac{x+1+5}{x+2} = \dfrac{x+6}{x+2}$

Take note that this rule only allows the sum of fractions that have the same denominator. In other words, two or more fractions can only be added if they have a common denominator. The rule for adding any two or more fractions will require the skills developed in the last two sections in addition to knowledge of combining like terms.

To add two or more fractions follow these steps:
Step 1 Find the least common denominator (LCD) for all fractions involved using the method developed in section 5-4.
Step 2 Change each fraction to an equivalent fraction having the least common denominator (section 5-5).
Step 3 Find the sum of the numerators and place this sum over the least common denominator.
Step 4 Simplify (or reduce) the fraction obtained in step 3.

These four steps should be used whenever you add fractions.

Example 3 Find the sum of $\dfrac{2}{x}$ and $\dfrac{3x}{5x^2 + x}$.

Solution **Step 1** First denominator: x
Second denominator: $x(5x + 1)$
Least common denominator: $x(5x + 1)$

Step 2 $\dfrac{2}{x} = \dfrac{2(5x + 1)}{x(5x + 1)}$ and $\dfrac{3x}{x(5x + 1)}$ needs no adjustment

Step 3 $\dfrac{2(5x + 1) + (3x)}{x(5x + 1)} = \dfrac{10x + 2 + 3x}{x(5x + 1)} = \dfrac{13x + 2}{x(5x + 1)}$

Step 4 $\dfrac{13x + 2}{x(5x + 1)}$ cannot be reduced

WARNING

Remember to multiply the *numerator* as well as the *denominator* by the same expression.

Example 4   Add: $\dfrac{x-4}{x^2-2x-15} + \dfrac{3}{x^2+5x+6}$

Solution $x^2 - 2x - 15 = (x+3)(x-5)$
$x^2 + 5x + 6 = (x+3)(x+2)$

so the least common denominator is $(x+3)(x-5)(x+2)$.

Again, don't forget to multiply the numerator by the same expression that you multiply the denominator by.

$$\dfrac{x-4}{(x+3)(x-5)} = \dfrac{(x-4)(x+2)}{(x+3)(x-5)(x+2)}$$

$$\dfrac{3}{(x+3)(x+2)} = \dfrac{3(x-5)}{(x+3)(x-5)(x+2)}$$

Hence, $\dfrac{(x-4)(x+2)}{(x+3)(x-5)(x+2)} + \dfrac{3(x-5)}{(x+3)(x-5)(x+2)}$

$$= \dfrac{(x-4)(x+2) + 3(x-5)}{(x+3)(x-5)(x+2)}$$

$$= \dfrac{x^2 - 2x - 8 + 3x - 15}{(x+3)(x-5)(x+2)} = \dfrac{x^2 + x - 23}{(x+3)(x-5)(x+2)}$$

This answer is in reduced form.

Example 5   Add: $\dfrac{3}{x} + \dfrac{5}{y} + \dfrac{x}{z}$

Solution   The least common denominator is xyz.

Whenever the denominators have no common factors, the LCD is the product of the denominators.

$$\dfrac{3}{x} = \dfrac{3yz}{xyz}$$

$$\dfrac{5}{y} = \dfrac{5xz}{xyz}$$

$$\dfrac{x}{z} = \dfrac{x^2y}{xyz}$$

The sum is $\dfrac{3yz + 5xz + x^2y}{xyz}$.

Example 6   Add: $\dfrac{2}{x-5} + \dfrac{x-23}{x^2-x-20}$

Here only the first fraction must be changed in form.

Solution   The least common denominator is $(x-5)(x+4)$.

$$\dfrac{2}{x-5} = \dfrac{2(x+4)}{(x-5)(x+4)}$$

$$\dfrac{x-23}{(x-5)(x+4)} \qquad \text{needs no adjustment}$$

The sum is

$$\frac{2(x + 4) + (x - 23)}{(x - 5)(x + 4)} = \frac{2x + 8 + x - 23}{(x - 5)(x + 4)}$$

$$= \frac{3x - 15}{(x - 5)(x + 4)}.$$

Now this sum can be reduced as in section 5–1.

$$\frac{3x - 15}{(x - 5)(x + 4)} = \frac{3(x - 5)}{(x - 5)(x + 4)} = \frac{3}{x + 4}$$

Therefore, $\dfrac{2}{x - 5} + \dfrac{x - 23}{x^2 - x - 20} = \dfrac{3}{x + 4}.$

Note that the numerator $3x - 15$ can be factored as $3(x - 5)$ and the factor $(x - 5)$ matches a factor in the denominator.

We can use fewer written steps if we note that "common denominator" means all the fractions have the same denominator, and if all have the same denominator, then it is necessary to write the denominator only once. To illustrate this we will re-work the preceding example.

Example 7   Add: $\dfrac{2}{x - 5} + \dfrac{x - 23}{x^2 - x - 20}$

Solution

$$\frac{2}{x - 5} + \frac{x - 23}{(x - 5)(x + 4)} = \frac{2(x + 4) + 1(x - 23)}{(x - 5)(x + 4)}$$

$$= \frac{2x + 8 + x - 23}{(x - 5)(x + 4)}$$

$$= \frac{3x - 15}{(x - 5)(x + 4)}$$

$$= \frac{3(x - 5)}{(x - 5)(x + 4)}$$

$$= \frac{3}{x + 4}$$

This shortcut is fine as long as you remember to multiply the numerators by the necessary factors.

Example 8   Add: $\dfrac{1}{x} + \dfrac{2}{x + 4} + \dfrac{x}{x + 2}$

Solution $\dfrac{1}{(x)} + \dfrac{2}{(x + 4)} + \dfrac{x}{(x + 2)}$

$$= \frac{1(x + 4)(x + 2) + 2x(x + 2) + x(x)(x + 4)}{x(x + 4)(x + 2)}$$

$$= \frac{x^2 + 6x + 8 + 2x^2 + 4x + x^3 + 4x^2}{x(x + 4)(x + 2)}$$

$$= \frac{x^3 + 7x^2 + 10x + 8}{x(x + 4)(x + 2)}$$

Again, the denominators have no common factors so the LCD is the product of all three denominators.

You may wish to review section 1–9.

In chapter 1 we established the order of operations. When more than one operation occurs in a problem involving fractions, this same rule of order applies.

Example 9 $\dfrac{2x + 1}{x + 3} + \dfrac{x^2 + x - 6}{x - 1} \cdot \dfrac{3}{x + 3}$

Multiplication and division are performed before addition and subtraction.

Solution   In the absence of parentheses we must multiply before we can add.

$$\frac{2x + 1}{x + 3} + \frac{\overset{1}{\cancel{(x + 3)}}(x - 2)}{x - 1} \cdot \frac{3}{\underset{1}{\cancel{(x + 3)}}}$$

$$= \frac{2x + 1}{x + 3} + \frac{3(x - 2)}{x - 1}$$

The LCD is $(x + 3)(x - 1)$.

$$= \frac{(2x + 1)(x - 1) + 3(x - 2)(x + 3)}{(x + 3)(x - 1)}$$

$$= \frac{2x^2 - x - 1 + 3x^2 + 3x - 18}{(x + 3)(x - 1)}$$

$$= \frac{5x^2 + 2x - 19}{(x + 3)(x - 1)}$$

Since the numerator will not factor, the answer cannot be reduced.

Example 10 $\left(\dfrac{2}{x + 1} + \dfrac{4x}{x^2 + 4x + 3} \right) \cdot \left(\dfrac{x + 3}{x - 4} \right)$

The parentheses indicate that the addition must be performed before the multiplication.

Solution $= \left(\dfrac{2}{x + 1} + \dfrac{4x}{(x + 1)(x + 3)} \right) \cdot \left(\dfrac{x + 3}{x - 4} \right)$

$$= \frac{2(x + 3) + 4x}{(x + 1)(x + 3)} \cdot \frac{x + 3}{x - 4}$$

$$= \frac{2x + 6 + 4x}{(x + 1)(x + 3)} \cdot \frac{x + 3}{x - 4}$$

$$= \frac{6x + 6}{(x + 1)(x + 3)} \cdot \frac{x + 3}{x - 4}$$

$$= \frac{6\overset{1}{\cancel{(x + 1)}}}{\underset{1}{\cancel{(x + 1)}}\underset{1}{\cancel{(x + 3)}}} \cdot \frac{\overset{1}{\cancel{(x + 3)}}}{(x - 4)}$$

$$= \frac{6}{x - 4}$$

▼ **EXERCISE 5-6-1**

Add.

1. $\dfrac{3}{7} + \dfrac{2}{7}$ $= \dfrac{5}{7}$

2. $\dfrac{2}{11} + \dfrac{5}{11}$ $= \dfrac{7}{11}$

3. $\dfrac{1}{8} + \dfrac{3}{8}$ $\dfrac{4}{8} = \dfrac{1}{2}$

4. $\dfrac{2}{9} + \dfrac{4}{9}$ $= \dfrac{6}{9} = \dfrac{2}{3}$

5. $\dfrac{x+3}{x-4} + \dfrac{4}{x-4}$ $= \dfrac{x+7}{x-4}$

6. $\dfrac{2x}{x+5} + \dfrac{2x-3}{x+5}$ $= \dfrac{4x-3}{x+5}$

7. $\dfrac{2x-3}{x-1} + \dfrac{1}{x-1}$ $= \dfrac{2x-2}{x-1}$

   $\dfrac{2(x-1)}{(x-1)} = 2$

8. $\dfrac{3x+2}{x+3} + \dfrac{7}{x+3}$

9. $\dfrac{1}{x} + \dfrac{1}{y}$ $=$

   $\dfrac{y}{xy} + \dfrac{x}{xy} = \dfrac{y+x}{xy}$

10. $\dfrac{3}{a} + \dfrac{1}{b}$ $= \dfrac{3b}{ab} + \dfrac{1a}{ab}$

    $\dfrac{3b+a}{ab}$

11. $\dfrac{1}{x} + \dfrac{1}{y} + \dfrac{1}{z}$

    $\dfrac{yz + xz + xy}{xyz}$

12. $\dfrac{a}{b} + \dfrac{1}{a} + \dfrac{b}{c}$

13. $\dfrac{1}{3} + \dfrac{2}{x+5}$

14. $\dfrac{2}{5} + \dfrac{1}{x-3}$

15. $\dfrac{2}{x} + \dfrac{1}{x+1}$

16. $\dfrac{3}{x} + \dfrac{x}{x+2}$

17. $\dfrac{1}{x+2} + \dfrac{2}{x^2-x-6}$

18. $\dfrac{2}{x-3} + \dfrac{4}{x^2+5x-24}$

19. $\dfrac{2}{x+1} + \dfrac{4x}{x^2+4x+3}$

20. $\dfrac{3}{2x-1} + \dfrac{x-14}{2x^2+7x-4}$

21. $\dfrac{2}{x+2} + \dfrac{1}{x-5}$

22. $\dfrac{1}{x+7} + \dfrac{3}{x-3}$

23. $\dfrac{5}{x+4} + \dfrac{2x-13}{x^2+x-12} + \dfrac{1}{x-3}$

24. $\dfrac{3}{x^2 - 9} + \dfrac{2}{x^2 + 6x + 9}$

25. $\dfrac{6}{5x + 10} + \dfrac{2x}{x^2 - 3x - 10}$

26. $\dfrac{1}{x + 2} + \dfrac{1}{x + 1} + \dfrac{1}{x - 3}$

27. $\dfrac{2}{x - 2} + \dfrac{3}{x + 1} + \dfrac{x - 8}{x^2 - x - 2}$

$\dfrac{(2x+2)+(3x-6)+(x-8)}{(x-2)(x+1)}$

$\dfrac{6x-12}{(x-2)(x+1)}$

$\dfrac{6(x-2)}{(x-2)(x+1)}$ = $\dfrac{6}{x+1}$

28. $\dfrac{3}{x + 1} + \dfrac{2}{x - 3} + \dfrac{5}{x + 4}$

29. $\dfrac{x + 3}{x - 6} + \dfrac{x - 1}{x^2 - 2x - 24}$

30. $\dfrac{x + 2}{x + 5} + \dfrac{x - 3}{x^2 + 3x - 10}$

31. $\dfrac{x + 3}{x^2 - x - 2} + \dfrac{2x - 1}{x^2 + 2x - 8}$

32. $\dfrac{2x}{x^2 + 6x + 9} + \dfrac{x - 1}{x^2 - 9}$

Sometimes more than one operation may appear in a problem. Recall the order of operations and the meaning of parentheses in the following problems.

33. $\dfrac{3}{x} + \dfrac{5}{y} \cdot \dfrac{4y}{3x}$

34. $\dfrac{1}{a} + \dfrac{2}{3b} \cdot \dfrac{3}{a^2}$

35. $\dfrac{3}{x - 2} + \dfrac{1}{x^2 - x - 2} \div \dfrac{1}{x + 1}$

36. $\dfrac{1}{x + 5} + \dfrac{2}{x^2 + 5x + 6} \div \dfrac{4}{x + 2}$

37. $\left(\dfrac{1}{x-4} + \dfrac{2x}{x^2-3x-4}\right) \cdot \left(\dfrac{x+1}{x-2}\right)$

38. $\left(\dfrac{x}{2x-1} + \dfrac{3x}{2x^2+11x-6}\right) \cdot \left(\dfrac{x+6}{x+1}\right)$

39. $\dfrac{1}{x^2-x-2} \div \left(\dfrac{1}{x+1} + \dfrac{3}{x-2}\right)$

40. $\dfrac{1}{x^2+2x-8} \div \left(\dfrac{2}{x+4} + \dfrac{1}{x+1}\right)$

◄ 5-7 SUBTRACTION OF ALGEBRAIC FRACTIONS

We know that subtraction is defined as "adding the negative," and we would expect the procedure for subtracting algebraic fractions to be the same as for addition. The fact is that the procedure is the same, but as you will see in this section there is enough cause for special caution to treat subtraction separately.

> The difference of any two fractions having the same denominator is the difference of their numerators over their common denominator.

Notice that this rule is the same as the rule for adding two fractions with the same denominator.

The steps for subtracting fractions are, therefore, the same as for adding fractions.

> To subtract fractions:
>
> **Step 1** Find the least common denominator of the two fractions.
> **Step 2** Change each fraction to an equivalent fraction having the least common denominator.
> **Step 3** Find the difference of the numerators and place this result over the least common denominator.
> **Step 4** Simplify (or reduce) the fraction obtained in step 3.

Special care must be taken when performing step 3.

The obvious question is, "If these two operations are the same, why study them separately?" The answer is that the subtraction gives rise to a very common error that the student must be prepared to avoid.

Example 1   Subtract: $\dfrac{3}{x+1} - \dfrac{x-4}{x+1}$

Solution

Notice we are subtracting the entire numerator of the second fraction. It will therefore be a good practice to place the entire numerator in parentheses with the subtraction sign in front of it.

$$\frac{3}{x+1} - \frac{x-4}{x+1} = \frac{3-(x-4)}{x+1}$$

$$= \frac{3-x+4}{x+1}$$

$$= \frac{7-x}{x+1}$$

The error referred to is often made by not recognizing that the minus sign affects the entire numerator of the second fraction and *NOT* just the first term.

This will happen if you don't use parentheses.

Do *not* do this!

$$\frac{3}{x+1} - \frac{x-4}{x+1} = \frac{3-x-4}{x+1}$$

The arrow points out the error most commonly made in subtraction of fractions. The best way to avoid this is to always use parentheses

$$\frac{3}{x+1} - \frac{x-4}{x+1} = \frac{3-(x-4)}{x+1}$$

and you are not so likely to fail to change the sign properly.

Example 2   Subtract: $\dfrac{x+4}{x-2} - \dfrac{2x+3}{x^2-4}$

Solution   The least common denominator is $(x-2)(x+2)$.

Notice we placed parentheses around the numerator of the second fraction.

$$\frac{x+4}{x-2} - \frac{2x+3}{(x-2)(x+2)} = \frac{(x+4)(x+2)-(2x+3)}{(x-2)(x+2)}$$

$$= \frac{x^2+6x+8-2x-3}{(x-2)(x+2)}$$

$$= \frac{x^2+4x+5}{(x-2)(x+2)}$$

Example 3   Subtract: $\dfrac{3x+2}{2x-1} - \dfrac{x-4}{x-3}$

Solution   The common denominator is $(2x-1)(x-3)$.

$$\frac{3x+2}{2x-1} - \frac{x-4}{x-3} = \frac{(3x+2)(x-3)-(x-4)(2x-1)}{(2x-1)(x-3)}$$

Notice we first multiplied $(x-4)$ $(2x-1)$ and then multiplied $(2x^2-9x+4)$ by -1. To multiply and change signs at the same time is to invite error.

$$= \frac{3x^2-7x-6-(2x^2-9x+4)}{(2x-1)(x-3)}$$

$$= \frac{3x^2-7x-6-2x^2+9x-4}{(2x-1)(x-3)}$$

$$= \frac{x^2+2x-10}{(2x-1)(x-3)}$$

▼ EXERCISE 5-7-1

Subtract.

1. $\dfrac{3}{5} - \dfrac{1}{5}$ $= \dfrac{2}{5}$

2. $\dfrac{8}{9} - \dfrac{2}{9}$ $= \dfrac{7}{9}$

3. $\dfrac{5}{6} - \dfrac{3}{4}$ $\quad \dfrac{10}{12} - \dfrac{9}{12} = \dfrac{1}{12}$

4. $\dfrac{4}{15} - \dfrac{1}{6}$

$\dfrac{8}{30} - \dfrac{5}{30} = \dfrac{3}{30} = \dfrac{1}{10}$

5. $\dfrac{3}{x} - \dfrac{1}{3}$

$\dfrac{9-x}{3x}$

6. $\dfrac{3}{5} - \dfrac{4}{x}$ $\quad \dfrac{3x-20}{5x}$

7. $\dfrac{2}{x} - \dfrac{4}{y}$

$\dfrac{2y-4x}{xy}$

8. $\dfrac{3}{a} - \dfrac{2a}{b}$

$\dfrac{3b-2a^2}{ab}$

9. $\dfrac{1}{x+1} - \dfrac{3}{4}$

$\dfrac{4-3x+3}{4(x+1)} = \dfrac{1-3x}{4(x+1)}$

10. $\dfrac{5}{x-3} - \dfrac{1}{2}$

11. $\dfrac{x}{x+1} - \dfrac{2}{3}$

$\dfrac{3x-2x+2}{(x+1)3} = \dfrac{x+2}{3(x+1)}$

12. $\dfrac{5}{8} - \dfrac{x}{x-2}$

$\dfrac{(5x-10)-(8x)}{(8)(x-2)} = \dfrac{-3x-10}{8x-16}$

13. $\dfrac{5}{6} - \dfrac{2x}{3x-12}$

$\dfrac{15x-60-12x}{6(3x-12)} = \dfrac{3x-60}{6(3x-12)}$

14. $\dfrac{x}{2} - \dfrac{3x}{2x-4}$

15. $\dfrac{4}{x} - \dfrac{6}{x+3}$

$\dfrac{(4x+12)-6x}{x(x+3)} \quad \dfrac{-2x+12}{x(x+3)}$

or $\dfrac{12-2x}{x(x+3)}$

16. $\dfrac{1}{2x-1} - \dfrac{2}{x}$

17. $\dfrac{1}{x+2} - \dfrac{1}{x+3}$

$\dfrac{x+3-x+2}{(x+2)(x+3)} =$

18. $\dfrac{2x}{x-3} - \dfrac{4}{x+5}$

19. $\dfrac{x+3}{x+4} - \dfrac{5}{x+2}$

$\dfrac{(x+3)(x+2)-(x+4)(5)}{(x+4)(x+2)} = \dfrac{x^2+5x+6-5x+9}{(x+4)(x+2)}$

$= \dfrac{x^2+10}{(x+4)(x+2)}$

20. $\dfrac{2x-1}{x-1} - \dfrac{x-4}{x+3}$

21. $\dfrac{3x-2}{2x+1} - \dfrac{x-1}{x-2}$ $\quad (x-1)(2x+1)$

$\dfrac{(3x-2)(x-2)-(x-1)}{(2x+1)(x-2)}\Big/(2x+1)$

$\dfrac{3x^2-8x+4-2x^2+x+1}{} = \dfrac{x^2-7x+5}{(2x+1)(x-2)}$

22. $\dfrac{x+5}{x-2} - \dfrac{x}{3x+2}$

23. $\dfrac{x}{x+1} - \dfrac{4}{x^2-1}$

$\dfrac{(x)(x^2-1)-(4)(x+1)}{(x+1)(x^2-1)} \quad \dfrac{(x^3-x)-4x-4}{}$

$\dfrac{x^3-5x-4}{(x+1)(x^2-1)}$

24. $\dfrac{6}{x+2} - \dfrac{x}{x^2-x-6}$

25. $\dfrac{2}{x-3} - \dfrac{x+7}{x^2-x-6}$

26. $\dfrac{2x}{x-1} - \dfrac{3x+2}{x^2+3x-4}$

27. $\dfrac{5}{x+3} - \dfrac{4x-13}{x^2+x-6}$

28. $\dfrac{x+2}{x-5} - \dfrac{5x+31}{x^2-2x-15}$

29. $\dfrac{x+1}{x^2-x-6} - \dfrac{x+2}{2x-6}$

30. $\dfrac{x+1}{3x^2-12} - \dfrac{x-3}{x+2}$

31. $\dfrac{x+4}{x^2-x-2} - \dfrac{x}{x^2+x-6}$

32. $\dfrac{2x}{x^2-1} - \dfrac{2x+4}{x^2+4x-5}$

Perform the indicated operations. Recall the order of operations.

33. $\dfrac{2}{x+5} + \dfrac{1}{x-4} - \dfrac{3}{x+3}$

$\dfrac{2x-8+x+5}{(x+5)+(x-4)} = \dfrac{x-3}{(x+5)+(x-4)} - \dfrac{3}{x+3}$

34. $\dfrac{x}{x^2-4x-5} - \dfrac{2}{x+1} + \dfrac{2x-1}{x^2-7x+10}$

35. $\dfrac{1}{x^2-x-6} \div \dfrac{1}{x+2} - \dfrac{1}{x-3}$

36. $\dfrac{2x}{x^2-9} \cdot \dfrac{2}{x+3} - \dfrac{1}{x-3}$

37. $\dfrac{1}{x^2-x-6} \div \left(\dfrac{1}{x+2} - \dfrac{1}{x-3} \right)$

38. $\dfrac{2x}{x^2-9} \cdot \left(\dfrac{2}{x+3} - \dfrac{1}{x-3} \right)$

39. $\dfrac{2}{x-1} \cdot \dfrac{3x-3}{x+5} + \dfrac{x+1}{x^2+4x-5} \div \dfrac{2}{x-1} - \dfrac{1}{2}$

40. $\dfrac{x-1}{x^2-2x-8} + \dfrac{3}{x+2} \cdot \dfrac{x+1}{x-4} - \dfrac{2}{x+3} \div \dfrac{1}{x^2-x-12}$

5-8 COMPLEX FRACTIONS

Not all algebraic fractions are composed of one polynomial over another. **Complex fractions** are those in which the numerator and denominator (or both) contains a fraction. We now wish to use procedures already learned to simplify complex fractions.

Examples of complex fractions are

$$\frac{\dfrac{a}{b}}{\dfrac{c}{d}}, \quad \frac{\dfrac{x}{y} + 1}{x}, \quad \frac{\dfrac{1}{x} + \dfrac{1}{y}}{\dfrac{a}{b} - \dfrac{c}{d}}.$$

We will discuss two methods of simplifying complex fractions. The student may prefer one method over the other and may be able to choose the easier method for a particular problem.

OBJECTIVES
Upon completing this section you should be able to:
1. Recognize a complex fraction.
2. Simplify a complex fraction.

Method 1 To simplify a complex fraction treat the fraction as a division problem.

Remember, $\dfrac{a}{b}$ means $a \div b$.

Example 1 Simplify: $\dfrac{\dfrac{1}{x} + \dfrac{1}{y}}{\dfrac{1}{x} - \dfrac{1}{y}}$

Solution We first write the problem as

$$\left(\frac{1}{x} + \frac{1}{y}\right) \div \left(\frac{1}{x} - \frac{1}{y}\right).$$

Numerator is divided by denominator.

Because of the parentheses, we have three problems to solve. First we must add the fractions in the first set of parentheses, then subtract the fractions in the second set of parentheses, and finally divide.

$$\left(\frac{1}{x} + \frac{1}{y}\right) \div \left(\frac{1}{x} - \frac{1}{y}\right) = \frac{y + x}{xy} \div \frac{y - x}{xy}$$

$$= \frac{y + x}{xy} \cdot \frac{xy}{y - x}$$

$$= \frac{y + x}{y - x}$$

We perform the operations in the parentheses first.

This solution is similar to problems in the previous section that involved more than one operation.

This method uses the fundamental principle of fractions.

$$\frac{a}{b} = \frac{ax}{bx}$$

A second method, which will be demonstrated several times, involves changing the complex fraction to a simple fraction and then reducing if possible. We will use the following rule.

Method 2   To simplify a complex fraction:

Step 1 Find the LCD (least common denominator) of all fractions in the expression.

Step 2 Multiply the numerator and denominator of the complex fraction by the common denominator.

Step 3 Simplify (reduce) the resulting fraction.

This is the same problem as in example 1. We will solve it here using rule 2.

Example 2   Simplify: $\dfrac{\dfrac{1}{x} + \dfrac{1}{y}}{\dfrac{1}{x} - \dfrac{1}{y}}$

Solution   **Step 1** The least common denominator of all the fractions is xy.

Step 2
$$\frac{\dfrac{1}{x} + \dfrac{1}{y}}{\dfrac{1}{x} - \dfrac{1}{y}} = \frac{xy\left(\dfrac{1}{x} + \dfrac{1}{y}\right)}{xy\left(\dfrac{1}{x} - \dfrac{1}{y}\right)}$$

Don't forget to multiply each term inside the parentheses by xy.

$$= \frac{xy\left(\dfrac{1}{x}\right) + xy\left(\dfrac{1}{y}\right)}{xy\left(\dfrac{1}{x}\right) - xy\left(\dfrac{1}{y}\right)}$$

$$= \frac{y + x}{y - x}$$

Step 3 The answer is simplified.

Example 3   Simplify: $\dfrac{\dfrac{1}{a^2} - \dfrac{1}{b^2}}{a + b}$

Solution   **Step 1** The least common denominator is a^2b^2.

Step 2
$$\frac{\dfrac{1}{a^2} - \dfrac{1}{b^2}}{a + b} = \frac{a^2b^2\left(\dfrac{1}{a^2} - \dfrac{1}{b^2}\right)}{a^2b^2(a + b)}$$

$$= \frac{b^2 - a^2}{a^2b^2(a + b)}$$

Also try solving this as $\left(\dfrac{1}{a^2} - \dfrac{1}{b^2}\right) \div (a + b)$.

Step 3
$$= \frac{(b - a)(b + a)}{a^2b^2(a + b)}$$

$$= \frac{b - a}{a^2b^2}$$

Example 4 Simplify: $\dfrac{\dfrac{3}{x^2 + 6x + 8} + \dfrac{1}{x + 2}}{\dfrac{2}{x + 4} - 1}$

Solution **Step 1** After factoring the first denominator $x^2 + 6x + 8$ into $(x + 2)(x + 4)$, we find that the LCD is $(x + 2)(x + 4)$.

Step 2 $\dfrac{\dfrac{3}{(x + 2)(x + 4)} + \dfrac{1}{x + 2}}{\dfrac{2}{x + 4} - 1}$

$= \dfrac{(x + 2)(x + 4)\left(\dfrac{3}{(x + 2)(x + 4)} + \dfrac{1}{x + 2}\right)}{(x + 2)(x + 4)\left(\dfrac{2}{x + 4} - 1\right)}$

Again, be very careful to multiply each term inside the parentheses by the LCD.

$= \dfrac{3 + (x + 4)}{2(x + 2) - (x + 2)(x + 4)}$

$= \dfrac{3 + x + 4}{2x + 4 - x^2 - 6x - 8}$

$= \dfrac{x + 7}{-x^2 - 4x - 4}$

Step 3 The answer is in simplest form but could be changed to

$$-\dfrac{x + 7}{x^2 + 4x + 4}.$$

We factored -1 from the denominator.

▼ **EXERCISE 5-8-1**

Simplify.

1. $\dfrac{\dfrac{1}{a}}{\dfrac{1}{a^2}}$ $\dfrac{a^2\left(\dfrac{1}{a}\right)}{a^2\left(\dfrac{1}{a^2}\right)}$ $\dfrac{\dfrac{a^2}{a}}{1}$

$= a$

2. $\dfrac{\dfrac{1}{a}}{\dfrac{1}{b}}$

3. $\dfrac{\dfrac{1}{x}}{\dfrac{1}{y}} \cdot \dfrac{\dfrac{1}{y}}{\dfrac{1}{y}} = \dfrac{\dfrac{1}{xy}}{1} = \dfrac{1}{xy}$

4. $\dfrac{\dfrac{1}{x} + \dfrac{1}{y}}{xy}$

5. $\dfrac{\dfrac{1}{a} + \dfrac{1}{b}}{\dfrac{1}{b} - \dfrac{1}{a}} = \dfrac{ab\left(\dfrac{1}{a}\right) + ab\left(\dfrac{1}{b}\right)}{ab\left(\dfrac{1}{b}\right) - ab\left(\dfrac{1}{a}\right)}$

$\dfrac{b + a}{a - b}$

6. $\dfrac{\dfrac{1}{x} - \dfrac{1}{y}}{\dfrac{1}{x} + \dfrac{1}{y}}$

7. $\dfrac{\dfrac{1}{a} + \dfrac{1}{b}}{a + b} = \dfrac{\dfrac{a+b}{ab}}{\dfrac{a+b}{1}}$

$\dfrac{a+b}{ab} \cdot \dfrac{1}{a+b} = \dfrac{1}{ab}$

8. $\dfrac{\dfrac{1}{x} + \dfrac{1}{y}}{\dfrac{1}{y} + \dfrac{1}{z}}$

9. $\dfrac{\dfrac{3}{a + 2}}{\dfrac{1}{a + 2}}$

$\dfrac{a+2\left(\dfrac{3}{a+2}\right)}{a+2\left(\dfrac{1}{a+2}\right)}$

$= \dfrac{3}{1} = 3$

10. $\dfrac{1 - \dfrac{1}{x - 1}}{\dfrac{2}{x - 1}}$

11. $\dfrac{1 + \dfrac{1}{a}}{ab} =$

$\dfrac{}{1}$

12. $\dfrac{\dfrac{1}{a} + \dfrac{1}{b}}{a^2 - b^2}$

13. $\dfrac{2 - \dfrac{1}{x}}{2x - 1}$

14. $\dfrac{\dfrac{1}{3} + \dfrac{1}{a}}{\dfrac{1}{b}}$

15. $\dfrac{x}{\dfrac{1}{x} + \dfrac{1}{x + 1}}$

16. $\dfrac{\dfrac{a}{a + 1}}{1 - \dfrac{1}{a + 1}}$

17. $\dfrac{\dfrac{a}{b} - \dfrac{b}{a}}{a + b}$

18. $\dfrac{\dfrac{1}{x + 2} - \dfrac{1}{x}}{2}$

19. $\dfrac{\dfrac{1}{x^2} - \dfrac{1}{y^2}}{\dfrac{1}{x^2 y^2}}$

20. $\dfrac{\dfrac{1}{x^2 - 1} + 1}{\dfrac{1}{x + 1} - \dfrac{1}{x - 1}}$

21. $\dfrac{\dfrac{1}{x - 3} + \dfrac{1}{x^2 - 2x - 3}}{\dfrac{1}{x + 1}}$

22. $\dfrac{\dfrac{1}{a+b} - \dfrac{1}{a-b}}{\dfrac{2b}{a^2-b^2}}$

23. $\dfrac{\dfrac{1}{a} + \dfrac{1}{a+b}}{\dfrac{2}{a+b} + 1}$

24. $\dfrac{2 + \dfrac{1}{x-2}}{\dfrac{1}{x+3} - 1}$

25. $\dfrac{\dfrac{1}{x+y} - \dfrac{1}{y}}{\dfrac{1}{x^2-y^2}}$

26. $\dfrac{\dfrac{1}{2x} - \dfrac{1}{6y}}{\dfrac{1}{3y} + \dfrac{1}{4x}}$

27. $\dfrac{\dfrac{6}{x^2+3x-10} - \dfrac{1}{x-2}}{\dfrac{1}{x-2} + 1}$

28. $\dfrac{\dfrac{1}{x+3} + \dfrac{6}{x^2-x-12}}{1 - \dfrac{1}{x+3}}$

29. $\dfrac{\dfrac{1}{x^2+4x+3} + \dfrac{1}{x-2}}{\dfrac{1}{x+3} + \dfrac{1}{x^2-x-2}}$

30. $\dfrac{\dfrac{1}{x+1} - \dfrac{1}{x^2+5x+4}}{\dfrac{1}{x^2-2x-3} + \dfrac{1}{x-3}}$

◣◢ 5-9 EQUATIONS HAVING ALGEBRAIC FRACTIONS

In chapter 2 we encountered equations that have fractions. However, those fractions all had numerical denominators. Now we will discuss equations that have fractions involving variables in the denominators.

The method of solving these equations will follow the same pattern as in chapter 2, but there are some additional cautions that you must be prepared to take.

To refresh your memory the steps for solving such equations are repeated here.

Step 1 Eliminate fractions by multiplying each term of the equation by the least common denominator of all fractions in the equation.

Step 2 Simplify by combining like terms on each side of the equation.

Step 3 Add or subtract the necessary quantities to get the unknown quantity on one side and the numbers of arithmetic on the other.

Step 4 Divide by the coefficient of the unknown quantity.

Step 5 Check your answer.

OBJECTIVES

Upon completing this section you should be able to:

1. Apply the method of solving fractional equations.
2. Determine when a fractional equation does not have a solution.

You may wish to look back at some examples in chapter 2 to refresh your memory.

The main difference in solving equations with arithmetic fractions and those with algebraic fractions comes in step 5. The checking process will not just be to find a possible error, but will also be to determine if the equation has an answer.

Remember, we may only multiply each side of an equation by a *nonzero* quantity.

This last possibility arises because with algebraic fractions we multiply by an unknown quantity. This unknown quantity could actually be zero, which would make all the work invalid. Therefore, checking is necessary.

Example 1 Solve for x: $\dfrac{2}{x^2 - 1} - \dfrac{1}{x + 1} = \dfrac{1}{x - 1}$

Solution **Step 1** The least common denominator is $(x - 1)(x + 1)$. So we multiply all terms by $(x - 1)(x + 1)$.

This means that neither $(x - 1)$ nor $(x + 1)$ can be zero.

$$\frac{(x - 1)(x + 1)(2)}{(x - 1)(x + 1)} - \frac{(x - 1)(x + 1)(1)}{(x + 1)} = \frac{(x - 1)(x + 1)1}{(x - 1)}$$

$$2 - (x - 1) = x + 1$$
$$2 - x + 1 = x + 1$$

Step 2 Simplify, getting $3 - x = x + 1$.

Step 3 Getting the unknown to one side and the numbers to the other gives $-2x = -2$.

Step 4 Divide both sides by (-2) and get $x = 1$.

Step 5 Now check to see if $x = 1$ makes the original equation true.

$$\frac{2}{x^2 - 1} - \frac{1}{x + 1} = \frac{1}{x - 1}$$

$$\frac{2}{(1)^2 - 1} - \frac{1}{(1) + 1} = \frac{1}{(1) - 1}$$

$\dfrac{2}{0}$ and $\dfrac{1}{0}$ are obtained.

In other words, $x = 1$ is not a solution, since it gives a statement with no meaning.

Since division by zero is not possible, we must conclude that $x = 1$ is not a solution. And since we made no error in the computations we must conclude that this equation has no solution.

The correct answer is "no solution."

Example 2 Solve for x: $\dfrac{2x}{x + 1} + \dfrac{1}{3x - 2} = 2$

Solution The least common denominator is $(x + 1)(3x - 2)$, so we multiply every term by $(x + 1)(3x - 2)$.

Don't forget, the 2 must also be multiplied by $(x + 1)(3x - 2)$ even though it is not a fraction.

$$\frac{(x + 1)(3x - 2)2x}{x + 1} + \frac{(x + 1)(3x - 2)(1)}{3x - 2} = (x + 1)(3x - 2)(2)$$

$$(3x - 2)(2x) + (x + 1)(1) = (x + 1)(3x - 2)(2)$$

$$6x^2 - 4x + x + 1 = 6x^2 + 2x - 4$$

$$6x^2 - 3x + 1 = 6x^2 + 2x - 4$$

$$-5x = -5$$

$$x = 1$$

Check: $\dfrac{2x}{x+1} + \dfrac{1}{3x-2} = 2$

$$\dfrac{2(1)}{(1)+1} + \dfrac{1}{3(1)-2} = 2$$

$$\dfrac{2}{2} + \dfrac{1}{1} = 2$$

$$1 + 1 = 2$$

$$2 = 2$$

Remember, the check is an extremely important step as it will determine whether there is a solution or not.

Thus, $x = 1$ is a solution.

Example 3 Solve for x: $\dfrac{2}{x} + \dfrac{x}{x+1} = 1$

Solution The least common denominator is $x(x+1)$.

$$(2)(x+1) + (x)(x) = x(x+1)(1)$$
$$2x + 2 + x^2 = x^2 + x$$
$$x = -2$$

Notice in these examples that when we have x^2 terms, they cancel out and we are left with a linear equation. If they did not cancel out, we would have an x^2 term in the equation. This type of equation (quadratic) will be dealt with in chapter 7.

Check: $\dfrac{2}{(-2)} + \dfrac{(-2)}{(-2)+1} = 1$

$$-1 + \dfrac{(-2)}{(-1)} = 1$$

$$-1 + 2 = 1$$

$$1 = 1$$

Thus $x = -2$ is a solution.

Example 4 The numerator of the fraction $\dfrac{13}{27}$ is increased by an amount so that the value of the resulting fraction is $\dfrac{8}{9}$. By what amount was the numerator increased?

Solution If we let x represent the number added to the numerator then we have

$$\dfrac{13+x}{27} = \dfrac{8}{9}.$$

Is $\dfrac{13+x}{27}$ the same as $\dfrac{13}{27} + x$? Why?

Solving for x, we use the least common denominator, 27, to multiply each term.

$$\dfrac{\cancel{27}(13+x)}{\cancel{27}} = \dfrac{\overset{3}{\cancel{27}}(8)}{\cancel{9}}$$

$$13 + x = 24$$

$$x = 11$$

$$Check: \quad \frac{13 + x}{27} = \frac{8}{9}$$

$$\frac{13 + 11}{27} = \frac{8}{9}$$

$$\frac{24}{27} = \frac{8}{9}$$

$$\frac{8}{9} = \frac{8}{9}$$

Therefore, 11 is the amount by which the numerator was increased.

A class of problems using algebraic fractions is **work problems.**

Example 5   A painter can do a certain job alone in 21 hours. His helper can do the same job alone in 28 hours. If both work together, how long would it take to complete the job?

Solution   The key to this class of problems is the amount of work done per unit of time. In this problem it is work per hour. We set up the following table.

	Time Alone	Fractional Part Done in 1 Hour
Painter	21	$\frac{1}{21}$
Helper	28	$\frac{1}{28}$
Together	x	$\frac{1}{x}$

The equation is obtained from the work done in one unit of time. In this problem the equation is

$$\begin{pmatrix} \text{Amount of work} \\ \text{done by painter} \\ \text{in 1 hour} \end{pmatrix} + \begin{pmatrix} \text{Amount of work} \\ \text{done by helper} \\ \text{in 1 hour} \end{pmatrix} = \begin{pmatrix} \text{Amount of work} \\ \text{done together} \\ \text{in 1 hour} \end{pmatrix}$$

$$\frac{1}{21} + \frac{1}{28} = \frac{1}{x}$$

The LCD is $84x$.

$$(84x)\left(\frac{1}{21}\right) + (84x)\left(\frac{1}{28}\right) = (84x)\left(\frac{1}{x}\right)$$

$$4x + 3x = 84$$

$$7x = 84$$

Check this answer.

$$x = 12$$

Together they would complete the job in 12 hours.

▼ **EXERCISE 5–9–1**

Solve.

1. $\dfrac{2}{x} + \dfrac{3}{2} = 2$

2. $\dfrac{5}{x} + \dfrac{1}{2} = 1$

3. $\dfrac{1}{x} + \dfrac{1}{2x} = 3$

$\left[\dfrac{(2x)(x)(1)+(2x)(x)}{x \qquad 2x}\right] =$

$(3)(2x)(x) \qquad (2x)+(x) = 6x^2$

$\boxed{x = \dfrac{1}{2}} \qquad 3x = 6x^2$

$x = \dfrac{6x^2}{3x} = 2x$

4. $\dfrac{1}{3x} + \dfrac{1}{x} = 2$

5. $\dfrac{2}{x} + \dfrac{1}{3} = \dfrac{1}{2x} - 1$

6. $\dfrac{x+1}{x} - \dfrac{x-2}{2x} = 1$

7. $\dfrac{1}{x} = \dfrac{3}{2x}$

8. $\dfrac{4}{x} + \dfrac{1}{2x} = \dfrac{9}{4}$

9. $\dfrac{3}{x-1} = 1$

10. $\dfrac{x}{x-2} = \dfrac{4}{3}$

11. $\dfrac{2}{x} = \dfrac{1}{x+1}$

12. $\dfrac{3}{x-5} = \dfrac{4}{x}$

13. $\dfrac{x}{x+1} = \dfrac{x-1}{x-2}$

14. $\dfrac{x+2}{x+3} = \dfrac{x}{x-4}$

15. $\dfrac{1}{x+1} = \dfrac{2}{1-x^2}$

16. $\dfrac{2}{x-5} = \dfrac{4}{x^2-25}$

17. $\dfrac{x+2}{x^2+7x} = \dfrac{1}{x+3}$

18. $\dfrac{1}{x^2-3x} = \dfrac{2}{x^2-9}$

19. $\dfrac{x}{x+1} + \dfrac{1}{2x+1} = 1$

20. $\dfrac{3}{x+5} = 1 - \dfrac{x-4}{2x+10}$

21. $\dfrac{1}{2} + \dfrac{2}{x-1} = \dfrac{x}{2x-4}$

22. $\dfrac{2}{x+4} + \dfrac{1}{3} = \dfrac{x}{3x-9}$

23. $\dfrac{x}{x+4} - \dfrac{x}{x-4} = \dfrac{x+18}{x^2-16}$

24. $\dfrac{2}{x-1} + \dfrac{x}{x+1} = 1 - \dfrac{1}{x^2-1}$

25. $\dfrac{2}{3x + 6} = \dfrac{1}{6} - \dfrac{1}{2x + 4}$

26. $\dfrac{x + 1}{x + 5} - \dfrac{2x + 1}{x - 2} = \dfrac{5 - x^2}{x^2 + 3x - 10}$

27. Three divided by a given number is equal to four divided by three more than the given number. Find the number.

28. In a fraction the numerator is three less than the denominator. If one is added to both the numerator and denominator, the value of the resulting fraction is $\dfrac{5}{6}$. Find the original fraction.

29. The numerator of a fraction is twelve less than the denominator. If the numerator is increased by three and the denominator is increased by seven, the resulting fraction has the same value as the original fraction. Find the original fraction.

30. The sum of the reciprocals of two consecutive even integers is equal to seven divided by four times the first number. Find the numbers.

31. A camera utilizes the formula from optics $\dfrac{1}{f} = \dfrac{1}{d_o} + \dfrac{1}{d_i}$, where f represents the focal length of the lens, d_o represents the distance of the object from the lens, and d_i represents the distance of the image from the lens. If a lens has a focal length of 12 centimeters, how far from the lens will the image appear when the object is 36 centimeters from the lens?

32. If an electric circuit is wired in parallel with two resistors, the total resistance R (measured in ohms) is given by the formula $\dfrac{1}{R} = \dfrac{1}{R_1} + \dfrac{1}{R_2}$, where R_1 and R_2 represent the resistances of the two resistors. If $R = 300$ ohms and $R_2 = 750$ ohms, find R_1.

33. One worker can complete a job in 36 hours. A second worker can do the same job in 45 hours. How long will it take them to complete the job if they work together?

34. One pipe can fill a swimming pool in 10 hours. If a second pipe is used together with the first pipe, the pool can be filled in 6 hours. How long would it take the second pipe alone to fill the pool?

Key Words

Section 5–1

- An **algebraic fraction** is the indicated ratio of two algebraic expressions.

- A fraction is in **simplified form** if the numerator and denominator have no common factor other than 1.

Section 5–4

- A **common denominator** for two or more fractions is an expression that contains all factors of the denominators of each fraction.

- A **least common denominator** contains the minimum number of factors to be a common denominator.

Section 5–5

- The **fundamental principle of fractions** is

$$\frac{x}{y} = \frac{ax}{ay}, \text{ when } a \neq 0.$$

Section 5–8

- **Complex fractions** are those in which the numerator or denominator (or both) contains a fraction.

Procedures

Section 5–1

- To simplify or reduce fractions to lowest terms factor the numerator and denominator and divide by all like factors.

Section 5–2

- To multiply fractions factor all numerators and denominators and divide by all like factors before multiplying.

Section 5–3

- To divide by a fraction invert the divisor and then multiply.

Section 5–4

- To find a least common denominator (LCD) first factor all denominators, then find a denominator that contains all factors of each denominator but does not contain any unnecessary factors.

Section 5–5

- To change a fraction to an equivalent fraction multiply numerator and denominator by the same nonzero expression.

Section 5–6

- To add fractions follow these steps:
  1. Find the least common denominator.
  2. Change each fraction to an equivalent fraction having the LCD as its denominator.
  3. Add the numerators and place over the LCD.
  4. Simplify or reduce the answer.

Section 5–7

- To subtract fractions proceed as in addition but combine the numerators by subtracting.

Section 5–8

- Complex fractions may be simplified by using either of the following two methods:
  1. Simplify the numerator and denominator to single fractions and then divide the numerator by the denominator.
  2. Multiply numerator and denominator of the complex fraction by the LCD of all fractions in the expression.

Section 5–9

- To solve equations having fractions first eliminate all fractions by multiplying the entire equation by the LCD of the fractions involved. The resulting equation is then solved, and the solution must be checked in the original equation.

C H A P T E R
5   **REVIEW**

A N S W E R S

Simplify.

1. _____

2. _____

3. _____

4. _____

5. _____

6. _____

7. _____

8. _____

9. _____

10. _____

11. _____

12. _____

13. _____

14. _____

15. _____

16. _____

1. $\dfrac{3x + 45}{12}$

2. $\dfrac{5x^2 - 10x}{15x}$

3. $\dfrac{2x + 8}{3x + 12}$

4. $\dfrac{3x - 9}{4x - 12}$

5. $\dfrac{x^2 - 2x - 15}{x^2 + 3x - 40}$

6. $\dfrac{x^2 + 7x + 12}{x^2 - 4x - 32}$

7. $\dfrac{x^2 - 49}{x^2 + 14x + 49}$

8. $\dfrac{4x^2 - 9}{4x^2 + 12x + 9}$

9. $\dfrac{6x + 30}{10x^2 + 40x - 50}$

10. $\dfrac{6x^2 - 50x + 56}{24x^2 - 44x + 16}$

Find the least common denominator.

11. $\dfrac{3}{x}, \dfrac{2}{x + 2}$

12. $\dfrac{1}{x + 4}, \dfrac{3}{x - 5}$

13. $\dfrac{4}{x^2 + 3x}, \dfrac{x}{x + 3}$

14. $\dfrac{2}{x^2 + 2x - 24}, \dfrac{5}{x - 4}$

15. $\dfrac{3}{x^2 - 5x}, \dfrac{x - 1}{x^2 - 4x - 5}$

16. $\dfrac{6}{2x^2 - 11x + 12}, \dfrac{1}{6x - 9}$

17. $\dfrac{4}{x+1}$, $\dfrac{1}{x^2-4}$, $\dfrac{5}{x+2}$

18. $\dfrac{5}{3x-4}$, $\dfrac{x-3}{9x^2-16}$, $\dfrac{3x}{3x+4}$

17. _____

19. $\dfrac{2x}{x^2-1}$, $\dfrac{3}{x^2-x-2}$

20. $\dfrac{2x+1}{x^2-x-30}$, $\dfrac{2x}{x^2-11x+30}$

18. _____

19. _____

Find the missing numerator.

20. _____

21. $\dfrac{2}{x-4}=\dfrac{?}{(x-4)(x+10)}$

22. $\dfrac{4}{2x-3}=\dfrac{?}{4x^2+4x-15}$

21. _____

23. $\dfrac{x+3}{2x-1}=\dfrac{?}{(2x-1)(x+7)}$

24. $\dfrac{x-3}{x+4}=\dfrac{?}{3x^2+10x-8}$

22. _____

23. _____

25. $\dfrac{4x-1}{3x+2}=\dfrac{?}{6x^2+4x}$

26. $\dfrac{3x+4}{2x-1}=\dfrac{?}{4x^2-1}$

24. _____

25. _____

27. $\dfrac{x-7}{3x}=\dfrac{?}{6x^2+9x}$

28. $\dfrac{2x-5}{x^2-x}=\dfrac{?}{x(x-5)(x-1)}$

26. _____

27. _____

29. $\dfrac{x+8}{4x^2+13x-12}=\dfrac{?}{(4x-3)(x+4)(2x+3)}$

28. _____

29. _____

30. $\dfrac{2x-3}{10x+23x-5}=\dfrac{?}{(5x-1)(2x-3)(2x+5)}$

30. _____

Perform the indicated operation.

31. _____

31. $\dfrac{3x}{4x - 12} \cdot \dfrac{2x - 6}{x^2 + x}$ 　　　**32.** $\dfrac{2x - 6}{3x} \div \dfrac{x^2 - 2x - 3}{x^2 + x}$

32. _____

33. $\dfrac{2}{x - 5} + \dfrac{4}{x}$ 　　　**34.** $\dfrac{4}{5} - \dfrac{x}{x + 2}$

33. _____

34. _____

35. $\dfrac{3x + 6}{x} \cdot \dfrac{5x}{21x + 42}$ 　　　**36.** $\dfrac{x + 1}{x^2 - 5x - 6} \div \dfrac{x^2 - 1}{x - 6}$

35. _____

36. _____

37. $\dfrac{2x}{2x^2 - 2} \div \dfrac{x^2 + 3x}{x^2 + 2x - 3}$ 　　　**38.** $\dfrac{2x + 1}{2x - 16} \cdot \dfrac{4x + 12}{4x + 2}$

37. _____

38. _____

39. $\dfrac{3}{x} + \dfrac{2}{x + 2}$ 　　　**40.** $\dfrac{3x}{x^2 - x - 12} + \dfrac{5}{x + 3}$

39. _____

40. _____

41. $\dfrac{x + 4}{x^2 - 6x + 8} \div \dfrac{x^2 + 12x + 32}{x^2 + 6x - 16}$

41. _____

42. $\dfrac{x^2 - 25}{x^2 + 7x + 10} \div \dfrac{x^2 - 9x + 20}{x^2 + x - 2}$

42. _____

43. _____

43. $\dfrac{3}{4} - \dfrac{x}{x + 5}$ 　　　**44.** $\dfrac{4}{x^2 + 3x} + \dfrac{x}{x + 3}$

44. _____

45. $\dfrac{x^2 + 9x + 18}{x^2 - 6x - 27} \cdot \dfrac{x^2 - 11x + 18}{x^2 + 8x + 12}$

45. _____

46. $\dfrac{x}{x^2 + 4x} + \dfrac{2}{x^2 - 3x}$

46. _____

47. $\dfrac{x^2 + 7x + 10}{x^2 + 4x - 5} \div \dfrac{x^2 + 7x + 12}{x^2 + 2x - 3}$

47. _____

48. _____

48. $\dfrac{3}{x + 4} - \dfrac{1}{x - 3}$

49. _____

49. $\dfrac{3}{x^2 - 5x} + \dfrac{x - 1}{x^2 - 4x - 5}$ **50.** $\dfrac{4x^2 - 4x - 3}{2x^2 + 5x - 12} \cdot \dfrac{2x^2 + 9x + 4}{6x^2 + 5x + 1}$

50. _____

51. $\dfrac{2x + 3}{x^2 + 3x + 2} - \dfrac{2}{x - 7}$ **52.** $\dfrac{x^2 - 9}{x^2 + 6x + 9} \div \dfrac{x^2 - 6x + 9}{x^2 - 8x - 33}$

51. _____

52. _____

53. $\dfrac{x^3 - 64}{2x^2 - 7x - 4} \cdot \dfrac{2x^2 + 7x + 6}{x^2 + 4x + 16}$ **54.** $\dfrac{4}{x^2 - 6x} - \dfrac{11}{x^2 - 3x - 18}$

53. _____

55. $\dfrac{9x^2 + 18x + 8}{6x^2 + 19x + 10} \div \dfrac{12x^2 + 13x - 4}{8x^2 + 10x - 3}$

54. _____

55. _____

56. $\dfrac{x^2 - 16}{2x^2 - 7x - 4} \cdot \dfrac{2x^2 + 7x + 6}{2x^2 + 11x + 12}$

56. _____

57. $\dfrac{5}{x-2} + \dfrac{3}{x^2+2x-8} + \dfrac{1}{x+3}$

57. _____

58. $\dfrac{2x}{x^2+5x+4} - \dfrac{x+3}{x^2-2x-3}$

58. _____

59. $\dfrac{x+6}{3x^2-12x+12} \div \dfrac{x+2}{3x^2-12}$ **60.** $\dfrac{3}{x+1} - \dfrac{2}{x-2}$

59. _____

61. $\dfrac{5}{x-1} - \dfrac{2x+6}{x^2+2x-3}$ **62.** $\dfrac{4}{x+1} + \dfrac{1}{x^2-4} + \dfrac{5}{x+2}$

60. _____

63. $\dfrac{3x^2-13x+12}{3x^2-8x-3} \cdot \dfrac{x^2+6x+9}{3x^2-13x+12}$

61. _____

62. _____

64. $\dfrac{x^2+10x+24}{2x^2-6x+18} \div \dfrac{x^2+9x+18}{2x^3+54}$

63. _____

65. $\dfrac{4x}{2x^2+x-6} - \dfrac{2x-1}{2x^2+3x-9}$

64. _____

66. $\dfrac{x^2-16}{x^2+5x+4} \div \dfrac{x^2-8x+16}{x^2-1}$

65. _____

66. _____

67. $\dfrac{x^2+3x+2}{x^2+5x+6} \cdot \dfrac{x^2-2x-15}{x^2-3x-10} \cdot \dfrac{2x^2-x-10}{x^2+5x+4}$

67. _____

68. $\dfrac{3x}{x^2 - 9} + \dfrac{4}{x^2 + 12x + 27}$

68. _____

69. $\dfrac{2x}{x^2 - 1} + \dfrac{3}{x^2 - x - 2}$

69. _____

70. $\dfrac{2x}{x^2 - 4} - \dfrac{x + 4}{x^2 + 4x - 12}$

70. _____

71. _____

Simplify.

71. $\dfrac{\dfrac{1}{a}}{\dfrac{3}{a}}$

72. $\dfrac{\dfrac{1}{a} + \dfrac{1}{b}}{a + b}$

73. $\dfrac{\dfrac{1}{2} - \dfrac{1}{x}}{\dfrac{1}{x}}$

72. _____

73. _____

74. _____

74. $\dfrac{\dfrac{2}{x} - 1}{\dfrac{1}{x} - 2}$

75. $\dfrac{1 - \dfrac{1}{a}}{a - 1}$

76. $\dfrac{\dfrac{1}{x} + \dfrac{1}{y}}{\dfrac{1}{x + y}}$

75. _____

76. _____

77. _____

77. $\dfrac{\dfrac{x}{x + 2} + 5}{\dfrac{2}{x + 2}}$

78. $\dfrac{\dfrac{4}{5} + \dfrac{3}{a - 3}}{\dfrac{1}{5} + \dfrac{1}{a - 3}}$

79. $\dfrac{\dfrac{1}{x + 3} - 1}{\dfrac{1}{x^2 - 9}}$

78. _____

79. _____

80. $\dfrac{\dfrac{-13}{x^2 - 2x - 35} - \dfrac{1}{x + 5}}{\dfrac{1}{x + 5} + 1}$

80. _____

81. _____

Solve for x.

82. _____

81. $\dfrac{3}{2x} = \dfrac{1}{x} + \dfrac{1}{2}$ **82.** $\dfrac{1}{x} + \dfrac{2}{3x} = \dfrac{1}{4}$

83. _____

83. $\dfrac{x}{x + 7} = \dfrac{x - 1}{x}$ **84.** $\dfrac{4x + 1}{x} - \dfrac{x}{x + 1} = 3$

84. _____

85. $\dfrac{1}{x - 2} + \dfrac{1}{x + 3} = \dfrac{2x}{x^2 - 4}$

85. _____

86. $\dfrac{2}{x + 3} = \dfrac{1}{x^2 - 4x - 21} + \dfrac{3}{x - 7}$

86. _____

87. $\dfrac{1}{x + 1} - \dfrac{1}{2} = \dfrac{1}{3x + 3}$

87. _____

88. _____

88. $\dfrac{2}{3} - \dfrac{1}{2x - 8} = \dfrac{1}{x - 4}$

89. _____

89. $\dfrac{x + 1}{x + 2} - \dfrac{x - 3}{x + 5} = \dfrac{4}{x^2 + 7x + 10}$

90. _____

SCORE: _____

90. $\dfrac{x + 2}{x - 3} - \dfrac{x}{x - 9} = \dfrac{2}{x^2 - 12x + 27}$

1. Simplify: $\dfrac{x^2 - 2x - 15}{x^2 - 25}$

2. Multiply:

$\dfrac{x^2 + 8x + 15}{x^2 + x - 6} \cdot \dfrac{x^2 + 3x - 10}{x^2 + 4x - 5}$

A N S W E R S

1. _____

3. Divide: $\dfrac{2x + 8}{3x} \div \dfrac{x^2 + 3x - 4}{x^2 - x}$

2. _____

3. _____

4. a. Find the LCD for the following fractions:

$\dfrac{2}{x + 6}, \dfrac{5}{x^2 - 36}, \dfrac{1}{x + 4}$

b. Find the missing numerator:

$\dfrac{x - 2}{x + 3} = \dfrac{?}{2x^2 + x - 15}$

4a. _____

4b. _____

5. Add:

$\dfrac{x - 1}{x^2 + 5x + 6} + \dfrac{3}{x^2 - x - 6}$

6. Subtract:

$\dfrac{3x}{x + 2} - \dfrac{x + 3}{x^2 + 6x + 8}$

5. _____

6. _____

7. Simplify: $\dfrac{1 - \dfrac{3}{x + 3}}{\dfrac{1}{x^2 - 9}}$

8. Solve:

$\dfrac{3}{x + 3} + \dfrac{1}{x - 1} = \dfrac{8}{x^2 + 2x - 3}$

7. _____

8. _____

SCORE: _____

325

C H A P T E R

6

PRETEST

Before beginning this chapter, answer as many of the following questions as you can. When you have finished the chapter, take the practice test at the end of the chapter and compare the scores of the two tests to see how much you have learned.

A N S W E R S

1. Simplify: **a.** $(xy^2)^4$

x^4y^8

b. $\left(\dfrac{x^2}{2}\right)^3\left(\dfrac{4}{x^3}\right)^2$

$\left(\dfrac{x^6}{8}\right)\left(\dfrac{16}{x^6}\right)$

1a. _____

1b. _____

2. Simplify using only positive exponents:

a. $\dfrac{x^{-2}y^{-5}}{x^{-4}y}$

b. $(a^{-1} + b^{-1})^{-2}$

2a. _____

2b. _____

3. Write 2,730,000 in scientific notation.

4. Compute, and leave the answer in scientific notation:

$$\dfrac{(5.1 \times 10^3)(2.0 \times 10^4)}{3.0 \times 10^{-2}}$$

3. _____

4. _____

5. Evaluate: $\sqrt[3]{-64}$

6. Evaluate: $(-27)^{2/3}$

5. _____

6. _____

7. Simplify: **a.** $\sqrt[3]{x^9 y^6}$ **b.** $\sqrt[5]{-64x^8 y^{16}}$

8. Multiply and simplify:
 a. $\sqrt{5}(3\sqrt{2} + \sqrt{3})$
  $+ \sqrt{2}(\sqrt{3} - 2\sqrt{5})$
 b. $(2\sqrt{3} + 5\sqrt{7})(2\sqrt{3} - 5\sqrt{7})$

9. Simplify: **a.** $\dfrac{3}{\sqrt[3]{x}}$ **b.** $\dfrac{4}{\sqrt{5} + \sqrt{2}}$

10. a. Remove fractional exponents from the denominator: $\dfrac{2}{x^{1/3} y^{1/2}}$

 b. Simplify:
 $\sqrt{20} - \sqrt{\dfrac{4}{45}} + \sqrt{\dfrac{5}{9}}$

6

Exponents and Radicals

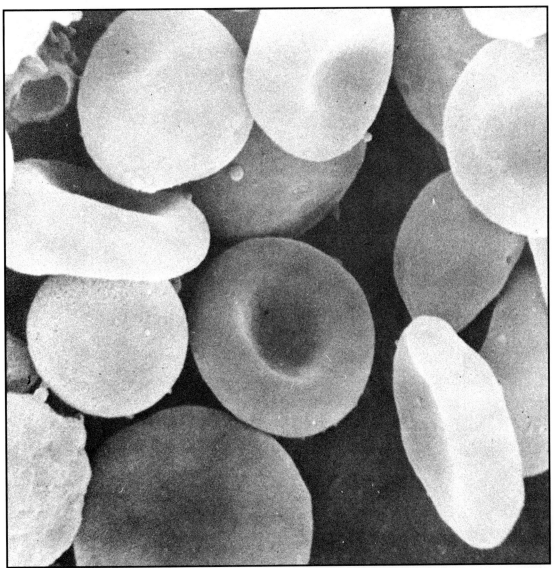

A red blood cell is approximately .001 cm in diameter. Express this length in scientific notation.

When we go beyond the four basic operations of addition, subtraction, multiplication, and division, a knowledge of radicals becomes necessary. In this chapter you will study techniques of simplifying radicals, the laws of exponents, and the relationship that exists between radicals and exponents.

◣ 6-1 THE LAWS OF EXPONENTS

OBJECTIVES

Upon completing this section you should be able to:

1. State the five laws of exponents.
2. Apply one or more of the laws to simplify an expression.

In chapter 1 we defined the positive whole number exponent and gave the first three laws of exponents.

Law 1 $x^a \cdot x^b = x^{a+b}$ *LIKE bASES AOO EXP.*

Law 2 $(x^a)^b = x^{ab}$ *Pwr to a Pwr   multiply*

Law 3 $\dfrac{x^a}{x^b} = x^{a-b}$, if $a > b$ *NUM > DENOM* $\dfrac{x^4}{x^2} = x^{4-2} = x^2$

$\dfrac{x^a}{x^b} = \dfrac{1}{x^{b-a}}$, if $a < b$ *NUM < DENOM* $\dfrac{x^2}{x^5} = \dfrac{1}{x^{5-2}} = \dfrac{1}{x^3}$

Here are some examples to refresh your memory.

Remember, y means y^1.

Example 1 $(3x^2y)(4x^3y^4) = 12x^5y^5$

Example 2 $(2a^2bc^2)(-7ab^4c^3) = -14a^3b^5c^5$

Example 3 $(2a^5)^2 = 4a^{10}$

Note that only the quantity inside the parentheses is raised to the indicated power.

Example 4 $5(a^4b^2)^3 = 5a^{12}b^6$

Example 5 $\dfrac{18x^2y^3}{6xy} = 3xy^2$

Example 6 $\dfrac{24x^2y^5}{-3x^5y^3} = -\dfrac{8y^2}{x^3}$

There are two more laws of exponents.

Law 4 $(xy)^a = x^a y^a$ $(xy)^3 = x^3 y^3$

Raising the product of factors to a power is the same as raising each factor to the power and then finding the product.

Example 7 $(xy)^3 = x^3y^3$

Example 8 $[(2)(3)]^2 = (2)^2(3)^2$

$[(2)(3)]^2 = 6^2 = 36$
$(2)^2(3)^2 = (4)(9) = 36$

Example 9 $(x^2y^3)^2 = x^4y^6$

Example 10 $[2(x^2y^3)^2]^3 = [2x^4y^6]^3 = 2^3x^{12}y^{18} = 8x^{12}y^{18}$

Example 11 $(-x^4y^3)^2 = x^8y^6$

Example 12 $(-2x^3y)^3 = (-2)^3x^9y^3 = -8x^9y^3$

Law 5 $\left(\dfrac{x}{y}\right)^a = \dfrac{x^a}{y^a}$

$\left(\dfrac{x^1}{y^2}\right)^3 = \dfrac{x^3}{y^6}$

Raising a quotient to a power is the same as raising the numerator and denominator to the power and then finding the quotient.

Example 13 $\left(\dfrac{x}{y}\right)^4 = \dfrac{x^4}{y^4}$

Example 14 $\left(\dfrac{3}{2}\right)^2 = \dfrac{3^2}{2^2}$

Example 15 $\left(\dfrac{x^2}{y^3}\right)^4 = \dfrac{x^8}{y^{12}}$

Example 16 $\left(\dfrac{-3x}{2y^2}\right)^3 = \dfrac{(-3)^3x^3}{2^3y^6} = \dfrac{-27x^3}{8y^6}$

> ▼ **WARNING**
>
> Do not attempt to use the fourth law of exponents on sums or differences, since
>
> $(x + y)^2 = x^2 + 2xy + y^2$
> NOT $x^2 + y^2$
>
> You can never distribute an exponent over a sum or difference.

▼ **EXERCISE 6-1-1**

Apply any one or a combination of the five laws of exponents to the following.

1. $(x^2)^5$ x^{10}

2. $(xy)^3$ x^3y^3

3. $(x^2y)^4$ x^8y^4

4. $\left(\dfrac{x}{y}\right)^2$ $\dfrac{x^2}{y^2}$

5. $\left(\dfrac{x^2}{y^3}\right)^3$ $\dfrac{x^6}{y^9}$

6. $\left(\dfrac{x}{2}\right)^3$ $\dfrac{x^3}{8}$

7. $(2x^2)^4$

$16x^8$

8. $(3xy^2)^3$

9. $\left(\dfrac{2}{x^2y}\right)^3$

10. $[(x^2)^3]^4$

$(x^2)^3 = [x^6]^4 = x^{24}$

11. $\left(\dfrac{xy^2}{x^2y}\right)^2$

12. $\left(\dfrac{2x^3}{xy^2}\right)^3$

13. $\left(2(xy)^2\right)^3$

14. $\left(\dfrac{1}{3x^2}\right)^2$ $\dfrac{1}{9x^4}$

15. $\left((2x^3y)^2\right)^3$

$2^2(x^3)^2(y)^2$

$(4x^6y^2)^3 = 64x^{18}y^6$

16. $\dfrac{(x^2)^3}{(x^3)^2}$

17. $\dfrac{(-3x)^3}{9x}$ $\dfrac{-27x^3}{9x} = -3x^2$

18. $\dfrac{-4x^2}{(2x)^3}$

19. $\dfrac{-5xy^3}{(-5x)^2}$ $= \dfrac{-5xy^3}{25x^2}$

$\dfrac{-1y^3}{5x}$

20. $\dfrac{(2x^2y^3)^3}{(2x^3y^2)^2}$ $\dfrac{8x^6y^9}{4x^6y^4}$

$\dfrac{8y^9}{4y^4} = 2y^5$

21. $2x^2y(x^2y)^3$

22. $(2xy)^3(2x^2y)^2$

23. $(-x^2y)^3(-2xy^3)^2$

24. $(x^2yz^2)^3(-2xy^3z)^2$

25. $[2(xy)^2]^5[3(x^2y)^3]^2$

26. $\left(\dfrac{x}{y}\right)^3\left(\dfrac{2x}{y}\right)^4$

27. $\left(\dfrac{-2}{x}\right)^3\left(\dfrac{x}{2}\right)^2$

28. $\left(\dfrac{x^2y}{3}\right)\left(\dfrac{3}{xy^2}\right)$

29. $\left(\dfrac{-2x}{5y^2}\right)^3\left(\dfrac{5y}{4x^2}\right)^2$

30. $\left(\dfrac{x^3}{8}\right)^2\left(\dfrac{4}{x^2}\right)^3$

$\left(\dfrac{-8x^3}{125y^6}\right)\left(\dfrac{25y^2}{16x^4}\right) = \dfrac{-1}{10xy^4}$

6-2 NEGATIVE AND ZERO EXPONENTS

The set of numbers used as exponents in our discussion thus far has been the set of positive integers. This is the only set that can be used when exponents are defined as they were in chapter 1. In this section, however, we would like to expand this set to include all integers (positive, negative, and zero) as exponents. This will, of course, require further definitions. These new definitions must be consistent with the system. Furthermore, we will expect all of the laws of exponents as well as other previously known facts to still be true.

> If $x \neq 0$, then $x^0 = 1$. (Any number, except zero, when raised to the zero power is 1.)
>
> $5^0 = 1 \qquad -6^0 = 1$

Example 1 $\quad 5^0 = 1$

Note that using this definition, 0^0 has no meaning.

Example 2 $\quad (-2)^0 = 1$

Example 3 $\quad (xyz)^0 = 1$, if $x, y, z \neq 0$

Example 4 $\quad \left(\dfrac{5x^2}{2y^3}\right)^0 = 1$, if $x, y \neq 0$

To check the consistency of the definition consider the product $x^0 \cdot x^5$, if $x \neq 0$.

$$x^0 \cdot x^5 = x^{0+5} = x^5$$

Since we know the product of x^5 and a number will be x^5 only if that number is 1, $x^0 = 1$ is consistent in this instance.

You may wish to verify that this definition is consistent with the other laws of exponents. Another definition becomes necessary if we are to allow negative numbers to be used as exponents.

We want to be able to use these laws for the integer zero and all the negative integers as well as the positive integers.

> $x^{-a} = \dfrac{1}{x^a}$, $x \neq 0$ and $a > 0$. (A negative exponent represents the reciprocal.)
>
> $x^{-2} = \dfrac{1}{x^2}$

Example 5 $\quad x^{-2} = \dfrac{1}{x^2}$

Example 6 $\quad x^{-2}y^{-3} = \dfrac{1}{x^2y^3}$

Example 7 $3^{-2} = \dfrac{1}{3^2} = \dfrac{1}{9}$

The negative sign of the exponent *never* affects the sign of the expression.

Example 8 $5^{-1} = \dfrac{1}{5}$

Example 9 $\dfrac{1}{x^{-2}} = \dfrac{1}{\dfrac{1}{x^2}} = x^2$

As mentioned earlier, the preceding definitions are consistent with the laws of exponents. That is, negative and zero exponents follow the same laws for positive exponents.

Law 1 $x^a x^b = x^{a+b}$

Remember, a negative exponent means reciprocal.

Example 10 $x^{-2}x^{-3} = \dfrac{1}{x^2 x^3} = \dfrac{1}{x^5} = x^{-5}$

Example 11 $x^5 x^{-2} = \dfrac{x^5}{x^2} = x^3$

Law 2 $(x^a)^b = x^{ab}$

Example 12 $(x^{-2})^3 = \left(\dfrac{1}{x^2}\right)^3 = \dfrac{1}{x^6} = x^{-6}$

Law 3   This law in section 6.1 could now be shortened to one rule stated as

Since we now allow negative exponents, there is no need for two parts to this law.

$$\dfrac{x^a}{x^b} = x^{a-b}$$

without regard to the relative size of a and b.

Example 13 $\dfrac{x^5}{x^2} = x^{5-2} = x^3$

Example 14 $\dfrac{x^3}{x^7} = x^{3-7} = x^{-4} = \dfrac{1}{x^4}$

Example 15 $\dfrac{x^3}{x^3} = x^{3-3} = x^0 = 1$

Law 4 $(xy)^a = x^a y^a$

Easiest way

Example 16 $(xy)^{-2} = \dfrac{1}{(xy)^2} = \dfrac{1}{x^2 y^2} = x^{-2} y^{-2}$

We will agree from this point on that in all the examples and problems the variables must never take on a value that will give a meaningless expression. That means we cannot have a denominator equal to zero or a value of a variable that gives 0^0.

Law 5 $\left(\dfrac{x}{y}\right)^a = \dfrac{x^a}{y^a}$

Example 17 $\left(\dfrac{x}{y}\right)^{-3} = \left(\dfrac{y}{x}\right)^{3} = \dfrac{y^3}{x^3} = \dfrac{x^{-3}}{y^{-3}}$

▼ **EXERCISE 6-2-1**

Rewrite using only positive exponents and leave the answers in simplest form.

1. x^{-3} $\dfrac{1}{x^3}$

2. x^{-5} $\dfrac{1}{x^5}$

3. $x^{-2}y^{-5}$ $\dfrac{1}{x^2 y^5}$

4. $a^{-3}b^{-1}$ $\dfrac{1}{a^3 b^1}$

5. $(ab)^{-3}$ $\dfrac{1}{(ab)^3}$ $\dfrac{1}{a^3 b^3}$

6. $(xy)^{-4}$ $\dfrac{1}{(xy)^4}$

7. 2^{-3} $\dfrac{1}{(2)^3} = \dfrac{1}{8}$

8. 4^{-2} $\dfrac{1}{4^2} = \dfrac{1}{16}$

9. $\dfrac{1}{3^{-2}}$ $\dfrac{1}{3^2} = \dfrac{1}{9}$

10. $\dfrac{1}{2^{-4}}$ $\dfrac{2^4}{1} = 16$

11. $\left(\dfrac{a}{b}\right)^{-5}$ $\left(\dfrac{b}{a}\right)^5$

12. $\left(\dfrac{x}{y}\right)^{-3}$ $\dfrac{x^{-3}}{y^{-3}}$ $\dfrac{y^3}{x^3}$

13. 3^0 $= 1$

14. 15^0 $= 1$

15. $(ab)^{-1}$ $a^{-1}b^{-1}$ $\dfrac{1}{ab}$

16. $(a + b)^{-1}$ $\dfrac{1}{a+b}$

17. $a^{-1} + b^{-1}$ $\dfrac{1}{a} + \dfrac{1}{b}$

18. $x^{-2}y^4z^{-1}$ $\dfrac{y^4}{x^2 z}$

19. $(x^3)^{-2}$ $\dfrac{1}{x^6}$

20. $(x^{-2}y^4)^{-4}$

21. $\dfrac{x^{-3}}{x^5}$ $\dfrac{1}{x^3} + \dfrac{1}{x^5}$ $\dfrac{1}{x^8}$

22. $x^5 x^{-3}$ $\dfrac{x^5}{x^3} = x^2$

23. $\dfrac{x^{-2}x^{-6}}{x^7}$ $\dfrac{1}{x^2 + x^6 \, x^7} = \dfrac{1}{x^{15}}$

24. $(x^2 y^{-3})^{-2}$

25. $(2x^2)^{-3}$ $(2^{-3})(x^2)^{-3}$ $= 8x^{-6}$ $\dfrac{1}{8x^6}$

26. $\dfrac{5^3 \cdot 5^0}{5^2}$ $= \dfrac{5^3}{5^2}$ 5

27. $\dfrac{2^{-5} \cdot 2^0}{2^{-2}}$ $\dfrac{2^2}{2^5} = \dfrac{1}{2^3}$

28. $\dfrac{x^{-3}y^2}{x^{-5}y^{-1}}$

29. $\dfrac{x^3 y^{-7}}{x^3 y^0}$ $\dfrac{x^3 \cdot 1}{x^3 y^7} = \dfrac{1}{y^7}$

30. $\left(\dfrac{3^{-1} \cdot 3^0}{3^5}\right)^0$ $= 1$

The practice you have gained from the previous exercise should now enable you to work slightly more involved problems.

Example 18   Simplify and rewrite $(-3a^2)^{-4}(b^{-1})^{-3}$ using only positive exponents.

Perform only one step at a time and be very careful.

    Solution   Using the fourth law of exponents, we may write

$$(-3)^{-4}(a^2)^{-4}(b^{-1})^{-3}.$$

Using the second law of exponents, we obtain

$$(-3)^{-4}a^{-8}b^3.$$

The definition of negative exponents enables us to write

$$\left(\frac{1}{(-3)^4}\right)\left(\frac{1}{a^8}\right)(b^3)$$

or

$$\frac{b^3}{81a^8}.$$

Example 19   Simplify and rewrite $\left(\dfrac{x^{-2}}{y}\right)^{-3}$ using only positive exponents.

There are several ways to approach this problem.

    Solution   We will examine this example very carefully to illustrate options that can make our work easier in more complicated problems.

We can first use the fifth law of exponents and obtain

Raise both numerator and denominator to the -3 power.

$$\left(\frac{x^{-2}}{y}\right)^{-3} = \frac{(x^{-2})^{-3}}{(y)^{-3}}.$$

Now using the second law, we get

Multiply exponents.

$$\frac{(x^{-2})^{-3}}{(y)^{-3}} = \frac{x^6}{y^{-3}}.$$

The definition of negative exponents now gives

Negative exponent means reciprocal.

$$\frac{x^6}{y^{-3}} = x^6 \div \frac{1}{y^3} = x^6 y^3.$$

A second possibility for the original problem $\left(\dfrac{x^{-2}}{y}\right)^{-3}$ is to note that the negative sign with (-3) designates reciprocal. We then can write

Note that we only inverted. Don't make the mistake of getting

$$\frac{y}{x^2}.$$

$$\left(\frac{x^{-2}}{y}\right)^{-3} = \left(\frac{y}{x^{-2}}\right)^3.$$

We now note that

$$\left(\frac{y}{x^{-2}}\right)^3 = (x^2y)^3 = x^6y^3$$

$$\text{or } \left(\frac{y}{x^{-2}}\right)^3 = \frac{y^3}{x^{-6}} = x^6y^3.$$

In other words, we can first eliminate negative exponents and then simplify or we can first simplify and then eliminate negative exponents or we can combine these techniques. As you obtain practice by working the exercises, you will probably find each of these methods useful.

The type of approach that is best to take will depend on the problem.

Example 20 Simplify and rewrite $\dfrac{y^{-1} - x^{-1}}{x - y}$ using only positive exponents.

Solution Using the definition of negative exponents, we write

$$\frac{\dfrac{1}{y} - \dfrac{1}{x}}{x - y}.$$

Here we have a complex fraction. To simplify it we need to multiply both the numerator and denominator by xy (the LCD of the fractions contained in the complex fraction).

Thus,

$$\frac{xy\left(\dfrac{1}{y} - \dfrac{1}{x}\right)}{xy(x - y)}.$$

Don't forget to multiply each term inside the parentheses by xy.

Simplifying, we obtain

$$\frac{x - y}{xy(x - y)} = \frac{1}{xy}.$$

Example 21 Simplify and rewrite $\dfrac{(x - y)^{-1}}{(2xy)^{-2}}$ using only positive exponents.

Solution

$$\frac{(x - y)^{-1}}{(2xy)^{-2}} = \frac{\dfrac{1}{(x - y)}}{\dfrac{1}{(2xy)^2}}$$

$$= \frac{1}{x - y} \div \frac{1}{4x^2y^2}$$

$$= \frac{4x^2y^2}{x - y}$$

WARNING

Be careful to note $(x - y)^{-1}$ is not the same as $x^{-1} - y^{-1}$. We can never distribute an exponent over a sum or difference of terms.

▼ EXERCISE 6-2-2

Simplify and rewrite using only positive exponents.

1. $(x^2)^{-5}(x^{-3})^2$

$(x^{-10})(x^{-6}) = x^{-16}$

$\dfrac{1}{x^{16}}$

2. $(3^{-2})^3(3^0)^{-1}$

$(3^{-6})(3^0)$

$\dfrac{1}{3^6}$

3. $(-2x^3)^{-3}(3x^{-1})^2$

$(-8x^{-9})(9x^{-2})$

$-\dfrac{1}{8x^9} \quad -\dfrac{9}{8x^{11}}$

4. $(2^{-3}x^5)^{-2}$

$2^6 x^{-10} \quad \dfrac{2^6}{x^{10}}$

5. $(3^{-1}x^{-4})^{-3}$

$3^{+3}x^{+12}$

$3^3 x^{12}$

6. $\left(\dfrac{-2}{x^{-3}}\right)^3 \quad \left(\dfrac{x^3}{2}\right)^3$

$\dfrac{-2^3}{x^{-9}} = \dfrac{x^9}{8}$

7. $(3^0 x^{-2}y^3)^{-4}$

$1\,x^8 y^{-12} \quad \dfrac{x^8}{y^{12}}$

8. $(2x^0 y^{-3})^{-2}$

$1\,y^{-6} \quad \dfrac{1}{y^6}$

9. $\left(\dfrac{x^{-3}}{y^{-1}}\right)^{-2}$

$\dfrac{x^{-6}}{y^{-2}} = \dfrac{y^2}{x^6}$

10. $\left(\dfrac{x^{-1}y^2}{y^{-3}}\right)^{-5}$

$\dfrac{x^{-5}y^{10}}{y^{-15}} = \dfrac{y^{15}y^{10}}{x^5}$

11. $\left(\dfrac{2^{-3}x^5y^{-1}}{2^2x^{-2}}\right)^2$

$\dfrac{2^{-6}x^{10}y^{-2}}{2^4x^{-4}} \quad \dfrac{x^{10}x^4}{2^6\,2^4 y^2}$

$\dfrac{x^{14}}{2^{10}y^2} \quad \boxed{\dfrac{x^{14}}{1024\,y^2}}$

12. $(2x^2y^{-3})^{-3}(-2x^{-3}y^2)^2$

13. $\left(\dfrac{x}{y}\right)^{-3}\left(\dfrac{x^2}{y^{-1}}\right)^0$

$\dfrac{x^{-3}}{y^{-3}} \cdot 1 = \dfrac{y^3}{x^3}$

14. $\left(\dfrac{2x}{y^2}\right)^{-2}\left(\dfrac{y^3}{x}\right)^2$

$\left(\dfrac{y^2}{2x}\right)^2 \left(\dfrac{y^3}{x}\right)^2$

$\left(\dfrac{y^4}{4x^2}\right)\left(\dfrac{y^6}{x^2}\right) = \dfrac{y^{10}}{4x^4}$

15. $\left(\dfrac{x^2y^{-1}}{x^{-5}}\right)\left(\dfrac{x^{-1}y^3}{x^5}\right)^{-3}$

16. $\left(\dfrac{2^{-3}}{x^4}\right)^{-2}\left(\dfrac{-2x^3}{x^{-1}}\right)^3$

17. $\dfrac{x^{-1}+y^{-1}}{xy}$

18. $(x^{-1}+2^{-2})^{-1}$

19. $(x^{-1} + y^{-1})^{-2}$

20. $\dfrac{x^{-1} + y^{-1}}{x + y}$ $\dfrac{1}{(x+y)^2}$

21. $\dfrac{(x + y)^{-1}}{(2x^2y)^{-3}}$ $\dfrac{8x^6y^3}{x+y}$

$\dfrac{2x^2y^3}{x+y} = \dfrac{8x^6y^3}{x+y}$

22. $\dfrac{(3xy^2)^{-3}}{(x - y)^{-2}}$ $= \dfrac{(x-y)^2}{(3xy^2)^3} = \dfrac{x^2-y^2}{27x^3y^6} =$

6-3 SCIENTIFIC NOTATION

Exponents are used in many fields of science to write numbers in what is called scientific notation. If a number is either very large or very small, this method of expressing it keeps it from being cumbersome and can make computations easier. Many calculators utilize scientific notation when numbers are too large or too small for the display.

> A number is in **scientific notation** if it is expressed as the product of a power of ten and a number equal to or greater than one and less than ten.

Example 1 3.6×10^3 is in scientific notation.

$3.6 \times 10^3 = 3,600$

Example 2   Earth is approximately 93,000,000 miles from the Sun. Express this number in scientific notation.

   Answer $93,000,000 = 9.3 \times 10^7$

Notice that the definition is very explicit. The product must be of a number equal to or greater than one and less than ten, and a power of ten.

$$93,000,000 = .93 \times 10^8$$

but this is *not* scientific notation because .93 is not equal to or greater than one.

OBJECTIVES

Upon completing this section you should be able to:

1. Identify a number that is in scientific notation.
2. Write a given number in scientific notation.
3. Change a number from scientific notation to one without exponents.
4. Perform computations using numbers in scientific notation.

Is 1×10^3 in scientific notation?

Is 2.3×10^0 in scientific notation?

Example 3   Write 2.5×10^6 without using exponents.

Answer $2.5 \times 10^6 = 2,500,000$

(Note that multiplying by 10^6 moves the decimal point six places to the right.)

Example 4   Write 2.5×10^{-6} without using exponents.

Answer $2.5 \times 10^{-6} = .0000025$

(Note that multiplying by 10^{-6} moves the decimal point six places to the left.)

Example 5   Write .0000000345 in scientific notation.

Solution   Immediately we see that part of the answer must be 3.45 (equal to or greater than one and less than ten always gives one digit to the left of the decimal point).

If we now look at 3.45 we must ask, "What power of ten will return the decimal point to its original position?" Counting, we get eight places to the left, so

$$.0000000345 = 3.45 \times 10^{-8}.$$

To summarize: If the exponent is *positive*, the decimal point is moved to the *right* that number of places.

If the exponent is *negative*, the decimal point is moved to the *left* that number of places.

▼   **EXERCISE 6–3–1**

State whether or not the given number is in scientific notation.

1. 3.6×10^5 **2.** $.05 \times 10^8$ **3.** 2.78×10^{16} **4.** 54.1×10^6 **5.** 8.2×10^{-3}

6. 7.8×10^{-4} **7.** 25×10^{-3} **8.** $.645 \times 10^4$ **9.** 5×10^8 **10.** 10×10^7

1.0×10^8

Write each number in scientific notation.

11. 5,000   **12.** 346,000,000   **13.** .000000235   **14.** 23,000,000,000,000   **15.** .0000000052

16. 5,280   **17.** 68   **18.** 728   **19.** 728,000   **20.** 728,000,000

7.28×10^8

Write each number without using exponents.

21. 3.201×10^5 **22.** 7.28×10^{-6} **23.** 6.23×10^{23}

     320100   .00000728   6230000000000000000000000

24. 1.07×10^{-9} **25.** 5.02×10^{10} **26.** 3.58×10^{-1}

   .00000000107   50200000000   .358

27. 5.3762×10^2 **28.** 4.07×10^{-5} **29.** 3.6×10^0 **30.** 9.9×10^1

     537.62   .0000407   3.6   99

31. Mars is approximately 49,000,000 miles from Earth. Express this distance in scientific notation.

32. An enormous cloud of hydrogen gas eight million miles in diameter was discovered around the comet Bennet by NASA in 1970. Express this distance in scientific notation.

33. A light year (the distance light travels in a year) is approximately 6.0×10^{12} miles. Express this distance without exponents.

     6000000000000

34. The mass of the Earth is 1.3×10^{25} pounds. Express this without exponents.

     F.0

35. An "Angstrom" is a unit of length. One Angstrom is approximately 1×10^{-8} cm. Express this length without exponents.

36. The diameter of the nucleus of an average atom is approximately 3.5×10^{-12} cm. Express this distance without exponents.

37. A red blood cell is approximately .001 cm in diameter. Express this length in scientific notation.

38. One electron volt produces approximately .0000000000016 ergs of energy. Express this number in scientific notation.

39. The Rubik's Cube puzzle has over 43,000,000,000,000,000,000 color combinations. Express this number in scientific notation.

40. The Milky Way galaxy has a diameter of approximately 590,000,000,000,000,000 miles. Express this distance in scientific notation.

Scientific notation is useful in other ways than making numbers less cumbersome. Computations involving very large or very small numbers can be simplified because of the laws of exponents.

Recall that $(10^4)(10^7) = 10^{11}$.

Example 6   Find the product: $(2.3 \times 10^4)(1.06 \times 10^7)$

Solution $(2.3 \times 10^4)(1.06 \times 10^7) = (2.3)(1.06)(10^{11})$
$= 2.438 \times 10^{11}$

Most calculators cannot handle this problem in this form.

Example 7   Multiply 728,000,000 by 12,000,000.

Solution   First change each number to scientific notation.

$$728,000,000 = 7.28 \times 10^8$$
$$12,000,000 = 1.2 \times 10^7$$
$$(7.28 \times 10^8)(1.2 \times 10^7) = (7.28)(1.2)(10^{15})$$
$$= 8.736 \times 10^{15}$$

Example 8   Compute: $\dfrac{(3.41 \times 10^5)(5.6 \times 10^7)}{2.4 \times 10^8}$

$(10^5)(10^7) = 10^{12}$

and $\dfrac{10^{12}}{10^8} = 10^{12-8} = 10^4$

Solution   By the laws of exponents $\dfrac{(10^5)(10^7)}{10^8} = 10^4$. So we have

$\dfrac{(3.41)(5.6)}{(2.4)} \times 10^4 = 7.96 \times 10^4$ (to two places).

Example 9   Compute: $(3.5 \times 10^{-5})(2.0 \times 10^8)$

Solution $(3.5 \times 10^{-5})(2.0 \times 10^8) = (3.5)(2.0)(10^3)$
$= 7.0 \times 10^3$

Example 10   Compute: $(4.2 \times 10^5)(3.5 \times 10^8)$

Solution $(4.2 \times 10^5)(3.5 \times 10^8) = (4.2)(3.5)(10^{13})$
$= 14.7 \times 10^{13}$

Also work this problem:

$\dfrac{2.0 \times 10^8}{5.0 \times 10^{11}}$

(Answer: 4.0×10^{-4})

Notice that 14.7 is not a number less than ten, so 14.7×10^{13} is *not* in scientific notation. We can write

$$14.7 \times 10^{13} = 1.47(10)(10^{13})$$
$$= 1.47 \times 10^{14}$$

which *is* in scientific notation.

▼ **EXERCISE 6–3–2**

Compute using the laws of exponents. Leave all answers in scientific notation.

1. $(3.2 \times 10^4)(1.5 \times 10^5)$ **2.** $(2.5 \times 10^9)(3.4 \times 10^4)$ **3.** $(2.6 \times 10^{-4})(3.0 \times 10^6)$

4. $(5.0 \times 10^8)(1.7 \times 10^{-11})$ **5.** $\dfrac{3.4 \times 10^{18}}{2.0 \times 10^{11}}$ **6.** $\dfrac{4.8 \times 10^{-3}}{2.4 \times 10^4}$

7. $(4.25 \times 10^{-2})(2.2 \times 10^{-5})$ **8.** $(3.8 \times 10^5)(1.75 \times 10^{-6})$ **9.** $\dfrac{9.45 \times 10^5}{2.5 \times 10^{11}}$

10. $\dfrac{8.51 \times 10^{-2}}{4.6 \times 10^{-3}}$ **11.** $(3.45 \times 10^3)(2.68 \times 10^4)$ **12.** $(5.6 \times 10^{-5})(3.0 \times 10^3)$

13. $(6.23 \times 10^{-23})(3.5 \times 10^3)$ **14.** $\dfrac{5.25 \times 10^5}{2.1 \times 10^3}$ **15.** $\dfrac{(3.0 \times 10^5)(2.73 \times 10^4)}{(3.5 \times 10^3)}$

16. $\dfrac{9 \times 10^4}{3 \times 10^{-4}}$ **17.** $\dfrac{1.2 \times 10^{-6}}{3.0 \times 10^4}$ **18.** $\dfrac{(5.0 \times 10^{10})(2.5 \times 10^3)}{(1.25 \times 10^{13})}$

19. $\dfrac{(7.5 \times 10^{-3})(2.0 \times 10^3)}{3.0 \times 10^0}$ **20.** $\dfrac{5.8 \times 10^4}{2.9 \times 10^4}$

◣◢ 6-4 RADICALS

[handwritten: PRINCIPAL SQ ROOT $\sqrt{4} = 2$; SQ ROOT $= 2$ or -2]

The symbol $\sqrt{}$ is called the *radical sign*. An expression containing the radical sign is called a *radical*. The radical $\sqrt[5]{32}$ is read "the principal fifth root of 32." The number under the radical sign (32 in this case) is called the *radicand* and the integer 5 is called the *index*. If the index is 2, it is usually omitted. Hence, both $\sqrt{8}$ and $\sqrt[2]{8}$ indicate the principal square root of 8. If the radical sign is used to indicate any root other than the square root, the index must be written. The index number is always a positive integer greater than 1.

Example 1 $\sqrt[4]{24}$ means the principal fourth root of 24.

Example 2 $\sqrt[3]{24}$ means the principal cube root of 24.

Example 3 $\sqrt{24}$ means the principal square root of 24.

O B J E C T I V E S

Upon completing this section you should be able to:

1. Give the principal root of a number.
2. Use a table to find the square root of a number.

Again, we do not write the index 2.

> If $x^n = y$, then x is an **nth root** of y.

Note there are two different square roots of 9.

Example 4 Since $3^2 = 9$, 3 is a square root of 9.

Example 5 Since $2^4 = 16$, 2 is a fourth root of 16.

Example 6 Since $(-3)^2 = 9$, -3 is a square root of 9.

Example 7 Since $(-2)^4 = 16$, -2 is a fourth root of 16.

Notice in the preceding examples that 9 has two square roots, 3 and -3. It can be shown by more advanced methods that every number has two square roots, three cube roots, four fourth roots, and so on. Not all of these roots, however, are in the set of real numbers.

We use the radical sign to indicate the *principal* root of a number.

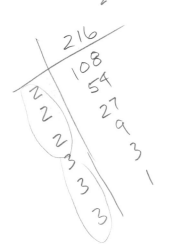

> The **principal nth root** of x ($\sqrt[n]{x}$) is
> a. zero, if $x = 0$.
> b. positive, if $x > 0$.
> c. negative, if $x < 0$ and n is an odd integer.
> d. not a real number, if $x < 0$ and n is an even integer.

Example 8 $\sqrt{16} = 4$

Example 9 $\sqrt[3]{8} = 2$

These are all *principal* roots.

Example 10 $\sqrt[3]{-8} = -2$

Example 11 $\sqrt[4]{-16}$ is not a real number.

▼ **EXERCISE 6-4-1**

Evaluate. If not a real number, so state.

1. $\sqrt{4}$ 2

2. $\sqrt{25}$ 5

3. $\sqrt[3]{27}$ 3

4. $\sqrt{1}$ 1

5. $\sqrt[3]{-1}$ -1

6. $\sqrt{9}$ 3

7. $\sqrt{100}$ 10

8. $\sqrt[3]{-27}$ -3

9. $\sqrt[4]{16}$ 2

10. $\sqrt{-169}$ NOT REAL

11. $\sqrt[5]{-32}$ _-2_   **12.** $\sqrt[3]{125}$ _5_   **13.** $\sqrt{225}$ _15_   **14.** $\sqrt[9]{-1}$ _-1_   **15.** $\sqrt{144}$ _12_

16. $\sqrt[3]{-216}$ _=6_   **17.** $\sqrt{625}$ _25_   **18.** $\sqrt[6]{-64}$ _Not Real_   **19.** $\sqrt[87]{-1}$ _-1_   **20.** $\sqrt[7]{-128}$

One of the most common uses of radicals is in finding the square root of a number. In the previous set of exercises we found square roots such as $\sqrt{4}$, $\sqrt{9}$, $\sqrt{100}$, and so on. Each of these values was a whole number. However, we will not always obtain a whole number or even an exact value as the principal square root of a given number. For example, consider $\sqrt{3}$. There is no exact value that when multiplied by itself yields 3 as a result. In such a case we must find an approximation of the square root of the number.

You may remember from arithmetic how to find the approximation of a square root of a number. To simplify our work we will use a square root table or a calculator. If you have a calculator with a square root key, you will want to use it in the following exercises. If you do not have such a calculator, the square root table on the inside back cover will aid you. Carefully read the following instructions on how to use the table.

A calculator will be more accurate in finding square roots of numbers not listed in the table.

To find the square root of a number between 1 and 200 simply read the value directly from the table. For example, if we want to find $\sqrt{3}$, we find 3 in the first column and look for the proper entry in the square root column. In this case, the value is 1.7321. Thus, $(1.7321)^2 \simeq 3$. Suppose we want to find $\sqrt{5.6}$ to three decimal places. The actual value 5.6 does not appear in the table. However, 5.6 is between 5 and 6, whose values are in the table. So the value of $\sqrt{5.6}$ is between $\sqrt{5}$ and $\sqrt{6}$, or between 2.2361 and 2.4495. The most accurate method of finding the square root of a number not in the table is to use a calculator. Using a calculator, we find $\sqrt{5.6} = 2.366$ correct to three decimal places.

The symbol $\simeq$ means "approximately equal to."

▼ **EXERCISE 6–4–2**

Using either a calculator or the table of square roots, evaluate the following to three decimal places.

1. $\sqrt{15}$ **2.** $\sqrt{96}$ **3.** $\sqrt{7}$ **4.** $\sqrt{35}$ **5.** $\sqrt{102}$

6. $\sqrt{120}$ **7.** $\sqrt{8.3}$ **8.** $\sqrt{6.2}$ **9.** $\sqrt{21.4}$ **10.** $\sqrt{15.6}$

11. $\sqrt{110.3}$ **12.** $\sqrt{134.2}$ **13.** $\sqrt{220}$ **14.** $\sqrt{244}$ **15.** $\sqrt{1,179}$

16. $\sqrt{747}$ **17.** $\sqrt{1,075}$ **18.** $\sqrt{2,300}$ **19.** $\sqrt{284.8}$ **20.** $\sqrt{254.8}$

◣◣ 6-5 FRACTIONAL EXPONENTS

OBJECTIVES

Upon completing this section you should be able to:

1. State the meaning of a fractional exponent.
2. Change an expression such as $a^{3/4}$ to radical form.
3. Change an expression such as $\sqrt[5]{x^2}$ to exponential form.
4. Evaluate expressions such as $\left(\dfrac{1}{8}\right)^{-1/3}$.

Our work with exponents thus far has involved only integers. Since exponents can be positive, zero, or negative, you might ask if they can also be fractions. The answer is *yes*. Whenever a new concept, such as fractional exponents, is introduced in a mathematical system, that new concept must be consistent with all previous laws or procedures. Fractional exponents will be given meaning in terms of radicals, which we discussed in the previous section.

$$x^{a/b} = \sqrt[b]{x^a} \text{ or } (\sqrt[b]{x})^a.$$

$8^{\frac{2}{3}} = \sqrt[3]{8^2} \quad \sqrt[3]{64} = 4$

$8^{\frac{2}{3}} = (\sqrt[3]{8})^2 \quad (2)^2 = 4$

This definition is consistent with the laws of exponents. For example,

$$x^{1/2} = \sqrt{x}.$$

Since

$$x^{1/2}x^{1/2} = x^{1/2 + 1/2} = x$$

we see that the definition is consistent in this instance.

The numerator is the power of the base and the denominator is the index.

Example 1 $x^{2/3} = \sqrt[3]{x^2}$

Example 2 $a^{4/5} = \sqrt[5]{a^4}$

Example 3 $2^{1/2} = \sqrt{2}$

Example 4 $x^{-1/3} = \dfrac{1}{\sqrt[3]{x}}$

Notice in these cases it is easier to take the root before raising to a power.

Example 5 $8^{2/3} = \sqrt[3]{8^2} = (\sqrt[3]{8})^2 = 2^2 = 4$

Example 6 $(-32)^{2/5} = (\sqrt[5]{-32})^2 = (-2)^2 = 4$

Example 7 $-32^{2/5} = -(\sqrt[5]{32})^2 = -(2)^2 = -4$

▼ **EXERCISE 6-5-1**

Change from exponential to radical form.

1. $x^{1/3}$ $\sqrt[3]{x^1}$

2. $x^{-1/2}$ $\sqrt[2]{x^{-1}}$ or $\dfrac{1}{\sqrt{x}}$

3. $a^{3/4}$ $\sqrt[4]{a^3}$

4. $6^{1/2}$ $\sqrt[2]{6} =$

5. $x^{-2/3}$ $\sqrt[3]{x^{-2}}$ or $\dfrac{1}{\sqrt[3]{x^2}}$

6. $2^{2/5}$ $\sqrt[5]{2^2}$

7. $(-3)^{2/7}$ $\sqrt[7]{-3^2}$

8. $(a^{-1})^{2/3}$

9. $(ab)^{-2/3}$ $\dfrac{1}{\sqrt[3]{ab^2}}$

10. $(2x)^{2/3}$ $\sqrt[3]{2x}^2$

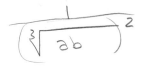 $\dfrac{1}{\left(\sqrt[3]{ab}\right)^2}$

Change from radical to exponential form.

11. $\sqrt{x}$ $x^{1/2}$

12. $\sqrt[5]{x^2}$ $x^{2/5}$

13. $\sqrt[4]{a^3}$ $a^{3/4}$

14. $(\sqrt[3]{a})^2$ $a^{2/3}$

15. $\dfrac{1}{\sqrt{x}}$ $\dfrac{1}{x^{1/2}}$ or $x^{-1/2}$

16. $(\sqrt[5]{x})^4$ $x^{4/5}$

17. $(\sqrt[3]{ab})^2$ $\left(ab^{1/3}\right)^2$ $ab^{2/3}$

18. $\sqrt[5]{\sqrt{x}}$ $\sqrt[5]{x^2}$ $x^{2/5}$

19. $(\sqrt[7]{x^2})^2$

$\left(x^{2/7}\right)^2 = x^{4/7}$

$\left(\dfrac{2}{7}\right)^2 = \dfrac{4}{7}$!!

20. $\dfrac{1}{(\sqrt[5]{x})^2}$ $\dfrac{1}{\left(x^{1/5}\right)^2} = \dfrac{1}{x^{2/5}}$

Evaluate.

21. $4^{1/2}$ $\sqrt{4} = 2$

22. $9^{1/2}$

23. $(-27)^{2/3}$ $\left(\sqrt[3]{-27}\right)^2$ $(-3)^2 = 9$

24. $-27^{2/3}$

25. $(64)^{2/3}$ $\left(\sqrt[3]{64}\right)^2$ $4^2 = 16$

26. $\left(\dfrac{1}{8}\right)^{-1/3}$

27. $\left(-\dfrac{1}{8}\right)^{-2/3}$ $\left(\sqrt[3]{\dfrac{-1}{8}}\right)^{-2}$ $-\dfrac{1}{2}$

28. $-\left(\dfrac{1}{8}\right)^{-2/3}$

29. $(-1)^{4/5}$ **30.** $\left(\dfrac{64}{27}\right)^{-2/3}$ **31.** $(\sqrt[3]{27})^2$ **32.** $(\sqrt[5]{-1})^3$

33. $(\sqrt[5]{-1})^2$ **34.** $(\sqrt[4]{16})^2$ **35.** $(\sqrt{25})^3$ **36.** $\sqrt{144^2}$

37. $\sqrt[3]{(-2)^3}$ **38.** $\sqrt[4]{(53)^4}$ **39.** $\sqrt[5]{(-21)^5}$ **40.** $\sqrt{\sqrt{16}}$

◤◣ 6–6 SIMPLIFICATION OF RADICALS

OBJECTIVES

▼

Upon completing this section you should be able to:
1. Simplify radicals by removing perfect roots.
2. Simplify radicals by reducing the index.

We have already noted that, in general, radicals have no exact numerical value unless they indicate roots of perfect powers. For this reason, when exact values are necessary, answers are left in radical form. In such cases, we wish to simplify the radical as much as possible.

A radical may be simplified if the radicand contains a factor that is raised to a power equal to or greater than the index.

For example, $\sqrt{48}$ may be simplified, since

$$\sqrt{48} = \sqrt{(4^2)(3)}$$

and the factor 4 is raised to a power equal to the index. We may then write

$$\sqrt{(4^2)(3)} = \sqrt{4^2}\sqrt{3}$$

since the fourth law of exponents gives

$$[(4^2)(3)]^{1/2} = (4^2)^{1/2}(3)^{1/2}.$$

$(4^2)^{1/2} = 4^{2/2} = 4$

Thus,

$$\begin{aligned}\sqrt{48} &= \sqrt{(4^2)(3)}\\ &= \sqrt{4^2}\sqrt{3}\\ &= 4\sqrt{3},\end{aligned}$$

which is the simplified form.

Example 1 Simplify: $\sqrt{75}$

Solution The question to ask is, "Does 75 have a factor that is raised to the second or greater power?" To answer this question we need only to look for factors of 75 that are perfect squares.

25 is a perfect square so we factor 75 as (25)(3).

$$\begin{aligned}\sqrt{75} &= \sqrt{(25)(3)}\\ &= \sqrt{(5^2)(3)}\\ &= \sqrt{5^2}\sqrt{3}\\ &= 5\sqrt{3}\end{aligned}$$

Example 2 Simplify: $\sqrt{8x^3y^4}$

Solution
$$\sqrt{8x^3y^4} = \sqrt{(2^2)(2)(x^2)(x)(y^4)}$$
$$= \sqrt{2^2}\sqrt{2}\sqrt{x^2}\sqrt{x}\sqrt{y^4}$$
$$= 2xy^2\sqrt{2x}$$

We factor 8 as $(4)(2) = (2^2)(2)$ since 4 is a perfect square.

Example 3 Simplify: $\sqrt[3]{16x^4y^5}$

Solution
$$\sqrt[3]{16x^4y^5} = \sqrt[3]{(2^3)(2)(x^3)(x)(y^3)(y^2)}$$
$$= 2xy\sqrt[3]{2xy^2}$$

We want a perfect cube factor of 16 so we factor 16 as $(8)(2) = (2^3)(2)$.

Example 4 Simplify: $\sqrt[3]{54x^4y^6}$

Solution Here we look for a perfect cube factor of 54, so we factor 54 as $(27)(2)$. 27 is a perfect cube.

$$\sqrt[3]{54x^4y^6} = \sqrt[3]{(27)(2)x^3xy^6}$$
$$= 3xy^2\sqrt[3]{2x}$$

▼ **WARNING**

Don't be careless and forget to write the index number if other than 2.

Example 5 Simplify: $\sqrt{54x^4y^6}$

Solution Here we look for a perfect square factor of 54, so we factor 54 as $(9)(6)$. 9 is a perfect square.

$$\sqrt{54x^4y^6} = \sqrt{(9)(6)x^4y^6}$$
$$= 3x^2y^3\sqrt{6}$$

We must always keep in mind the root we are seeking.

▼ **EXERCISE 6–6–1**

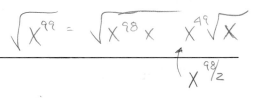

Simplify.

1. $\sqrt{8}$ $\sqrt{4 \cdot 2} = 2\sqrt{2}$ **2.** $\sqrt{18}$ $9 \cdot 2$ **3.** $\sqrt{27}$ $\sqrt{3 \cdot 9}$ **4.** $\sqrt{125}$ 5

$3\sqrt{3}$

5. $\sqrt[3]{16}$ 2.519 **6.** $\sqrt[3]{54}$ 3.779 **7.** $\sqrt{x^3}$ $\sqrt{x^2 \cdot x^1}$ **8.** $\sqrt[3]{x^7}$ $x^{7/3}$

$x\sqrt{x}$

9. $\sqrt[3]{x^6y^9}$ x^2y^3 **10.** $\sqrt{25x^3}$ **11.** $\sqrt{18x^5}$ $9 \cdot 2 \, x^4 \cdot x$ **12.** $\sqrt{27x^3y^5}$ $\sqrt{9 \cdot 3 \cdot x^2 \cdot x \cdot y^4 \cdot y}$

$5x^2\sqrt{x}$ $3x^2\sqrt{2x}$ $3xy^2\sqrt{3xy}$

13. $\sqrt{8xy^6}$ $4 \cdot 2 \, x(y^3)^2$ **14.** $\sqrt{32x^5y^7}$ **15.** $\sqrt[3]{16x^5}$ $8 \cdot 2 \, x^3 \cdot x^2$ **16.** $\sqrt[3]{54x^4y^6}$

$2y^3\sqrt{2x}$ $2x\sqrt[3]{2x^2}$

17. $\sqrt[5]{64x^{11}}$ $2^5 \cdot 2 \, x^{10} \cdot x^1$ **18.** $\sqrt{16x^8}$ **19.** $\sqrt{18x^5y^3}$ **20.** $\sqrt[3]{-8x^7y^6}$

$2x^2\sqrt[5]{2x}$

21. $\sqrt[4]{162x^9y^5}$ **22.** $\sqrt[3]{64x^4y^6}$ **23.** $\sqrt[5]{-64x^{10}y^2}$ **24.** $\sqrt{144x^{10}y^5}$

25. $\sqrt[3]{81x^{14}y^7}$ **26.** $\sqrt[5]{64x^7y^{12}}$

Radicals are in simplest form when

     1. no factor of the radicand is raised to a power equal to or greater than the index and

     2. the index cannot be smaller.

Finding the smallest possible index can best be accomplished by first changing the expression to one involving fractional exponents.

Example 6   Simplify: $\sqrt[6]{x^2y^4}$

     Solution   **Step 1** $\sqrt[6]{x^2y^4} = (x^2y^4)^{1/6} = x^{2/6}y^{4/6}$
> We apply the fourth law of exponents.

                          The fractional exponents are then reduced to simplest form.

               Step 2 $x^{2/6}y^{4/6} = x^{1/3}y^{2/3}$

               Step 3   If the fractional exponents do not have a common denominator, they are changed to a common denominator. In this case, they already have a common denominator. Therefore, $x^{1/3}y^{2/3} = (xy^2)^{1/3}$.

> We apply the fourth law of exponents in reverse.

               Step 4   Change back to radical form.
                          $(xy^2)^{1/3} = \sqrt[3]{xy^2}$

> We have reduced the index from 6 to 3.

                          Therefore, $\sqrt[6]{x^2y^4} = \sqrt[3]{xy^2}$.

Example 7   Simplify: $\sqrt[12]{x^6y^8}$

     Solution   **Step 1** $\sqrt[12]{x^6y^8} = (x^6y^8)^{1/12} = x^{6/12}y^{8/12}$

> The LCD of 2 and 3 is 6.

               Step 2 $= x^{1/2}y^{2/3}$
               Step 3 $= x^{3/6}y^{4/6}$

> The new index is 6.

                            $= (x^3y^4)^{1/6}$
               Step 4 $= \sqrt[6]{x^3y^4}$

To simplify an expression we sometimes need to remove factors as well as make the index smaller.

Example 8   Simplify: $\sqrt[6]{25x^4y^8}$

Solution
$$\sqrt[6]{25x^4y^8} = \sqrt[6]{5^2x^4y^8}$$
$$= 5^{2/6}x^{4/6}y^{8/6}$$
$$= 5^{1/3}x^{2/3}y^{4/3}$$
$$= \sqrt[3]{5x^2y^4}$$

Change to fractional exponents.

The index has been reduced but we still have a factor in the radicand raised to a power that is higher than the index.

When the index is made smaller, we still have a factor (y) raised to a power higher than the index number (3). We simplify as discussed earlier, obtaining

$$\sqrt[3]{5x^2y^4} = \sqrt[3]{5x^2y^3y}$$
$$= y\sqrt[3]{5x^2y}.$$

Reducing the index number does not arise very often. It will only be possible when the index number and all exponents have a common divisor.

▼ EXERCISE 6-6-2

Simplify.

1. $\sqrt[6]{x^2}$

2. $\sqrt[6]{y^4}$

3. $\sqrt[9]{y^{12}}$

4. $\sqrt[4]{x^{10}}$

5. $\sqrt[4]{x^2y^4}$

6. $\sqrt[8]{x^2y^4}$

7. $\sqrt[10]{x^6y^2}$

8. $\sqrt[10]{x^8y^4}$

9. $\sqrt[12]{x^{10}y^6}$

10. $\sqrt[12]{x^2y^8}$

11. $\sqrt[8]{4x^4}$

12. $\sqrt[10]{16x^8}$

13. $\sqrt[6]{9x^2}$

14. $\sqrt[6]{9x^2y^4}$

15. $\sqrt[4]{4x^2y^8}$

16. $\sqrt[8]{16x^6y^{10}}$

17. $\sqrt[6]{25x^2y^{14}}$

18. $\sqrt[10]{16x^6y^8}$

19. $\sqrt[12]{81x^{16}y^8}$

20. $\sqrt[6]{512x^9y^3}$

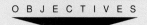 **6-7 OPERATIONS WITH RADICALS**

Since all radicals can be written in exponential form and we already have rules for adding, subtracting, multiplying, and dividing algebraic expressions containing exponents, it is not really necessary to have separate rules for these operations with radicals. However, it is sometimes more convenient to operate with the radicals, so we will develop such rules. In this section we will develop addition, subtraction, and multiplication rules. Division will be discussed in the next section.

Let us look first at the problem of simplifying. For example,

$$2\sqrt{x} + 3\sqrt{x}.$$

Changing to exponents, we have

$$2\sqrt{x} + 3\sqrt{x} = 2(x)^{1/2} + 3(x)^{1/2}.$$

We see a common factor of $x^{1/2}$. Thus,

$$2(x)^{1/2} + 3(x)^{1/2} = (2 + 3)(x)^{1/2} = 5x^{1/2} = 5\sqrt{x}.$$

Hence,

$$2\sqrt{x} + 3\sqrt{x} = 5\sqrt{x}.$$

Recall how to combine similar terms.

This example, plus the facts we already know about operating on algebraic expressions, leads us to the following definition and rule.

Similar (or **like**) **radicals** are simplified radicals that have the same radicand and index number.

Only similar radicals can be added or subtracted.

Note that these radicals all have the same radicand and index number.

Example 1   Simplify: $4\sqrt[3]{2} + 3\sqrt[3]{2} - \sqrt[3]{2}$

Solution $4\sqrt[3]{2} + 3\sqrt[3]{2} - \sqrt[3]{2} = (4 + 3 - 1)\sqrt[3]{2}$
$$= 6\sqrt[3]{2}$$

Example 2   Simplify: $3\sqrt{2} + \sqrt{50} + \sqrt{32}$

At first glance, you might not realize that these radicals can be expressed as similar radicals.

Solution   We first simplify $\sqrt{50}$ and $\sqrt{32}$, obtaining

$$3\sqrt{2} + \sqrt{50} + \sqrt{32} = 3\sqrt{2} + 5\sqrt{2} + 4\sqrt{2}$$
$$= (3 + 5 + 4)\sqrt{2}$$
$$= 12\sqrt{2}.$$

Example 3 Simplify: $\sqrt[3]{a^4} + \sqrt[3]{8a^4} - \sqrt[6]{4a^2}$

Solution Again, we first simplify the radicals, obtaining

$$\sqrt[3]{a^4} + \sqrt[3]{8a^4} - \sqrt[6]{4a^2} = a\sqrt[3]{a} + 2a\sqrt[3]{a} - \sqrt[3]{2a}$$
$$= 3a\sqrt[3]{a} - \sqrt[3]{2a}.$$

We use the techniques learned in the previous section.

Note that $\sqrt[3]{a}$ and $\sqrt[3]{2a}$ are not similar radicals and therefore cannot be combined.

▼ **EXERCISE 6-7-1**

Simplify by combining similar radicals.

1. $\sqrt{2} + 3\sqrt{2}$ $= 4\sqrt{2}$
wow!

2. $2\sqrt{5} + 4\sqrt{5}$ $6\sqrt{5}$

3. $3\sqrt{x} + 4\sqrt{x}$ $7\sqrt{x}$

4. $2\sqrt{x} + 5\sqrt{x}$ $3\sqrt{x}$

5. $3\sqrt{3} - 2\sqrt{3}$ $\sqrt{3}$

6. $8\sqrt{7} - 9\sqrt{7}$ $-\sqrt{7}$

7. $\sqrt{2} + 3\sqrt{2} + 5\sqrt{2}$ $9\sqrt{2}$

8. $4\sqrt{3} + \sqrt{3} + 2\sqrt{3}$ $7\sqrt{3}$

9. $\sqrt{3} + 4\sqrt{3} - 2\sqrt{3}$ $3\sqrt{3}$

10. $5\sqrt{2} - 7\sqrt{2} + \sqrt{2}$

11. $3\sqrt{5} - 2\sqrt{3} + 4\sqrt{5} - \sqrt{3}$
$7\sqrt{5} - 3\sqrt{3}$

12. $2\sqrt{6} + 3\sqrt[3]{6} - \sqrt{6} + 2\sqrt[3]{6}$

13. $2\sqrt{2} - 3\sqrt[3]{2} - \sqrt{8}$

14. $3\sqrt{5} - 6\sqrt{5} + 2\sqrt[3]{5}$

[handwritten: $2 \cdot 3\sqrt{2}$ $5\sqrt{25 \cdot 2}$]

15. $\sqrt{8} - 2\sqrt{18} + \sqrt{50}$ *[handwritten: AS IS]*

[handwritten: $2\sqrt{2} - 6\sqrt{2} + 5\sqrt{2} = \sqrt{2}$]

16. $\sqrt{9} + \sqrt{18} - \sqrt{36}$

17. $\sqrt[3]{16} - \sqrt[3]{54} + \sqrt[3]{3}$

18. $\sqrt[3]{24} + \sqrt[3]{3} - \sqrt[3]{40}$

19. $\sqrt{50} + \sqrt[3]{2} - \sqrt{32} - \sqrt[3]{16}$

20. $\sqrt{a} + \sqrt{8a^3} + \sqrt[4]{16a^{10}} - 3\sqrt[6]{8a^9}$

Multiplication with radicals can also be accomplished by changing the radicals to exponential form and using the laws of exponents. However, the following examples will show that working with radical forms may be more convenient.

$$\sqrt[n]{a}\sqrt[n]{b} = \sqrt[n]{ab}$$

This rule is an application of the fourth law of exponents. Since

Definition of fractional exponents.

$$\sqrt[n]{a} = a^{1/n} \text{ and } \sqrt[n]{b} = b^{1/n}$$

then

Fourth law of exponents.

$$\sqrt[n]{a}\sqrt[n]{b} = a^{1/n}b^{1/n} = (ab)^{1/n} = \sqrt[n]{ab}.$$

Example 4 $\sqrt{3}\sqrt{7} = \sqrt{(3)(7)} = \sqrt{21}$

Example 5 $(2\sqrt{5})(3\sqrt{2}) = (2)(3)\sqrt{(5)(2)} = 6\sqrt{10}$

Notice how the product in examples 6 and 7 can be simplified.

Example 6 $\sqrt[3]{4}\sqrt[3]{2} = \sqrt[3]{8} = 2$

Example 7 $\sqrt[5]{x^2}\sqrt[5]{x^4} = \sqrt[5]{x^6} = x\sqrt[5]{x}$

This rule can be expanded to simplify the following example.

Example 8   Multiply: $2\sqrt{2}(3\sqrt{3} + 4\sqrt{5})$

Solution   Here we multiply each term in the parentheses by $2\sqrt{2}$. Hence,

$$2\sqrt{2}(3\sqrt{3} + 4\sqrt{5}) = 6\sqrt{6} + 8\sqrt{10}.$$

We use the distributive property.

Example 9   Simplify: $2\sqrt[3]{3}(\sqrt[3]{7} + \sqrt[3]{5}) - 4\sqrt[3]{5}(\sqrt[3]{3} + \sqrt[3]{2})$

Solution $2\sqrt[3]{3}(\sqrt[3]{7} + \sqrt[3]{5}) - 4\sqrt[3]{5}(\sqrt[3]{3} + \sqrt[3]{2})$
$= 2\sqrt[3]{21} + 2\sqrt[3]{15} - 4\sqrt[3]{15} - 4\sqrt[3]{10}$
$= 2\sqrt[3]{21} - 2\sqrt[3]{15} - 4\sqrt[3]{10}$

Combine similar radicals.

The methods of multiplying polynomial expressions that were covered in chapter 4 can also be transferred to the multiplication of some radical expressions.

Example 10   Expand: $(\sqrt{3} + 4\sqrt{2})(2\sqrt{3} - \sqrt{2})$

Solution   In chapter 4 we learned that

$$(x + 4y)(2x - y) = 2x^2 + 7xy - 4y^2.$$

The product of two binomials

Using the same pattern,

$$(\sqrt{3} + 4\sqrt{2})(2\sqrt{3} - \sqrt{2}) = 2(\sqrt{3})^2 + 7(\sqrt{3})(\sqrt{2}) - 4(\sqrt{2})^2$$
$$= 6 + 7\sqrt{6} - 8$$
$$= 7\sqrt{6} - 2.$$

Example 11   Expand: $(3\sqrt{2} + 5\sqrt{3})(3\sqrt{2} - 5\sqrt{3})$

Solution   Using the fact that

$$(a + b)(a - b) = a^2 - b^2,$$

we have

$$(3\sqrt{2} + 5\sqrt{3})(3\sqrt{2} - 5\sqrt{3}) = (3\sqrt{2})^2 - (5\sqrt{3})^2$$
$$= 9(2) - 25(3)$$
$$= -57.$$

Notice that the result has no radical. This idea will be useful in the next section.

Example 12   Expand: $(4\sqrt{5} + \sqrt{6})^2$

Solution   Using the fact that

$$(x + y)^2 = x^2 + 2xy + y^2,$$

we obtain

$$(4\sqrt{5} + \sqrt{6})^2 = (4\sqrt{5})^2 + 2(4\sqrt{5})(\sqrt{6}) + (\sqrt{6})^2$$
$$= 80 + 8\sqrt{30} + 6$$
$$= 86 + 8\sqrt{30}.$$

Squaring a binomial.
Of course, you could write the problem as $(4\sqrt{5} + \sqrt{6})(4\sqrt{5} + \sqrt{6})$ and proceed as in example 10.

▼ **EXERCISE 6-7-2**

Multiply and simplify when possible.

1. $\sqrt{2}\sqrt{5}$

2. $\sqrt[3]{3}\sqrt[3]{11}$ = $\sqrt[3]{33}$

3. $(2\sqrt{7})(3\sqrt{2})$ = $6\sqrt{14}$

4. $\sqrt{8}\sqrt{2}$

5. $(4\sqrt{a})(2\sqrt{a^3})$ $8\sqrt{a^4}$
$8a^2$

6. $(\sqrt[3]{16})(2\sqrt[3]{5})$

7. $\sqrt[4]{x^3}\sqrt[4]{x^2}$ $\sqrt[4]{x^6}$

8. $(3\sqrt{12a})(2\sqrt{18a})$

9. $(4\sqrt{20x^3})(-\sqrt{50x})$

10. $\sqrt[3]{54a^2}\sqrt[3]{-16a^2}$

11. $\sqrt{2}(\sqrt{3}+\sqrt{5})$

12. $\sqrt{3}(2\sqrt{2}-\sqrt{7})$

13. $3\sqrt{3}(2\sqrt{5}-3\sqrt{7})$ $6\sqrt{15}-9\sqrt{21}$

14. $\sqrt{2}(3\sqrt{3}-2\sqrt{2})$

15. $5\sqrt{3}(2\sqrt{6}+\sqrt{15})$ $10\sqrt{18}+5\sqrt{75}$
$10\sqrt{9\cdot2}+5\sqrt{25\cdot3}=30\sqrt{2}+25\sqrt{3}$

16. $\sqrt[3]{2}(4\sqrt[3]{4}-2\sqrt[3]{32})$

17. $4\sqrt{3}(\sqrt{6}-\sqrt{3}+\sqrt{18})$

18. $2\sqrt{2x}(5\sqrt{2x}-\sqrt{6x^3})$

19. $\sqrt{3}(\sqrt{2} + \sqrt{7}) + \sqrt{2}(\sqrt{5} - \sqrt{3})$

$\sqrt{6} + \sqrt{21} + \sqrt{10} - \sqrt{6}$

$\sqrt{21} + \sqrt{10}$

20. $\sqrt{2}(\sqrt{6} + \sqrt{2}) - \sqrt{3}(\sqrt{6} + \sqrt{3})$

$\sqrt{12} + \sqrt{4} - \sqrt{18} + \sqrt{9}$

21. $\sqrt{2}(3\sqrt{6} - \sqrt{10}) - 2\sqrt{3}(\sqrt{15} + 2\sqrt{12})$

22. $2\sqrt{5}(2\sqrt{3} - \sqrt{10}) - \sqrt{3}(3\sqrt{5} - \sqrt{6})$

23. $\sqrt[3]{3}(2\sqrt[3]{9} + \sqrt[3]{2}) + 3\sqrt[3]{2}(\sqrt[3]{3} - 3\sqrt[3]{4})$

24. $2\sqrt[3]{5}(\sqrt[3]{50} - \sqrt[3]{2}) - \sqrt[3]{2}(2\sqrt[3]{5} - \sqrt[3]{4})$

25. $(\sqrt{2} + \sqrt{5})(\sqrt{3} + \sqrt{2})$

26. $(\sqrt{6} + \sqrt{2})(\sqrt{3} - \sqrt{2})$

27. $(2\sqrt{3} - \sqrt{5})(3\sqrt{5} - \sqrt{2})$

28. $(3\sqrt{6} + 2\sqrt{3})(2\sqrt{3} - \sqrt{2})$

29. $(\sqrt{3} + \sqrt{5})(\sqrt{3} - \sqrt{5})$

30. $(2\sqrt{2} + \sqrt{6})(2\sqrt{2} - \sqrt{6})$

31. $(2\sqrt{2} - 5\sqrt{7})(2\sqrt{2} + 5\sqrt{7})$

32. $(2\sqrt{3} + \sqrt{2})(2\sqrt{3} - 5\sqrt{2})$

33. $(\sqrt{3} + \sqrt[3]{2})(2\sqrt{6} - 3\sqrt{10})$

34. $(\sqrt[3]{2} - 2\sqrt{3})(\sqrt{2} + 3\sqrt{6})$

35. $(\sqrt{2} + 3\sqrt{5})^2$

36. $(3\sqrt{5} - 4)^2$

37. $(2\sqrt{3} - 5\sqrt{7})^2$

38. $(3\sqrt{2} + 2\sqrt{5})^2$

 **6-8 SIMPLIFYING RADICAL EXPRESSIONS**

In the previous section we noted that radicals should be left in simplified form. The same holds true for algebraic expressions that contain radicals.

An algebraic expression containing radicals is in simplest form when:

1. Each individual radical is in simplest form.
2. No radical appears in the denominator of a fraction.

Simplifying the individual radicals was discussed in section 6–6, so we will concern ourselves here with the second requirement. Removing all radicals from the denominator of an expression is referred to as "rationalizing the denominator."

Problems of rationalizing the denominator fall into two categories:

1. Those with only one term in the denominator.
2. Those with two or more terms in the denominator.

Two different techniques are used so we will discuss these separately.

Example 1 Simplify: $\dfrac{1}{\sqrt{3}}$

Solution Our problem is to change the given fraction to another fraction with no radical in the denominator. Multiplying the numerator and denominator of any fraction by the same nonzero number will yield an equal fraction. We must multiply by a number that will give a perfect square in the denominator. In this case, the number is $\sqrt{3}$. Thus,

We use $\sqrt{3}$ since
$$\sqrt{3} \cdot \sqrt{3} = \sqrt{9} = 3.$$

$$\frac{1}{\sqrt{3}} = \frac{1}{\sqrt{3}} \cdot \frac{\sqrt{3}}{\sqrt{3}}$$
$$= \frac{\sqrt{3}}{\sqrt{9}}$$
$$= \frac{\sqrt{3}}{3}.$$

This form satisfies the requirement that there is no radical in the denominator. That is to say, we have "rationalized the denominator."

Example 2 Simplify: $\dfrac{3}{\sqrt{8}}$

Solution Here the requirement is to find a number that when multiplied by $\sqrt{8}$ yields the square root of a perfect square. Since $(\sqrt{8})(\sqrt{2}) = \sqrt{16}$, the desired number is $\sqrt{2}$. Hence,

Don't forget to multiply the numerator by $\sqrt{2}$ also.

$$\frac{3}{\sqrt{8}} = \frac{3}{\sqrt{8}} \cdot \frac{\sqrt{2}}{\sqrt{2}}$$
$$= \frac{3\sqrt{2}}{\sqrt{16}}$$
$$= \frac{3\sqrt{2}}{4}.$$

Example 3 Simplify: $\dfrac{2}{\sqrt[3]{5}}$

Solution We must now find a number that when multiplied by $\sqrt[3]{5}$ yields the cube root of a perfect cube. Since

$$(\sqrt[3]{5})(\sqrt[3]{25}) = \sqrt[3]{125}$$

the desired number is $\sqrt[3]{25}$. Thus,

$$\begin{aligned}
\frac{2}{\sqrt[3]{5}} &= \frac{2}{\sqrt[3]{5}} \cdot \frac{\sqrt[3]{25}}{\sqrt[3]{25}} \\
&= \frac{2\sqrt[3]{25}}{\sqrt[3]{125}} \\
&= \frac{2\sqrt[3]{25}}{5}.
\end{aligned}$$

Be careful here. The denominator has a cube root.

Again, make sure you also multiply the numerator by the radical.

Example 4 Simplify: $\dfrac{x}{\sqrt[3]{x^2 y}}$

Solution Remember that perfect cube exponents are multiples of three.

$$\begin{aligned}
\frac{x}{\sqrt[3]{x^2 y}} &= \frac{x}{\sqrt[3]{x^2 y}} \cdot \frac{\sqrt[3]{x y^2}}{\sqrt[3]{x y^2}} \\
&= \frac{x\sqrt[3]{x y^2}}{\sqrt[3]{x^3 y^3}} \\
&= \frac{x\sqrt[3]{x y^2}}{xy} \\
&= \frac{\sqrt[3]{x y^2}}{y}
\end{aligned}$$

We need an x^3 and y^3 under the radical.

Here we can reduce the fraction.

Example 5 Simplify: $\dfrac{3}{\sqrt[5]{16 x^3 y^2}}$

Solution

$$\begin{aligned}
\frac{3}{\sqrt[5]{16 x^3 y^2}} &= \frac{3}{\sqrt[5]{16 x^3 y^2}} \cdot \frac{\sqrt[5]{2 x^2 y^3}}{\sqrt[5]{2 x^2 y^3}} \\
&= \frac{3\sqrt[5]{2 x^2 y^3}}{\sqrt[5]{32 x^5 y^5}} \\
&= \frac{3\sqrt[5]{2 x^2 y^3}}{2xy}
\end{aligned}$$

We need $2^5 x^5 y^5$ under the radical.

Example 6   Simplify: $\sqrt{\dfrac{x + y}{x}}$

Solution   Remember that $\left(\dfrac{a}{b}\right)^n = \dfrac{a^n}{b^n}$. Thus,

Here we have radicals in both numerator and denominator. It is only the radical in the denominator that we need eliminate.

$$\sqrt{\dfrac{x + y}{x}} = \dfrac{\sqrt{x + y}}{\sqrt{x}}$$

$$= \dfrac{\sqrt{x + y}}{\sqrt{x}} \cdot \dfrac{\sqrt{x}}{\sqrt{x}}$$

$$= \dfrac{\sqrt{x(x + y)}}{\sqrt{x^2}}$$

$$= \dfrac{\sqrt{x^2 + xy}}{x}.$$

Example 7   Simplify: $\dfrac{a - b}{\sqrt{a + b}}$

Solution

In this example there are two terms under the radical, but the radical itself is still only one term.

$$\dfrac{a - b}{\sqrt{a + b}} = \dfrac{a - b}{\sqrt{a + b}} \cdot \dfrac{\sqrt{a + b}}{\sqrt{a + b}}$$

$$= \dfrac{(a - b)\sqrt{a + b}}{\sqrt{(a + b)^2}}$$

$$= \dfrac{(a - b)\sqrt{a + b}}{a + b}$$

▼ **EXERCISE 6-8-1**

Simplify.

1. $\dfrac{1}{\sqrt{2}}$ **2.** $\dfrac{1}{\sqrt{5}}$ **3.** $\dfrac{1}{\sqrt{7}}$ **4.** $\dfrac{2}{\sqrt{3}}$ **5.** $\dfrac{1}{\sqrt{12}}$

6. $\dfrac{3}{\sqrt{20}}$ **7.** $\dfrac{1}{\sqrt[3]{2}}$ **8.** $\dfrac{1}{\sqrt[3]{4}}$ **9.** $\dfrac{1}{\sqrt[5]{8}}$ **10.** $\dfrac{1}{\sqrt[5]{9}}$

11. $\dfrac{1}{\sqrt[3]{x^2}}$ 12. $\dfrac{2}{\sqrt[5]{x^3}}$ 13. $\dfrac{3}{\sqrt[3]{2x}}$ 14. $\dfrac{x}{\sqrt[3]{x}}$ 15. $\dfrac{2y}{\sqrt[3]{4x^2y}}$

16. $\dfrac{2xy}{\sqrt[5]{8x^3y^2}}$ 17. $\sqrt{\dfrac{1}{2}}$ 18. $\sqrt{\dfrac{3}{5}}$ 19. $\sqrt{\dfrac{1}{8}}$ 20. $\sqrt{\dfrac{x+2}{x}}$

21. $\dfrac{1}{\sqrt{x+3}}$ 22. $\sqrt{\dfrac{x}{x+1}}$ 23. $\sqrt{\dfrac{8}{x-2}}$ 24. $\dfrac{x+3}{\sqrt{x-5}}$ 25. $\dfrac{x+y}{\sqrt{x+y}}$

26. $\dfrac{x-3}{\sqrt{x-3}}$

When the denominator of a fraction contains two or more terms and one or more of these contains a radical, the problem of rationalizing the denominator becomes more complex. We will discuss only the simplest type of these problems, which are those problems with two terms, in which one or both contain radicals with the index of two. Problems containing radicals with higher indices and more than two terms in the denominator are left for more advanced courses.

Example 8   Simplify: $\dfrac{3}{\sqrt{5}+1}$

This denominator contains two terms, $\sqrt{5}$ and 1.

Solution $(\sqrt{5}+1)$ is a binomial. Multiplication of a binomial by another binomial will yield at least a trinomial in all cases except the special case of the product of the sum and difference of the same two numbers.

$$(a-b)(a+b) = a^2 - b^2$$

If $(\sqrt{5} + 1)$ is multiplied by any number other than $(\sqrt{5} - 1)$, a middle term still containing a radical will result. Therefore, to rationalize the denominator in this example we will multiply numerator and denominator by $(\sqrt{5} - 1)$. Thus,

Again, don't forget to also multiply the numerator by the same quantity.

$$\frac{3}{\sqrt{5} + 1} = \frac{3}{(\sqrt{5} + 1)} \cdot \frac{(\sqrt{5} - 1)}{(\sqrt{5} - 1)}$$

$$= \frac{3(\sqrt{5} - 1)}{\sqrt{25} - 1}$$

$\sqrt{25} - 1 = 5 - 1 = 4$

$$= \frac{3\sqrt{5} - 3}{4}.$$

To simplify an algebraic expression having a binomial of the form $(\sqrt{a} \pm \sqrt{b})$ in the denominator multiply the numerator and denominator by a binomial having the same terms but whose second term has the opposite sign.

Example 9 Simplify: $\dfrac{5}{\sqrt{2} + \sqrt{3}}$

Since the denominator is $\sqrt{2} + \sqrt{3}$, the rule says to multiply by $\sqrt{2} - \sqrt{3}$.

Solution

$$\frac{5}{\sqrt{2} + \sqrt{3}} = \frac{5}{(\sqrt{2} + \sqrt{3})} \cdot \frac{(\sqrt{2} - \sqrt{3})}{(\sqrt{2} - \sqrt{3})}$$

$$= \frac{5(\sqrt{2} - \sqrt{3})}{2 - 3}$$

$$= \frac{5(\sqrt{2} - \sqrt{3})}{-1}$$

$$= -5(\sqrt{2} - \sqrt{3})$$

This is the same as $-5\sqrt{2} + 5\sqrt{3}$.

or $= 5\sqrt{3} - 5\sqrt{2}$

Example 10 Simplify: $\dfrac{\sqrt{2} - 1}{\sqrt{2} + 1}$

We have two binomials to multiply in the numerator as well as in the denominator.

Solution

$$\frac{\sqrt{2} - 1}{\sqrt{2} + 1} = \frac{(\sqrt{2} - 1)}{(\sqrt{2} + 1)} \cdot \frac{(\sqrt{2} - 1)}{(\sqrt{2} - 1)}$$

$$= \frac{2 - 2\sqrt{2} + 1}{2 - 1}$$

$$= 3 - 2\sqrt{2}$$

Example 11 Simplify: $\dfrac{2\sqrt{x} + y}{3\sqrt{x} - \sqrt{y}}$

Solution

$$\frac{2\sqrt{x} + y}{3\sqrt{x} - \sqrt{y}} = \frac{(2\sqrt{x} + y)}{(3\sqrt{x} - \sqrt{y})} \cdot \frac{(3\sqrt{x} + \sqrt{y})}{(3\sqrt{x} + \sqrt{y})}$$

$$= \frac{6x + 3y\sqrt{x} + 2\sqrt{xy} + y\sqrt{y}}{9x - y}$$

There are a lot of radicals here—be careful.

▼ **Exercise 6–8–2**

Simplify.

1. $\dfrac{1}{\sqrt{3} + \sqrt{2}}$

2. $\dfrac{1}{\sqrt{2} + \sqrt{5}}$

3. $\dfrac{1}{\sqrt{2} + 3}$

4. $\dfrac{1}{\sqrt{3} + 1}$

5. $\dfrac{1}{\sqrt{2} - 1}$

6. $\dfrac{1}{\sqrt{5} - 3}$

7. $\dfrac{1}{\sqrt{x} + \sqrt{y}}$

8. $\dfrac{1}{\sqrt{x} - \sqrt{y}}$

9. $\dfrac{2}{\sqrt{5} + 1}$

10. $\dfrac{4}{3 - \sqrt{5}}$

11. $\dfrac{6}{\sqrt{x} + y}$

12. $\dfrac{x}{\sqrt{x} - 2}$

13. $\dfrac{3}{\sqrt{2} + \sqrt{3}}$

14. $\dfrac{4}{\sqrt{3} + \sqrt{5}}$

15. $\dfrac{1}{\sqrt{2} - \sqrt{3}}$

16. $\dfrac{1}{\sqrt{3} - \sqrt{5}}$

17. $\dfrac{\sqrt{2}}{\sqrt{2} + 1}$

18. $\dfrac{\sqrt{2}}{\sqrt{6} - 2}$

19. $\dfrac{\sqrt{3} + 1}{\sqrt{3} - 1}$

20. $\dfrac{1 - \sqrt{2}}{1 + \sqrt{2}}$

21. $\dfrac{a - 1}{\sqrt{a} - 1}$

22. $\dfrac{x - 4}{\sqrt{x} + 2}$

23. $\dfrac{5}{\sqrt{7} - \sqrt{2}}$

24. $\dfrac{y}{\sqrt{x} - \sqrt{y}}$

25. $\dfrac{\sqrt{3} + \sqrt{2}}{\sqrt{3} - \sqrt{2}}$

26. $\dfrac{\sqrt{7} - \sqrt{3}}{\sqrt{7} + \sqrt{3}}$

27. $\dfrac{2\sqrt{3} - 5}{3\sqrt{3} + 1}$

28. $\dfrac{3\sqrt{2} + 4}{5\sqrt{2} - 1}$

29. $\dfrac{3\sqrt{2} + \sqrt{3}}{\sqrt{2} - 2\sqrt{3}}$

30. $\dfrac{2\sqrt{5} - \sqrt{2}}{\sqrt{5} + 3\sqrt{2}}$

◣◣ 6-9 EXERCISES ON EXPONENTS AND RADICALS

In future work you may encounter many problems in which exponents and radicals of every type occur. We now wish to present exercises that combine the new concepts learned in this chapter with skills you have previously developed.

In section 6-5 we noted that fractional exponents follow the same laws for integral exponents.

OBJECTIVES

Upon completing this section you should be able to:

1. Simplify expressions containing fractional exponents.
2. Simplify various radical expressions.

Example 1   Simplify: $x^{1/2}x^{-1/3}$

Solution
$$x^{1/2}x^{-1/3} = x^{3/6}x^{-2/6}$$
$$= x^{3/6 - 2/6} = x^{1/6}$$

Example 2   Simplify: $(x^{1/3} + y^{1/3})(x^{1/3} - y^{1/3})$

Solution
$(x^{1/3} + y^{1/3})(x^{1/3} - y^{1/3})$
$= x^{1/3}x^{1/3} - x^{1/3}y^{1/3} + x^{1/3}y^{1/3} - y^{1/3}y^{1/3}$
$= x^{2/3} - y^{2/3}$

These are multiplied the same as any other two binomials.

A fractional exponent in the denominator of a fraction is the same as a radical in the denominator. Simplifying such an expression requires that fractional exponents be removed from the denominator.

Example 3   Simplify: $\dfrac{y}{x^{2/3}}$

Solution   We could change the problem to radical form and proceed as in section 6-8.

If we work with the expression as is, we must multiply the numerator and denominator by a number that will give an integer exponent in the denominator. In this case, we will multiply the numerator and denominator by $x^{1/3}$.

$$\dfrac{y}{x^{2/3}} = \dfrac{y}{x^{2/3}} \cdot \dfrac{x^{1/3}}{x^{1/3}}$$
$$= \dfrac{x^{1/3}y}{x^{2/3 + 1/3}}$$
$$= \dfrac{x^{1/3}y}{x}$$

$$\dfrac{y}{x^{2/3}} = \dfrac{y}{\sqrt[3]{x^2}}$$

Work the problem in this form as you did in section 6-8. Then change the final answer to exponential form.

Example 4   Simplify: $\dfrac{x}{x^{1/2} - y^{1/2}}$

Note we are using the same rule from section 6–8.

Solution

$$\frac{x}{x^{1/2} - y^{1/2}} = \frac{x}{x^{1/2} - y^{1/2}} \cdot \frac{x^{1/2} + y^{1/2}}{x^{1/2} + y^{1/2}}$$

$$= \frac{x(x^{1/2} + y^{1/2})}{x - y}$$

Combining radicals sometimes requires obtaining a common denominator.

Example 5   Simplify: $\sqrt{5} + \sqrt{\dfrac{2}{5}}$

The fifth law of exponents.

Solution   We use the law $\left(\dfrac{a}{b}\right)^n = \dfrac{a^n}{b^n}$ to obtain

We need to eliminate the radical from the denominator.

$$\sqrt{5} + \sqrt{\frac{2}{5}} = \sqrt{5} + \frac{\sqrt{2}}{\sqrt{5}}$$

$$= \sqrt{5} + \frac{\sqrt{2}(\sqrt{5})}{\sqrt{5}(\sqrt{5})}$$

Finally, we obtain a common denominator.

$$= \sqrt{5} + \frac{\sqrt{10}}{5}$$

$$= \frac{5\sqrt{5} + \sqrt{10}}{5}.$$

The following exercise will provide practice in using the skills developed in this chapter.

▼ EXERCISE 6–9–1

Simplify. Use only positive exponents in the answers.

1. $x^{1/2}x^{1/3}$

2. $x^{3/4}x^{1/3}$

3. $x^{-1/5}x^{1/2}$

4. $x^{1/3}x^{-1/4}$

5. $x^{2/3}(x^{1/2} + x^{1/4})$

6. $x^{-1/2}(x^{3/4} - x^{4/5})$

7. $x^{1/4}(x^{-1/8} - x^{2/3})$

8. $(x^{1/2} + y^{1/2})(x^{1/2} - y^{1/2})$

9. $(x^{2/3} + y^{2/3})(x^{2/3} - y^{2/3})$

10. $(x^{1/2} + y^{1/2})^2$

11. $x^{-1} + y^{-1}$

12. $x^{-2} - y^{-2}$

13. $\left(\dfrac{x^{1/2}}{x}\right)^2$

14. $\left(\dfrac{x^{1/2}}{x^{1/3}}\right)^6$

15. $\dfrac{1}{x^{1/2}}$

16. $\dfrac{1}{x^{1/3}}$

17. $\dfrac{2}{x^{3/2}}$

18. $\dfrac{3}{x^{2/5}}$

19. $\dfrac{2}{x^{5/6}}$

20. $\dfrac{y^{2/3}}{2x^{4/3}}$

21. $\dfrac{2}{x^{1/3}y^{3/5}}$

22. $\dfrac{1}{x^{1/2} - 2}$

23. $\dfrac{x}{x^{1/2} + 5}$

24. $\dfrac{x}{x^{1/2} + 3^{1/2}}$

Simplify the following radical expressions.

25. $\sqrt{2} + \sqrt{\dfrac{1}{2}}$

26. $\sqrt{3} - \sqrt{\dfrac{1}{3}}$

27. $\sqrt{\dfrac{1}{8}} + \sqrt{2}$

28. $\sqrt{\dfrac{1}{50}} - \sqrt{8}$

29. $\sqrt{\dfrac{1}{12}} - \sqrt{27}$

30. $\sqrt{\dfrac{1}{2}} + \sqrt{\dfrac{1}{8}} - 3\sqrt{2}$

31. $\sqrt{\dfrac{1}{12}} - \sqrt{75} + 2\sqrt{3}$

32. $\sqrt{28} + \sqrt{\dfrac{1}{63}} - 3\sqrt{7}$

33. $\sqrt{\dfrac{1}{24}} - \sqrt{\dfrac{1}{54}} + \sqrt{\dfrac{1}{150}}$

34. $\sqrt{20} - \sqrt{\dfrac{4}{45}} + \sqrt{\dfrac{5}{9}}$ **35.** $\sqrt{\dfrac{2}{3}} + \sqrt{6}$ **36.** $\sqrt{\dfrac{2}{5}} + \sqrt{10}$

37. $\sqrt{6} + \sqrt{\dfrac{2}{75}}$ **38.** $\sqrt{\dfrac{5}{8}} - \sqrt{10}$ **39.** $\sqrt{\dfrac{3}{5}} + \sqrt{\dfrac{3}{20}} - \sqrt{\dfrac{12}{45}}$

40. $\sqrt{\dfrac{2}{3}} - \sqrt{\dfrac{2}{27}} + \sqrt{\dfrac{8}{75}}$

SUMMARY

Key Words

Section 6–2

- A **zero exponent** is defined as

$$x^0 = 1, \text{ if } x \neq 0.$$

- A **negative exponent** is defined as

$$x^{-a} = \frac{1}{x^a}, \text{ if } x \neq 0.$$

Section 6–3

- A number is in **scientific notation** if it is expressed as the product of a number n ($1 \leq n < 10$) and a power of ten.

Section 6–4

- x is an ***n*th root** of y if $x^n = y$.

- The **principal *n*th root** of x is indicated by the symbol $\sqrt[n]{x}$, where x is called the **radicand** and n is called the **index.**

Section 6–5

- A **fractional exponent** is defined by the statement

$$x^{a/b} = \sqrt[b]{x^a} \text{ or } (\sqrt[b]{x})^a.$$

Section 6–6

- A radical is in **simplest form** when
  1. no power of a factor under the radical is greater than or equal to the index and
  2. the index cannot be smaller.

Section 6–7

- **Similar** or **like radicals** are simplified radicals that have the same radicand and index.

Procedures

Section 6–2

- The laws of exponents for integers are
  1. $x^a x^b = x^{a+b}$
  2. $(x^a)^b = x^{ab}$
  3. $\dfrac{x^a}{x^b} = x^{a-b}$
  4. $(xy)^a = x^a y^a$
  5. $\left(\dfrac{x}{y}\right)^a = \dfrac{x^a}{y^a}$

Section 6–4

- The principal nth root of x (written $\sqrt[n]{x}$) is
  a. zero, if $x = 0$.
  b. positive, if $x > 0$.
  c. negative, if $x < 0$ and n is odd.
  d. not a real number, if $x < 0$ and n is even.

Section 6–5

- To interchange exponential and radical form use the definition of fractional exponents.

Section 6–7

- Only similar radicals can be combined.

- To multiply radicals with the same index use the rule

$$\sqrt[n]{a}\,\sqrt[n]{b} = \sqrt[n]{ab}.$$

Section 6–8

- To simplify an algebraic expression containing radicals:
  1. Simplify each radical in the expression.
  2. Rationalize all denominators in the expression.

NAME: _____

CLASS / SECTION: _____   DATE: _____

Apply the laws of exponents to each of the following.

1. $(2x^2y^3)^5$

2. $\left(\dfrac{x}{y^2}\right)^3$

3. $\left(\dfrac{2x^3}{3y}\right)^3$

4. $[-3(x^2y)^2]^3[2xy^2]^2$

5. $\left(\dfrac{-2}{x^2}\right)^3\left(\dfrac{x}{4}\right)^2$

6. $\left(\dfrac{x^2y^3}{3}\right)^4\left(\dfrac{-9}{xy^5}\right)^3$

Simplify.

7. x^8x^{-2}

8. $\dfrac{x^2 + 3^0}{x^{-1}}$

9. $(x^3)^{-4}(x^{-2})^3$

10. $\left(\dfrac{2x}{y}\right)^{-2}\left(\dfrac{x^3}{y^2}\right)^3$

11. $\dfrac{(x^2 - y^2)^{-1}}{(x + y)^{-2}}$

12. $\dfrac{a^{-1} - b^{-1}}{a - b}$

13. Write 37,623,000 in scientific notation.

14. Write .00000571 in scientific notation.

15. Write 4.12×10^4 without exponents.

16. Write 3.5×10^{-4} without exponents.

1. _____

2. _____

3. _____

4. _____

5. _____

6. _____

7. _____

8. _____

9. _____

10. _____

11. _____

12. _____

13. _____

14. _____

15. _____

16. _____

A N S W E R S

17. _____

18. _____

19. _____

20. _____

21. _____

22. _____

23. _____

24. _____

25. _____

26. _____

27. _____

28. _____

29. _____

30. _____

31. _____

32. _____

33. _____

34. _____

35. _____

Compute, and leave answers in scientific notation.

17. $\dfrac{6.8 \times 10^5}{3.4 \times 10^{-5}}$

18. $\dfrac{(2.14 \times 10^8)(3.2 \times 10^{-7})}{(4.0 \times 10^{10})}$

Evaluate.

19. $\sqrt{16}$

20. $\sqrt[3]{-125}$

21. $\sqrt[5]{243}$

22. $\sqrt{175}$

23. $\sqrt{408}$

24. $\sqrt{657}$

Evaluate.

25. $27^{2/3}$

26. $\left(\dfrac{4}{9}\right)^{-1/2}$

27. $\left(-\dfrac{125}{8}\right)^{-2/3}$

28. $-64^{-2/3}$

29. $\sqrt[3]{(216)^2}$

30. $\sqrt[5]{(-1)^3}$

Simplify.

31. $\sqrt{50}$

32. $\sqrt[3]{32}$

33. $\sqrt[5]{x^6 y^{12}}$

34. $\sqrt{75x^3 y^6}$

35. $\sqrt[3]{-40x^3 y^7}$

Simplify.

36. $\sqrt{75} + 3\sqrt{24} - 4\sqrt{27}$

37. $\sqrt{2a^3} - 3\sqrt{4a^2} + 5a\sqrt{288a}$

36. _____

37. _____

38. $(2\sqrt{12x^3})(3\sqrt{48x^5})$

39. $\sqrt{5}(3\sqrt{20} - 2\sqrt{3})$

38. _____

39. _____

40. $(2\sqrt{6} - \sqrt{2})^2$

40. _____

41. _____

Simplify.

41. $\dfrac{1}{\sqrt{18}}$

42. $\dfrac{3}{\sqrt[3]{4}}$

43. $\dfrac{a - b}{\sqrt{a - b}}$

42. _____

43. _____

44. $\dfrac{x - 2}{\sqrt{x} - \sqrt{2}}$

45. $\dfrac{\sqrt{3} - 1}{2\sqrt{3} + 3}$

44. _____

45. _____

Simplify.

46. $x^{2/5}x^{1/3}$

47. $(x^{1/2} - 3^{1/2})^2$

48. $\dfrac{3}{x^{1/2}y^{2/3}}$

49. $\sqrt{\dfrac{1}{32}} - 2\sqrt{8} + \sqrt{\dfrac{1}{50}}$

50. $\sqrt{\dfrac{2}{27}} + \sqrt{\dfrac{3}{50}}$

46. _____

47. _____

48. _____

49. _____

50. _____

SCORE: _____

1. Simplify: **a.** $[(x^2y^3)^4]^3$ **b.** $\left(\dfrac{2x}{y^2}\right)^3\left(\dfrac{y}{2x^2}\right)^5$

A N S W E R S

1a. _____

2. Simplify using only positive exponents:

   a. $\dfrac{x^{-1}y^{-4}}{x^3y^{-2}}$ **b.** $(a + b^{-1})^{-1}$

1b. _____

2a. _____

2b. _____

3. Write 2.56×10^{-5} without exponents.

4. Compute, and leave the answer in scientific notation.
$$\frac{(6.6 \times 10^{15})(7.0 \times 10^{-8})}{(2.2 \times 10^2)}$$

3. _____

4. _____

5. Evaluate: $\sqrt[3]{-216}$ **6.** Evaluate: $\left(\dfrac{49}{25}\right)^{-1/2}$

5. _____

6. _____

7. Simplify: **a.** $\sqrt[5]{x^6y^{12}}$ **b.** $\sqrt[3]{-27ab^8}$

7a. _____

7b. _____

8. Multiply and simplify:
   a. $\sqrt{3}(\sqrt{2} + 3\sqrt{5})$
   $+ \sqrt{2}(\sqrt{5} - 2\sqrt{3})$ **b.** $(\sqrt{3} - 4\sqrt{2})(\sqrt{3} + 4\sqrt{2})$

8a. _____

8b. _____

9. Simplify: **a.** $\dfrac{1}{\sqrt[3]{2}}$ **b.** $\dfrac{5}{\sqrt{2} - \sqrt{x}}$

9a. _____

9b. _____

10a. _____

10. a. Remove fractional exponents from the denominator: $\dfrac{1}{x^{2/3}y^{1/5}}$ **b.** Simplify:
$$\sqrt{2} - \sqrt{\frac{1}{2}} + \sqrt{\frac{3}{8}}$$

10b. _____

SCORE: _____

CUMULATIVE TEST

NAME:

CLASS / SECTION:

DATE:

A N S W E R S

1. Evaluate: $\left|-(+14)\right|$

2. Find the product:
   $2a^2b(4a^3b^2 - ab^4 + 6a^3b)$

1. _____

2. _____

3. Evaluate $3x^3y + x^2y^2 - 4xy^3$
   if $x = -3, y = -1$.

4. Solve for x:
   $2x - (4x - 2)$
   $= \dfrac{1}{2}(x + 1) - 1$

3. _____

4. _____

5. Solve for x and graph the solution on the number line:
   $5x + 3 < 7x + 5$

   0

5. _____

6. Solve for x: $\left|2x - 5\right| = 9$

6. _____

7. The sum of two numbers is 48. Write expressions for the two numbers.

8. Tom is two years younger than Jean and four years older than Alice. The sum of their ages is 64. Find Tom's age.

7. _____

8. _____

9. A 20% salt solution is added to a 45% salt solution to obtain ten liters of a 30% salt solution. How many liters of each is to be used?

10. Multiply:
    $(3x - 1)(2x^2 - 3x + 4)$

9. _____

10. _____

11. Factor: $3x - 3y - ax + ay$

12. Factor: $8x^2 - 10xy + 3y^2$

13. Factor: $x^2 + 4x + 4 - y^2$

14. Simplify: $\dfrac{2x^2 - 8x}{x^2 - 16}$

15. Subtract: $\dfrac{2x}{x - 1} - \dfrac{3x + 1}{x^2 - 1}$

16. Simplify: $\dfrac{\dfrac{2}{x - 3} + 1}{\dfrac{x - 1}{x^2 - 2x - 3}}$

17. Solve for x:

$$\frac{2}{x + 2} + \frac{1}{2} = \frac{x}{2x - 4}$$

18. Write 24,500,000 in scientific notation.

19. Simplify: $\dfrac{3}{\sqrt{5} + \sqrt{2}}$

20. Remove fractional exponents from the denominator: $\dfrac{2}{x^{1/2}y^{1/5}}$

11. _____

12. _____

13. _____

14. _____

15. _____

16. _____

17. _____

18. _____

19. _____

20. _____

SCORE: _____

C H A P T E R

7

PRETEST

Before beginning this chapter, answer as many of the following questions as you can. When you have finished the chapter, take the practice test at the end of the chapter and compare the scores of the two tests to see how much you have learned.

A N S W E R S

1. Solve: $3x^2 + 11x - 4 = 0$

1. _____

2a. _____

2. Solve: **a.** $3x^2 = 5x$ **b.** $5x^2 - 10 = 0$

2b. _____

3. Solve by completing the square: $5x^2 + 3 = 20x$

4. Use the quadratic formula to solve: $3x^2 + 8x = 5$

3. _____

4. _____

5. Give the nature of the roots by evaluating the discriminant:
a. $3x^2 - 7x + 4 = 0$ **b.** $x^2 + 5x + 7 = 0$

5a. _____

5b. _____

6. Solve: $2x^2 - 2x + 5 = 0$

7. Solve: $3x^4 - 8x^2 + 4 = 0$

6. _____

7. _____

8. _____

8. Solve:
$$\sqrt{x + 2} + \sqrt{2x + 11} = 4$$

9. One number is five greater than another number. The sum of their squares is 97. Find the numbers.

9. _____

SCORE: _____

7

Quadratic Equations

A farm manager has 200 meters of fence on hand and wishes to enclose a rectangular field so that it will contain 2,400 square meters in area. What should the dimensions of the field be?

Solving equations is the central theme of algebra. All skills learned lead eventually to the ability to solve equations and simplify the solutions. In previous chapters we have solved equations of the first degree. You now have the necessary skills to solve equations of the second degree, which are known as *quadratic equations*.

◤◢ 7–1 QUADRATICS SOLVED BY FACTORING

O B J E C T I V E S

Upon completing this section you should be able to:

1. Identify a quadratic equation.
2. Place a quadratic equation in standard form.
3. Solve a quadratic equation by factoring.

A **quadratic equation** is a polynomial equation that contains the second degree, but no higher degree, of the variable.

The **standard form** of a quadratic equation is

$$ax^2 + bx + c = 0 \text{ when } a \neq 0.$$

All quadratic equations can be put in standard form, and any equation that can be put in standard form is a quadratic equation. In other words, the standard form represents all quadratic equations.

The *solution* to an equation is sometimes referred to as the *root* of the equation.

This theorem is proved in most college algebra books.

An important theorem, which cannot be proved at the level of this text, states "Every polynomial equation of degree *n* has exactly *n* roots." Using this fact tells us that quadratic equations will always have two solutions. It is possible that the two solutions are equal.

A quadratic equation will have two solutions because it is of degree two.

The simplest method of solving quadratics is by factoring. This method cannot always be used, because not all polynomials are factorable, but it is used whenever factoring is possible.

The method of solving by factoring is based on a simple theorem.

In other words, if the product of two factors is zero then at least one of the factors is zero.

If $AB = 0$, then either $A = 0$ or $B = 0$.

We will not attempt to prove this theorem but note carefully what it states. We can never multiply two numbers and obtain an answer of zero unless at least one of the numbers is zero. Of course, both of the numbers can be zero since $(0)(0) = 0$.

Example 1   Solve for *x* if $x^2 - 5x = 6$.

Solution   **Step 1** Put the equation in standard form.

We must subtract 6 from both sides.

$$x^2 - 5x = 6$$
$$x^2 - 5x - 6 = 0$$

Step 2 Factor completely.

$$x^2 - 5x - 6 = 0$$
$$(x - 6)(x + 1) = 0$$

Recall how to factor trinomials.

Step 3 Set each factor equal to zero and solve for x. Since we have $(x - 6)(x + 1) = 0$, we know that $x - 6 = 0$ or $x + 1 = 0$, in which case $x = 6$ or $x = -1$.

This applies the theorem on page 382, which says that at least one of the factors must have a value of zero.

Step 4 Check the solution in the original equation. If $x = 6$, then $x^2 - 5x = 6$ becomes $(6)^2 - 5(6) = 6$.

$$36 - 30 = 6$$
$$6 = 6$$

Checking your solutions is a sure way to tell if you have solved the equation correctly.

Therefore, $x = 6$ is a solution. If $x = -1$, then $x^2 - 5x = 6$ becomes $(-1)^2 - 5(-1) = 6$.

$$1 + 5 = 6$$
$$6 = 6$$

Therefore, -1 is a solution.

The solutions can be indicated either by writing $x = 6$ and $x = -1$ or by using set notation and writing $\{6, -1\}$, which we read "the solution set for x is 6 and -1." In this text we will use set notation.

In this example 6 and -1 are called the *elements* of the set.

Example 2 Solve for x if $3x^2 + 7x - 6 = 0$.

Note in this example that the equation is already in standard form.

Solution

$$3x^2 + 7x - 6 = 0$$
$$(3x - 2)(x + 3) = 0$$

If $3x - 2 = 0$,   If $x + 3 = 0$,
$$3x = 2 \qquad\qquad x = -3.$$
$$x = \frac{2}{3}.$$

Check: $3\left(\dfrac{2}{3}\right)^2 + 7\left(\dfrac{2}{3}\right) - 6 = 0$

Again, checking the solutions will assure you that you did not make an error in solving the equation.

$$3\left(\frac{4}{9}\right) + 7\left(\frac{2}{3}\right) - 6 = 0$$
$$\frac{12}{9} + \frac{14}{3} - 6 = 0$$
$$\frac{12}{9} + \frac{42}{9} - 6 = 0$$
$$\frac{54}{9} - 6 = 0$$
$$6 - 6 = 0$$

Hence, $\dfrac{2}{3}$ is a solution.

$$3(-3)^2 + (7)(-3) - 6 = 0$$
$$3(9) - 21 - 6 = 0$$
$$27 - 21 - 6 = 0$$
$$27 - 27 = 0$$

$\dfrac{2}{3}$ and -3 are also called *roots* of the equation.

Hence, -3 is a solution. The solution set is $\left\{ \dfrac{2}{3}, -3 \right\}$.

Example 3 Solve for x if $\dfrac{2}{x + 1} + 3x = 4$.

$(x + 1)$ is the least common denominator of all the fractions in the equation. Note that $x + 1 \neq 0$.

Solution We clear fractions by multiplying by $x + 1$.

$$(x + 1)\left(\dfrac{2}{x + 1} + 3x \right) = 4(x + 1)$$
$$2 + 3x(x + 1) = 4(x + 1)$$
$$2 + 3x^2 + 3x = 4x + 4$$
$$3x^2 - x - 2 = 0$$

Remember, *every term* of the equation must be multiplied by $(x + 1)$.

$$X^2 - 3 = 5X$$
$$X^2 - 5X - 3 = 0$$

Now factoring we have

$$(3x + 2)(x - 1) = 0.$$

If $3x + 2 = 0,$	If $x - 1 = 0,$
$3x = -2$	$x = 1.$

Then $x = -\dfrac{2}{3}$.

Check the solutions in the original equation.

Check in the original equation to make sure you do not obtain a denominator with a value of zero.

If $x = -\dfrac{2}{3}: \dfrac{2}{\left(-\dfrac{2}{3}\right) + 1} + 3\left(-\dfrac{2}{3}\right) = 4$

$$\dfrac{2}{\dfrac{1}{3}} - 2 = 4$$
$$6 - 2 = 4$$
$$4 = 4$$

$$X^2 = 4$$
$$X^2 - 4 = 0$$

PRIME
$$X^2 + 16 = 0 \quad ?$$

If $x = 1: \dfrac{2}{(1) + 1} + 3(1) = 4$
$$1 + 3 = 4$$
$$4 = 4$$

$$(X + 2)(X - 2) = 0$$
$$X = -2 \quad X = +2$$

The solution set is $\left\{ -\dfrac{2}{3}, 1 \right\}$.

▼ **EXERCISE 7-1-1**

Arrange each equation in standard form.

1. $x^2 + 5x = -6$

$x^2 + 5x + 6 = 0$

$(x + 2)(x - 3) = 0$

$x = -2 \quad x = -3$

2. $x^2 + 3 = 4x$

$x^2 + 3 - 4x = 0$

$x -$

3. $x^2 = 3x - 2$

$x^2 - 3x + 2 = 0$

$(x - 2)(x - 1) = 0$

$x = 2 \quad x = 1$

4. $2x = 4 - x^2$

$2x - 4 + x^2 = 0$

$x -$

5. $6x^2 + 1 = 5x$

$6x^2 + 1 - 5x = 0$

6. $4x^2 = 3x$

$4x^2 - 3x = 0$

7. $\dfrac{1}{x} + 2x = 5$

$\dfrac{1}{x} + 2x - 5 = 0$

8. $\dfrac{2}{x} + \dfrac{1}{2} = \dfrac{7}{x} - 3x$

9. $\dfrac{1}{(x + 1)^2} = \dfrac{3}{x + 1} - 5$

10. $\dfrac{2}{x + 2} + 3 = \dfrac{1}{x - 2}$

Solve by factoring.

11. $x^2 + 3x + 2 = 0$

12. $x^2 + 8x + 15 = 0$

$(x + 5)(x + 3)$

$x = -5 \quad x = -3$

13. $x^2 + 8x + 7 = 0$

14. $x^2 - 4x + 3 = 0$

15. $x^2 + 2x = 8$

16. $x^2 = 3x + 10$

17. $x^2 + 18 = 9x$

18. $x^2 + 8 = 6x$

19. $x^2 + 4x - 21 = 0$

20. $x^2 - 5x - 24 = 0$

21. $x^2 + 2x = 3$

22. $x^2 = 2x + 15$

$x^2 - 2x - 15 = 0$

$(x - 5)(x + 3) = 0$

$x = +5 \quad x = -3$

23. $x^2 + 2x + 1 = 0$

24. $x^2 - 6x + 9 = 0$

$x^2 - 6x + 9 = 0$

$(x - 3)(x - 3) = 0$

$x = 3 \quad x = 3$

$x = 3$

25. $x^2 + 4x = 0$

26. $x^2 = 6x$

27. $2x^2 = x$

$2x^2 - x = 0$

$x(2x - 1) = 0$

$x = 0 \quad 2x - 1 = 0$

$2x = 1 \quad x = \dfrac{1}{2}$

28. $3x^2 + 2x = 0$

29. $x + 7 + \dfrac{10}{x} = 0$

30. $x = 2 + \dfrac{35}{x}$

31. $2x^2 + 5x + 3 = 0$

32. $6x^2 + 5x + 1 = 0$

33. $10x^2 + 19x + 6 = 0$

34. $2x^2 + 5x = 12$

35. $6x^2 - 13x = 5$

36. $24x^2 + 66x = 63$

37. $10x^2 + 19x - 15 = 0$

38. $6x^2 - 13x + 6 = 0$

39. $x + \dfrac{17}{6} + \dfrac{5}{3x} = 0$

40. $5x + \dfrac{4}{3x} = \dfrac{23}{3}$

◣ 7-2 INCOMPLETE QUADRATICS

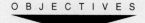

OBJECTIVES

Upon completing this section you should be able to:

1. Identify an incomplete quadratic equation.
2. Solve an incomplete quadratic equation.

x is a common factor. The product of two factors is zero. We therefore use the theorem from the previous section.

Check these solutions.

$$X = \frac{-b \pm \sqrt{b^2 - 4ac}}{2a}$$

$a =$
$b =$

$c =$

Note that there are *two* values that when squared will equal A.

If, when an equation is placed in standard form $ax^2 + bx + c = 0$, either $b = 0$ or $c = 0$, the equation is an *incomplete quadratic*.

Example 1 $5x^2 - 10 = 0$ is an incomplete quadratic, since the middle term is missing and therefore $b = 0$.

When you encounter an incomplete quadratic with $c = 0$ (third term missing), it can still be solved by factoring.

Example 2   Solve for x if $3x^2 - 2x = 0$.

Solution   Factor $x(3x - 2) = 0$. Then $x = 0$ or $3x - 2 = 0$.

$$3x = 2$$
$$x = \frac{2}{3}$$

The solution set is $\left\{0, \dfrac{2}{3}\right\}$.

Notice that if the c term is missing, you can always factor x from the other terms. This means that in all such equations, zero will be one of the solutions.

An incomplete quadratic with the b term missing must be solved by another method, since factoring will be possible only in special cases.

Example 3   Solve for x if $x^2 - 12 = 0$.

Solution   Since $x^2 - 12$ has no common factor and is not the difference of squares, it cannot be factored into rational factors. But, from previous observations, we have the following theorem.

> If $A^2 = b$, then $A = \pm \sqrt{b}$.

Using this theorem, we have

$$x^2 - 12 = 0$$
$$x^2 = 12$$
$$x = \pm\sqrt{12}.$$

Since all answers should be left in simplified form, we will simplify $\pm\sqrt{12}$ and obtain $x = \pm 2\sqrt{3}$. Notice that this gives two solutions, $+2\sqrt{3}$ and $-2\sqrt{3}$. We can write the solutions using set notation as $\{2\sqrt{3}, -2\sqrt{3}\}$ or simply as $\{\pm 2\sqrt{3}\}$.

$\pm\sqrt{12} = \pm\sqrt{4 \cdot 3} = \pm 2\sqrt{3}$

Check these solutions.

Example 4   Solve for x if $2x^2 - 10 = 0$.

Solution
$$2x^2 - 10 = 0$$
$$2x^2 = 10$$
$$x^2 = 5$$
$$x = \pm\sqrt{5}$$

Add 10 to each side.

The solution set is $\{\pm\sqrt{5}\}$.

Check these solutions.

Example 5   Solve for x if $7x^2 + 14x = 0$.

Solution
$$7x(x + 2) = 0$$
$$7x = 0 \qquad x + 2 = 0$$
$$x = 0 \qquad\quad x = -2$$

Here $7x$ is a common factor.

The solution set is $\{-2, 0\}$.

Check these solutions.

Example 6   Solve for x if $x^2 + 25 = 0$.

Solution
$$x^2 + 25 = 0$$
$$x^2 = -25$$

No real solution

Why?

Note that in this example we have the square of a number equal to a negative number. This can never be true in the real number system and, therefore, we have no real solution.

▼ **EXERCISE 7-2-1**

Solve.

1. $x^2 + 3x = 0$

$x(x+3) = 0$
$x = 0 \quad -3$

2. $x^2 - 5x = 0$

$x(x-5) = 0$
$x = 0 \quad x = 5$

3. $2x^2 - 3x = 0$

$x(2x-3)$
$x = 0 \quad 2x-3 = 0$
$x = 1.5$

4. $3x^2 + x = 0$

$3x = -1$
$x = \dfrac{-1}{3}$

$x(3x+1)$
$x = 0 \quad x = -\dfrac{1}{3}$

5. $x^2 = 8x$

$x^2 - 8x = 0$
$x(x-8)$
$x = 0 \quad x = 8$

6. $x^2 + x = 0$

$x(x+1) = 0$
$x = 0 \quad x = -1$

7. $3x^2 = 5x$

$3x^2 - 5x = 0$
$x(3x-5) = 0$
$x = 0 \quad x = 5/3$

8. $4x^2 + 12x = 0$

$4x(x+3) = 0$
$x = 0 \quad x = -3$

(handwritten top margin: $X^2 - 9 = 0$)

(handwritten: $X = 0 \quad X = 2$)

9. $5x^2 = 10x$

$5x(x-2) = 0$

$X = 0 \quad X = 2$

10. $2x^2 + 3x = 7x$

$2x^2 + 3x - 7x = 0$
$2x^2 - 4x = 0$
$2x(x-2) = 0$

11. $x^2 = 4$

$X^2 = 4$
$X = \pm 2$

12. $x^2 = 9$

$X^2 = 9$
$X = \pm 3$

13. $x^2 = 5$

$X^2 = 5$
$X = \pm\sqrt{5}$

14. $x^2 = 13$

$X = \pm\sqrt{13}$

15. $x^2 = 20$

$X = \pm\sqrt{20}$
$X = \sqrt{4}\sqrt{5}$
$X = \pm 2\sqrt{5}$

16. $x^2 = 32$

$X = \pm\sqrt{32}$
$X = \pm\sqrt{16}\sqrt{2}$
$X = \pm 4\sqrt{2}$

17. $x^2 - 14 = 0$

$X^2 - 14 = 0$
$X = \pm\sqrt{14}$

18. $x^2 + 16 = 0$

$X^2 + 16 = 0$
$X + 4 = 0$
$X = -4$

19. $x^2 + 6 = 42$

$X^2 + 6 - 42 = 0$
$X^2 - 36 = 0$
$X - 6 = 0 \quad X = 6$

20. $5x^2 = 45$

$X = \pm\sqrt{9}$
$X = \pm 3$

21. $2x^2 = 32$

$X^2 = \sqrt{16}$
$X = \pm 4$

22. $3x^2 = 108$

$X^2 = 36$
$X = \pm\sqrt{6}$

23. $2x^2 = 4$

$X^2 = 2$
$X = \pm\sqrt{2}$

24. $3x^2 = 9$

$X^2 = 3$
$X = \pm\sqrt{3}$

25. $2x^2 = 16$

$X^2 = 8$
$X = \pm\sqrt{8}$
$X = \pm 2\sqrt{2}$

26. $3x^2 = 54$

$X^2 = 18$
$X = \pm\sqrt{9} + \pm\sqrt{2}$
$X = \pm 3\sqrt{2}$

27. $5x^2 - 15 = 0$

$X^2 - 3 = 0$
$X^2 - 3 = 0$
$X^2 = 3$
$X = \pm\sqrt{3}$

28. $4x^2 + 3 = 103$

$4x^2 + 3 - 103 = 0$
$4x^2 - 100 = 0$
$X^2 - 25 = 0$
$X^2 = 25$
$X = \pm 5$

29. $6x^2 + 19 = -5$

30. $x = \dfrac{147}{3x}$

◥◣ 7-3 COMPLETING THE SQUARE

O B J E C T I V E S

Upon completing this section you should be able to:

1. Identify a perfect square trinomial.
2. Complete the third term to make a perfect square trinomial.
3. Solve a quadratic equation by completing the square.

Remember, squaring a binomial means multiplying it by itself.

In other words, the first and third terms are perfect squares.

From your experience in factoring you already realize that not all polynomials are factorable. Therefore, we need a method for solving quadratics that are not factorable. The method needed is called "completing the square."

First let us review the meaning of "perfect square trinomial." When we square a binomial we obtain a perfect square trinomial. The general form is $(a + b)^2 = a^2 + 2ab + b^2$.

Example 1 $(2x + 3)^2 = (2x + 3)(2x + 3)$
$$= 4x^2 + 12x + 9$$

Example 2 $(5x - 4)^2 = (5x - 4)(5x - 4)$
$$= 25x^2 - 40x + 16$$

From the general form and these examples we can make the following observations concerning a perfect square trinomial.

1. Two of the three terms are perfect squares. $4x^2$ and 9 in the first example, $25x^2$ and 16 in the second example, and a^2 and b^2 in the general form.

2. The other term is either plus or minus two times the product of the square roots of the other two terms.

In the first example $12x = (2)(2x)(3)$.

In the second example $-40x = -2(5x)(4)$.

In the general form $2ab = (2)(a)(b)$.

$+2$ times the square roots of $4x^2$ and 9

-2 times the square roots of $25x^2$ and 16

$+2$ times the square roots of a^2 and b^2

Now consider the polynomial $x^2 + 6x - 7$. Is this polynomial $x^2 + 6x - 7$ a perfect square trinomial? It is not, because two terms are not perfect squares. The -7 term immediately says this cannot be a perfect square trinomial. The task in completing the square is to find a number to replace the -7 such that there will be a perfect square.

$x^2 + 6x \underbrace{- 7}$ not a perfect square

Consider this problem: Fill in the blank so that "$x^2 + 6x + \underline{\quad}$" will be a perfect square trinomial. From the two conditions for a perfect square trinomial we know that the blank must contain a perfect square and that $6x$ must be twice the product of the square root of x^2 and the number in the blank. Since x is already present in $6x$ and is a square root of x^2, then 6 must be twice the square root of the number we place in the blank. In other words, if we first take half of 6 and then square that result, we will obtain the necessary number for the blank.

$x^2 + 6x + \underline{\quad}$

and

$6x = 2(x)\sqrt{}$

$$\left(\frac{1}{2}\right)(6) = 3 \text{ and } 3^2 = 9$$

$6x = 2(x)\sqrt{(9)}$

Therefore $x^2 + 6x + 9$ is a perfect square trinomial.

Example 3 Fill in the blank so that $x^2 + 12x + \underline{\quad}$ will be a perfect square, then factor.

Solution Since $12x$ already contains x, which is a square root of x^2, we take half of 12 and square it.

$12x = 2(x)\sqrt{(\ )}$
$\quad = 2(x)\sqrt{(36)}$

$$\frac{1}{2}(12) = 6 \quad \text{and} \quad 6^2 = 36.$$

The perfect square trinomial is $x^2 + 12x + 36$ and factors as $(x + 6)^2$.

Example 4 Fill the blank so that $x^2 - 10x + \underline{\quad}$ is a perfect square trinomial, then factor.

$$\frac{1}{2}(-10) = -5 \quad \text{and} \quad (-5)^2 = 25.$$

Solution So, $x^2 - 10x + 25$ is a perfect square trinomial and factors as $(x - 5)^2$.

The following exercise will give you practice in finding trinomials that are perfect squares.

▼ **EXERCISE 7-3-1**

Fill the blank in each of the following to obtain a perfect square trinomial. Then factor the remaining perfect square.

1. $x^2 + 4x +$ ___4___
$(x+2)(x+2)$

2. $x^2 + 10x +$ ___25___
$(x+5)(x+5)$

3. $x^2 - 6x +$ ___9___
$(x-3)(x-3)$

4. $x^2 - 4x +$ ___4___
$(x-2)(x-2)$

5. $x^2 + 8x +$ ___16___
$(x+4)(x+4)$

6. $x^2 + 20x +$ ___106___

7. $x^2 + 22x +$ ___121___

8. $x^2 - 18x +$ ___81___

9. $x^2 - 32x +$ ___256___

10. $x^2 + 40x +$ ___400___

11. $x^2 + x +$ ___.25___
$(x+.5)(x+.5)$

12. $x^2 + 3x +$ ___2.25___
$(x+1.5)(x+1.5)$

13. $x^2 - 24x +$ ___144___

14. $x^2 + 100x +$ ___2500___
$(x+50)(x+50)$

15. $x^2 - 2x +$ ___1___
$(x-1)(x-1)$

16. $x^2 - 3x +$ ___2.25___
$(x-1.5)(x-1.5)$

17. $x^2 - 5x +$ ___6.25___

18. $x^2 + 9x +$ ___20.25___
$(x+4.5)(x+4.5)$

19. $x^2 + 7x +$ ___12.25___
$(x+3.5)(x+3.5)$

20. $x^2 - 25x +$ ___156.25___
$(x-12.5)(x+12.5)$

Now let's consider how we can use completing the square to solve quadratic equations.

Recall that instead of -7, a $+9$ would make the expression a perfect square.

Example 5 Solve $x^2 + 6x - 7 = 0$ by completing the square.

Solution First we notice that the -7 term must be replaced if we are to have a perfect square trinomial, so we will rewrite the equation, leaving a blank for the needed number.

$$x^2 + 6x - 7 = 0$$
$$x^2 + 6x + \underline{\quad} = 7 + \underline{\quad}$$

At this point, be careful not to violate any rules of algebra. For instance, note that the second form came from adding $+7$ to both sides of the equation. Never add something to one side without adding the same thing to the other side.

Now we find half of $6 = 3$ and $3^2 = 9$, to give us the number for the blank. Again, if we place a 9 in the blank we must also add 9 to the right side as well.

Remember, if 9 is added to the left side of the equation, it must also be added to the right side.

$$x^2 + 6x + 9 = 7 + 9.$$

Now factor the perfect square trinomial, which gives

$$x^2 + 6x + 9 = 7 + 9$$
$$(x + 3)^2 = 16.$$

Now $x^2 + 6x + 9$ may be written as $(x + 3)^2$.

We are now able to use the theorem "if $A^2 = b$ then $A = \pm\sqrt{b}$" and obtain

$$(x + 3)^2 = 16$$
$$x + 3 = \pm\sqrt{16}$$
$$x + 3 = \pm 4.$$

Solving for x, we have

Add -3 to both sides.

$$x = -3 \pm 4.$$

Now to obtain the two necessary values of x remember that $\pm$ means plus or minus and get

$$x = -3 + 4 \quad \text{or} \quad x = -3 - 4$$
$$x = 1 \quad \text{or} \quad x = -7.$$

Thus, 1 and -7 are solutions or roots of the equation.

The solution set is $\{1, -7\}$.

Example 6   Solve $2x^2 + 12x - 4 = 0$ by completing the square.

Solution   This problem brings in another difficulty. The first term, $2x^2$, is not a perfect square.

We will correct this by dividing all terms of the equation by 2 and obtain

In other words, obtain a coefficient of 1 for the x^2 term.

$$\frac{2x^2}{2} + \frac{12x}{2} - \frac{4}{2} = 0$$
$$x^2 + 6x - 2 = 0.$$

$$x^2 + \textcircled{6}x + \underline{} = 2 + \underline{}$$

Take half of this coefficient and square it.

We now add 2 to both sides, giving

$$x^2 + 6x + \underline{} = 2 + \underline{}$$

$$\left(\frac{1}{2}\right)(6) = 3 \quad \text{and} \quad 3^2 = 9, \text{ so we have}$$

$$x^2 + 6x + 9 = 2 + 9$$
$$x^2 + 6x + 9 = 11$$
$$(x + 3)^2 = 11$$
$$x + 3 = \pm\sqrt{11}$$
$$x = -3 \pm \sqrt{11}.$$

Again, this is more concise.

Notice that since $\sqrt{11}$ cannot be simplified, the solution set can be stated as $\{-3 + \sqrt{11}, -3 - \sqrt{11}\}$, or simply left as $\{-3 \pm \sqrt{11}\}$.

Example 7   Solve $3x^2 + 7x - 9 = 0$ by completing the square.

Again, obtain a coefficient of 1 for x^2 by dividing by 3.

Solution   **Step 1**   Divide all terms by 3.

$$\frac{3x^2}{3} + \frac{7x}{3} - \frac{9}{3} = 0$$

$$x^2 + \frac{7}{3}x - 3 = 0$$

Step 2   Rewrite the equation, leaving a blank for the term necessary to complete the square.

$$x^2 + \frac{7}{3}x + \underline{} = 3 + \underline{}$$

$$x^2 + \boxed{\frac{7}{3}}x + \underline{} = 3$$

Take half of this coefficient and square it.

Step 3   Find the square of half of the coefficient of x and add to both sides.

$$\left(\frac{1}{2}\right)\left(\frac{7}{3}\right) = \frac{7}{6} \quad \text{and} \quad \left(\frac{7}{6}\right)^2 = \frac{49}{36}$$

It looks complex, but we are following the same exact rules as before.

$$x^2 + \frac{7}{3}x + \frac{49}{36} = 3 + \frac{49}{36}$$

Step 4   Factor the completed square.

$$x^2 + \frac{7}{3}x + \frac{49}{36} = 3 + \frac{49}{36}$$

$$\left(x + \frac{7}{6}\right)^2 = 3 + \frac{49}{36}$$

The factoring should *never* be a problem since we know we have a perfect square trinomial, which means we find the square roots of the first and third terms and use the sign of the middle term.

You should review the arithmetic involved in adding the numbers on the right at this time if you have any difficulty.

We now have $\left(x + \dfrac{7}{6} \right)^2 = \dfrac{157}{36}$.

Step 5 Take the square root of each side of the equation.

$$\left(x + \frac{7}{6} \right)^2 = \frac{157}{36}$$

$$x + \frac{7}{6} = \frac{\pm \sqrt{157}}{6}$$

Step 6 Solve for x (two values).

$$x = -\frac{7}{6} \pm \frac{\sqrt{157}}{6} \text{ or } \frac{-7 \pm \sqrt{157}}{6}$$

The solution set is $\left\{ \dfrac{-7 + \sqrt{157}}{6}, \dfrac{-7 - \sqrt{157}}{6} \right\}$.

$$3 + \frac{49}{36} = \frac{3(36)}{36} + \frac{49}{36}$$
$$= \frac{108}{36} + \frac{49}{36}$$
$$= \frac{108 + 49}{36}$$
$$= \frac{157}{36}$$

$\sqrt{157}$ cannot be simplified. We could also write the solution to this problem in a more condensed form as

$$\left\{ \frac{-7 \pm \sqrt{157}}{6} \right\}.$$

Example 8 Solve $x^2 + 2x + 8 = 0$ by completing the square.

Solution
$$x^2 + 2x + 8 = 0$$
$$x^2 + 2x + \underline{\hspace{1cm}} = -8 + \underline{\hspace{1cm}}$$
$$x^2 + 2x + 1 = -8 + 1$$
$$(x + 1)^2 = -7$$

$x^2 + ②x + \underline{\hspace{0.5cm}} = -8 + \underline{\hspace{0.5cm}}$

Take half of this coefficient and square it.

Follow the steps in the previous computation and then note especially the last line. What is the conclusion when the square of a quantity is equal to a negative number? "No real solution."

What real number can we square and obtain -7?

In summary, to solve a quadratic equation by completing the square follow this step-by-step method.

These steps will help in solving the equations in the following exercise.

Step 1 If the coefficient of x^2 is not 1, divide all terms by that coefficient.

Step 2 Rewrite the equation in the form of $x^2 + bx + \underline{\hspace{1cm}}$
$= c + \underline{\hspace{1cm}}$

Step 3 Find the square of one-half of the coefficient of the x term and add this quantity to both sides of the equation.

Step 4 Factor the completed square and combine the numbers on the right-hand side of the equation.

Step 5 Find the square root of each side of the equation.

Step 6 Solve for x and simplify.

If step 5 is not possible, then the equation has no real solution.

▽ **EXERCISE 7-3-2**

Solve by completing the square.

1. $x^2 + 3x + 2 = 0$

2. $x^2 + 6x + 8 = 0$

3. $x^2 + x - 6 = 0$

4. $x^2 + 3x - 4 = 0$

5. $x^2 - 7x + 6 = 0$

6. $x^2 - 5x + 6 = 0$

7. $x^2 + 6x + 9 = 0$

8. $x^2 - 4x + 4 = 0$

9. $x^2 + 4x - 3 = 0$

10. $x^2 + x - 2 = 0$

11. $x^2 + 4x - 5 = 0$

12. $x^2 + 6x - 16 = 0$

13. $x^2 - 3 = 2x$

14. $x^2 - 8x + 7 = 0$

15. $x^2 + 20 = 4x$

NO SOLUTION

$a = 1$
$b = -4$
$c = 20$

$\dfrac{-(-4) \pm \sqrt{16 - (4)(20)(1)}}{2 \cdot 1}$

16. $x^2 + 3x - 1 = 0$

17. $x^2 - 3 = 5x$

18. $x^2 - 2 = x$

$\dfrac{4 \pm \sqrt{-64}}{2}$ $\dfrac{4 \pm 8i}{2}$

$\boxed{2 \pm 4i}$

19. $x^2 + x + 5 = 0$

20. $x^2 + 6x + 2 = 0$

21. $3x^2 + 8x + 3 = 0$

22. $2x^2 - 3x - 5 = 0$

23. $2x^2 - 5x - 4 = 0$

24. $2x^2 - 6x + 3 = 0$

25. $2x^2 + 3x - 2 = 0$

26. $2x^2 - 5x + 2 = 0$

27. $3x^2 + 6x - 4 = 0$

28. $5x^2 + 3 = 20x$

29. $2x^2 + 3x + 4 = 0$

30. $4x^2 + 1 = 4x$

◣▬ 7-4 THE QUADRATIC FORMULA

The standard form of a quadratic equation is $ax^2 + bx + c = 0$. This means that every quadratic equation can be put in this form. In a sense then $ax^2 + bx + c = 0$ represents *all* quadratics. If you can solve this equation, you will have the solution to all quadratic equations.

We will solve the general quadratic equation by the method of completing the square.

$$ax^2 + bx + c = 0$$

Step 1 Divide by a: $x^2 + \dfrac{b}{a}x + \dfrac{c}{a} = 0$

Step 2 Rearrange: $x^2 + \dfrac{b}{a}x + \underline{\quad} = -\dfrac{c}{a} + \underline{\quad}$

Step 3 Find $\left[\dfrac{1}{2}\left(\dfrac{b}{a}\right)\right]^2$ and add to both sides.

$$x^2 + \frac{b}{a}x + \frac{b^2}{4a^2} = \frac{b^2}{4a^2} - \frac{c}{a}$$

Step 4 Factor and combine.

$$\left(x + \frac{b}{2a}\right)^2 = \frac{b^2 - 4ac}{4a^2}$$

Step 5 Find the square root of each side of the equation.

$$x + \frac{b}{2a} = \frac{\pm\sqrt{b^2 - 4ac}}{2a}$$

Step 6 Solve for x and simplify.

$$x = \frac{-b \pm \sqrt{b^2 - 4ac}}{2a}$$

This form $x = \dfrac{-b \pm \sqrt{b^2 - 4ac}}{2a}$ is called the **quadratic formula** and represents the solution to all quadratic equations.

To use the quadratic formula you must identify a, b, and c. To do this the given equation must *always* be placed in standard form.

Example 1   Solve for x if $3x^2 - 5x = 7$.

Solution $3x^2 - 5x = 7$ in standard form is

$$3x^2 - 5x - 7 = 0$$

$a = 3$, $b = -5$, and $c = -7$.

OBJECTIVES

Upon completing this section you should be able to:

1. Solve the general quadratic equation by completing the square.
2. Solve any quadratic equation by using the quadratic formula.

This we did in the previous section many times.

We must add $-\dfrac{b}{2a}$ to each side.

Memorize this formula.

Add -7 to both sides.

Carefully substitute the values of a, b, and c in the formula.

Therefore by the formula

$$x = \frac{-(-5) \pm \sqrt{(-5)^2 - (4)(3)(-7)}}{(2)(3)}$$

$$x = \frac{5 \pm \sqrt{25 + 84}}{6}$$

$$x = \frac{5 \pm \sqrt{109}}{6}.$$

Or simply $\left\{ \dfrac{5 \pm \sqrt{109}}{6} \right\}$

The solution set is $\left\{ \dfrac{5 + \sqrt{109}}{6}, \dfrac{5 - \sqrt{109}}{6} \right\}.$

Example 2 Solve for x: $x^2 + 3x - 10 = 0$

Solution $a = 1$, $b = 3$, and $c = -10$.

$$x = \frac{-3 \pm \sqrt{(3)^2 - 4(1)(-10)}}{2(1)}$$

$$x = \frac{-3 \pm \sqrt{49}}{2}$$

$\sqrt{49}$ may be simplified.

$$x = \frac{-3 \pm 7}{2}$$

So

$$x = \frac{-3 + 7}{2} = \frac{4}{2} = 2$$

or

$$x = \frac{-3 - 7}{2} = \frac{-10}{2} = -5.$$

The solution set is $\{2, -5\}$.

This equation is already in standard form.

Example 3 Solve for x: $2x^2 - 7x + 12 = 0$

Solution $a = 2$, $b = -7$, and $c = 12$.

$$x = \frac{-(-7) \pm \sqrt{(-7)^2 - 4(2)(12)}}{(2)(2)}$$

$$x = \frac{7 \pm \sqrt{49 - 96}}{4}$$

$$x = \frac{7 \pm \sqrt{-47}}{4}$$

$\sqrt{-47}$: We are in trouble!

There is no real solution since -47 has no real square root.

Example 4 Solve for x: $x^2 + 4x - 3 = 0$

Solution $a = 1$, $b = 4$, and $c = -3$.

Again, this equation is in standard form.

$$x = \frac{-4 \pm \sqrt{(4)^2 - 4(1)(-3)}}{2(1)}$$

$$x = \frac{-4 \pm \sqrt{16 + 12}}{2}$$

$$x = \frac{-4 \pm \sqrt{28}}{2}$$

$\sqrt{28}$ can be simplified.
$$\sqrt{28} = \sqrt{4 \cdot 7} = 2\sqrt{7}$$

This solution should now be simplified.

$$x = \frac{-4 \pm 2\sqrt{7}}{2} \text{ since } \sqrt{28} = \sqrt{(4)(7)} = 2\sqrt{7}$$

$$x = \frac{2(-2 \pm \sqrt{7})}{2} \text{ (Factor a 2.)}$$

$$x = -2 \pm \sqrt{7}$$

The solution set is $\{-2 + \sqrt{7}, -2 - \sqrt{7}\}$.

Or simply $\{-2 \pm \sqrt{7}\}$.

▼ **EXERCISE 7-4-1**

Solve by the quadratic formula. If the radicand is negative, then indicate no real solution.

1. $x^2 + 2x - 15 = 0$

$A = 1$
$b = 2$
$c = -15$
$(5)(-3)$

2. $x^2 - 9x + 20 = 0$

$A = 1$
$B = -9$
$C = +20$

3. $x^2 + 5x + 3 = 0$

$A = 1$
$b = 5$
$c = 3$
$\frac{-5 \pm \sqrt{13}}{2}$

4. $x^2 + 3x - 5 = 0$

$A = 1$
$b = 3$
$C = -5$
$\frac{-3 \pm \sqrt{29}}{2}$

5. $x^2 - 5x + 1 = 0$

$A = 1$
$B = -5$
$C = +1$
$\frac{5 \pm \sqrt{21}}{2}$

6. $x^2 - 7x + 3 = 0$

$A = 1$
$B = -7$
$C = +3$

7. $5x^2 - 7x - 6 = 0$

$A = 5$
$b = -7$
$C = -6$
$(2)(-3/5)$

8. $6x^2 = x + 2$

9. $2x^2 - 3x + 4 = 0$

$A = 2$
$B = -3$
$C = 4$
$\frac{3 \pm \sqrt{-9}}{4}$ NO SOLU

10. $2x^2 - 4x - 5 = 0$

11. $3x^2 + 5x + 2 = 0$

12. $4x^2 + 8x + 1 = 0$

13. $3x^2 - x - 1 = 0$

14. $2x^2 - 10x + 9 = 0$

15. $x^2 + 3x + 1 = 0$

16. $2x^2 = x + 3$

17. $2x^2 + 1 = 3x$

18. $4x^2 + 5x + 1 = 0$

19. $3x^2 = 2x + 1$

20. $x^2 - 3x + 4 = 0$

21. $3x^2 + 5x + 2 = 0$

22. $5x^2 + 7x + 1 = 0$

23. $2x^2 - 6x + 3 = 0$

24. $x^2 + 2x = 7$

25. $x^2 + 25 = 10x$

26. $2x^2 - 3x + 5 = 0$

27. $9x^2 + 12x + 4 = 0$

28. $3x^2 + 8x = 5$

29. $5x^2 - 6x + 1 = 0$

30. $3x^2 + 4x - 5 = 0$

◥◣ 7-5 THE NATURE OF THE ROOTS OF A QUADRATIC EQUATION

OBJECTIVES

Upon completing this section you should be able to:

1. Compute the value of the discriminant $b^2 - 4ac$.
2. Determine the nature of the roots of a quadratic equation from the value of the discriminant.

When we examine the solution of the general quadratic equation (the formula), we discover some important facts.

$$x = \frac{-b \pm \sqrt{b^2 - 4ac}}{2a}$$

First we notice that the radicand $b^2 - 4ac$ will determine whether or not we have solutions that are real numbers. If $b^2 - 4ac$ is negative, and we are working with real numbers, we have no solution.

The radicand $b^2 - 4ac$ is called the **discriminant** of a quadratic equation because it determines the nature of the roots (solutions).

Example 1 In the equation $x^2 + 2x + 5 = 0$ $a = 1$, $b = 2$, and $c = 5$.

$$\text{So } b^2 - 4ac = (2)^2 - 4(1)(5)$$
$$= 4 - 20$$
$$= -16$$

We will have a negative number under the radical.

Since the discriminant is negative, there will be no real roots.

The discriminant $b^2 - 4ac$ can give further information about the roots. Since $\sqrt{b^2 - 4ac}$ is added to give one solution and subtracted to give the other, then the only way both solutions could be the same is for the discriminant to be zero. This leads us to the second property.

If $b^2 - 4ac = 0$, the solutions of the equation are equal.

Both solutions of a quadratic equation can be the same only if the quadratic is a perfect square trinomial.

Example 2 In the equation $x^2 - 6x + 9 = 0$ $a = 1$, $b = -6$, and $c = 9$.

$$\text{So } b^2 - 4ac = (-6)^2 - 4(1)(9)$$
$$= 36 - 36$$
$$= 0.$$

Thus the solutions to this equation will be equal.

Notice that this is a perfect square trinomial.

We will have a zero under the radical.

If $b^2 - 4ac$ is positive, we will have two different solutions for the equation. Furthermore, if the value of $b^2 - 4ac$ is positive and not a perfect square (such as 4, 9, 25, and so on) the solution must contain a radical.

If the solution to a quadratic equation does not contain a radical, the roots can be obtained by factoring. Thus the value of $b^2 - 4ac$ becomes a sure test for the factorability of a trinomial.

The following table summarizes the facts discussed.

Roots of a Quadratic

Value of the Discriminant	Nature of the Roots
$b^2 - 4ac < 0$	No real roots
$b^2 - 4ac = 0$	Roots are real and equal. Two identical roots. The trinomial is a perfect square.
$b^2 - 4ac > 0$ but not a perfect square.	Roots are irrational and unequal. Two different roots that contain a radical.
$b^2 - 4ac > 0$ and a perfect square.	Roots are unequal and rational. Two different roots containing no radical. The trinomial is factorable.

As you can see from this table, the value of the discriminant gives you a good idea of what to expect when finding the roots of an equation.

▼ **EXERCISE 7–5–1**

Compute $b^2 - 4ac$ for each of the following and give the nature of their roots.

1. $x^2 + 4x + 4 = 0$ **2.** $x^2 + 2x + 1 = 0$ **3.** $x^2 + 6x - 9 = 0$

4. $x^2 + 7x - 3 = 0$

5. $x^2 + 7x + 12 = 0$

6. $x^2 + 3x - 18 = 0$

7. $x^2 - 12x - 12 = 0$

8. $x^2 - 4x - 5 = 0$

9. $5x^2 - 3x + 1 = 0$

10. $2x^2 + 4x + 3 = 0$

11. $3x^2 - 7x - 1 = 0$

12. $2x^2 + 5x + 3 = 0$

13. $4x^2 - 20x + 25 = 0$

14. $x^2 + 8x + 16 = 0$

15. $x^2 - 8x + 11 = 0$

16. $x^2 - 6x + 8 = 0$

17. $2x^2 + 13x - 7 = 0$

18. $3x^2 - 13x - 10 = 0$

19. $x^2 - 4x + 4 = 0$

20. $x^2 - 3x + 1 = 0$

21. $2x^2 - 7x + 3 = 0$

22. $x^2 - 3x + 5 = 0$

23. $4x^2 + 4x + 1 = 0$

24. $2x^2 + x + 1 = 0$

25. $3x^2 + 5x + 2 = 0$ **26.** $2x^2 + 5x - 3 = 0$ **27.** $x^2 + 5x + 7 = 0$

28. $x^2 - 11x + 31 = 0$ **29.** $2x^2 - 7x + 4 = 0$ **30.** $3x^2 - 8x + 4 = 0$

◥◣ 7-6 COMPLEX NUMBERS

In preceding sections many problems have had no real solution or no real roots. To the observant student this may have implied that there is a set of numbers other than the real numbers in which such equations do have solutions. This is, in fact, the case.

To introduce this new set of numbers we define the imaginary unit "i" as the square root of negative 1.

$$i = \sqrt{-1} \quad \text{or} \quad i^2 = -1.$$

OBJECTIVES

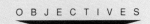

Upon completing this section you should be able to:

1. Add, subtract, multiply, and divide complex numbers.
2. Solve a quadratic equation for complex roots.

Accepting this definition of i makes it possible to find values for the square roots of negative numbers, or at least to indicate such values.

Example 1   Find the value of $\sqrt{-4}$.

    Solution

$$\sqrt{-4} = \sqrt{(-1)(4)}$$
$$= \sqrt{-1}\sqrt{4}$$
$$= i\sqrt{4}$$
$$= 2i$$

We substitute i for $\sqrt{-1}$.

To check $\sqrt{-4} = 2i$ by the definition of the square root we evaluate $(2i)^2$.

$$(2i)^2 = (2i)(2i)$$
$$= (2)(2)(i)(i)$$
$$= 4i^2$$

but since

$$i^2 = -1$$

then

$$4i^2 = -4.$$

$4i^2 = 4(-1) = -4$

Example 2   Find the value of $\sqrt{-10}$.

Solution
$$\sqrt{-10} = \sqrt{(-1)(10)}$$
$$= \sqrt{-1}\sqrt{10}$$
$$= i\sqrt{10}$$

To avoid confusion it is best to write the *i* in front of the radical.

(The answer is left in this form since $\sqrt{10}$ is not rational.)

A number such as *2i* or $i\sqrt{10}$ is called an **imaginary number.**

▽ **EXERCISE 7-6-1**

Express the following as imaginary numbers.

1. $\sqrt{-9}$ 2. $\sqrt{-16}$ 3. $\sqrt{-2}$ 4. $\sqrt{-3}$ 5. $\sqrt{-25}$ 6. $\sqrt{-49}$

   $3i$ $4i$ $i\sqrt{2}$ $i\sqrt{3}$ $5i$ $7i$

   $\sqrt{-1\cdot 9}$

7. $\sqrt{-7}$ 8. $\sqrt{-11}$ 9. $\sqrt{-30}$ 10. $\sqrt{-6}$ 11. $\sqrt{-8}$ 12. $\sqrt{-20}$

   $i\sqrt{7}$ $i\sqrt{11}$ $i\sqrt{30}$ $i\sqrt{6}$ $i\sqrt{8}$ $2i\sqrt{5}$

13. $\sqrt{-100}$ 14. $\sqrt{-45}$ 15. $\sqrt{-18}$ 16. $\sqrt{-200}$ 17. $\sqrt{-75}$ 18. $\sqrt{-27}$

   $10i$ $3i\sqrt{5}$ $i\sqrt{18}$ $10i\sqrt{2}$ $5i\sqrt{3}$ $3i\sqrt{3}$

19. $\sqrt{-80}$ 20. $\sqrt{-28}$

   $4i\sqrt{5}$ $2i\sqrt{7}$

The indicated sum of a real number and an imaginary number, (*a* + *bi*) where *a* and *b* are real, is called a **complex number.**

The real number 5 can be expressed as 5 + 0*i*.

It should first be noted that the set of real numbers is a subset of the set of complex numbers. A real number *x* can be expressed as *x* + 0*i* and by the definition is complex. Also, the set of imaginary numbers is a subset of the set of complex numbers since an imaginary number *bi* can be expressed as 0 + *bi*.

Complex numbers can be added, subtracted, multiplied, divided, raised to powers, and so on, as you would expect of any set of numbers. Rules for these operations follow.

To add or subtract complex numbers combine the real parts and the imaginary parts separately. (Note that this follows a previous rule that only like terms can be combined.)

Example 3   Add: $(7 + 6i) + (3 + 2i)$

Solution
$$(7 + 6i) + (3 + 2i) = (7 + 3) + (6i + 2i)$$
$$= 10 + 8i$$

You could also add them in column form.
$$\begin{array}{r} 7 + 6i \\ 3 + 2i \\ \hline 10 + 8i \end{array}$$

Example 4   Add: $(3 + 4i) + (2 - 6i)$

Solution
$$(3 + 4i) + (2 - 6i) = (3 + 2) + (4i - 6i)$$
$$= 5 - 2i$$

To multiply complex numbers consider them as binomials and use the distributive law. (We can use the pattern developed in chapter 4.)

Example 5   Multiply: $(3 + 2i)(4 + 6i)$

Solution
$$(3 + 2i)(4 + 6i) = 12 + 26i + 12i^2$$
$$= 12 + 26i + 12(-1)$$
$$= 26i$$

Recall $i^2 = -1$.

Example 6   Multiply: $5(3 + 4i)$

Solution
$$5(3 + 4i) = 15 + 20i$$

(Note that we could write 5 as $5 + 0i$ and use the preceding pattern. However, a simple use of the distributive law is all that is needed.)

Example 7   Find: $(3 + i)^2$

Solution
$$(3 + i)^2 = (3 + i)(3 + i)$$
$$= 9 + 6i + i^2$$
$$= 9 + 6i - 1$$
$$= 8 + 6i$$

Square the binomial.

The next operation we will discuss is division. Division is always defined as multiplying by the inverse. This is actually rationalizing the denominator since i is just a symbol for a radical.

Example 8   Divide: $(3 + 4i) \div (2 + i)$

Solution
$$(3 + 4i) \div (2 + i) = \frac{3 + 4i}{2 + i}$$

We want this in a form that does not have an i in the denominator.

If we write this as $\dfrac{3 + 4\sqrt{-1}}{2 + \sqrt{-1}}$, you will recognize it as a type of problem found in chapter 6 (exercise 6–8–2). We should multiply the numerator and denominator by $2 - i$.

$4 - i^2 = 4 - (-1)$
$= 4 + 1 = 5$

$$\frac{3 + 4i}{2 + i} \cdot \frac{2 - i}{2 - i} = \frac{6 + 5i - 4i^2}{4 - i^2}$$

$$= \frac{10 + 5i}{5}$$

This solution can be reduced to

Reduce when possible.

$$\frac{10 + 5i}{5} = \frac{5(2 + i)}{5} = 2 + i.$$

▼ **EXERCISE 7–6–2**

Perform the following operations.

1. $(2 + 3i) + (5 + 4i)$

$7 + 7i$

2. $(1 + 2i) + (7 - 5i)$

$8 - 3i$

3. $(6 + 5i) + (1 - 6i)$

$7 - i$

4. $(3 + 6i) + (4 + i)$

5. $(5 - 4i) + 2(4 + 2i)$

$5 - 4i + 8 + 4i$

13

6. $(2 + 5i) + 3(2 - 8i)$

$2 + 5i + 6 - 24i$

$8 - 21i$

7. $(5 - 2i) - (8 - i)$

8. $(2 + 9i) - (5 - 2i)$

$2 + 9i - 5 + 2i$

$11i - 3$

9. $(11 + 7i) - (3 + 4i)$

$8 + 3i$

10. $(24 - 11i) - (15 + 6i)$

11. $(11 + i) - 3(2 - 5i)$

12. $(7 - i) - 5(2 + 3i)$

$(7 - i) - (10 + 15i)$

$7 - i - 10 - 15i$

$-3 - 16i$

13. $(2 + 3i)(1 + 4i)$

$2 + 8i + 3i + 12i^2$

$2 + 11i + 12i^2 \quad i^2 = -1$

$2 + 11i + 12(-1)$

$-10 + 11i$

14. $(3 - i)(5 + 5i)$

15. $(x + iy)(x - iy)$

$x^2 - xiy + xiy - iy^2$

$x^2 - iy^2$

$x^2 - y^2$

16. $(6 - 8i)(5 - 2i)$

$30 - 12i - 40i - (16 \cdot -1)$

$30 + 16 - 52i$

$46 - 52i$

17. $(5 + 4i)^2$

$(5 + 4i)(5 + 4i)$

$25 + 20i + 20i + 16i^2$

$25 + 16 + 40i$

$91 + 40i$

18. $(3 - 2i)^2$

$(3 - 2i)(3 - 2i)$

$9 - 6i - 6i + 4i^2$

$9 - 12i - 4i^2$

19. $(6 - 5i)^2$

$(6 - 5i)(6 - 5i)$

$36 - 10i + 25i^2$

$36 - 60i + 25 \cdot -1$

$11 - 60i$

20. $(10 + 7i)^2$

21. $(2 + 3i) \div (1 - 2i)$

22. $(1 - 4i) \div (2 + 3i)$

23. $5 \div (4 + 3i)$

24. $7 \div (3 - 5i)$

25. $2i \div (6 + i)$

26. $8i \div (4 - 3i)$

27. $(3 + i) \div i$

28. $(13 - 4i) \div 2i$

29. $(1 - 4i) \div (2 - 3i)$

30. $(3 - i) \div (6 - 5i)$

The most common use of complex numbers at this level of algebra is in expressing solutions to quadratic equations that have no real solution.

Example 9 Solve and check: $x^2 - 2x + 5 = 0$

Solution Since the expression on the left will not factor, we will use the quadratic formula with $a = 1$, $b = -2$, and $c = 5$.

Remember to be careful when substituting.

$$x = \frac{2 \pm \sqrt{(-2)^2 - 4(1)(5)}}{2}$$

$$= \frac{2 \pm \sqrt{-16}}{2}$$

$$= \frac{2 \pm 4i}{2}$$

$$= 1 \pm 2i$$

Make sure to check each answer.

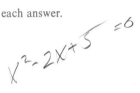

Check: $x = 1 + 2i$

$$(1 + 2i)^2 - 2(1 + 2i) + 5 = 1 + 4i + 4i^2 - 2 - 4i + 5$$
$$= 4 + 4i^2$$
$$= 4 + 4(-1)$$
$$= 0$$

Check: $x = 1 - 2i$

$$(1 - 2i)^2 - 2(1 - 2i) + 5 = 1 - 4i + 4i^2 - 2 + 4i + 5$$
$$= 4 + 4i^2$$
$$= 4 + 4(-1)$$
$$= 0$$

We see that both answers check. The solution set is therefore $\{1 \pm 2i\}$.

▼   **EXERCISE 7-6-3**

Solve and check.

1. $x^2 - 2x + 2 = 0$

a 1
b −2 $(x-1)$
c 2

2. $x^2 - 2x + 10 = 0$

a 1
b −2
c 10

3. $x^2 - 4x + 8 = 0$

a 1
b −4 $2 \pm 2i$
c 8

4. $x^2 - 6x + 10 = 0$

5. $x^2 + 2x + 5 = 0$

$a = 1$
$b = 2$ $-1 \pm 2i$
$c = 5$

6. $x^2 - 4x + 5 = 0$

7. $x^2 + 4x + 13 = 0$

$a = 1$
$b = 4$ $\dfrac{-4 \pm \sqrt{36i}}{2}$
$c = 13$

$\boxed{-2 \pm 3i}$

8. $x^2 + 2x + 17 = 0$

9. $2x^2 - 2x + 1 = 0$

10. $4x^2 - 8x + 5 = 0$

11. $8x^2 - 4x + 1 = 0$

12. $9x^2 - 6x + 2 = 0$

13. $2x^2 + 2x + 5 = 0$

14. $4x^2 + 4x + 5 = 0$

15. $x^2 + x + 1 = 0$

16. $x^2 - x + 1 = 0$

17. $2x^2 + x + 1 = 0$

18. $x^2 + x + 2 = 0$

19. $2x^2 - 3x + 2 = 0$

20. $3x^2 - 2x + 1 = 0$

21. $x^2 - 2x + 4 = 0$

22. $x^2 - 2x + 3 = 0$

23. $x^2 - 4x + 6 = 0$

24. $x^2 - 6x + 14 = 0$

25. $3x^2 + x + 1 = 0$

26. $3x^2 - 2x + 4 = 0$

27. $5x^2 - 2x + 3 = 0$

28. $2x^2 - 5x + 7 = 0$

29. $x^2 + 3x + 4 = 0$

30. $x^2 + 3x + 6 = 0$

 **7-7 EQUATIONS QUADRATIC IN FORM**

O B J E C T I V E S

Upon completing this section you should be able to:

1. Determine if an equation is quadratic in form.
2. Solve equations that are quadratic in form.

Solving equations of degree higher than two is, in general, a topic beyond the level of this text. However, certain special cases can be solved by the methods we have already covered. We will now discuss those special cases that are in the form of a quadratic.

An equation is quadratic in form if a suitable substitution for the unknown can be found so that the resulting equation is a quadratic equation.

For example,

$$x^4 - 5x^2 + 6 = 0$$

is quadratic in form since the substitution $y = x^2$ would give

$$y^2 - 5y + 6 = 0,$$

which is a quadratic equation. Such equations may often be solved by solving the resulting quadratic equation and then using these solutions and the substitution to obtain solutions to the original equation.

Example 1   Solve: $x^4 - 5x^2 + 6 = 0$

Solution   Let $y = x^2$.

This trinomial factors.

Then
$$y^2 - 5y + 6 = 0$$
$$(y - 3)(y - 2) = 0$$
$$y - 3 = 0 \quad \text{or} \quad y - 2 = 0$$
$$y = 3 \qquad\qquad y = 2.$$

Make sure you don't stop here in this type problem.

Since $y = x^2$, we have

$$x^2 = 3 \quad \text{or} \quad x^2 = 2$$
$$x = \pm\sqrt{3} \qquad x = \pm\sqrt{2}.$$

There are four solutions to this equation.

Checking these values in the original equation, we find they all satisfy the equation. Therefore, the solution set is $\{\sqrt{3}, -\sqrt{3}, \sqrt{2}, -\sqrt{2}\}$.

Example 2   Solve: $2x^{1/2} - 7x^{1/4} + 6 = 0$

Solution   Let $y = x^{1/4}$.

$x^{1/2} = (x^{1/4})^2 = y^2$

Then
$$2y^2 - 7y + 6 = 0$$
$$(2y - 3)(y - 2) = 0$$
$$2y - 3 = 0 \quad \text{or} \quad y - 2 = 0$$
$$2y = 3 \qquad\qquad y = 2$$
$$y = \frac{3}{2}.$$

Since $y = x^{1/4}$, then

We substitute $x^{1/4}$ for y.

$$x^{1/4} = \frac{3}{2} \text{ or } x^{1/4} = 2.$$

Raising both sides to the fourth power, we obtain

$$\left(x^{1/4}\right)^4 = \left(\frac{3}{2}\right)^4 \quad \text{or} \quad \left(x^{1/4}\right)^4 = 2^4$$

$$x = \frac{81}{16} \qquad\qquad x = 16.$$

Check these answers in the original equation.

Checking these values, we find the solution set to be $\left\{\frac{81}{16}, 16\right\}$.

▼ EXERCISE 7-7-1

Solve for real or complex roots.

1. $x^4 - 5x^2 + 4 = 0$

2. $x^4 - 10x^2 + 9 = 0$

3. $x^4 - 6x^2 + 8 = 0$

4. $x^4 - 7x^2 + 12 = 0$

5. $x^4 - 7x^2 + 10 = 0$

6. $2x^4 - 7x^2 + 3 = 0$

7. $3x^4 - 8x^2 + 4 = 0$

8. $x^4 - 3x^2 - 4 = 0$

9. $x^4 - 7x^2 - 18 = 0$

10. $x^4 + 8x^2 + 15 = 0$

11. $x^6 - 9x^3 + 8 = 0$

12. $x^6 - 7x^3 - 8 = 0$

13. $x^6 + 28x^3 + 27 = 0$

14. $x^{1/2} - 5x^{1/4} + 6 = 0$

15. $x^{1/2} - 4x^{1/4} - 5 = 0$

16. $2x^{1/2} + 7x^{1/4} + 3 = 0$ **17.** $x - 6x^{1/2} + 8 = 0$ **18.** $x - 4x^{1/2} - 21 = 0$

19. $x^{2/3} + 2x^{1/3} - 3 = 0$ **20.** $2x^{2/3} + 5x^{1/3} + 2 = 0$

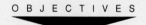 **7-8 EQUATIONS WITH RADICALS**

OBJECTIVES

Upon completing this section you should be able to:

1. Solve equations involving radicals.
2. Identify extraneous roots.

In chapter 2 we found that certain operations will always yield an equivalent equation. This fact was used to solve first-degree equations. We are now faced with equations that will require something more than the four basic operations to obtain an equivalent equation. In fact, you will find that it is not always possible to obtain an equivalent equation.

Consider the two equations $x = 2$ and $x^2 = 4$. These equations certainly have something in common but it is also obvious that they are not equivalent, since 2 is the only solution to $x = 2$, but 2 and -2 are both solutions to $x^2 = 4$.

If both sides of an equation are squared, the resulting equation is not always equivalent to the original. However, any solution to the original equation will be a solution of the resulting equation.

> If both sides of an equation are raised to the same power, the resulting equation will contain all solutions of the original equation. However, the resulting equation may also contain solutions that are not solutions of the original equation. These solutions are called **extraneous roots.** It is therefore necessary to check all solutions in the original equation.

Example 1   Solve: $x - \sqrt{2x - 5} = 4$

Solution   First we isolate the radical on one side of the equation.

We subtract x from each side.

$$-\sqrt{2x - 5} = 4 - x$$

Squaring each side yields

Be careful. Make sure you square $(4 - x)$.

$$(-\sqrt{2x - 5})^2 = (4 - x)^2$$
$$2x - 5 = 16 - 8x + x^2$$
$$x^2 - 10x + 21 = 0$$
$$(x - 7)(x - 3) = 0$$
$$x - 7 = 0 \quad \text{or} \quad x - 3 = 0$$
$$x = 7 \quad\quad\quad\quad x = 3.$$

Check if $x = 7$.

$$x - \sqrt{2x - 5} = 7 - \sqrt{2(7) - 5}$$
$$= 7 - \sqrt{9}$$
$$= 7 - 3 = 4$$

In this type problem checking is extremely important.

Therefore, $x = 7$ is a solution.

If $x = 3$,

$$x - \sqrt{2x - 5} = 3 - \sqrt{2(3) - 5}$$
$$= 3 - \sqrt{1}$$
$$= 3 - 1 = 2.$$

To be a solution the value should be 4.

Therefore, $x = 3$ is *not* a solution. It is extraneous. The solution set is $\{7\}$.

Example 2   Solve: $\sqrt{5x - 1} - \sqrt{2x} = \sqrt{x - 1}$

Solution   Here we have more than one radical in the equation. Since there is a single radical isolated on one side of the equation, it will disappear when we square both sides.

$$(\sqrt{5x - 1} - \sqrt{2x})^2 = (\sqrt{x - 1})^2$$
$$5x - 1 - 2\sqrt{5x - 1}\sqrt{2x} + 2x = x - 1$$
$$-2\sqrt{5x - 1}\sqrt{2x} = -6x$$
or
$$-2\sqrt{10x^2 - 2x} = -6x$$

The left side of this equation is a binomial.

Dividing each side by -2 gives $\sqrt{10x^2 - 2x} = 3x$.
Again, squaring each side, we obtain

We must now square again.

$$(\sqrt{10x^2 - 2x})^2 = (3x)^2$$
$$10x^2 - 2x = 9x^2$$
$$x^2 - 2x = 0$$
$$x(x - 2) = 0$$
$$x = 0 \quad \text{or} \quad x - 2 = 0$$
$$x = 2.$$

Check if $x = 0$.

$$\sqrt{5x - 1} - \sqrt{2x} = \sqrt{-1} - \sqrt{0} = \sqrt{-1}$$
$$\sqrt{x - 1} = \sqrt{0 - 1} = \sqrt{-1}$$

Make sure you check the values in the original equation.

Therefore, $x = 0$ is a solution.

If $x = 2$,

$$\sqrt{5x - 1} - \sqrt{2x} = \sqrt{9} - \sqrt{4} = 3 - 2 = 1$$
$$\sqrt{x - 1} = \sqrt{2 - 1} = \sqrt{1} = 1$$

Thus, the solution set is $\{0,2\}$.

▼ EXERCISE 7-8-1

Solve. Check each answer carefully.

1. $x + \sqrt{x} = 6$

2. $x + \sqrt{x} = 12$

3. $x - 2\sqrt{x} = 3$

4. $2x - \sqrt{x} = 6$

5. $x + \sqrt{x - 1} = 7$

6. $x + \sqrt{x - 3} = 9$

7. $x - \sqrt{x + 2} = 10$

8. $x - \sqrt{x + 5} = 1$

9. $x + \sqrt{2x - 5} = 10$

10. $x + \sqrt{3x + 1} = 9$

11. $2x - \sqrt{3x - 2} = 8$

12. $3x - \sqrt{5x - 4} = 8$

13. $x - 2\sqrt{2x + 1} = -2$

14. $x - 2\sqrt{3x - 2} = -2$

15. $5 - \sqrt{5x - 1} = x$

16. $8 - \sqrt{2x - 1} = 5$

17. $6 + \sqrt{3x + 1} = 2x$

18. $5 + \sqrt{5x + 1} = 3x$

19. $\sqrt{x + 7} = 1 + \sqrt{2x}$

20. $\sqrt{x + 5} = \sqrt{4x} - 1$

21. $\sqrt{2x - 1} + \sqrt{x + 3} = 3$

22. $\sqrt{2x - 1} + \sqrt{x + 4} = 6$

23. $\sqrt{x + 2} + \sqrt{2x + 11} = 4$

24. $\sqrt{x + 6} + \sqrt{3x + 7} = 3$

25. $\sqrt{x + 4} + \sqrt{2x + 10} = 3$

26. $\sqrt{x + 10} + \sqrt{2x + 21} = 5$

27. $\sqrt{x + 1} + \sqrt{x - 7} = \sqrt{2x}$

28. $\sqrt{x + 3} + \sqrt{x - 5} = \sqrt{6x}$

29. $\sqrt{2x + 5} - \sqrt{x + 2} = \sqrt{3x - 5}$

30. $\sqrt{x + 12} - \sqrt{2x + 7} = \sqrt{x + 7}$

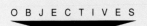 **7-9 WORD PROBLEMS**

O B J E C T I V E S

Upon completing this section you should be able to:

1. Identify word problems that require a quadratic equation for their solution.
2. Solve word problems involving quadratic equations.

Certain types of word problems can be solved by quadratic equations. The process of outlining and setting up the problem is the same as taught in chapter 3, but with those problems that are solved by quadratics you must be very careful to check the solutions in the problem itself. The physical restrictions within the problem can eliminate one or both of the solutions.

Example 1   If the length of a rectangle is 1 unit more than twice the width, and the area is 55 square units, find the length and width.

Solution   The formula for the area of a rectangle is Area = Length $\times$ Width. Let x = width, $2x + 1$ = length.

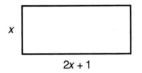

If x represents the width, then $2x$ represents twice the width and $2x + 1$ represents one more than twice the width.

Substituting these values in the formula, we have

$$55 = (2x + 1)(x)$$
$$55 = 2x^2 + x$$
$$2x^2 + x - 55 = 0$$
$$(2x + 11)(x - 5) = 0$$

Place the quadratic equation in standard form.

This quadratic can be solved by factoring.

$$2x + 11 = 0 \quad \text{or} \quad x - 5 = 0$$
$$2x = -11 \qquad\qquad x = 5$$
$$x = -\frac{11}{2}$$

A measurement cannot be a negative value.

At this point, you can see that the solution $x = -\frac{11}{2}$ is not valid since x represents a measurement of the width and negative numbers are not used for such measurements. Therefore, the solution is

$$\text{width} = x = 5, \text{length} = 2x + 1 = 11.$$

Example 2   A number added to its reciprocal is $2\frac{9}{10}$. Find the number.

The reciprocal of x is $\frac{1}{x}$.

Remember LCD means *lowest common denominator*.

Solution   Let x = the number. Then $x + \frac{1}{x} = 2\frac{9}{10}$ or $x + \frac{1}{x} = \frac{29}{10}$.

The LCD is $10x$.

$$10x\left(x + \frac{1}{x}\right) = 10x\left(\frac{29}{10}\right)$$

Every term must be multiplied by $10x$.

$$10x^2 + 10 = 29x$$
$$10x^2 - 29x + 10 = 0$$

Again, this quadratic can be factored.

$$(5x - 2)(2x - 5) = 0$$

$$5x - 2 = 0 \quad \text{or} \quad 2x - 5 = 0$$
$$5x = 2 \qquad\qquad 2x = 5$$
$$x = \frac{2}{5} \qquad\qquad x = \frac{5}{2}$$

Both solutions check. Therefore, the solution set is $\left\{\frac{5}{2}, \frac{2}{5}\right\}$.

There are two solutions to this problem.

Example 3   If a certain integer is subtracted from 6 times its square, the result is 15. Find the integer.

Solution   Let $x =$ the integer. Then $6x^2 - x = 15$ or $6x^2 - x - 15 = 0$.

$$(3x - 5)(2x + 3) = 0$$
$$3x - 5 = 0 \quad \text{or} \quad 2x + 3 = 0$$
$$3x = 5 \qquad\qquad 2x = -3$$
$$x = \frac{5}{3} \qquad\qquad x = -\frac{3}{2}$$

You might be tempted to give these values as a solution unless you paid close attention to the fact that the problem asked for an integer.

Since neither solution is an integer, the problem has *no solution*.

Example 4   A farm manager has 200 meters of fence on hand and wishes to enclose a rectangular field so that it will contain 2,400 square meters in area. What should the dimensions of the field be?

Solution   Here there are two formulas involved. $P = 2\ell + 2w$ for the perimeter and $A = \ell w$ for the area.

First using $P = 2\ell + 2w$, we get $200 = 2\ell + 2w$

Divide each term by 2.

or

$100 = \ell + w$

$w = 100 - \ell$

ℓ

or

$w = 100 - \ell$.

We can now use the formula $A = \ell w$ and substitute the $(100 - \ell)$ for w, giving

We could just as well solve for ℓ obtaining $\ell = 100 - w$. Then

$$A = \ell w$$
$$= (100 - w)w$$
$$= 100w - w^2$$

$$2,400 = \ell(100 - \ell)$$
$$2,400 = 100\ell - \ell^2$$

or

$$2{,}400 = 100w - w^2$$

or

$$w^2 - 100w + 2{,}400 = 0$$
$$(w - 40)(w - 60) = 0$$
$$w - 40 = 0 \text{ or } w - 60 = 0$$
$$w = 40 \qquad w = 60$$

or

$$\ell^2 - 100\ell + 2{,}400 = 0$$
$$(\ell - 40)(\ell - 60) = 0$$
$$\ell - 40 = 0 \quad \text{or} \quad \ell - 60 = 0$$
$$\ell = 40 \qquad\qquad \ell = 60.$$

If $\ell = 60$, then $w = 100 - \ell = 40$. If $\ell = 40$, then $w = 100 - \ell = 60$.

The field must be 40 meters wide by 60 meters long.

Note that in this problem we actually use a system of equations

$$P = 2\ell + 2w$$
$$A = \ell w.$$

In general, a system of equations in which a quadratic is involved will be solved by the *substitution* method. (See chapter 9.)

▼ **EXERCISE 7-9-1**

Solve. If one or more answers will not check in the word problem itself, state that fact.

1. The sum of two numbers is 25; the product is 144. What are the numbers?

2. The sum of a number and twice its reciprocal is $\dfrac{17}{6}$. What is the number?

3. The denominator of a certain positive fraction is three more than its numerator. The sum of the fraction and its reciprocal is $2\dfrac{9}{10}$. What is the fraction?

4. There are three consecutive integers such that the square of the second, increased by the product of the other two, is 161. What are these three integers?

5. In an auditorium there are 720 seats. If the number of rows of seats is six less than the number of seats in each row, find the number of rows and the number of seats in each row.

6. The length of a rectangle is two more than twice its width. Find the dimensions of the rectangle if its area is 24.

7. A triangle of area 35 has an altitude that is three less than its base. Find the base and altitude.

8. Find two consecutive positive odd integers whose product is 143.

9. One positive number is five greater than the other. Find the numbers if their product is 176.

10. Find three consecutive positive integers if the sum of their squares is 77.

11. The sum of a number and its reciprocal is $2\frac{1}{12}$. Find the number.

12. The diagonal of a rectangle is ten centimeters and the width is two centimeters less than the length. Find the dimensions of the rectangle. (Recall from geometry that in a right triangle the sum of the squares of the two legs is equal to the square of the hypotenuse, i.e., $a^2 + b^2 = c^2$.)

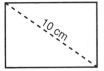

13. The sum of the reciprocals of two consecutive even integers is $\frac{9}{40}$. Find the integers.

14. One number is four greater than the other. The sum of their squares is 106. Find the numbers.

15. One number is five more than twice the other. Find the numbers if the difference of their squares is 153 and both numbers are positive.

16. Find the number that is six greater than its positive square root.

17. One leg of a right triangle is two centimeters longer than the other. If the hypotenuse is 10 centimeters, find the lengths of the two legs. (See hint in problem 12.)

18. One side of a right triangle is four centimeters. The other side is seven centimeters less than twice the length of the hypotenuse. Find the length of the hypotenuse.

19. The area of a square is 36 square meters. If the area is to be increased by 28 square meters, how many meters should each side be increased by?

20. The length of a rectangle is twice the width. If each dimension is increased by three, the new area would be 104 square meters. Find the original dimensions.

21. A polygon of n sides has $\dfrac{n(n-3)}{2}$ diagonals. How many sides does a polygon have if it has 54 diagonals?

22. An object dropped from the top of a 45-meter tower falls according to the formula $s = -5t^2 + 45$, where s represents the distance of the object above the ground at any time t in seconds. How long will it take the object to reach the ground?

23. A ball is thrown upward with a velocity of 15 meters per second. The distance s of the ball above the ground in t seconds is given by $s = 15t - 5t^2$. How long will it take the ball to be within 10 meters of the ground? Why are there two answers?

24. A small motorboat can travel 10 kilometers per hour in still water. The boat travels 8 kilometers upstream and returns in one hour and forty minutes. What is the current's speed?

25. A merchant bought some calculators for a total price of $225.00. All but one of the calculators was sold at a profit of $10.00 per calculator for a total return of $220.00. How many calculators did the merchant buy?

26. The numerator of a certain positive fraction is three less than the denominator. If the numerator and denominator are each increased by one, the new fraction will exceed the original fraction by $\dfrac{1}{10}$. Find the original fraction.

27. A certain lawn is 20 meters wide and 40 meters long. How wide a border must be cut all around the edge of the lawn so that $\dfrac{3}{8}$ of the area will be left uncut?

28. Bill and Bob working together can mow a lawn in 2 hours. Bill can mow the lawn by himself in 3 hours less time than it takes Bob. How many hours would it take each of them alone?

Key Words

Section 7–1

- A **quadratic equation** is a polynomial equation that contains the second degree, but no higher degree, of the variable.

- The **standard form** of a quadratic equation is

$$ax^2 + bx + c = 0, \text{ when } a \neq 0.$$

Section 7–2

- An **incomplete quadratic** equation is of the form

$$ax^2 + bx + c = 0, \text{ and either } b = 0$$
$$\text{or } c = 0.$$

Section 7–4

- The **quadratic formula** is

$$x = \frac{-b \pm \sqrt{b^2 - 4ac}}{2a}.$$

Section 7–5

- The **discriminant** of a quadratic equation is

$$b^2 - 4ac.$$

Section 7–6

- The **imaginary unit** i is defined as

$$i = \sqrt{-1} \text{ or } i^2 = -1.$$

- A **complex number** is in the form of $a + bi$, where a and b are real numbers.

Section 7–7

- Equations that are **quadratic in form** are quadratics in some power of the variable.

Section 7–8

- **Extraneous roots** sometimes occur when both sides of an equation are raised to a power.

Procedures

Section 7–1

- The most direct and generally easiest method of finding the solutions to a quadratic equation is factoring. This method is based on the theorem: if $AB = 0$, then $A = 0$ or $B = 0$. To use this theorem we put the equation in standard form, factor, and set each factor equal to zero.

Section 7–3

- To solve a quadratic equation by completing the square, follow these steps:

  Step 1 If the coefficient of x^2 is not 1, divide all terms by that coefficient.

  Step 2 Rewrite the equation in the form of
  $x^2 + bx + \underline{\quad} = c + \underline{\quad}$.

  Step 3 Find the square of one-half of the coefficient of the x term and add this quantity to both sides of the equation.

  Step 4 Factor the completed square and combine the numbers on the right-hand side of the equation.

  Step 5 Find the square root of each side of the equation.

  Step 6 Solve for x and simplify.

  If step 5 is not possible, then the equation has no real solution.

Section 7–4

- The method of completing the square is used to derive the quadratic formula.

- To use the quadratic formula write the equation in standard form, identify a, b, and c, and substitute these values into the formula. All solutions should be simplified.

Section 7–5

- The nature of the roots of a quadratic equation is dependent on the value of the discriminant.
  1. If $b^2 - 4ac \geq 0$, the roots are real.
  2. If $b^2 - 4ac < 0$, the roots are complex.

Section 7–6

- To add or subtract two complex numbers combine the real parts of each and then the imaginary parts of each.

- To multiply or divide two complex numbers treat the numbers as binomials.

Section 7–7

- The solution of an equation that is quadratic in form may be found by using the following steps:

  Step 1 Solve for the power of the variable as if solving a quadratic equation.

  Step 2 Find the solutions of the original equation by either raising both sides to a power or taking a root of both sides.

Section 7–8

- To solve equations containing radicals isolate the radical when possible and then raise both sides of the equation to the power necessary to eliminate the radical. If the equation contains more than one radical, the process of raising both sides to a power must continue until all radicals have been eliminated. All answers should be checked in the original equation to identify extraneous roots.

Solve by factoring.

1. $x^2 + 10x + 21 = 0$

2. $x^2 + x - 30 = 0$

3. $x^2 - 11x + 24 = 0$

4. $2x^2 + 3x - 2 = 0$

5. $6x^2 + 7x - 3 = 0$

Solve.

6. $x^2 = 36$

7. $x^2 = 17$

8. $5x^2 = 100$

9. $2x^2 + 32 = 0$

10. $3x^2 - 27 = 0$

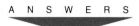

A N S W E R S

1. _____

2. _____

3. _____

4. _____

5. _____

6. _____

7. _____

8. _____

9. _____

10. _____

Solve by completing the square.

11. $x^2 + 8x + 15 = 0$ **12.** $x^2 - 10x + 8 = 0$

11. _____

12. _____

13. $x^2 + 3x - 1 = 0$ **14.** $3x^2 - 6x + 2 = 0$

13. _____

15. $2x^2 + 5x + 4 = 0$

14. _____

15. _____

Solve by using the quadratic formula.

16. _____

16. $x^2 + 3x - 28 = 0$ **17.** $x^2 - 4x + 3 = 0$

17. _____

18. $x^2 - x + 1 = 0$ **19.** $x^2 + 10x + 26 = 0$

18. _____

19. _____

20. $3x^2 - 2x - 1 = 0$

20. _____

Compute $b^2 - 4ac$ for each of the following and give the nature of their roots.

21. $x^2 + 3x - 2 = 0$

22. $x^2 - 3x - 18 = 0$

21. _____

23. $x^2 + 6x + 9 = 0$

24. $2x^2 - 3x + 2 = 0$

22. _____

25. $3x^2 + 5x + 2 = 0$

23. _____

24. _____

Solve for real or complex roots.

26. $x^4 - 10x^2 + 9 = 0$

27. $x^4 + 3x^2 - 10 = 0$

25. _____

26. _____

28. $x^{1/2} - 7x^{1/4} + 10 = 0$

29. $x - 9x^{1/2} + 20 = 0$

27. _____

28. _____

30. $x^{2/3} + 3x^{1/3} - 4 = 0$

29. _____

30. _____

Solve for x.

31. $x + \sqrt{x} = 12$

32. $x - 3\sqrt{x} = -2$

31. _____

32. _____

33. $4 + \sqrt{2x} = x$

34. $\sqrt{3x + 1} + \sqrt{2x - 1} = 7$

33. _____

35. $\sqrt{x + 7} - \sqrt{x + 6} = \sqrt{2x + 13}$

34. _____

35. _____

Solve.

36. _____

36. The area of a rectangle is 105 square meters. Find the dimensions if the length is one meter longer than twice the width.

37. The sum of the squares of two consecutive odd integers is 74. Find the integers.

37. _____

38. The base of a triangle is four units longer than the altitude. Find the length of the base if the area of the triangle is 96.

39. The sum of a number and twice its reciprocal is $\dfrac{27}{5}$. Find the number.

38. _____

39. _____

40. One number is three greater than another number. The square of the smaller number is one less than five times the larger number. Find the numbers.

40. _____

SCORE: _____

1. Solve: $2x^2 + 5x - 3 = 0$

1. _____

2. Solve: **a.** $2x^2 = 7x$ **b.** $x^2 = 10$

2a. _____

3. Solve by completing the square: $9x^2 - 1 = 12x$

4. Use the quadratic formula to solve: $2x^2 + 3x = 4$

2b. _____

3. _____

5. Give the nature of the roots by evaluating the discriminant:
a. $2x^2 - 3x - 1 = 0$ **b.** $x^2 + 6x + 9 = 0$

4. _____

5a. _____

6. Solve: $x^2 - 3x + 4 = 0$ **7.** Solve: $x^4 - 3x^2 - 10 = 0$

5b. _____

6. _____

8. Solve:
$$\sqrt{x + 3} + \sqrt{2x - 1} = 3$$

9. The area of a rectangle is 84 square meters. If the length is 5 meters longer than the width, find the width.

7. _____

8. _____

9. _____

SCORE: _____

C H A P T E R

8

PRETEST

Before beginning this chapter, answer as many of the following questions as you can. When you have finished the chapter, take the practice test at the end of the chapter and compare the scores of the two tests to see how much you have learned.

A N S W E R S

1. Give the domain (values of x) and range (values of y) of each of the following relations. In each case state if the relation is a function of x.
   a. $y = x^2 + 2$
   b. $x^2 + y^2 = 16$

2. Complete the table of values and sketch the graph of $2x - y = 7$.

x	2	4	5
y			

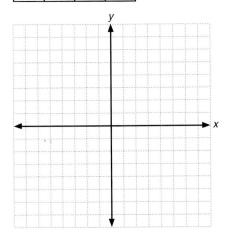

3. Write the equation, in standard form, of the straight line through the points $(-2, 4)$ and $(3, -1)$.

1a. _____

1b. _____

2. _____

3. _____

4. _____

5. _____

6. _____

7. _____

8. _____

4. Sketch the graph of the straight line through the point (3,4) if

$$m = -\frac{2}{3}.$$

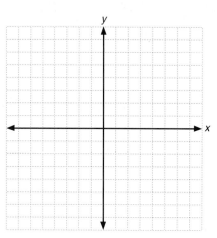

5. Write the equation of the circle $x^2 + y^2 + 4x - 8y = 5$ in standard form. Give the center and radius.

6. Sketch the graph of

$$\frac{(x-2)^2}{25} + \frac{(y+3)^2}{9} = 1.$$

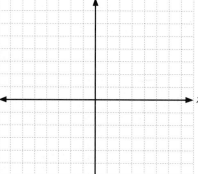

7. Sketch the graph of $4y = x^2 - 2x - 7$.

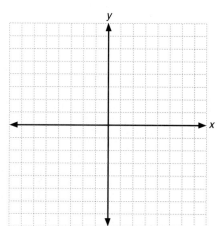

8. Sketch the graph of the linear inequality $2x + y \le 5$.

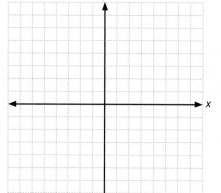

SCORE: _____

428

8

Graphing: Functions, Relations, and Inequalities

Find the vertex, focus, and directrix of the parabola $(x - 4)^2 = 8(y - 3)$.

A graph is a pictorial representation of number facts. There are many types of graphs, such as bar graphs, circular graphs, line graphs, and so on. You can usually find examples of these graphs in the financial section of a newspaper. Graphs are used because a picture usually makes the number facts more easily understood.

In this chapter we will discuss the method of graphing an equation in two variables. In other words, we will sketch a picture of an equation in two variables.

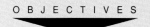 **8-1 DEFINITIONS**

O B J E C T I V E S

Upon completing this section you should be able to:
1. Find the domain and range of a relation.
2. Determine if a relation is a function.

Before we can proceed to our goal of graphing equations with two variables, we must introduce some new ideas and notations that we will be using.

If we are given two sets of numbers, we can establish some rule that shows a relation between elements of the first set and elements of the second set.

For instance, given the two sets

$$A: \{1, 3, 5\}$$
$$B: \{2, 4, 6, 8, 10\}$$

the rule "Relate a number in set A to a number in set B so that the number in set B is twice as large as the number in set A" would give

1 is related to 2
3 is related to 6
5 is related to 10.

We can express this more compactly as a set of ordered pairs

$$\{(1,2),(3,6),(5,10)\}$$

where the first number (abscissa) in each pair is from set A and the second number (ordinate) is from set B.

The rule "Relate a number in set A to a number in set B that is larger" would give

Note in this example that each number in set A is related to several numbers in set B.

1 is related to 2, 4, 6, 8, and 10
3 is related to 4, 6, 8, and 10
5 is related to 6, 8, and 10.

Writing this as a set of ordered pairs, we have

$$\{(1,2),(1,4),(1,6),(1,8),(1,10),(3,4),(3,6),(3,8),(3,10),(5,6),(5,8),(5,10)\}.$$

Both of the examples above are relations.

A **relation** is a set of ordered pairs.

The set of all first numbers in the ordered pairs of a relation is called the **domain.** In the first example on page 430 the domain is $\{1,3,5\}$. It is also the domain of the second example.

The set of second numbers in the ordered pairs of a relation is called the **range.** In the first example the range is $\{2,6,10\}$. The range in the second example is $\{2,4,6,8,10\}$. Sometimes the numbers in the range are referred to as *images* and the range is called the *image set.*

A **function** is a relation such that each element of the domain has exactly one image.

Make sure you understand this definition.

From this definition note that the first example is a function whereas the second example is not a function.

Do you see in the second example that each element of the domain has more than one image?

In mathematics the rule relating the elements of the domain to elements of the range is usually given as an algebraic expression. For instance, in the first example if x represents an element of set A, then $2x$ represents the image of x in set B.

The notation $f(x)$ is read "f of x" or "f at x" and means "the image of x."

$f(x) = 2x$ means "the image of x is a number that is twice x."

$f(x) = x^2 + 1$ means "the image of x is a number that is one more than the square of x."

The notation is only used when the rule actually gives a relation that is a function.

The domain of a function can be the set of real numbers or any subset of the real numbers and is sometimes limited by the algebraic expression of the rule. For instance,

$$f(x) = \frac{1}{x}$$

$f(x)$ would be meaningless if $x = 0$.

has a domain of all real numbers except zero.

At this point it will be helpful to introduce a special type of symbolism called *interval notation.* We will use brackets [ ] and parentheses ( ) to indicate intervals of numbers.

To denote an interval on the number line when the endpoints are included, we will use the bracket. For example, $[-5, 3]$ designates the interval from -5 to 3 including the numbers -5 and 3. To say x is in $[-5, 3]$, sometimes written $x \in [-5, 3]$, would mean the same as $-5 \le x \le 3$.

When the endpoints of the interval are not included, we will use parentheses. For example, $(-5, 3)$ designates the interval from -5 to 3 excluding the numbers -5 and 3. To say x is in $(-5, 3)$ would mean the same as $-5 < x < 3$.

We need to be careful here to distinguish when we are referring to an interval or an ordered pair.

When the endpoints of an interval are included in the set, the interval is said to be a *closed interval*. When the endpoints are excluded, the interval is called an *open interval*. Of course we can have "half-open" or "half-closed" intervals. In this case one endpoint is included and the other is not.

Refer to section 2–7.

This is the same type of notation we used when graphing the solution to an inequality on the number line.

Note use of parentheses to exclude the number zero.

Therefore the domain of the function $f(x) = \dfrac{1}{x}$ could be expressed as $(-\infty,0)$ or $(0,+\infty)$.

The symbol ∞ (read as "infinity") is *not* to be considered a real number. $(-\infty,+\infty)$ designates the set of all real numbers, $(-\infty,0)$ the negative real numbers and $(0,+\infty)$ the positive real numbers.

The symbol $\cup$ is sometimes used in place of the logical "or." Thus the domain could also be written as $(-\infty,0) \cup (0,+\infty)$.

Also

If x was negative, $f(x)$ would not be a real number.

$$f(x) = \sqrt{x}$$

has a domain of all nonnegative real numbers, that is, $[0,+\infty)$.

If we are given an algebraic equation in two variables, it may be important to know if one variable is a function of the other.

Example 1   Is y a function of x in the equation $x^2 + 2x - y = 5$? If so, determine the domain and range.

Solve $x^2 + 2x - y = 5$ for y.

Solution   From the definition y is a function of x if for each value of x there is exactly one value of y (image). Therefore to answer the question we must first solve the equation for y in terms of x and then determine if the definition of a function is satisfied. Thus

$$y = x^2 + 2x - 5$$

and with the equation in this form we can see that any value for x will yield exactly one value of y. Thus the answer to the question is "yes."

Would any value of x not yield a real number?

To determine the domain and range we need to think negatively. In other words, "Is any real number *eliminated* from the domain or range for any reason?" Inspecting $y = x^2 + 2x - 5$ shows that any real number substituted for x will give a real value for y. No real number is eliminated as a possible value of x so the domain is the set of all real numbers, that is, $(-\infty,+\infty)$.

To determine the range we will solve the equation for x in terms of y by use of the quadratic formula. This gives

Solve $x^2 + 2x - (5 + y) = 0$ using the quadratic formula.

$$x = -1 \pm \sqrt{6 + y},$$

which implies that

$$6 + y \geq 0$$

is a condition imposed on y so that x will be real. Thus the range is the set of all reals y such that $y \geq -6$, which in interval notation is $[-6,+\infty)$.

Example 2 Is y a function of x in the expression $y^2 = x - 1$?

Solution First we solve for y obtaining

$$y = \pm\sqrt{x - 1}. \qquad\qquad x \in [1, +\infty)$$

A value of x will yield *two* values of y. Therefore y is not a function of x.

▼ **EXERCISE 8-1-1**

In each of the relations consider the replacement set for x as the domain and the replacement set for y as the range. Find (a) the domain, (b) the range, (c) determine if y is a function of x.

1. $y = x^2 - 1$ **2.** $y = x^2 + 1$ **3.** $\dfrac{1}{y} = x$

4. $y = \dfrac{3}{x}$ **5.** $y^2 = x + 3$ **6.** $y^2 = x - 2$

7. $y = x^3$ **8.** $y = x^4$ **9.** $x = y^3$

10. $x = y^4$ **11.** $x^2 + y^2 = 9$ **12.** $x^2 + y^2 = 1$

13. $y = \sqrt{9 - x^2}$ **14.** $y = \sqrt{1 - x^2}$ **15.** $y = \dfrac{1}{x - 1}$

16. $y = \dfrac{1}{2x + 1}$ **17.** $x = \sqrt{y^2}$ **18.** $x = \sqrt{y^3}$

19. $y = |x^2 + 3|$ **20.** $y = \dfrac{1}{|x + 4|}$ **21.** $y = \sqrt{x^2 - 9}$

22. $y = \sqrt{x^2 - 1}$ **23.** $y = \dfrac{1}{\sqrt{x - 2}}$ **24.** $y = \dfrac{2}{\sqrt{x + 5}}$

◤◢ 8-2 POINTS ON THE PLANE

Perpendicular means that two lines are at right angles to each other.

We have already used the number line on which we have represented numbers as points on a line.

Note that this concept contains elements from two fields of mathematics, the line from geometry and the numbers from algebra. René Descartes (1596–1650) devised a method of relating points on a plane to algebraic numbers. This scheme is called the **Cartesian coordinate system** (for Descartes) and is sometimes referred to as the *rectangular coordinate system*.

This system is composed of two number lines that are perpendicular at their zero points.

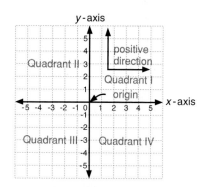

Study the diagram carefully as you note each of the following facts.

The number lines are called *axes*. The horizontal line is the *x-axis* and the vertical is the *y-axis*. The zero point at which they are perpendicular is called the *origin*.

Positive is to the *right* and *up;* negative is to the *left* and *down*.

The plane is divided into four parts called *quadrants*. These are numbered in a counterclockwise direction starting at the upper right.

Points on the plane are designated by **ordered pairs** of numbers written in parentheses with a comma between them, such as (5,7). This is called an *ordered pair* because the order in which the numbers are written is important. The ordered pair (5,7) is *not* the same as the ordered pair (7,5). Points are located on the plane in the following manner.

First, start at the origin and count left or right the number of spaces designated by the first number of the ordered pair. Second, from the point on the x-axis given by the first number count up or down the number of spaces designated by the second number of the ordered pair. Ordered pairs are always written with x first and then y, (x,y). The numbers represented by x and y are called the *coordinates* of the point (x,y).

Example 1 On the following Cartesian coordinate system the points A (3,4), B (0,5), C (−2,7), D (−4,1), E (−3,−4), F (4,−2), G (0,−5), and H (−6,0) are designated. Check each one to determine how they are located.

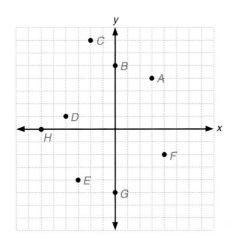

Axes is plural. Axis is singular.

The arrows indicate the number lines extend indefinitely. Thus, the plane extends indefinitely in all directions.

This is important. The first number of the ordered pair always refers to the horizontal direction and the second number always refers to the vertical direction.

What are the coordinates of the origin?

▽ **EXERCISE 8-2-1**

1. Locate each of the following points on the coordinate system. Label each point with the letter and the ordered pair. *A* is given as an example. *A* (−3,5), *B* (5,2), *C* (5,−4), *D* (−4,−5), *E* (0,4), *F* (5,0), *G* (−3,0), and *H* (0,−3)

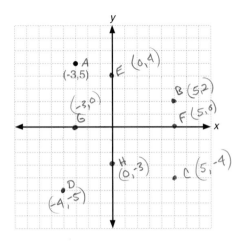

2. Note the points indicated on the coordinate system. Give the ordered pair associated with each point.

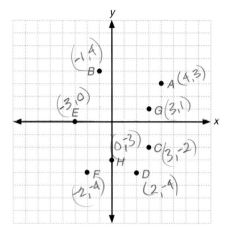

3. Points in quadrant I correspond to ordered pairs with both coordinates positive, (+,+). What are the coordinates in the other three quadrants?

$$\text{I} = +\ +$$
$$\text{II} = -\ +$$
$$\text{III} = -\ -$$
$$\text{IV} = +\ -$$

 **8-3 THE STRAIGHT LINE**

OBJECTIVES

▼

Upon completing this section you should be able to:

1. Sketch the graph of a linear equation.
2. Find the equation of a line given two points on the line.
3. Find the equation of a line given its slope and a point on the line.
4. Write the equation of a line in slope-intercept form.

In the chapter introduction we stated that we would graph algebraic equations with two variables. We have now covered the new concepts and notations necessary to proceed. In this section we will graph equations in two variables when both are of the first-degree.

Consider the equation $x + y = 7$ and note that we can easily find many solutions. For instance, if $x = 5$ then $y = 2$, since $5 + 2 = 7$. Also, if $x = 3$ then $y = 4$, since $3 + 4 = 7$. If we represent these answers as ordered pairs (x,y), then we have (5,2) and (3,4) as two points on the plane that represent answers to the equation $x + y = 7$.

All possible answers to this equation, located as points on the plane, will give us the graph (or picture) of the equation.

Of course we could never find all numbers x and y such that $x + y = 7$, so we must be content with a sketch of the graph. A sketch can be described as the "curve of best fit." In other words, it is necessary to locate enough points to give a reasonably accurate picture of the equation.

Example 1   Sketch the graph of $2x + y = 3$.

Solution   We wish to find several pairs of numbers that will make this equation true. We will accomplish this by choosing a number for x and then finding a corresponding value for y. A *table of values* is used to record the data.

x							
y							

In the top line (x) we will place numbers that we have chosen for x. Then in the bottom line (y) we will place the corresponding value of y derived from the equation.

In this example we will allow x to take on the values $-3, -2, -1, 0, 1, 2, 3$.

Of course, we could also start by choosing values for y, and then find the corresponding values for x.

x	-3	-2	-1	0	1	2	3
y							

These values are arbitrary. We could choose any values at all.

If $x = -3$, we have
$$2x + y = 3$$
$$2(-3) + y = 3$$
$$-6 + y = 3$$
$$\boxed{y = 9}.$$

Notice that once we have chosen a value for x, the value for y is determined by the condition of the equation.

If $x = -2$, we have
$$2x + y = 3$$
$$2(-2) + y = 3$$
$$-4 + y = 3$$
$$\boxed{y = 7}.$$

If $x = -1$, we have
$$2x + y = 3$$
$$2(-1) + y = 3$$
$$-2 + y = 3$$
$$\boxed{y = 5}.$$

These values of x give integers for values of y. Thus, they are good choices. Suppose we chose $x = \dfrac{2}{5}$. Then

If $x = 0$, we have
$$2x + y = 3$$
$$2(0) + y = 3$$
$$0 + y = 3$$
$$\boxed{y = 3}.$$

$$2x + y = 3$$
$$2\left(\frac{2}{5}\right) + y = 3$$
$$\frac{4}{5} + y = 3$$
$$y = \frac{11}{5} = 2\frac{1}{5}.$$

If $x = 1$, we have
$$2x + y = 3$$
$$2(1) + y = 3$$
$$2 + y = 3$$
$$\boxed{y = 1}.$$

The ordered pair $\left(\dfrac{2}{5}, 2\dfrac{1}{5}\right)$ satisfies the equation but would be very difficult to locate on the coordinate system. Hence, $x = \dfrac{2}{5}$ is not a very good choice.

If $x = 2$, we have
$$2x + y = 3$$
$$2(2) + y = 3$$
$$4 + y = 3$$
$$\boxed{y = -1}.$$

If $x = 3$, we have
$$2x + y = 3$$
$$2(3) + y = 3$$
$$6 + y = 3$$
$$\boxed{y = -3}.$$

These facts give us the following table of values:

x	-3	-2	-1	0	1	2	3
y	9	7	5	3	1	-1	-3

We now locate the ordered pairs $(-3,9)$, $(-2,7)$, $(-1,5)$, $(0,3)$, $(1,1)$, $(2,-1)$, $(3,-3)$ on the coordinate plane and connect them with a line.

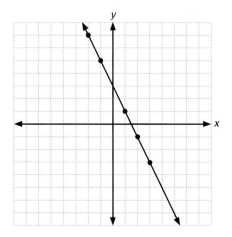

The line indicates that all points on the line satisfy the equation, as well as the points from the table. The arrows indicate the line continues indefinitely.

We now have the graph of $2x + y = 3$.

Thus, any equation of the form $ax + by = c$ where a, b, and c are real numbers is a linear equation.

The graphs of all first-degree equations in two variables will be straight lines. This fact will be used here even though it will be much later in mathematics before you can prove this statement. Such first-degree equations are called **linear equations.**

Since the graph of a first-degree equation in two variables is a straight line, it is only necessary to have two points. However, your work will be more consistently accurate if you find at least three points. Mistakes can be located and corrected when the points found do not lie on a line. We thus refer to the third point as a "checkpoint."

This is important. Don't try to shorten your work by finding only two points. You will be surprised how often you will find an error by locating all three points.

Example 2   Sketch the graph of $3x - 2y = 7$.

Solution   First make a table of values and decide on three numbers to substitute for x. We will try 0, 1, 2.

x	0	1	2
y			

If $x = 0$, then we have $3x - 2y = 7$

$$3(0) - 2y = 7$$
$$y = -3\frac{1}{2}.$$

Again, you could also have started with arbitrary values of y.

The answer $y = -3\frac{1}{2}$ is not as easy to locate on the graph as an integer would be. So it seems that $x = 0$ was not a very good choice. Sometimes it is possible to look ahead and make better choices for x.

If $x = 1$, we have $3x - 2y = 7$

$$3(1) - 2y = 7$$
$$3 - 2y = 7$$
$$-2y = 4$$
$$y = -2.$$

The point $(1, -2)$ will be easier to locate. If $x = 2$, we will have another fraction.

Since both x and y are integers, $x = 1$ was a good choice.

If $x = 3$, we have $3x - 2y = 7$

$$3(3) - 2y = 7$$
$$9 - 2y = 7$$
$$-2y = -2$$
$$y = 1.$$

$x = 3$ was another good choice.

The point $(3,1)$ will be easy to locate.

If $x = -1$, then we have $3x - 2y = 7$

$$3(-1) - 2y = 7$$
$$-3 - 2y = 7$$
$$-2y = 10$$
$$y = -5.$$

We will readjust the table of values and use the points that gave integers. This may not always be feasible, but trying for integral values will give a more accurate sketch. We now have the table for $3x - 2y = 7$.

We can do this since the choices for x were arbitrary.

x	1	3	-1
y	-2	1	-5

Locating the points $(1,-2)$, $(3,1)$, $(-1,-5)$ gives the graph of $3x - 2y = 7$.

How many ordered pairs satisfy this equation?

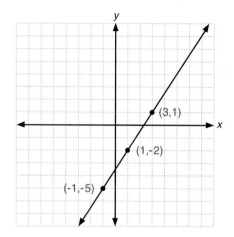

EXERCISE 8–3–1

Sketch the graphs.

x	4	-2	0
y	3	9	7

1. $x + y = 7$

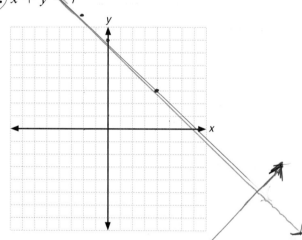

2. $x - y = 4$

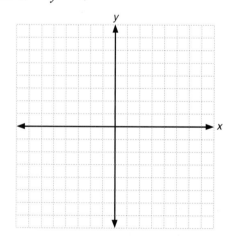

3. $y - x = 3$

x	4	-6	0	7
y	7	-3	3	10

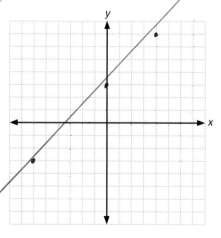

4. $y = x$

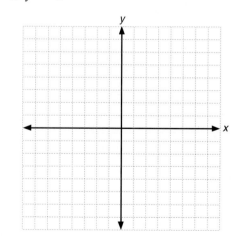

5. $y = 4 - x$

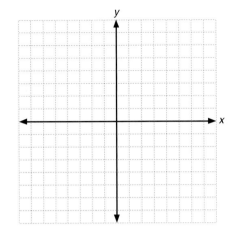

6. $2x - y = 3$

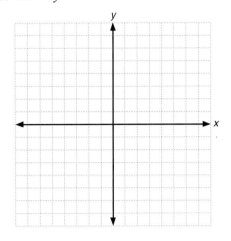

7. $2x + y = 3$

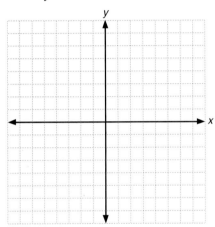

8. $x + 3y = 2$

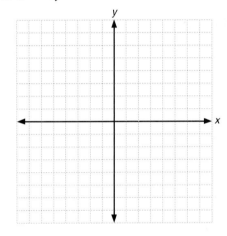

9. $2x + 3y = 8$

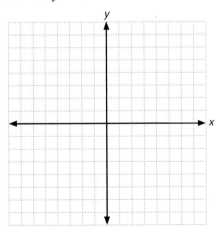

10. $3x - 4y = 1$

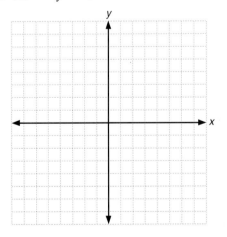

11. $2x + 3y = 4$

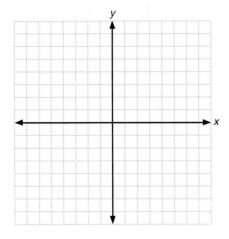

12. $2x - 3y = 6$

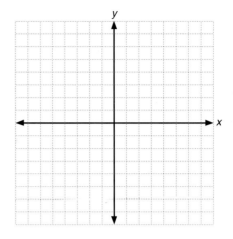

13. $x - 4y = 4$

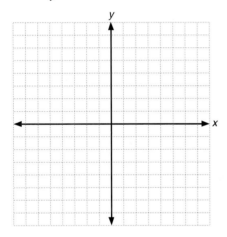

14. $5x - y = 5$

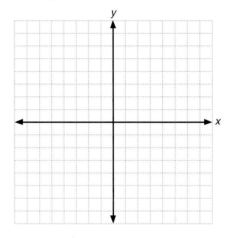

15. $x + 2y = -6$

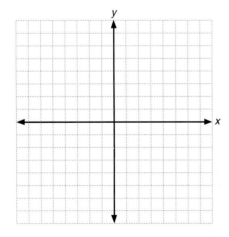

16. $3x - y = -3$

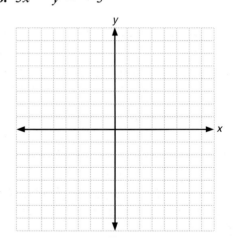

$$y_2 - y_1$$

$$\frac{4 - 4}{5 - (-3)} = \frac{0}{8}$$

$$x_2 - x_1$$

17. $3x - 2y = -9$

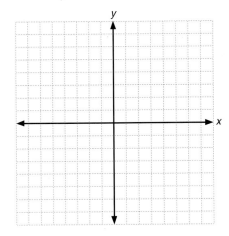

18. $3x + 4y = 12$

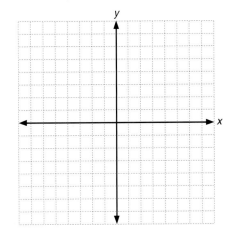

19. $2x + 5y = 5$

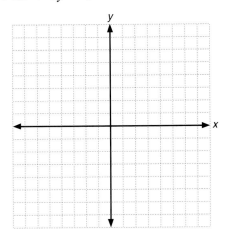

20. $4x - 3y = 9$

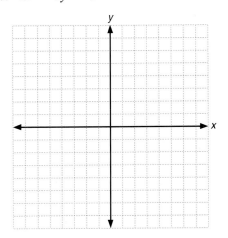

We now wish to examine the straight line as a function. The graph of a straight line parallel to the y-axis does not represent a function since each x value has many y values. All other straight lines represent functions. Remember that the graph of any first-degree equation in two variables will form a straight line.

We will call

$$ax + by = c$$

the **standard form** of the equation of a straight line. An important concept, related to the equation of a straight line, is that of slope. Intuitively we think of the slope of a line as the "steepness" of the line. The following definition gives us a more precise meaning.

The **slope** (m) of a line through the two distinct points (x_1, y_1) and (x_2, y_2) is given by the ratio

$$m = \frac{y_2 - y_1}{x_2 - x_1}, \quad x_1 \neq x_2.$$

Note that the denominator cannot be zero.

Example 3   Find the slope of a line through the points $(1, -5)$ and $(4, 0)$.

Solution   If we let $(1, -5)$ be (x_1, y_1) and $(4, 0)$ be (x_2, y_2), then applying the formula, we obtain

Be careful when substituting for x_1, y_1, x_2, and y_2.

$$m = \frac{0 + 5}{4 - 1} = \frac{5}{3}.$$

The ratio $\dfrac{y_2 - y_1}{x_2 - x_1}$ is not dependent on the points chosen as long

It doesn't matter which two points are chosen on the line, the slope will be the same.

as they are two different points on the line. A given line has only one slope.

The definition of slope and the fact that the slope of a given line is constant is the basis for the solution of many problems concerning linear functions.

For instance, if we are given that a line contains the points (x_1, y_1) and (x_2, y_2) and are asked for the equation of the line, we proceed in the following manner.

First we choose any other point on the line and call it (x, y). The slope is now

given by the ratio $\dfrac{y_2 - y_1}{x_2 - x_1}$ *or* by the ratio $\dfrac{y - y_1}{x - x_1}$, and since the slope is con-

stant we can equate these two ratios giving

The slopes are equal.

$$\frac{y - y_1}{x - x_1} = \frac{y_2 - y_1}{x_2 - x_1},$$

which becomes

We multiplied each side by $(x - x_1)$.

$$y - y_1 = \left(\frac{y_2 - y_1}{x_2 - x_1}\right)(x - x_1), \; x_1 \neq x_2.$$

This is known as the **two-point form** of the equation of a line.

Example 4   Write the equation, in standard form, of the line through the points $(3, 5)$ and $(-2, 7)$.

Solution   Using the two-point form, we have

Let $(x_1, y_1) = (3, 5)$ and $(x_2, y_2) = (-2, 7)$

$$y - 5 = \left(\frac{7 - 5}{-2 - 3}\right)(x - 3).$$

Simplifying, we obtain

$$y - 5 = -\frac{2}{5}(x - 3)$$

or

This is in standard form.

$$2x + 5y = 31.$$

Check the points here in the margin.

Check to see that both points are on this line by substituting into the equation.

If we are given the slope m of a line and a point (x_1,y_1) and asked to write the equation, we proceed as follows.

We first choose some other point on the line and call it (x,y). The definition of the slope gives us the ratio $\dfrac{y - y_1}{x - x_1}$. But since we are given the slope to be m, we can write the equation

$$m = \frac{y - y_1}{x - x_1}$$

or

$$y - y_1 = m(x - x_1).$$

This is the **point-slope form** of the equation of a straight line.

Example 5   Write the equation in standard form of the line through the point $(2,5)$ and having slope of $\dfrac{2}{5}$.

Solution   Using the point-slope form gives us

$$y - 5 = \frac{2}{5}(x - 2),$$

which in standard form is

$$2x - 5y = -21.$$

Check to see if $(2,5)$ satisfies this equation.

$(x_1,y_1) = (2,5)$

$m = \dfrac{2}{5}$

The **y-intercept** of a straight line is the ordinate of the point where the line intersects the y-axis. In other words, it is the value of y when $x = 0$. We will designate the y intercept as b.

If we are given the slope m of a line and the point $(0,b)$ on the line, then the point-slope form gives us

$$y - b = m(x - 0)$$

or

$$y = mx + b.$$

m is the slope and b is the y-intercept.

This very useful form of the equation of a straight line is called the **slope-intercept form.**

$m = -\dfrac{5}{8}$ and $b = 6$.

Example 6   If the slope of a line is $-\dfrac{5}{8}$ and the y-intercept is 6, write the equation in standard form.

Solution   The slope-intercept form gives us

$$y = -\frac{5}{8}x + 6,$$

which in standard form is

$$5x + 8y = 48.$$

Change the standard form $5x + 8y = 48$ to slope-intercept form.

The slope-intercept form of the equation of a straight line is useful in graphing. We can use the equation from the last example to illustrate.

$$y = -\frac{5}{8}x + 6$$

We know that the y-intercept is 6. Thus the point whose coordinates are $(0,6)$ is on the graph.

The slope is $-\dfrac{5}{8}$ and since it is a ratio, it does not matter whether we place the negative sign in the numerator or denominator. If we put it in the numerator we obtain $\dfrac{-5}{8}$, which indicates that the change in y is -5 while the change in x is 8.

Using these values, we begin at $(0,6)$ and move 8 places in the positive x direction. We then move 5 places in the negative y direction. The point we arrive at is also on the line. Since we now have two points, we may draw the graph of the line as shown.

Draw the graph again using

$$m = \frac{5}{-8}.$$

Do you get the same line?

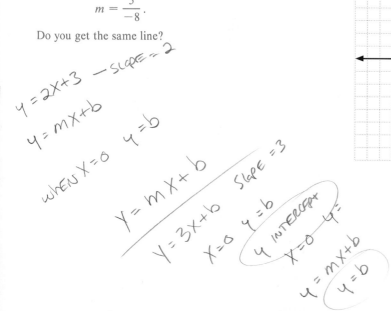

Example 7 Find the slope and the y-intercept of the line given by the equation $3x - 2y = 9$.

Solution Solving for y in terms of x gives us

$$y = \frac{3}{2}x - \frac{9}{2},$$

This is slope-intercept form.

so the slope is $m = \dfrac{3}{2}$ and the y-intercept is $b = -\dfrac{9}{2}$.

We now wish to state two special properties involving slopes of lines.

> If two distinct lines have the same slope, they are parallel.

Example 8 The lines given by the equations

$$y = \frac{1}{3}x + 5 \quad \text{and} \quad y = \frac{1}{3}x - 7$$

The slope of each line is $\dfrac{1}{3}$.

are parallel since their slopes are equal and their y-intercepts are different.

> If the slope of one line is the negative reciprocal of another line, the lines are perpendicular.

Perpendicular lines intersect at an angle of 90°.

Example 9 The lines given by the equations

$$y = \frac{2}{3}x + 4 \quad \text{and} \quad y = -\frac{3}{2}x + 1$$

are perpendicular since $\dfrac{2}{3}$ and $-\dfrac{3}{2}$ are negative reciprocals.

X INDEPENDENT EQUATION

// INCONSISTENT EQUATION

↗ DEPENDENT EQUATION

▼ EXERCISE 8-3-2

Find the equation, in standard form, of the line through each of the pairs of points.

1. (2,1) and (−3,9)   **2.** (3,5) and (−1,4)   **3.** (−6,2) and (4,−3)

4. (−2,−1) and (5,−7)   **5.** (16,−5) and (4,0)   **6.** (3,0) and (−9,4)

7. (3,2) and (−4,2)   **8.** (−5,−1) and (0,−1)   **9.** (−8,−1) and (3,−5)

10. (−3,7) and (−1,−4)

Use the information given to sketch the graph and then find the equation, in standard form, of the line represented.

11. (1,7), $m = 3$ Slope $(m) = 3$ **12.** (2,5), $m = 2$

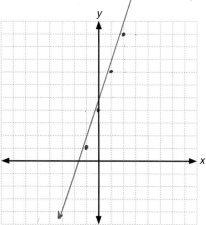

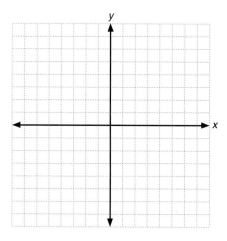

Slope = −1.6

$$\#1$$

$$y - y_1 = \frac{y_2 - y_1}{x_2 - x_1}(x - x_1)$$

$$y - (1) = \frac{9 - 1}{(-3) - (2)}(x - 2)$$

$$y - 1 = \frac{8}{-5}(x - 2)$$

$$-5y + 5 = 8x - 16 \qquad 8x + 5y = 21$$

$$\#3$$

$$y - y_1 = \frac{y_2 - y_1}{x_2 - x_1}(x - x_1)$$

$$y - (2) = \frac{-3 - (2)}{4 - (-6)}(x - (-6))$$

Slope = −.5

$$y - 2 = \frac{-5}{10}(x + 6)$$

$$10y - 20 = -5x - 30$$

$$5x + 10y = -10$$

$$x + 2y = -2$$

13. $(-2,0), m = \dfrac{1}{2}$

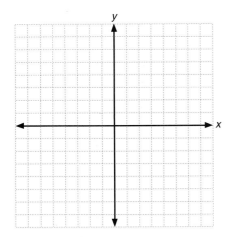

14. $(0,4), m = 5$

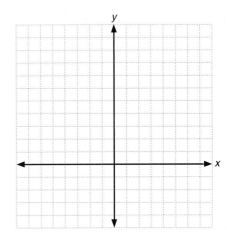

15. $(-5,9), m = -8$

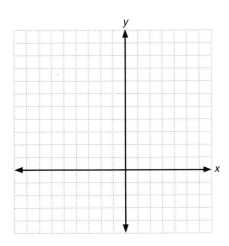

16. $(3,-1), m = -\dfrac{2}{3}$

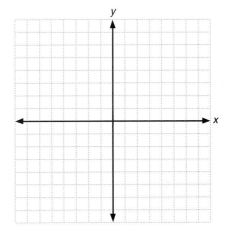

17. $(-2,3)$, parallel to the y-axis.

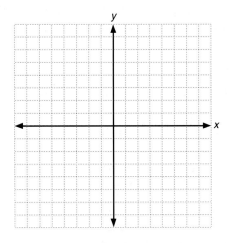

18. $(3,-1)$, parallel to the y-axis

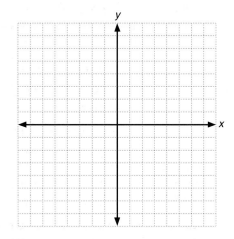

19. $(2,5)$, $m = 0$

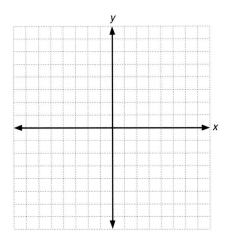

20. $(4,-3)$, $m = 0$

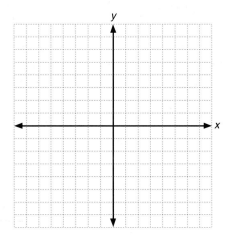

Write each equation in slope-intercept form. Specify the slope and y-intercept.

21. $2x + y = 5$

22. $3x + y = -1$

23. $3x + 4y = 1$

24. $2y - x = 3$

25. $2x - 5y = 9$

26. $3y - 6x = 4$

27. $5x - 2y = 0$

28. $2y - 3x = 0$

29. $x - 8y = 1$

30. $x - 3y = 6$

31. A line has an x-intercept of $x = -2$ and a y-intercept of $y = 7$. Find the equation of the line in standard form.

32. A line has an x-intercept of $x = 3$ and a y-intercept of $y = 5$. Find the equation of the line in standard form.

Write the equation, in standard form, of the line passing through the given point and parallel to the given line.

33. $(3,1)$, $y = 2x + 1$

34. $(2,0)$, $3x + y = 5$

35. $(5,-1)$, $2y - 5x = 1$

36. $(3,-10)$, $2x - y = 4$

37. $(-8,1)$, $y = -3$

38. $(2,6)$, $y = 1$

39. $(4,-9)$, $x = 2$

40. $(-3,2)$, $x = -5$

Write the equation, in standard form, of the line passing through the given point and perpendicular to the given line.

41. $(3,2)$, $y = \dfrac{2}{3}x + 1$ **42.** $(0,-5)$, $y = \dfrac{1}{2}x - 3$

43. $(-1,4)$, $y = -\dfrac{1}{5}x + 1$ **44.** $(2,5)$, $y = -2x + 3$

 8-4 THE CONIC SECTIONS

O B J E C T I V E S

Upon completing this section you should be able to:

1. Identify the conic sections from their equations in standard form.
2. Sketch a conic from its equation in standard form.

Equations in two unknowns that are higher than first-degree will graph as curves other than straight lines. We will limit our discussion here to second-degree equations.

The curves formed by second-degree equations in two variables are the *circle, ellipse, hyperbola,* and *parabola.* These curves are studied in detail in the subject of analytic geometry, but because their equations are also important in algebra, we will briefly discuss them in this section.

These curves mentioned are called **conic sections** or **conics** because they result from the intersection of a plane and a right circular cone as illustrated.

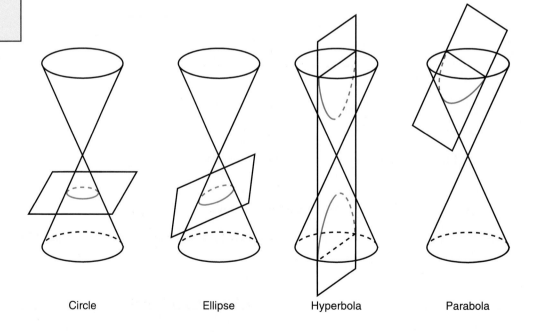

Circle   Ellipse   Hyperbola   Parabola

We will first consider the circle.

A **circle** is the set of all points in a plane that are a given distance from a fixed point. The given distance is the length of any line segment from the fixed point (called the center) to a point on the circle. All of these line segments are called radii (singular radius). Often r is used to represent the radius and the center is called (h,k).

The *standard form* of the equation of a circle having center (h,k) and radius r is

$$(x - h)^2 + (y - k)^2 = r^2.$$

Notice that if the center is at the origin $(0,0)$, this equation reduces to

$$x^2 + y^2 = r^2.$$

The graph of a circle is readily sketched when we know its center and radius.

Example 1 Sketch the graph of a circle with center $(2,-1)$ and a radius of 3. Write its equation in standard form.

We are looking for all points on the plane that are 3 units from the point $(2,-1)$.

Solution

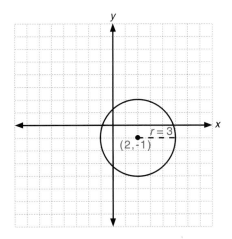

The equation in standard form is

$$(x - 2)^2 + [y - (-1)]^2 = 3^2$$

We substitute $h = 2$, $k = -1$, and $r = 3$.

or

$$(x - 2)^2 + (y + 1)^2 = 9.$$

Note this equation is not in standard form.

Example 2 $x^2 + y^2 + 4x + 8y = -4$ is the equation of a circle. Find its center and radius and sketch its graph.

Solution   First we wish to put the equation in standard form. To do this we complete the square on each part by adding the correct number to both sides of the equation.

Recall how to complete the square.

$$x^2 + 4x + 4 + y^2 + 8y + 16 = -4 + 4 + 16$$

or

$$(x + 2)^2 + (y + 4)^2 = 16$$

In this form we see that the center is $(-2,-4)$ and the radius is 4.

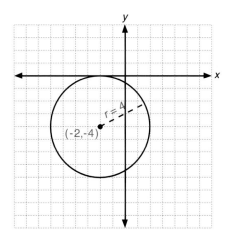

The domain is $[-6,2]$ and the range is $[-8,0]$. Is y a function of x?

Does each value of x give only one value of y?

  EXERCISE 8-4-1

1. Find the equation of a circle in standard form with center $(-2,3)$ and radius 6.

2. Find the equation of a circle in standard form with center $(0,0)$ and radius 7.

Each of the following is the equation of a circle. Put the equation in standard form and give the center and radius. Sketch the graph.

3. $x^2 + y^2 + 2x - 12y = 44$

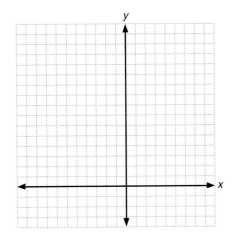

4. $x^2 + y^2 - 4x + 6y = 3$

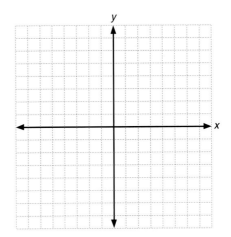

5. $x^2 + y^2 - 6x - 4y = 36$

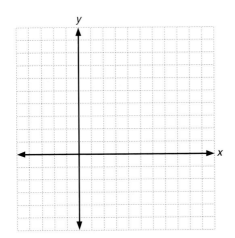

6. $x^2 + y^2 - 2x - 10y = -22$

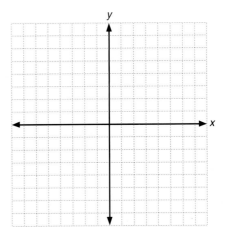

7. $x^2 + y^2 + 4x + 10y = -28$

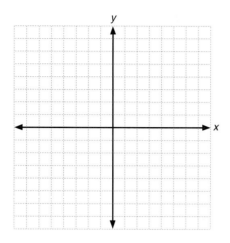

8. $x^2 + y^2 + 6x + 12y = -9$

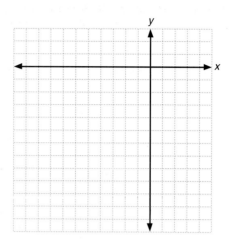

9. $4x^2 + 4y^2 - 4x + 32y = 35$

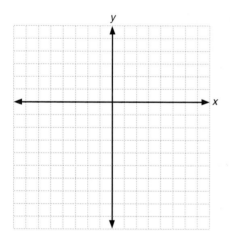

10. $16x^2 + 16y^2 + 64x - 8y = 79$

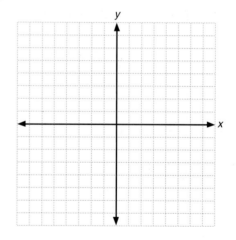

An **ellipse** is the set of all points in a plane, each such that the sum of its distances from two fixed points, called foci (singular focus), is a constant.

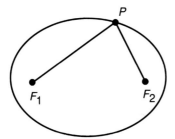

F_1 and F_2 are the foci of the ellipse.

$PF_1 + PF_2$ is a constant.

The *standard form* of the equation of an ellipse with center at (h,k) is

$$\frac{(x - h)^2}{a^2} + \frac{(y - k)^2}{b^2} = 1$$

where $2a$ is the length of the horizontal axis and $2b$ is the length of the vertical axis.

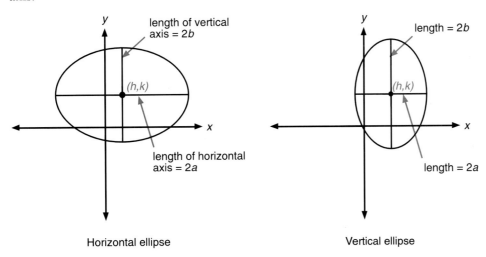

Horizontal ellipse

Vertical ellipse

Note that if $a > b$, the ellipse is a *horizontal ellipse* and if $a < b$, the ellipse is a *vertical ellipse*.

This implies that if $a^2 > b^2$, we have a horizontal ellipse and if $a^2 < b^2$, we have a vertical ellipse.

If we denote the distance from the center of an ellipse to its foci as *c*, then we have

$$c^2 = b^2 - a^2 \text{ for a vertical ellipse and}$$
$$c^2 = a^2 - b^2 \text{ for a horizontal ellipse.}$$

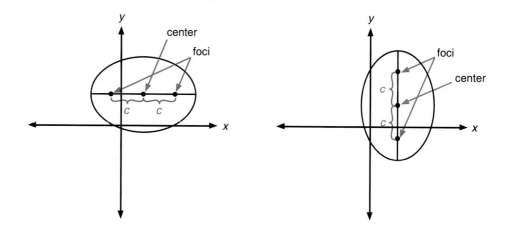

Vertical ellipse

Since *c* represents the distance from the center to the foci, it is a positive value.

Example 3   Sketch the graph of $\dfrac{(x-2)^2}{9} + \dfrac{(y+1)^2}{25} = 1$.

Which is larger, a^2 or b^2?

Solution   We immediately note that the center is $(2, -1)$, $a = 3$ and $b = 5$. First we plot the center. Then we move horizontally three units to the left and three units to the right of the center to find the endpoints of the horizontal axis. We then move vertically five units up and five units down from center to establish the endpoints of the vertical axis. These endpoints are sometimes referred to as *vertices*. We note that since $a < b$ we have a vertical ellipse and can now easily sketch the graph.

$2a = 2(3) = 6$ is the length of the horizontal axis. $2b = 2(5) = 10$ is the length of the vertical axis.

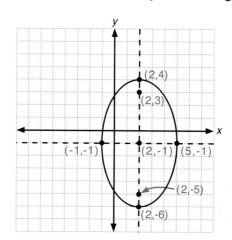

All we need to sketch the graph is the center and the values of *a* and *b*. These are easily obtained from the equation in standard form.

Notice that it is not necessary to know the foci to sketch the graph of the ellipse. However, we know that the foci are a distance *c* from the center. Since for a vertical ellipse

$$a^2 + c^2 = b^2$$

or

$$c^2 = b^2 - a^2,$$

then since $a = 3$ and $b = 5$

$$c^2 = 25 - 9$$

or

$$c = 4.$$

Thus the foci are $(2,3)$ and $(2,5)$.

The domain of the relation is $[-1,5]$, and the range is $[-6,4]$.

Is y a function of x?

Example 4 $9x^2 + 18x + 16y^2 - 64y = 71$ is the equation of an ellipse. Find its center, foci, and vertices (endpoints of the axes). Sketch the graph.

Solution   We first wish to get the equation in standard form. To do this we factor 9 from the x terms and 16 from the y terms.

$$9(x^2 + 2x + \quad) + 16(y^2 - 4y + \quad) = 71$$

We next complete the squares.

When completing the squares, don't forget to also add these quantities to the right side of the equation.

$$9(x^2 + 2x + 1) + 16(y^2 - 4y + 4) = 71 + 9 + 64$$
$$9(x + 1)^2 + 16(y - 2)^2 = 144$$

or

$$\frac{(x + 1)^2}{16} + \frac{(y - 2)^2}{9} = 1$$

The equation is now in standard form and we see that the center is $(-1,2)$. We also note that it is a horizontal ellipse since $a = 4$ is greater than $b = 3$. Since it is a horizontal ellipse

$$a^2 - c^2 = b^2$$

or

$$c^2 = a^2 - b^2$$
$$= 16 - 9$$
$$= 7.$$

Thus

Note that in either the horizontal or vertical ellipse the value of c^2 is always the larger denominator minus the smaller denominator when the equation is in standard form.

$$c = \sqrt{7}$$

and the coordinates of the foci are $(-1 - \sqrt{7}, 2)$ and $(-1 + \sqrt{7}, 2)$. The vertices of the horizontal axis are $(-5,2)$ and $(3,2)$ and the vertices of the vertical axis are $(-1,-1)$ and $(-1,5)$.

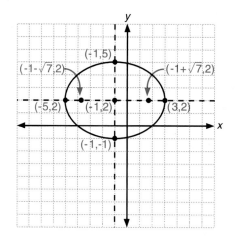

State the domain and range of this relation.

EXERCISE 8-4-2

Each of the following is the equation of an ellipse. Find the center, vertices, and foci. Sketch the graph.

1. $\dfrac{(x-3)^2}{25} + \dfrac{(y-1)^2}{9} = 1$

2. $\dfrac{(x-1)^2}{100} + \dfrac{(y-2)^2}{36} = 1$

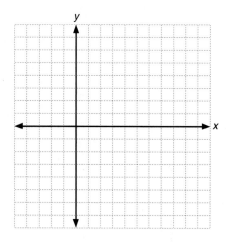

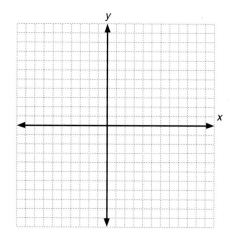

3. $\dfrac{(x + 1)^2}{16} + \dfrac{(y - 4)^2}{36} = 1$

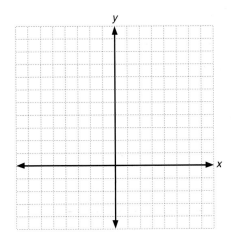

4. $\dfrac{(x + 2)^2}{16} + \dfrac{(y - 3)^2}{25} = 1$

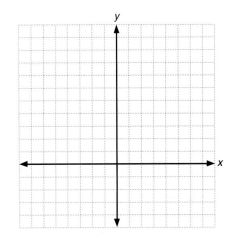

5. $\dfrac{x^2}{9} + \dfrac{y^2}{25} = 1$

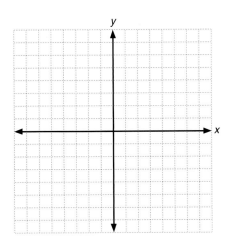

6. $\dfrac{x^2}{49} + \dfrac{(y + 1)^2}{64} = 1$

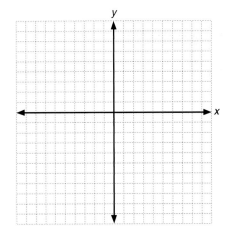

7. $4x^2 + 16y^2 - 16x + 160y = -352$

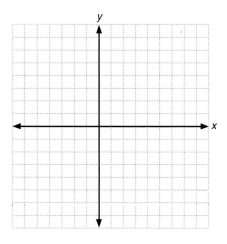

8. $9x^2 + 4y^2 - 36x + 24y = -36$

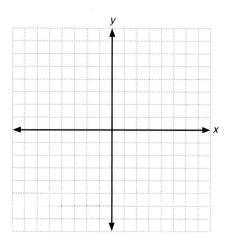

9. Write the equation of the ellipse whose vertices are $(-10,2)$, $(4,2)$, $(-3,-1)$, and $(-3,5)$.

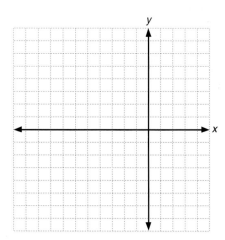

A **hyperbola** is the set of all points in a plane such that the absolute value of the differences of its distance to two fixed points (foci) is a constant.

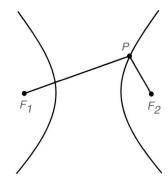

$\overline{PF_1} \cdot \overline{PF_2}$ is a constant.

Note that there are two "branches" of this curve.

The *standard form* for the equation of the hyperbola is

$$\frac{(x - h)^2}{a^2} - \frac{(y - k)^2}{b^2} = \pm 1,$$

where (h,k) is the *center*. The foci are a distance c from the center and $a^2 + b^2 = c^2$. If the right side of the equation is $+1$, we have a *horizontal* hyperbola, and if the right side of the equation is -1, the hyperbola is *vertical*. The hyperbola is one of the easiest curves to sketch. Using (h,k) as the center, we construct a rectangle that is $2a$ units wide and $2b$ units high. We then draw and extend the diagonals of the rectangle. These extended diagonals are the *asymptotes* for the branches of the hyperbola. The *asymptote* has the characteristic that as we move out on any branch of the hyperbola, the curve continually approaches the asymptote but never reaches it. The hyperbola can then be easily sketched as shown.

If the equation is

$$\frac{(x - h)^2}{a^2} - \frac{(y - k)^2}{b^2} = 1,$$

the graph would appear as follows.

This is a horizontal hyperbola since we have a positive 1.

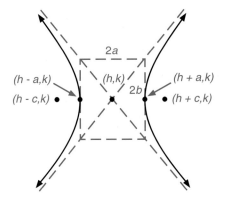

Note that all we need to sketch the curve is the center (h,k) and a and b. These values are easily obtained from the standard form of the equation.

If the equation is

$$\frac{(x - h)^2}{a^2} - \frac{(y - k)^2}{b^2} = -1,$$

the graph would be a vertical hyperbola.

In the hyperbola the relative size of a and b has no bearing on whether the curve is horizontal or vertical.

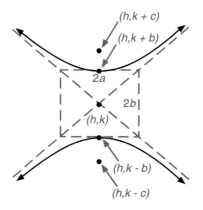

Example 5   Sketch the graph of $\dfrac{(x - 2)^2}{9} - \dfrac{(y + 1)^2}{16} = 1.$

Solution   We first note that it is a horizontal hyperbola. (Why?) We next locate the center at $(2, -1)$. Using this as the center, we construct a rectangle six units ($2a$) long and eight units ($2b$) high. We extend the diagonals of the rectangle to form the asymptotes for the hyperbola. Using $(-1, -1)$ and $(5, -1)$ as the vertices of the hyperbola, we may now sketch the graph.

Since $a = 3$, $2a = 6$. Since $b = 4$, $2b = 8$.

$$c^2 = a^2 + b^2$$
$$c^2 = 9 + 16$$
$$c^2 = 25$$
$$c = 5$$

Therefore, the distance from the center $(2, -1)$ to the foci is 5.

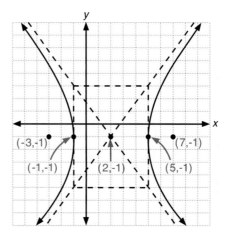

Is y a function of x?

The foci of the hyperbola are $(-3, -1)$ and $(7, -1)$. $(-1, -1)$ and $(5, -1)$ are the *vertices* of the hyperbola. What is the domain of this relation? What is the range?

▼ **EXERCISE 8-4-3**

For each hyperbola find the center, vertices, and foci. Draw the asymptotes and sketch the graph.

1. $\dfrac{(x-2)^2}{9} - \dfrac{(y-3)^2}{16} = 1$

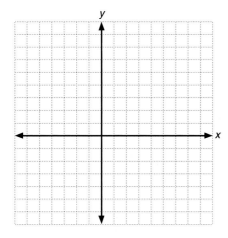

2. $\dfrac{(x-1)^2}{16} - \dfrac{(y+4)^2}{25} = 1$

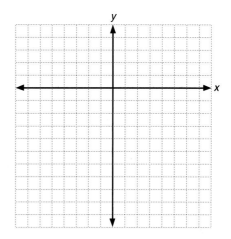

3. $\dfrac{(x+1)^2}{4} - \dfrac{(y-2)^2}{16} = -1$

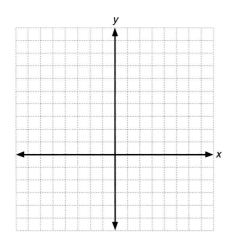

4. $\dfrac{(x+3)^2}{49} - \dfrac{(y-2)^2}{16} = -1$

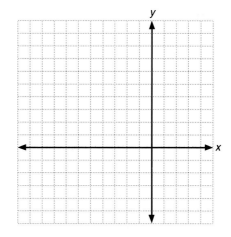

5. $y^2 - 4x^2 = 16$ (*Hint:* Put in standard form.)

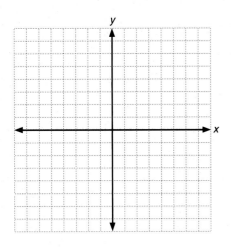

6. $4x^2 + 36 = 9y^2$

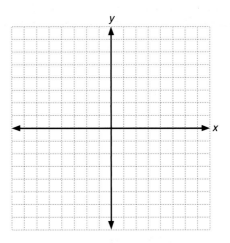

7. $9x^2 - 4y^2 - 54x - 32y = 19$

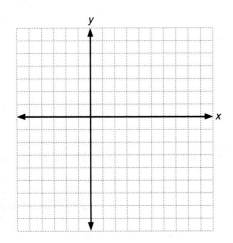

8. $4x^2 - 9y^2 - 16x - 90y = 65$

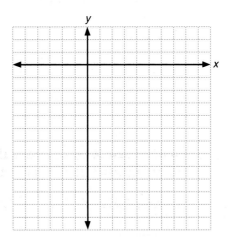

A **parabola** is the set of all points in a plane equidistant from a fixed point (focus) and a fixed line (directrix) not containing the fixed point.

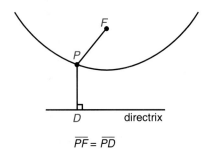

$$\overline{PF} = \overline{PD}$$

Satellite dishes are parabolic in shape because incoming rays are reflected to the fixed point called the focus.

It should be mentioned immediately that a parabola is *not* one branch of a hyperbola, as might be thought if they are carelessly drawn. They have different definitions and a parabola does not have any asymptotes.

The *standard form* of the equation for a vertical parabola is

$$(x - h)^2 = 4p(y - k)$$

where (h,k) is the vertex of the parabola, the focus is $(h, k + p)$, and the directrix is $y = k - p$. The parabola opens upward if $p > 0$ and downward if $p < 0$.

Example 6 Graph $(x - 4)^2 = 8(y - 3)$.

Solution This is a vertical parabola with vertex $(4,3)$, $p = 2$ and therefore the parabola opens upward. (Why?) The focus is $(4,5)$ and the directrix is $y = 1$.

If the equation was $(x - 4)^2 = -8(y - 3)$, the parabola would open downward.

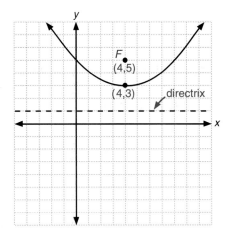

When sketching the graph, remember that each point on the curve is equidistant from the focus and the directrix.

The *standard form* of the equation for a horizontal parabola is

$$(y - k)^2 = 4p(x - h)$$

Notice you will have a y^2 term in this case.

where (h,k) is the vertex. The focus is $(h + p, k)$ and the directrix is $x = h - p$. The parabola opens to the right if $p > 0$ and to the left if $p < 0$.

Example 7   Graph $(y + 1)^2 = -12(x - 3)$.

Which way would the parabola open if the equation was $(y + 1)^2 = 12(x - 3)$?

Solution   This is a horizontal parabola with vertex $(3, -1)$, $p = -3$. The parabola opens to the left. (Why?) The focus is $(0, -1)$ and the directrix is $x = 6$.

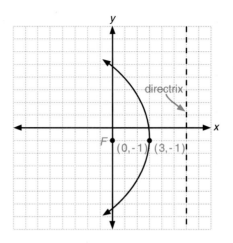

▼ **EXERCISE 8-4-4**

For each parabola find the vertex, focus, and directrix. Sketch the graph.

1. $(x - 3)^2 = 12(y + 1)$

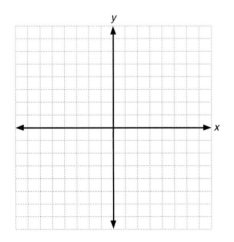

2. $(x - 1)^2 = 8(y - 4)$

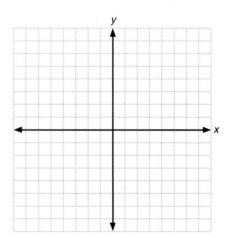

3. $(y - 2)^2 = 4(x + 3)$

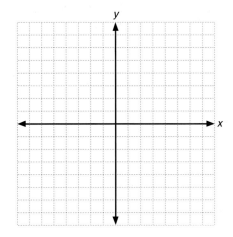

4. $(y - 3)^2 = 16(x + 1)$

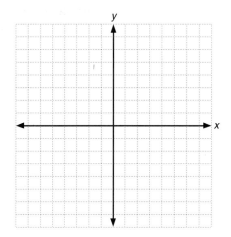

5. $(x + 1)^2 = -8(y - 3)$

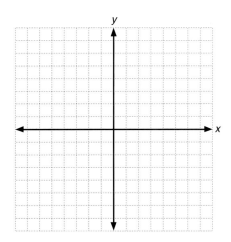

6. $(x + 2)^2 = -12(y - 5)$

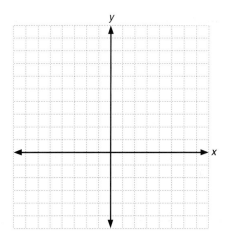

7. $y^2 - 2y + 8x - 15 = 0$

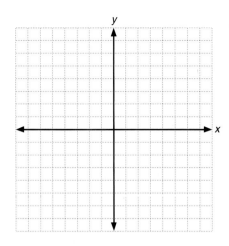

8. $y^2 + 20x - 8y = 24$

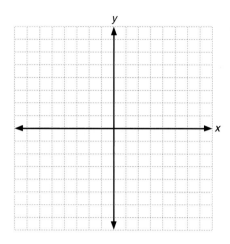

Put each equation in standard form. Identify the conic and give important data such as center, vertices, foci, directrix, and so on. Sketch the graph.

9. $x^2 + 9y^2 - 10x + 36y + 52 = 0$

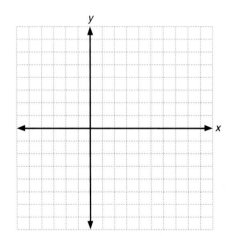

10. $4x^2 + 4y^2 + 24x - 4y = 63$

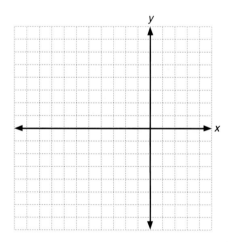

11. $x^2 - 8x - 8y + 8 = 0$

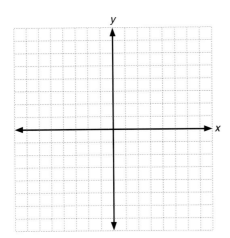

12. $25x^2 + 4y^2 + 100x - 40y + 100 = 0$

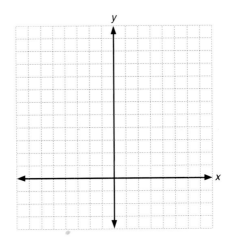

13. $x^2 + y^2 + 8x - 6y - 5 = 0$

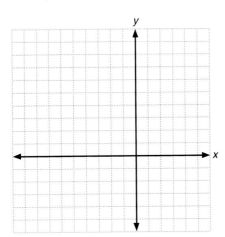

14. $4x^2 - 16y^2 - 8x - 96y = 204$

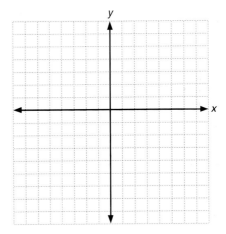

15. $9x^2 - 16y^2 + 72x + 96y + 144 = 0$

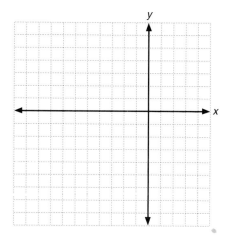

16. $y^2 + 8x + 6y = 23$

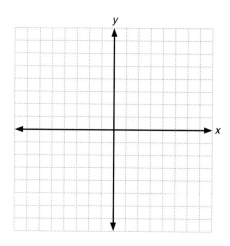

8-5 GRAPHING LINEAR INEQUALITIES

In chapter 2 we constructed line graphs of inequalities such as

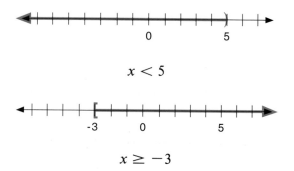

$$x < 5$$

$$x \geq -3$$

These were inequalities involving only one variable. We found that in all such cases the graph was some portion of the number line. Since an equation in two variables gives a graph on the plane, it seems reasonable to assume that an inequality in two variables would graph as some portion or region of the plane. This is in fact the case. The solution of the inequality $x + y < 5$ is the set of all ordered pairs of numbers (x,y) such that their sum is less than 5. ($x + y < 5$ is a linear inequality since $x + y = 5$ is a linear equation.)

You may want to review section 2–7.

Example 1   Are each of the following pairs of numbers in the solution set of $x + y < 5$? $(2,1)$, $(3,-4)$, $(5,6)$, $(3,2)$, $(0,0)$, $(-1,4)$, $(-2,8)$.

The solution set consists of all ordered pairs that make the statement true.

     Solution

$$
\begin{aligned}
(2,1) \quad & x + y < 5 \\
& 2 + 1 < 5 \\
& \quad\ 3 < 5 \qquad \text{Yes}
\end{aligned}
$$

$$(3,-4) \quad x + y < 5$$
$$3 + (-4) < 5$$
$$-1 < 5 \qquad \text{Yes}$$

$$(5,6) \quad x + y < 5$$
$$5 + 6 < 5$$
$$11 < 5 \qquad \text{No}$$

$$(3,2) \quad x + y < 5$$
$$3 + 2 < 5$$
$$5 < 5 \qquad \text{No}$$

$$(0,0) \quad x + y < 5$$
$$0 + 0 < 5$$
$$0 < 5 \qquad \text{Yes}$$

$$(-1,4) \quad x + y < 5$$
$$(-1) + 4 < 5$$
$$3 < 5 \qquad \text{Yes}$$

$$(-2,8) \quad x + y < 5$$
$$(-2) + 8 < 5$$
$$6 < 5 \qquad \text{No}$$

To summarize, the following ordered pairs give a *true* statement.

$$(2,1)$$
$$(3,-4)$$
$$(0,0)$$
$$(-1,4)$$

The following ordered pairs give a *false* statement.

$$(5,6)$$
$$(3,2)$$
$$(-2,8)$$

Following is a graph of the line $x + y = 5$. The points from example 1 are indicated on the graph with answers to the question "Is $x + y < 5$?"

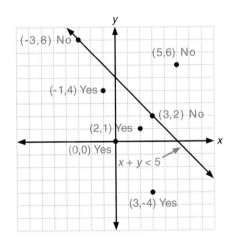

Notice that all the points that satisfy the equation are to the left and below the line while all the points that will not are above and to the right.

Observe that all "yes" answers lie on the same side of the line $x + y = 5$, and all "no" answers lie on the other side of the line or on the line itself.

The graph of the line $x + y = 5$ divides the plane into three parts: the line itself and the two sides of the lines (called *half-planes*).

$$x + y < 5 \text{ is a } half\text{-}plane$$
$$x + y \leq 5 \text{ is a } line \text{ and a } half\text{-}plane.$$

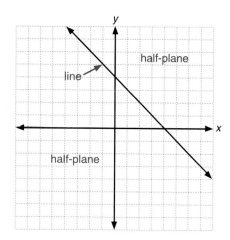

If one point of a half-plane is in the solution set of a linear inequality, then all points in that half-plane are in the solution set. This gives us a convenient method for graphing linear inequalities.

Why do we need to check only one point?

To graph a linear inequality
1. Replace the inequality symbol with an equal sign and graph the resulting line.
2. Check *one* point that is obviously in a particular half-plane of that line to see if it is in the solution set of the inequality.
3. If the point chosen *is* in the solution set, then that entire half-plane is the solution set. If the point chosen *is not* in the solution set, then the other half-plane is the solution set.

Example 2   Sketch the graph of $2x + 3y > 7$.

Step 1

One possible table of values is

x	5	2	-1
y	-1	1	3

Solution   First sketch the graph of the line $2x + 3y = 7$ using a table of values.

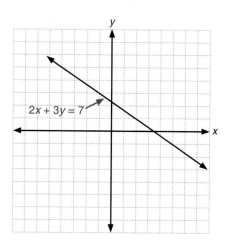

Next choose a point that is *not* on the line $2x + 3y = 7$. [If the line does not go through the origin, then the point $(0,0)$ is always a good choice.] Now turn to the inequality $2x + 3y > 7$ to see if the chosen point is in the solution set.

Step 2
Why do you think $(0,0)$ is a good choice?

$$(0,0) \quad 2x + 3y > 7$$
$$2(0) + 3(0) > 7$$
$$0 > 7 \qquad \text{No!}$$

The point $(0,0)$ is *not* in the solution set, therefore the half-plane containing $(0,0)$ is not the solution set. Hence, the other half-plane determined by the line $2x + 3y = 7$ is the solution set.

Step 3

Since the line itself is not a part of the solution, it is shown as a dashed line and the half-plane is shaded to show the solution set.

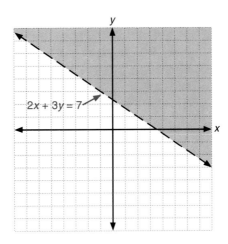

The solution set is the half-plane above and to the right of the line.

Example 3 Graph the solution for the linear inequality $2x - y \geq 4$.

Solution First graph $2x - y = 4$.

Step 1

Since the line graph for $2x - y = 4$ does not go through the origin $(0,0)$, check that point in the linear inequality.

$$(0,0) \quad 2x - y \geq 4$$
$$2(0) - (0) \geq 4$$
$$0 \geq 4 \qquad \text{No!}$$

Step 2

Since the point $(0,0)$ is not in the solution set, the half-plane containing $(0,0)$ is not in the set. Hence, the solution is the other half-plane. Notice, however, that the line $2x - y = 4$ is included in the solution set. Therefore, draw a solid line to show that it is part of the graph.

Step 3

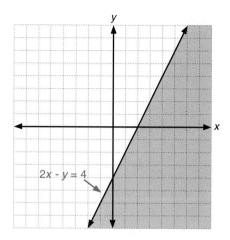

The solution set is the line and the half-plane below and to the right of the line.

Example 4   Graph $x < y$.

Solution   First graph $x = y$.

Next check a point not on the line. Notice that the graph of the line contains the point $(0,0)$, so we cannot use it as a checkpoint. To determine which half-plane is the solution set use any point that is obviously not on the line $x = y$. The point $(-2,3)$ is such a point.

$$(-2,3) \quad x < y$$
$$-2 < 3 \qquad \text{Yes!}$$

Using this information, graph $x < y$.

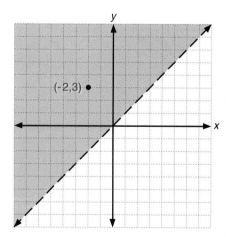

When the graph of the line goes through the origin, any other point on the x- or y-axis would also be a good choice.

▼ **EXERCISE 8–5–1**

Graph the linear inequalities.

1. $x + y > 3$

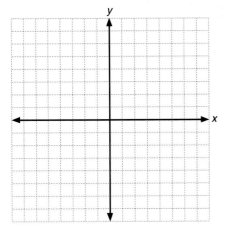

2. $x + y < 0$

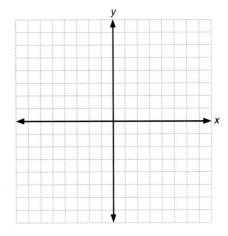

3. $x - y < 5$

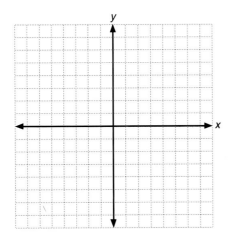

4. $x - y < 4$

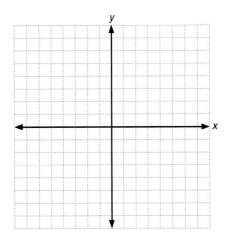

5. $x + 2y < 5$

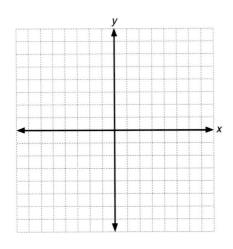

6. $2x + 3y > 1$

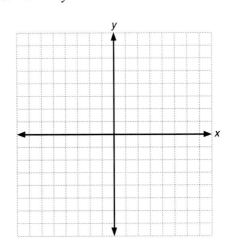

7. $x > y$

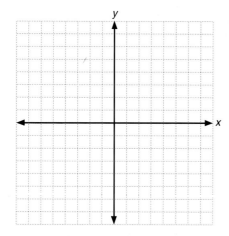

8. $x \leq y$

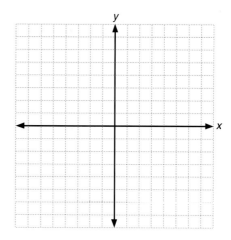

9. $x + 3y \leq 5$

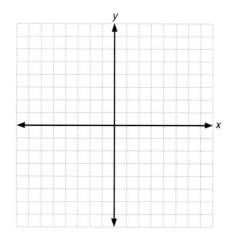

10. $x + 2y \geq 3$

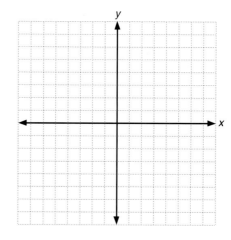

11. $x - 2y \leq 4$

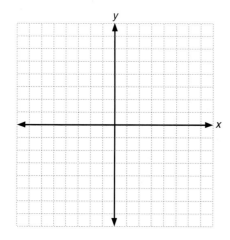

12. $3x \geq y$

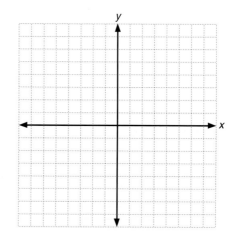

13. $3x - y > 0$

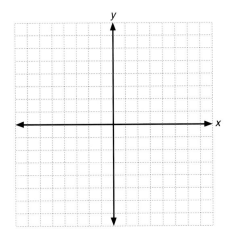

14. $x - 3y \leq 0$

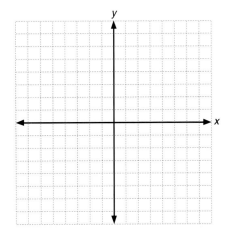

15. $x + 3y > 3$

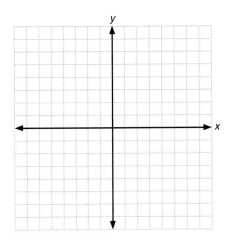

16. $x + 3y \geq 2$

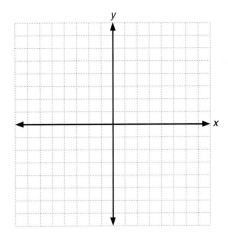

17. $2x + y \leq 3$

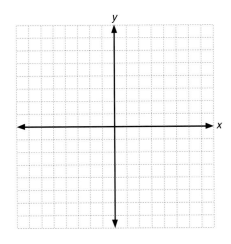

18. $2x + 3y > 8$

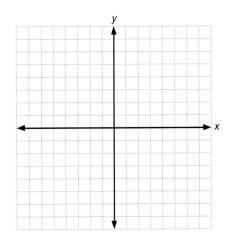

19. $3x - 4y \leq 2$

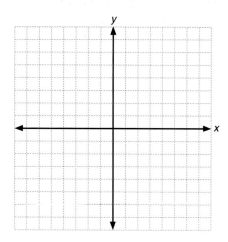

20. $2x - y \geq 3$

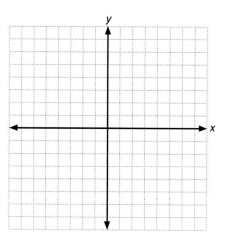

CHAPTER
8
SUMMARY

Key Words

Section 8–1

- A **relation** is a set of ordered pairs. The set of all first numbers is the **domain** and the set of all second numbers is the **range.**

- A **function** is a relation such that each element of the domain has exactly one image in the range.

Section 8–2

- The **Cartesian coordinate system** is a method of naming points on a plane.

- **Ordered pairs** of numbers are used to designate points on a plane.

Section 8–3

- A **linear equation** is a first-degree equation in two variables.

- The **standard form** of the equation of a straight line is

$$ax + by = c.$$

- The **slope** of a line is given by

$$m = \frac{y_2 - y_1}{x_2 - x_1}, \quad x_1 \neq x_2.$$

- The **two-point form** of the equation of a line is

$$y - y_1 = \left(\frac{y_2 - y_1}{x_2 - x_1}\right)(x - x_1), \quad x_1 \neq x_2.$$

- The **point-slope form** of the equation of a line is

$$y - y_1 = m(x - x_1).$$

- The **slope-intercept form** of the equation of a line is

$$y = mx + b.$$

Section 8–4

- Standard forms of the **conic sections:**
  1. **circle** $(x - h)^2 + (y - k)^2 = r^2$
  2. **ellipse** $\dfrac{(x - h)^2}{a^2} + \dfrac{(y - k)^2}{b^2} = 1$
  3. **hyperbola** $\dfrac{(x - h)^2}{a^2} - \dfrac{(y - k)^2}{b^2} = \pm 1$
  4. **parabola** vertical: $(x - h)^2 = 4p(y - k)$
     horizontal: $(y - k)^2 = 4p(x - h)$

Section 8–5

- A **linear inequality** is a first-degree inequality in two variables.

Procedures

Section 8–3

- To sketch the graph of a linear equation find ordered pairs of numbers that are solutions to the equation. Locate these points on the Cartesian coordinate system and connect them with a line.

Section 8–5

- To graph a linear inequality:
  1. Replace the inequality symbol with an equal sign and graph the resulting line.
  2. Check *one* point that is obviously in a particular half-plane of that line to see if it is in the solution set of the inequality.

3. If the point chosen *is* in the solution set, then that entire half-plane is the solution set. If the point chosen *is not* in the solution set, then the other half-plane is the solution set.

NAME:

CLASS / SECTION:

DATE:

A N S W E R S

Give the ordered pair associated with each point indicated on the coordinate system.

1. _____

1. *A*

2. *B*

2. _____

3. *C*

3. _____

4. *D*

5. *E*

4. _____

6. *F*

5. _____

7. *G*

8. *H*

6. _____

9. *I*

7. _____

10. *J*

8. _____

In each of the following relations consider the replacement set for *x* as the domain and the replacement set for *y* as the range. Find (a) the domain, (b) the range, (c) determine if *y* is a function of *x*.

9. _____

11. $y = 2x^2$

12. $y = \dfrac{1}{3x}$

10. _____

11. _____

12. _____

13. $x = y^2 + 1$

14. $x = \dfrac{3}{y}$

13. _____

15. $y = \sqrt{4 - x^2}$

16. $y = \dfrac{x - 1}{x + 5}$

14. _____

15. _____

17. $y = \dfrac{1}{\sqrt{x^2 - 1}}$

18. $y = x^2 - 1$

16. _____

Complete the table of values and sketch each of the following.

17. _____

19. $y = 2x - 1$

20. $y = x - 3$

x	-1	0	1
y			

x	-2	0	2
y			

18. _____

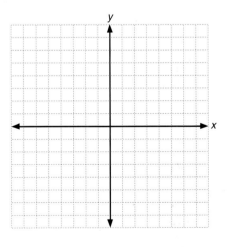

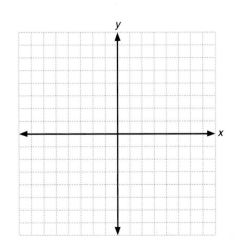

19. _____

20. _____

A N S W E R S

21. _____

22. _____

23. _____

24. _____

25. _____

26. _____

27. _____

28. _____

29. _____

30. _____

21. $3x + y = 7$

x	0	2	3
y			

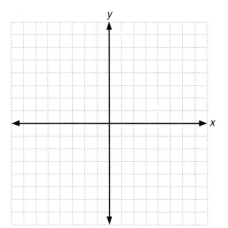

22. $x + 4y = 6$

x	-2	2	6
y			

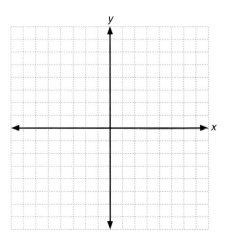

23. Find the equation, in standard form, of the line through the points $(4, -6)$ and $(-3, 5)$.

24. Find the equation, in standard form, of the line through the points $(3, 5)$ and $(-1, -3)$.

25. Find the equation, in standard form, of the line through the point $(2, -9)$ and having a slope of $-\dfrac{3}{5}$.

26. Find the equation, in standard form, of the line through the point $(-3, 7)$ and having a slope of $-\dfrac{2}{3}$.

27. Write the equation $2x - 5y = 10$ in slope-intercept form.

28. Write the equation $3x + 4y = 6$ in slope-intercept form.

29. Find the equation of a line, in standard form, whose slope is $\dfrac{1}{2}$ and has a y-intercept of 3.

30. Find the equation of a line, in standard form, whose slope is 5 and has a y-intercept of -3.

31. Sketch the graph of the straight line through the point (2,3) if $m = \dfrac{1}{2}$.

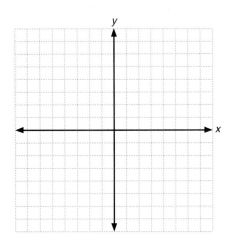

32. Sketch the graph of the straight line through the point $(-2,4)$ if $m = -\dfrac{2}{3}$.

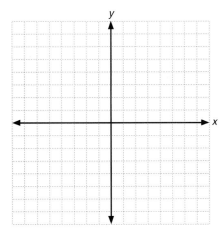

33. Write the equation, in standard form, of the line passing through the point $(-2,7)$ that is parallel to the line $3x + 8y = 1$.

34. Write the equation, in standard form, of the line passing through the point $(1,5)$ that is parallel to the line $7x + 3y = 10$.

A N S W E R S

31. _____

32. _____

33. _____

34. _____

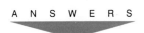

35. Put the equation of the circle $x^2 + y^2 - 6x + 2y = 15$ in standard form. Give the center and radius.

36. Put the equation of the circle $x^2 + y^2 + 8y = 0$ in standard form. Give the center and radius.

35. _____

For each of the following write the equation in standard form and classify the curve. Sketch the graph.

37. $25x^2 + 9y^2 + 150x - 90y + 225 = 0$

38. $4x^2 + 4y^2 - 4x - 48y + 45 = 0$

36. _____

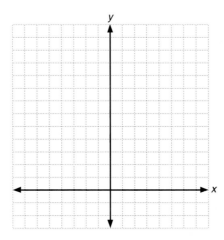

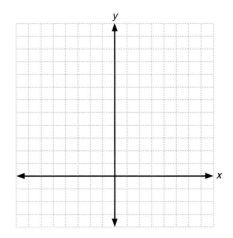

37. _____

38. _____

39. $25y^2 - 16x^2 + 50y + 64x$
$- 439 = 0$

40. $y^2 - 8y - 2x + 6 = 0$

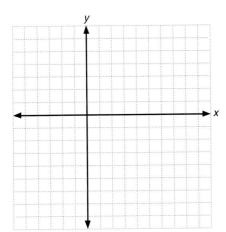

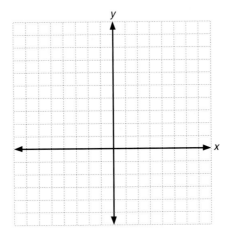

39. _____

40. _____

Sketch the graph of the linear inequalities.

41. $x + y < 6$

42. $2x + y > 4$

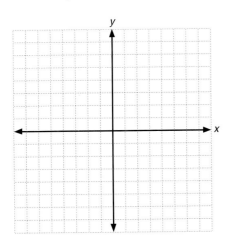

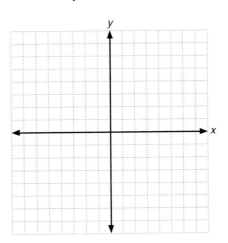

41. _____

42. _____

A N S W E R S

43. _____

44. _____

43. $3x - y \geq 4$

44. $2x - 3y \leq 8$

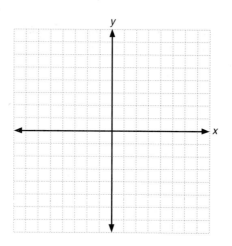

SCORE: _____

1. Give the domain (values of x) and range (values of y) of each of the following relations. In each case state if the relation is a function of x.

a. $y = \sqrt{2x - 5}$

b. $x^2 + y^2 = 9$

A   N   S   W   E   R   S

1a. _____

2. Complete the table of values and sketch the graph of $x + 2y = 5$.

x	-1	1	5
y			

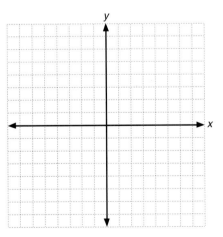

3. Write the equation, in standard form, of the straight line through the points $(-1, 5)$ and $(3, 3)$.

1b. _____

2. _____

4. Sketch the graph of the straight line through the point $(-2, -3)$ if $m = \dfrac{3}{4}$.

5. Write the equation of the circle $x^2 + y^2 - 10x + 4y = 7$ in standard form. Give the center and radius.

3. _____

4. _____

5. _____

6. _____

7. _____

8. _____

SCORE: _____

6. Sketch the graph of
$$\frac{(x-2)^2}{9} + \frac{(y+3)^2}{25} = 1.$$

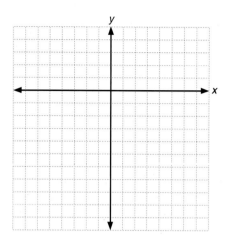

7. Sketch the graph of
$4y = x^2 + 2x + 9.$

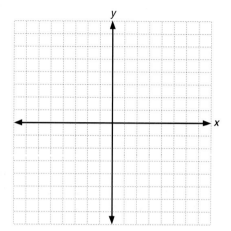

8. Sketch the graph of the linear inequality $x + 3y \geq -2$.

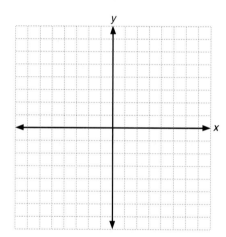

C H A P T E R

9

PRETEST

Before beginning this chapter, answer as many of the following questions as you can. When you have finished the chapter, take the practice test at the end of the chapter and compare the scores of the two tests to see how much you have learned.

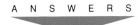

A   N   S   W   E   R   S

1. Solve the system by graphing:
$$\begin{cases} x + 2y = 1 \\ 2x - 3y = 9 \end{cases}$$

2. Solve the system of inequalities by graphing:
$$\begin{cases} 3x - 2y \le 6 \\ x + 2y \ge 4 \end{cases}$$

1. _____

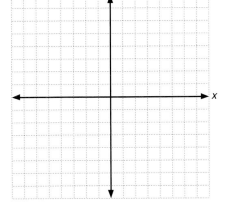

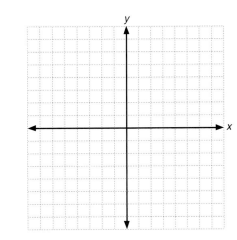

2. _____

3. Solve the system algebraically:
$$\begin{cases} 3x - 2y = -8 \\ 2x + 5y = 1 \end{cases}$$

4. Solve algebraically:
$$\begin{cases} \dfrac{1}{3}x + \dfrac{1}{2}y = 1 \\ 2y = x + 11 \end{cases}$$

3. _____

4. _____

5. Solve algebraically:
$$\begin{cases} 3x + 4y + z = -2 \\ y + z = 1 \\ 2x - y - z = -5 \end{cases}$$

6. Evaluate the determinants:

a. $\begin{vmatrix} 6 & 3 \\ 5 & -1 \end{vmatrix}$ **b.** $\begin{vmatrix} 4 & 0 & -3 \\ 1 & 2 & 5 \\ 3 & 1 & 6 \end{vmatrix}$

5. _____

6. _____

7. Solve by determinants:
$$x + y = -2$$
$$3x - z = 1$$
$$2x + y + z = 1$$

8. Solve the following using two unknowns: How many milliliters each of 6% acid solution and 10% acid solution must be used to produce 200 milliliters of 7% solution?

7. _____

8. _____

SCORE: _____

Systems of Equations and Inequalities

A boat travels 15 miles per hour downstream and 9 miles per hour upstream. Find the speed of the current and the speed of the boat in still water.

n previous chapters we have dealt with the solutions and graphs of individual equations and inequalities. A very important concept in mathematics is the simultaneous solutions of two or more equations. In this chapter we will discuss several ways of arriving at such a solution set.

 **9–1 GRAPHICAL SOLUTION OF A SYSTEM OF TWO EQUATIONS**

Example 1   The pair of equations $\begin{cases} 2x - y = 2 \\ x + 2y = 11 \end{cases}$ is called a **system of linear equations.**

We have observed that each of these equations has infinitely many solutions and each will form a straight line when we graph it on the Cartesian coordinate system.

We now wish to find solutions to the system. In other words, we want all points (x,y) that will be on the graph of both equations.

Solution   We reason in this manner: If all solutions of $2x - y = 2$ lie on one straight line and all solutions of $x + 2y = 11$ lie on another straight line, then a solution to both equations will be their points of intersection (if the two lines intersect).

Table of values for $2x - y = 2$:

x	0	1	2
y	-2	0	2

In this table we let x take on the values 0, 1, and 2. We then find the values for y by using the equation. Do this before going on.

Table of values for $x + 2y = 11$:

x	11	5	-1
y	0	3	6

In this table we let y take on the values 0, 3, and 6. We then find x by using the equation. Check these values also.

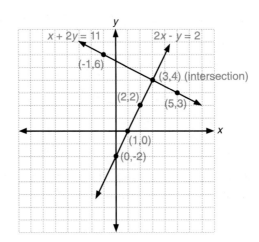

The two lines intersect at the point (3,4).

Note that the point of intersection appears to be (3,4). We must now check the point (3,4) in *both* equations to see that it is a solution to the system.

Check:
$$2x - y = 2$$
$$2(3) - 4 = 2$$
$$6 - 4 = 2$$
$$2 = 2$$

As a check we substitute the ordered pair (3,4) in each equation to see if we get a true statement.

(3,4) is a solution to the first equation.

Check:
$$x + 2y = 11$$
$$(3) + 2(4) = 11$$
$$3 + 8 = 11$$
$$11 = 11$$

Are there any other points that would satisfy both equations? Why?

(3,4) is a solution to the second equation.

Therefore, (3,4) is a solution to the system.

Not all pairs of equations will give a unique solution, as in this example. There are, in fact, three possibilities and you should be aware of them.

Since we are dealing with equations that graph as straight lines, we can examine these possibilities by observing graphs.

1. **Independent equations**   The two lines intersect in a single point. In this case there is a unique solution.

The example above was a system of independent equations.

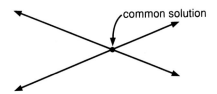

common solution

2. **Inconsistent equations**   The two lines are parallel. In this case there is no solution.

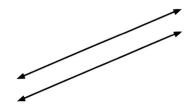

No matter how far these lines are extended, they will never intersect.

3. **Dependent equations**   The two equations give the same line. In this case any solution of one equation is a solution of the other.

In this case there will be infinitely many common solutions.

> To solve a system of two linear equations by graphing
> 1. Make a table of values and sketch the graph of each equation on the same coordinate system.
> 2. Find the values of (x,y) that name the point of intersection of the lines.
> 3. Check this point (x,y) in *both* equations.

Example 2   Solve by graphing: $\begin{cases} x + y = 5 \\ 2x + y = 8 \end{cases}$

Solution   Table of values for $x + y = 5$:   Table of values for $2x + y = 8$:

Again, these are possible tables of values. You could have chosen any values you wanted.

x	-2	0	5
y	7	5	0

x	0	2	4
y	8	4	0

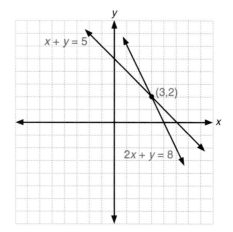

The apparent solution is $(3,2)$.

We say "apparent" because we have not yet checked the ordered pair in both equations. Once it checks it is then definitely the solution.

$$\begin{array}{lll} \textit{Check:} & x + y = 5 & 2x + y = 8 \\ & 3 + 2 = 5 & 2(3) + 2 = 8 \\ & 5 = 5 & 6 + 2 = 8 \\ & & 8 = 8 \end{array}$$

Since $(3,2)$ checks in both equations, it is the solution to the system.

Systems of equations can also involve graphs of curves other than straight lines.

Example 3 Solve by graphing: $\begin{cases} x - y = -1 \\ y = x^2 - 2x + 1 \end{cases}$

This system involves a first-degree equation and a second-degree equation.

Solution These equations represent a straight line and a parabola, respectively. Again set up tables of values and graph each equation.

$x - y = -1$

x	-3	1	5
y	-2	2	6

$y = x^2 - 2x + 1$

x	-2	-1	0	1	2	3	4
y	9	4	1	0	1	4	9

You could also graph the parabola using the standard form of the equation. See section 8–4.

In the second table we substituted each value of x in the equation to find the corresponding value of y. For instance, if we let $x = -2$ then

$$y = (-2)^2 - 2(-2) + 1 = 9.$$

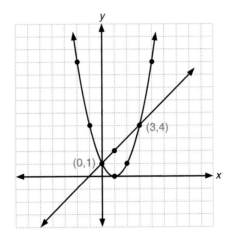

Check both of these points in both equations.

The solutions appear to be (0,1) and (3,4). These check in both equations.

▼ **EXERCISE 9-1-1**

Solve the systems by graphing.

1. $\begin{cases} x + y = 2 \\ 2x - y = 1 \end{cases}$

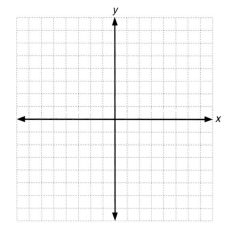

2. $\begin{cases} 3x + y = 0 \\ 2x - y = -5 \end{cases}$

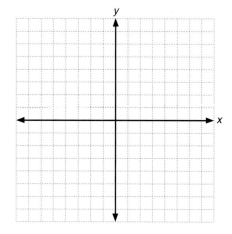

3. $\begin{cases} 3x + y = -9 \\ x - 2y = 4 \end{cases}$

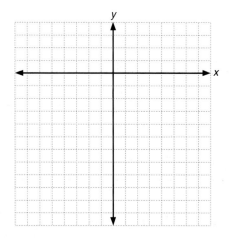

4. $\begin{cases} 3x + y = 8 \\ 2x - y = 7 \end{cases}$

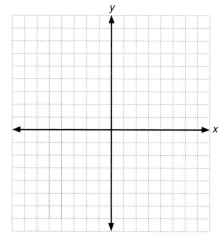

5. $\begin{cases} 3x + 2y = 7 \\ 2x - 3y = -4 \end{cases}$

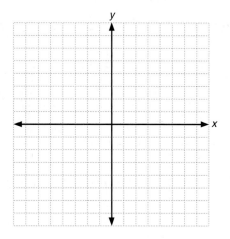

6. $\begin{cases} 2x - y = 8 \\ x + 3y = -3 \end{cases}$

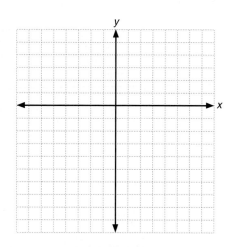

7. $\begin{cases} x + 4y = -13 \\ 2x - 7y = 4 \end{cases}$

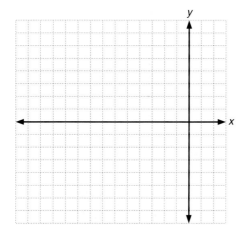

8. $\begin{cases} 5x - 2y = 4 \\ 3x + 4y = -8 \end{cases}$

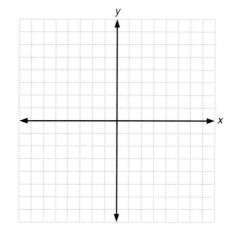

9. $\begin{cases} 2x + 3y = 1 \\ 3x - y = -15 \end{cases}$

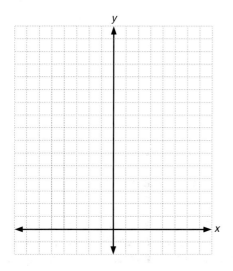

10. $\begin{cases} 3x - 5y = 9 \\ 2x - 3y = 5 \end{cases}$

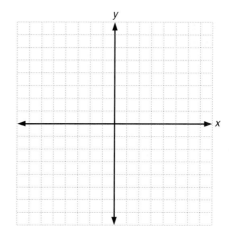

11. $\begin{cases} x + 2y = -4 \\ 3x - y = 9 \end{cases}$

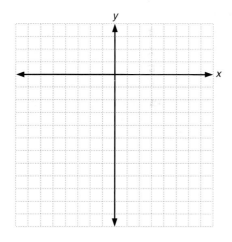

12. $\begin{cases} x + y = 3 \\ 3x + y = 9 \end{cases}$

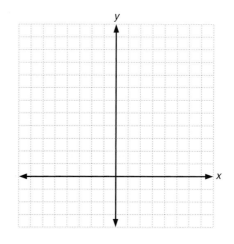

13. $\begin{cases} 2x - y = -2 \\ 5x + 3y = -5 \end{cases}$

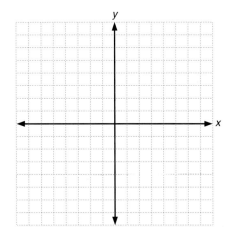

14. $\begin{cases} 4x + 3y = 0 \\ x + y = 1 \end{cases}$

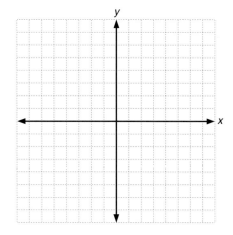

15. $\begin{cases} x + 3y = -1 \\ 2x + 5y = 0 \end{cases}$

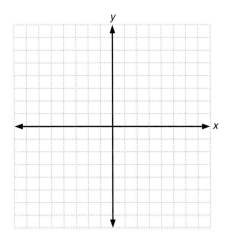

16. $\begin{cases} 4x - y = -3 \\ x + 3y = -4 \end{cases}$

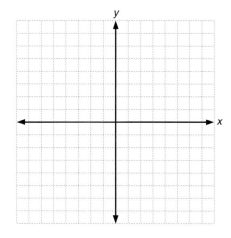

17. $\begin{cases} 2x - 5y = 2 \\ x - 3y = 2 \end{cases}$

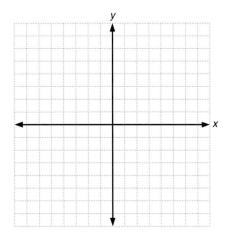

18. $\begin{cases} 2x + 3y = -3 \\ x + 2y = -3 \end{cases}$

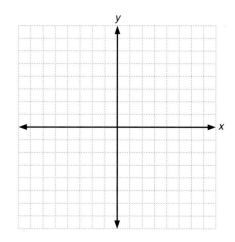

19. $\begin{cases} 2x - y = 4 \\ 5x + 3y = -12 \end{cases}$

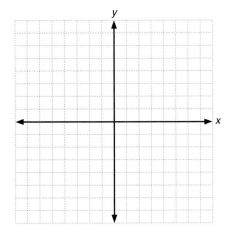

20. $\begin{cases} 3x - 2y = -6 \\ 2x + 3y = 9 \end{cases}$

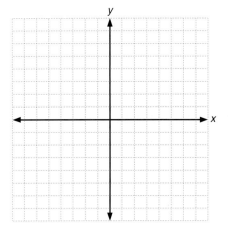

21. $\begin{cases} x - y = -2 \\ y = x^2 \end{cases}$

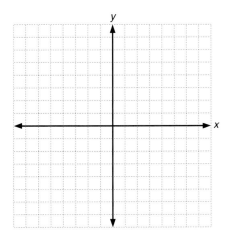

22. $\begin{cases} x + y = -1 \\ y = x^2 + x - 1 \end{cases}$

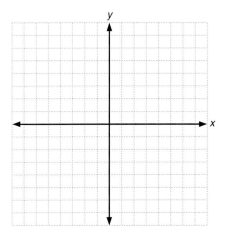

23. $\begin{cases} y = x^2 - 2x + 3 \\ x + y = 5 \end{cases}$

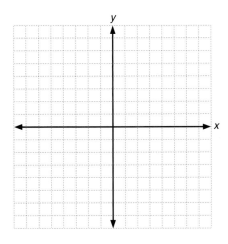

24. Given a system of two linear equations, discuss the possibilities for the solution set.

25. Given a system of one straight line and one parabola, discuss the possibilities for the solution set.

◥◣ 9-2 GRAPHICAL SOLUTION OF A SYSTEM OF LINEAR INEQUALITIES

O B J E C T I V E S

▼

Upon completing this section you should be able to:

1. Graph two or more linear inequalities on the same set of coordinate axes.
2. Determine the region of the plane that is the solution of the system.
3. Solve applications involving systems of linear inequalities.

Systems of inequalities and their solutions are the basis for a branch of mathematics called *linear programming*. Linear programming is very useful in business as well as in some scientific fields, and a student should have in her/his background the ability to solve systems of inequalities.

You found in the previous section that the solution to a system of linear equations is the intersection of the solutions to each of the equations. In the same manner the solution to a **system of linear inequalities** is the intersection of the half-planes (and perhaps lines) that are solutions to each individual linear inequality.

In other words, $x + y > 5$ has a solution set and $2x - y < 4$ has a solution set. Therefore, the system

$$\begin{cases} x + y > 5 \\ 2x - y < 4 \end{cases}$$

has as its solution set the region of the plane that is in the solution set of both inequalities.

To graph the solution to this system we graph each linear inequality on the same set of coordinate axes and indicate the intersection of the two solution sets.

Note that the solution to a system of linear inequalities will be a collection of points.

Example 1   Graph the solution: $\begin{cases} x + y > 5 \\ 2x - y < 4 \end{cases}$

Solution   First graph the lines $x + y = 5$ and $2x - y = 4$.

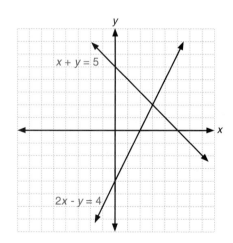

Checking the point $(0,0)$ in the inequality $x + y > 5$ indicates that the point $(0,0)$ is not in its solution set. We indicate the solution set of $x + y > 5$ with a screen to the right of the dashed line.

This region is to the right and above the line $x + y = 5$.

Checking the point $(0,0)$ in the inequality $2x - y < 4$ indicates that the point $(0,0)$ is in its solution set. We indicate this solution set with a screen to the left of the dashed line.

This region is to the left and above the line $2x - y = 4$.

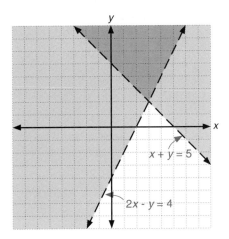

The intersection of the two solution sets is that region of the plane in which the two screens intersect. This region is shown in the graph.

Note again that the solution does not include the lines.

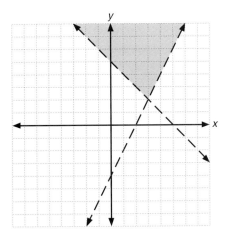

The results indicate that all points in the shaded section of the graph would be in the solution sets of $x + y > 5$ *and* $2x - y < 4$ at the same time.

If, for example, we were asked to graph the solution of the system

$$\begin{cases} x + y \geq 5 \\ 2x - y < 4 \end{cases}$$

we would obtain

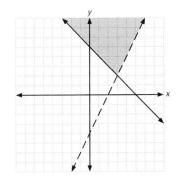

which indicates the solution includes points on the line $x + y = 5$.

This system contains four
inequalities.

Example 2 Graph the system: $\begin{cases} x + y \geq 3 \\ x + y \leq 5 \\ \quad\;\; x \geq 0 \\ \quad\;\; y \geq 0 \end{cases}$

Solution Graphing the equations $x + y = 3$, $x + y = 5$, $x = 0$, and $y = 0$ gives us two parallel lines and the two axes.

Substitution of coordinates in the original inequalities will then lead to the following graph of the solution set of the system.

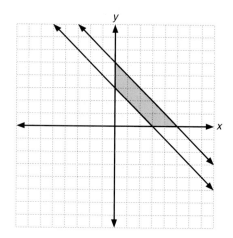

All points in the region of the solution must satisfy all four inequalities.

Notice that this system has a solution that is the trapezoid region having vertices at $(0,3)$, $(0,5)$, $(3,0)$, and $(5,0)$.

Sometimes a linear equation is used in conjunction with inequalities involving the same variables. This occurs in the branch of mathematics known as **linear programming.** The following example illustrates the process.

Example 3 A small manufacturing company produces two models of fishing reels. Model A sells for $25 and model B sells for $40. The company cannot sell more than 100 reels a week. It takes 1 hour to produce a model A reel and 3 hours to produce a model B reel. The total number of labor hours available in a week is 150. How many reels of each model should the company produce in a week to maximize income?

Solution First we shall let

$$x = \text{number of model } A \text{ reels to be produced}$$
$$y = \text{number of model } B \text{ reels to be produced.}$$

Then since the total number of reels produced cannot exceed 100 we have

$$x + y \leq 100.$$

Do you see why we use the
symbol $\leq$?

It would take a total of x hours to produce the model A reels and $3y$ hours to produce the model B reels. Thus

$$x + 3y \leq 150.$$

Can you find four inequalities
indicated in this problem?

We therefore have the following system of inequalities:

$$\begin{cases} x + y \le 100 \\ x + 3y \le 150 \\ x \ge 0 \\ y \ge 0. \end{cases}$$

$x \ge 0$ and $y \ge 0$ since we cannot produce a negative number of reels.

The graph of the system follows.

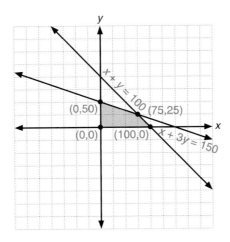

Now the income (I) for the company can be expressed as

$$I = 25x + 40y.$$

Here we have a linear equation.

This expression has certain restrictions or *constraints* on the values of x and y as provided for by the system of inequalities. We wish to find which values for x and y under the given constraints will produce a maximum income.

At this point we must introduce an important theorem from linear programming. It states that when we have a region such as in this example, the maximum and minimum values (if they exist) for the linear expression will be found at the vertices of the region.

Thus if there is a maximum income, it must occur at one of the vertices. We therefore try the coordinates for each vertex in the expression for income.

$$
\begin{array}{llll}
(0,0): & I = 25(0) & + 40(0) & = \$0 \\
(0,50): & I = 25(0) & + 40(50) & = \$2,000 \\
(75,25): & I = 25(75) & + 40(25) & = \$2,875 \\
(100,0): & I = 25(100) & + 40(0) & = \$2,500
\end{array}
$$

This is the maximum income.

We see that the vertex (75,25) produces the maximum income. Therefore the company should produce 75 model *A* reels and 25 model *B* reels.

▼ **EXERCISE 9–2–1**

Solve the systems by graphing.

1. $\begin{cases} x + y > 2 \\ 2x - y < 1 \end{cases}$

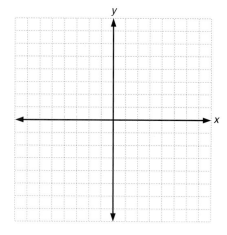

2. $\begin{cases} 3x + y > 0 \\ 2x - y < -4 \end{cases}$

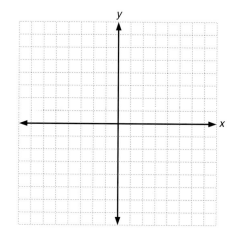

3. $\begin{cases} 3x + y > 6 \\ x - 2y < 4 \end{cases}$

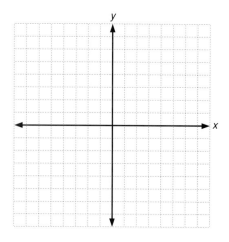

4. $\begin{cases} 2x + y < 3 \\ x - y < 5 \end{cases}$

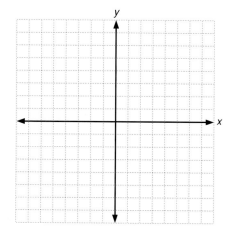

5. $\begin{cases} x + y > 0 \\ x - 3y > 3 \end{cases}$

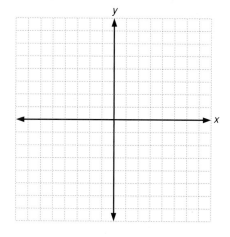

6. $\begin{cases} 2x - y \geq 2 \\ x + y \leq 3 \end{cases}$

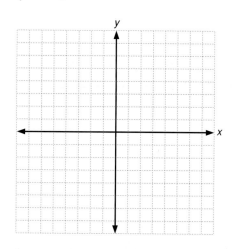

7. $\begin{cases} x + 2y \le 4 \\ 2x - y \ge 6 \end{cases}$

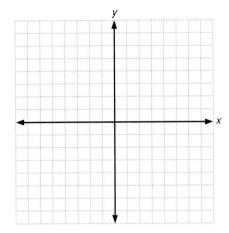

8. $\begin{cases} 3x - y > 6 \\ 2x + y \le 4 \end{cases}$

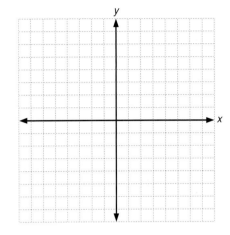

9. $\begin{cases} 4x - y < 4 \\ x + 2y \le 2 \end{cases}$

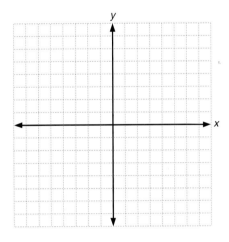

10. $\begin{cases} 5x - 2y \le 10 \\ x + 3y > 9 \end{cases}$

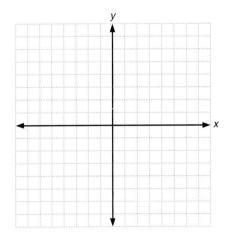

11. $\begin{cases} x - 2y < 3 \\ 2x + y > 8 \end{cases}$

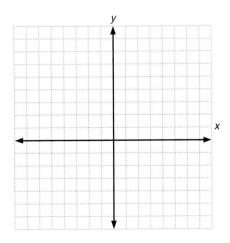

12. $\begin{cases} 2x - y \ge -2 \\ x + 5y \le -5 \end{cases}$

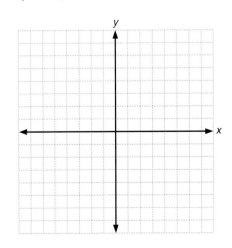

13. $\begin{cases} 4x - 3y \le -12 \\ x + 4y > 6 \end{cases}$

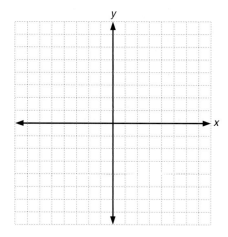

14. $\begin{cases} 5x - 3y > 15 \\ 2x + 3y \le -6 \end{cases}$

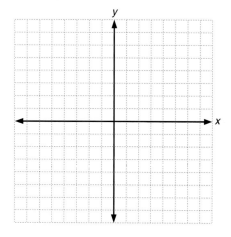

15. $\begin{cases} x - y \le 0 \\ x + y \ge 0 \end{cases}$

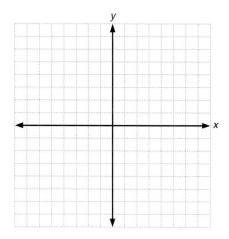

16. $\begin{cases} x + y \le 0 \\ 2x - y > 4 \end{cases}$

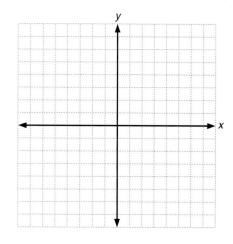

17. $\begin{cases} 3x + y \ge -9 \\ x - 2y \ge 4 \end{cases}$

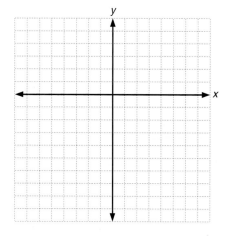

18. $\begin{cases} 2x - 3y \le -6 \\ x + 4y \le 4 \end{cases}$

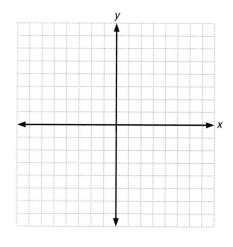

19. $\begin{cases} x - 3y < 5 \\ 2x + 3y \le 8 \end{cases}$

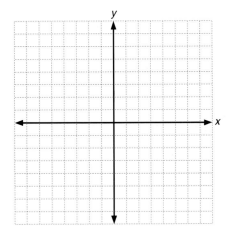

20. $\begin{cases} 3x + 5y \le -15 \\ 2x - 3y \ge 12 \end{cases}$

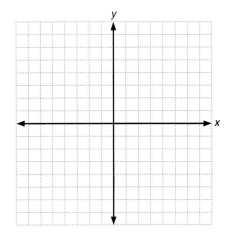

21. $\begin{cases} y \ge 0 \\ x \ge -1 \\ x + y \ge 1 \end{cases}$

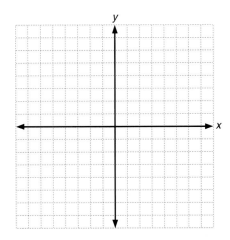

22. $\begin{cases} 3x + y \le 0 \\ 2x - y \ge -5 \\ y \ge 0 \end{cases}$

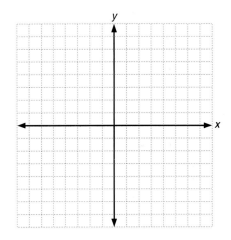

23. $\begin{cases} 3x - 4y \geq -12 \\ x + y \geq 1 \\ 3x + 4y \leq 12 \\ 2x - y \leq 2 \end{cases}$

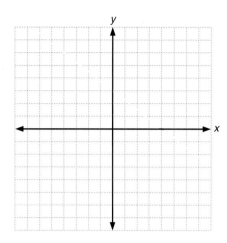

24. Two numbers are such that the first number is not less than 10 and the sum of twice the first number and three times the second number is not more than 60. Also twice the first number decreased by the second number is not greater than 40. Find the two numbers satisfying the constraints such that their sum will be minimum. Find the two numbers whose sum will be maximum.

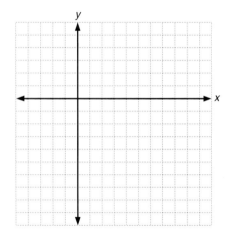

25. A company wishes to purchase advertising time on a television network. The network offers a package containing one-half minute commercials and one minute commercials. The company must purchase a total of at least 100 commercial spots of air time. The number of one minute commercials must be at least one and one-half that of the half minute commercials. The number of one minute commercials cannot exceed 120. If each half minute commercial costs $30,000 and each one minute commercial costs $50,000, how many of each should the company buy to minimize their expenses?

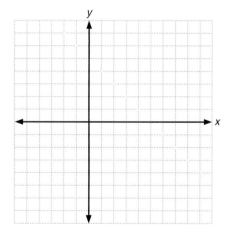

9-3 THE ALGEBRAIC SOLUTION OF A SYSTEM OF TWO LINEAR EQUATIONS

In section 9–1 we solved a system of two equations with two unknowns by graphing. The graphical method is very useful, but it would not be practical if the solutions were fractions. The actual point of intersection could be very difficult to determine. There are algebraic methods of solving systems and these methods give a much more precise solution. In this section we will discuss the *substitution* method and the *addition* method. We first look at the *substitution* method.

OBJECTIVES

Upon completing this section you should be able to:

1. Solve a system of two linear equations by the substitution method.
2. Solve a system of two linear equations by the addition method.
3. Determine whether a system of two linear equations is independent, inconsistent, or dependent.

Example 1   Solve by the substitution method: $\begin{cases} 2x + 3y = 1 \\ x - 2y = 4 \end{cases}$

Solution   **Step 1**   We must solve for one unknown in one equation. We can choose either x or y in either the first or second equation. Our choice can be based on obtaining the simplest expression. In this case we will solve for x in the second equation, obtaining $x = 4 + 2y$, because any other choice would have resulted in a fraction.

Step 2   Substitute the value of x into the other equation. In this case the equation is

$$2x + 3y = 1.$$

Substituting $(4 + 2y)$ for x, we obtain $2(4 + 2y) + 3y = 1$, an equation with only one unknown.

The reason for this is that if $x = 4 + 2y$ in one of the equations, then x must equal $4 + 2y$ in the other equation.

Step 3   Solve for the unknown.

$$2(4 + 2y) + 3y = 1$$
$$8 + 4y + 3y = 1$$
$$8 + 7y = 1$$
$$7y = -7$$
$$y = -1$$

Remember, first remove parentheses.

Step 4   Substitute $y = -1$ into either equation to find the corresponding value for x. Since we have already solved the second equation for x in terms of y, we may use it.

We may substitute $y = -1$ in either equation since y has the same value in both.

$$x = 4 + 2y$$
$$= 4 + 2(-1)$$
$$= 4 - 2$$
$$= 2$$

Thus, we have the solution $(2, -1)$.

Remember, x is written first in the ordered pair.

Step 5 Check the solution in both equations. Remember that the solution for a system must be true for each equation in the system. Since

This checks: $2x + 3y = 1$.

$$2x + 3y = 2(2) + 3(-1) = 4 - 3 = 1$$
$$\text{and}$$
$$x - 2y = 2 - 2(-1) = 2 + 2 = 4,$$

This checks: $x - 2y = 4$.

the solution $(2, -1)$ does check.

Example 2 Solve by substitution: $\begin{cases} 2x + 3y = 7 \\ 4x + 3y = 8 \end{cases}$

Neither of these equations has a variable with a coefficient of 1. In this case solving by substitution is not the best method, but we will do it that way just to show it can be done. The addition method will give us an easier method.

Solution **Step 1** We will obtain a fractional expression in any case, so one choice is as easy as another. We will solve for x in the first equation.

$$x = \frac{7 - 3y}{2}$$

Step 2 Substitute the expression $\left(\dfrac{7 - 3y}{2}\right)$ for x in the second equation.

$$4x + 3y = 8$$
$$4\left(\frac{7 - 3y}{2}\right) + 3y = 8$$

Step 3 Solve the equation.

$$4\left(\frac{7 - 3y}{2}\right) + 3y = 8$$
$$2(7 - 3y) + 3y = 8$$
$$14 - 6y + 3y = 8$$
$$-3y = -6$$
$$y = 2$$

Try substituting $y = 2$ in the second equation.

$$4x + 3y = 8$$

Step 4 Substitute $y = 2$ in either of the equations. If we use the first equation, we obtain

$$2x + 3y = 7$$
$$2x + 3(2) = 7$$
$$2x + 6 = 7$$
$$2x = 1$$
$$x = \frac{1}{2}.$$

Step 5 Checking, we find that the ordered pair $\left(\frac{1}{2}, 2\right)$ satisfies both equations and is thus the solution to the system.

Check this ordered pair in both equations.

▼ EXERCISE 9–3–1

Solve by the substitution method.

1. $\begin{cases} x + y = 3 \\ 2x + y = 5 \end{cases}$

2. $\begin{cases} 2x + y = 1 \\ x + 2y = 5 \end{cases}$

3. $\begin{cases} x - 2y = 5 \\ 2x + y = 10 \end{cases}$ $\quad X = 2y + 5$

$2(2y+5) + y = 10$

$X = 5$

$y = 0$ $(5,0)$

$pa.$

$4y + 10 + y = 10$

$5y = 0 \qquad X = 5$

$X - y = 5 \quad 5 - 0 = 5$

4. $\begin{cases} 2x + y = 4 \\ x - 2y = 7 \end{cases}$

5. $\begin{cases} 3x - y = 4 \\ x + 5y = -4 \end{cases}$

6. $\begin{cases} 2x + y = 4 \\ 3x - 2y = -1 \end{cases}$

7. $\begin{cases} x + 5y = 2 \\ 2x + 3y = -3 \end{cases}$

8. $\begin{cases} x + y = 5 \\ 2x - y = -5 \end{cases}$

9. $\begin{cases} 2x + y = 5 \\ 6x + 2y = 11 \end{cases}$

10. $\begin{cases} 4x + 3y = -3 \\ 8x - y = -6 \end{cases}$

11. $\begin{cases} 2x - y = 1 \\ 5x + 2y = -11 \end{cases}$

12. $\begin{cases} x + 6y = 7 \\ 3x - 4y = 32 \end{cases}$

13. $\begin{cases} x - 2y = 1 \\ x + 4y = 22 \end{cases}$

14. $\begin{cases} x - 5y = 3 \\ 3x - 2y = -17 \end{cases}$

15. $\begin{cases} 2x - y = 16 \\ 3x + 2y = 3 \end{cases}$

16. $\begin{cases} 3x + 8y = 4 \\ x - 6y = -3 \end{cases}$

17. $\begin{cases} 5x + y = 5 \\ 3x - 2y = 29 \end{cases}$

18. $\begin{cases} 3x + 2y = 0 \\ x + 4y = 15 \end{cases}$

19. $\begin{cases} x - 2y = 5 \\ 2x - 3y = 1 \end{cases}$

20. $\begin{cases} 2x + 3y = 1 \\ 4x + 5y = 3 \end{cases}$

The *addition* method for solving a system of linear equations is based on two facts that we have used previously.

First we know that the solutions to an equation do not change if every term of that equation is multiplied by a nonzero number. Second we know that if we add the same or equal quantities to both sides of an equation, the results are still equal.

Example 3   Solve by addition: $\begin{cases} 2x + y = 5 \\ 3x + 2y = 6 \end{cases}$

Note that we could solve this system by the substitution method, by solving the first equation for y. Solve this system by the substitution method and compare your solution with that obtained by the addition method.

Solution   **Step 1** Our purpose is to add the two equations and eliminate one of the unknowns so that we can solve the resulting equation in one unknown. If we add the equations as they are, we will not eliminate an unknown. This means we must first multiply each side of one or both of the equations by a number or numbers that will lead to the elimination of one of the unknowns when the equations are added.

After carefully looking at the problem, we note that the easiest unknown to eliminate is y. This is done by first multiplying each side of the first equation by -2.

$$\begin{cases} 2x + y = 5 \leftarrow \text{Multiply by } (-2). \\ 3x + 2y = 6 \end{cases}$$

obtaining the equivalent system

$$\begin{cases} -4x - 2y = -10 \\ 3x + 2y = 6. \end{cases}$$

Note that each term must be multiplied by (-2).

Step 2 Add the equations.

$$\begin{array}{r} -4x - 2y = -10 \\ \underline{3x + 2y = 6} \\ -x = -4 \end{array}$$

Step 3 Solve the resulting equation.

$$\begin{array}{r} -x = -4 \\ x = 4 \end{array}$$

In this case we simply multiply each side by (-1).

Step 4 Find the value of the other unknown by substituting this value into one of the original equations. Using the first equation,

$$\begin{array}{l} 2x + y = 5 \\ 2(4) + y = 5 \qquad \text{Value of } x \\ 8 + y = 5 \\ y = -3. \end{array}$$

Substitute $x = 4$ in the second equation and see if you get the same value for y.

Step 5 If we check the ordered pair $(4, -3)$ in both equations, we see that it is a solution of the system.

$$\begin{array}{l} 2x + y = 5 \\ 2(4) + (-3) = 8 - 3 = 5 \\ 3x + 2y = 6 \\ 3(4) + 2(-3) = 12 - 6 = 6 \end{array}$$

Example 4 Solve by addition: $\begin{cases} 2x + 3y = 7 \\ 3x + 2y = 3 \end{cases}$

Note that in this system no variable has a coefficient of 1. Therefore the best method of solving it is the addition method.

Solution **Step 1** Both equations will have to be changed to eliminate one of the unknowns. Neither equation will be easier than the other, so choose to eliminate either x or y.

If you choose to eliminate y, multiply the first equation by -2 and the second equation by 3. Do this and solve the system. Compare your solution with the one obtained in the example.

To eliminate x multiply each side of the first equation by 3 and each side of the second equation by -2.

$$\begin{cases} 2x + 3y = 7 & \longleftarrow \text{Multiply by 3.} \\ 3x + 2y = 3 & \longleftarrow \text{Multiply by } -2. \end{cases}$$

obtaining the equivalent system

$$\begin{cases} 6x + 9y = 21 \\ -6x - 4y = -6. \end{cases}$$

Step 2 Adding the equations, we obtain

$$\begin{aligned} 6x + 9y &= 21 \\ \underline{-6x - 4y} &= \underline{-6} \\ 5y &= 15. \end{aligned}$$

Step 3 Solving for y yields

$$y = 3.$$

Step 4 Using the first equation in the original system to find the value of the other unknown gives

$$\begin{aligned} 2x + 3y &= 7 \\ 2x + 3(3) &= 7 \\ 2x + 9 &= 7 \\ 2x &= -2 \\ x &= -1. \end{aligned}$$

The check is left up to you.

Step 5 Check to see that the ordered pair $(-1,3)$ is a solution of the system.

▼ **EXERCISE 9-3-2**

Solve by the addition method.

1. $\begin{cases} x + y = 2 \\ 4x - y = 13 \end{cases}$ $\quad 3+(-1)=2$

$\qquad \underline{5x \quad = 15} \quad (3,-1)$

$\qquad X = 3$

2. $\begin{cases} 3x + y = 8 \\ 2x - y = -3 \end{cases}$ $\quad \begin{aligned} 3+y&=8 \\ y&=5 \end{aligned}$

$\qquad \underline{5x \quad = 5} \quad (1,5)$

$\qquad X \quad = 1$

3. $\begin{cases} 2x - y = 0 \\ x + y = 6 \end{cases}$ $\quad \begin{aligned} 2\cdot2-4&=6 \\ 4-4&=0 \end{aligned}$

$\qquad \underline{3x \quad = 6}$

$\qquad X \quad = 2$

$\qquad \begin{aligned} X &= 2 \\ y &= 4 \end{aligned} \quad (2,4)$

4. $\begin{cases} x + 2y = 1 \\ x - 2y = 9 \end{cases}$

5. $\begin{cases} 3x - y = -7 \\ x + 2y = 7 \end{cases}$

6. $\begin{cases} x + 7y = 28 \\ 2x - y = -4 \end{cases}$

7. $\begin{cases} 5x - 2y = 54 \\ 2x + 3y = 14 \end{cases}$

8. $\begin{cases} 4x + y = 13 \\ 3x + y = 10 \end{cases}$

9. $\begin{cases} 8x - 3y = 13 \\ 5x - 4y = 6 \end{cases}$

10. $\begin{cases} 2x + 5y = 8 \\ 5x + 2y = 20 \end{cases}$

11. $\begin{cases} 2x + y = -3 \\ 7x + 2y = -18 \end{cases}$

12. $\begin{cases} 6x - y = 24 \\ 3x - 2y = 21 \end{cases}$

13. $\begin{cases} 2x - 5y = 44 \\ 10x + 3y = -4 \end{cases}$

14. $\begin{cases} 4x + 9y = -6 \\ 5x + 2y = 11 \end{cases}$

15. $\begin{cases} 2x + 5y = 21 \\ 4x - 3y = -10 \end{cases}$

16. $\begin{cases} 5x + 6y = 9 \\ 2x - 9y = -4 \end{cases}$

17. $\begin{cases} 4x + 7y = 27 \\ 3x - 2y = -16 \end{cases}$

18. $\begin{cases} 5x + 3y = 10 \\ 10x + 5y = 16 \end{cases}$

19. $\begin{cases} 8x + 3y = -5 \\ 9x + 7y = -2 \end{cases}$

20. $\begin{cases} 5x - 4y = 2 \\ 7x - 11y = 19 \end{cases}$

As mentioned in section 9–1, a system of two linear equations does not always have a unique solution and we should be aware of the other possibilities.

Recall that linear equations graph as straight lines.

Since we are dealing with linear equations, we may think of the lines represented by the equations. Two linear equations will represent one of the following three possibilities.

You may want to refer back to section 9–1.

1. **Independent equations:**   The two lines intersect in a point. In this case we have a unique solution.
2. **Inconsistent equations:**   The two lines are parallel. In this case we have no solution.
3. **Dependent equations:**   The two equations give the same line. In this case any solution of one equation is a solution of the other.

We can recognize any of the three situations by proceeding to work the problem either by the substitution or addition method.

If the equations are *independent,* we will obtain a *unique solution.*
If the equations are *inconsistent,* we will obtain a *contradiction.*
If the equations are *dependent,* we will obtain an *identity.*

Example 5   Solve: $\begin{cases} x + y = 6 \\ x + y = 8 \end{cases}$

$x + y = 6$
$\underline{-x - y = -8}$
$0 = -2$

This is not a true statement. Therefore, we have a contradiction.

Solution   If we multiply both sides of the second equation by -1 and add, we obtain

$$0 = -2.$$

Since this is a *contradiction,* the equations are *inconsistent* and their graphs will be parallel lines. This system has no solution.

Example 6   Solve: $\begin{cases} x + y = 6 \\ 2x + 2y = 12 \end{cases}$

$-2x - 2y = -12$
$\underline{2x + 2y = 12}$
$0 = 0$

This is an identity.

Solution   Multiplying both sides of the first equation by -2 and adding yields

$$0 = 0.$$

Since this is an *identity* the equations are *dependent* and represent the same line on the coordinate plane. Therefore any ordered pair that satisfies one equation will also satisfy the other. There are infinitely many solutions.

▼ **EXERCISE 9-3-3**

Classify each system as independent, inconsistent, or dependent. If the system is independent, find its solution.

1. $\begin{cases} 2x + y = 1 \\ 3x - y = 9 \end{cases}$

2. $\begin{cases} x + y = 4 \\ x - y = 6 \end{cases}$

3. $\begin{cases} x + 2y = 9 \\ 3x - y = -1 \end{cases}$

4. $\begin{cases} 2x - y = 1 \\ 6x - 3y = 3 \end{cases}$

5. $\begin{cases} 3x - y = 2 \\ 2x + 3y = -6 \end{cases}$

6. $\begin{cases} 3x - 2y = 3 \\ 6x - 4y = 1 \end{cases}$

7. $\begin{cases} 6x - 4y = 2 \\ 3x - 2y = 1 \end{cases}$

8. $\begin{cases} x - 2y = 15 \\ 3x + y = 3 \end{cases}$

9. $\begin{cases} 3x + 2y = 5 \\ 2x + y = 1 \end{cases}$

10. $\begin{cases} 2x + y = 5 \\ 10x + 5y = 10 \end{cases}$

11. $\begin{cases} 2x + 3y = 10 \\ 6x - 2y = -3 \end{cases}$

12. $\begin{cases} x + 3y = 1 \\ 2x - 9y = -8 \end{cases}$

13. $\begin{cases} 2x - 4y = 6 \\ 3x - 6y = 9 \end{cases}$

14. $\begin{cases} 6x + 3y = 3 \\ 10x + 5y = 15 \end{cases}$

15. $\begin{cases} 5x + 2y = -1 \\ 4x + 3y = -12 \end{cases}$

9-4 STANDARD FORM OF LINEAR EQUATIONS

Equations in the preceding sections have all had no fractions, both unknowns on the left of the equation, and unknowns in the same order.

Such equations are said to be in *standard form*. That is, they are in the form $ax + by = c$, where a, b and c are integers. Equations must be changed to the standard form before solving by the addition method.

Note that in example 1 the term $4y$ is out of place. We therefore wish to eliminate it from the right side of the equation by adding $-4y$ to both sides.

Example 1 Change $3x = 5 + 4y$ to standard form.

Solution $3x = 5 + 4y$ is not in standard form because one unknown is on the right. If we add $-4y$ to both sides, we have $3x - 4y = 5$, which is in standard form.

Example 2 Change $\dfrac{2}{3}x + \dfrac{5}{8}y = 3$ to standard form.

Solution Here we have fractions that must be cleared by multiplying by a common denominator. Multiplying each term of $\dfrac{2}{3}x + \dfrac{5}{8}y = 3$ by the least common denominator 24, we have

Be careful here. Many students forget to multiply the right side of the equation by 24.

$$24\left(\frac{2}{3}x\right) + 24\left(\frac{5}{8}y\right) = 24(3).$$

Simplifying, we have $16x + 15y = 72$, which is in standard form.

Example 3 Change $\dfrac{3}{4}y = 2x - \dfrac{5}{6}$ to standard form.

Solution First multiply by 12.

Again, make sure each term is multiplied by 12.

$$12\left(\frac{3}{4}y\right) = 12(2x) - 12\left(\frac{5}{6}\right)$$
$$9y = 24x - 10$$

Instead of saying "the first term is positive," we sometimes say "the leading coefficient is positive."

Now add $-24x$ to both sides, giving $-24x + 9y = -10$, which is in standard form. Usually, equations are written so the first term is positive. Thus we multiply each term of this equation by (-1).

$$(-1)(-24x) + (-1)(9y) = (-1)(-10)$$
$$\text{or}$$
$$24x - 9y = 10$$

Change to standard form.

1. $\dfrac{2}{3}x + \dfrac{1}{5}y = 3$

2. $\dfrac{3}{4}x - \dfrac{5}{8}y = 3$

3. $2x = 9 - 4y$

4. $5y = x - 4$

5. $\dfrac{2}{3}y = \dfrac{1}{2}x - \dfrac{1}{5}$

6. $\dfrac{1}{4}x = \dfrac{5}{6} + \dfrac{1}{7}y$

Change to standard form and solve the systems by the best method.

7. $\begin{cases} 2x = 6 + 3y \\ 2y = 3 - x \end{cases}$

8. $\begin{cases} x = 2y - 4 \\ 3y = 5 + x \end{cases}$

9. $\begin{cases} \dfrac{3}{8}x + \dfrac{1}{4}y = -1 \\ \dfrac{1}{5}x = \dfrac{1}{10} + \dfrac{1}{2}y \end{cases}$

10. $\begin{cases} x = -\dfrac{1}{2}y \\ \dfrac{1}{3}x + \dfrac{1}{4}y = \dfrac{1}{2} \end{cases}$

11. $\begin{cases} y = x + 3 \\ x = \dfrac{2}{3}y - 2 \end{cases}$

12. $\begin{cases} x = 2y + 6 \\ \dfrac{x}{3} = \dfrac{5}{6} - \dfrac{y}{2} \end{cases}$

 **9-5 WORD PROBLEMS WITH TWO UNKNOWNS**

<table>
<tr><td>

O B J E C T I V E S

Upon completing this section you should be able to:

1. Determine when a word problem can be solved using two unknowns.
2. Determine the equations and solve the word problem.

</td></tr>
</table>

Many word problems can be outlined and worked more easily by using two unknowns.

Example 1   The sum of two numbers is 5. Three times the first number added to five times the second number is 9. Find the numbers.

Three times the first number added to five times the second number is 9.

Solution   Let x = first number

y = second number.

The first statement gives us the equation

$$x + y = 5.$$

The second statement gives us the equation

$$3x + 5y = 9.$$

We now have the system

$$\begin{cases} x + y = 5 \\ 3x + 5y = 9, \end{cases}$$

Solve the system by substitution.

which we can solve by either method we have learned, to give

$$x = 8 \text{ and } y = -3.$$

Example 2   Two workers receive a total of $136 for 8 hours work. If one worker is paid $1.00 per hour more than the other, find the hourly rate for each.

Note that it is very important to say what x and y represent.

Solution   Let x = hourly rate of one worker

y = hourly rate of other worker.

The first statement gives the equation

$$8x + 8y = 136$$

and the second statement gives the equation

$$x = y + 1.$$

We now have the system (in standard form)

Solve this system by the addition method.

$$\begin{cases} 8x + 8y = 136 \\ x - y = 1. \end{cases}$$

Solving gives $x = 9$ and $y = 8$. One worker's rate is $9.00 per hour and the other's is $8.00 per hour.

◢ **EXERCISE 9–5–1**

Solve using two unknowns.

1. The sum of two numbers is 57 and their difference is 21. Find the numbers.

2. The sum of a number and twice another number is 38. The difference of the two numbers is 7. Find the numbers.

3. The total number of students in a class is 48. If there are six more men than women, find the number of men and women in the class.

4. A total of 40 points was scored in a football game, with the Pittsburgh Steelers beating the Houston Oilers by 14 points. Find the score for each team.

5. The perimeter of a rectangular lot is 450 feet. If the length is twice the width, find the length and width.

6. The total population of Dallas and Houston is approximately 2,265,000. Find the population of each city if Houston has 575,000 more people than Dallas.

7. The sum of the lengths of the Golden Gate Bridge and the Brooklyn Bridge is 5,795 feet. The Golden Gate Bridge is 1,010 feet longer than twice the length of the Brooklyn Bridge. Find the length of each bridge.

8. Jim worked eight hours and Tom worked six hours. They received a total of $135 for their work. Find the hourly rate for each if Tom is paid $1.50 per hour more than Jim.

9. The sum of two angles is 90°. If one angle is 15° more than twice the other, find the angles.

10. The sum of two angles is 180°. Find each angle if one is 12° less than three times the other.

11. The length of a rectangular lot is eight feet less than three times the width. Find the length and width if the perimeter of the lot is 600 feet.

12. A total of 5,000 tickets were sold for a football game. The price was $5.00 per adult and $2.50 per child. If the total receipts were $21,915, find the number of each type of ticket sold.

13. Sue invested $15,000, part of it at 11% and the rest at 14%. She receives $1,890 in interest at the end of a year. How much did she invest at each rate?

14. Sarah borrowed a total of $3,700 from two sources. One source charges 14% interest and the other charges 18%. If she pays $578 interest a year, how much did she borrow at each rate?

15. A boat travels twelve kilometers upstream in two hours. The return trip takes one hour. Find the rate of the boat in still water and the rate of the current.

16. A plane travels 1,600 kilometers with the wind in two hours. The return trip against the wind takes one-half hour longer. Find the speed of the plane (airspeed) and the speed of the wind.

17. Mike has two more dimes than nickels. The total value of the coins is $2.60. How many of each coin does he have?

18. A man has five more dimes than quarters. The total value of the coins is $4.70. How many of each kind of coin does he have?

19. A 40% copper alloy is to be used with a 70% copper alloy to produce 180 kilograms of a 60% alloy. How much of each alloy must be used?

20. How many milliliters each of a 5% acid solution and a 12% acid solution must be used to produce 140 milliliters of a 10% acid solution?

9–6 THE ALGEBRAIC SOLUTION OF THREE EQUATIONS WITH THREE VARIABLES

OBJECTIVES

Upon completing this section you should be able to:

1. Solve a system of three equations having three variables.
2. Solve applications using three equations and three variables.

A first-degree polynomial equation in three variables is the equation of a plane in three-dimensional space. Three such planes in space might intersect in a single point. Of course, there are other possibilities (such as all three parallel, and so on) and you may wish to list them. In this section we are interested only in those that intersect in a single point (that is, those that have a unique solution) and in an algebraic method of finding the solution.

A point in three space is represented by an **ordered triple** of numbers. We use x, y, and z as the three variables and the ordered triple is (x,y,z). The ordered triple $(2,3,-1)$ represents a point such that $x = 2$, $y = 3$, and $z = -1$.

The method we will use reduces a system of three equations in three variables to two equations in two variables by addition. This method is best illustrated by example.

Example 1 Solve the system: $\begin{cases} 2x + 3y - z = 11 & (1) \\ x + 2y + z = 12 & (2) \\ 3x - y + 2z = 5 & (3) \end{cases}$

We have numbered the equations for easy reference.

Solution **Step 1** Choose any two equations and eliminate any one of the three variables by addition.

Note that this gives a wide range of choices. We can choose on the basis of the variable which is easiest to eliminate. In this case we will choose to eliminate z by adding equations (1) and (2).

Equation (1) added to equation (2) yields

$$3x + 5y = 23.$$

$$\begin{array}{rcl} 2x + 3y - z &=& 11 \\ x + 2y + z &=& 12 \\ \hline 3x + 5y &=& 23 \end{array}$$

Step 2 Choose two *other* equations and eliminate the *same* variable by addition.

Eliminate the variable z by combining any other two equations.

The choices here are not as broad because we must eliminate the same variables as in step 1, but we cannot use the same two equations as before. In this example we must eliminate z and can use equations (1) and (3) or equations (2) and (3). We will choose equations (1) and (3) and eliminate z by multiplying both sides of equation (1) by 2 and adding the result to equation (3).

Twice equation (1) added to equation (3) yields

$$7x + 5y = 27.$$

$$\begin{array}{rcl} 4x + 6y - 2z &=& 22 \\ 3x - y + 2z &=& 5 \\ \hline 7x + 5y &=& 27 \end{array}$$

Step 3 Solve the system of two equations in two variables that results from step 1 and step 2.

$$\begin{cases} 3x + 5y = 23 \\ 7x + 5y = 27 \end{cases}$$

We now proceed as in section 8–6. Solve this system here in the margin.

Using either substitution or addition to solve this system, we find

$$x = 1$$
$$y = 4.$$

Step 4 Use the values from step 3 in any one of the original equations to find the other variable.

Using equation (2) and $x = 1$, $y = 4$, we have

Try the same values in equation (1).

$$x + 2y + z = 12$$
$$1 + 2(4) + z = 12$$
$$z = 3.$$

Step 5 Check the solution in all three of the original equations to see that each equation is satisfied.

This is a very important step.

In our example the solution (1,4,3) checks in all three equations.

Example 2 Solve the system: $\begin{cases} x - y + 2z = 6 & (1) \\ 2x + 3y - z = -3 & (2) \\ 3x + 2y + 2z = 5 & (3) \end{cases}$

Do this in the margin.

Solution **Step 1** Eliminate y by multiplying both sides of equation (1) by 3 and adding the result to equation (2).

Three times equation (1) added to equation (2) gives

$$5x + 5z = 15.$$

Do this in the margin.

Step 2 Eliminate y by multiplying both sides of equation (1) by 2 and adding the result to equation (3). This yields

$$5x + 6z = 17.$$

Step 3 Solve the system

Solve this system here.

$$\begin{cases} 5x + 5z = 15 \\ 5x + 6z = 17. \end{cases}$$

We find the solution to be

$$x = 1$$
$$z = 2.$$

Show work here.

Step 4 Substitute $x = 1$ and $z = 2$ in equation (1) to solve for y. We find the solution to be

$$y = -1.$$

Step 5 The ordered triple $(1, -1, 2)$ checks in all three equations.

If one of the variables is missing from one or more of the equations, finding the solution becomes easier.

Show the work for this example here in the margin

Example 3 Solve the system: $\begin{cases} x + z = 3 & (1) \\ 2x + y + z = 3 & (2) \\ 3x - y + 2z = 8 & (3) \end{cases}$

Solution Note that equation (1) contains only two variables. We can eliminate y using equations (2) and (3) and then we will have two equations in two variables (x and z).

Equation (2) added to equation (3) gives

$$5x + 3z = 11.$$

Now we solve the system

$$\begin{cases} x + z = 3 \\ 5x + 3z = 11. \end{cases}$$

We obtain as our solution

$$x = 1$$
$$z = 2.$$

Substituting these values into equation (2) yields

$$y = -1.$$

The solution $(1,-1,2)$ checks in all three equations.

Example 4 The equation of a vertical parabola can be put in the form $y = ax^2 + bx + c$. Find the equation of the vertical parabola that passes through the points $(-3,10)$, $(-2,3)$, and $(1,6)$.

Solution We must find the values of a, b, and c. Since each point must satisfy the equation, we have

for $(-3,10)$:	$10 = a(-3)^2 + b(-3) + c$	We substitute the values of x and
for $(-2,3)$:	$3 = a(-2)^2 + b(-2) + c$	y in the three points to obtain
and for $(1,6)$:	$6 = a(1)^2 + b(1) + c$	three equations.

These simplify to the system

$$\begin{cases} 9a - 3b + c = 10 \\ 4a - 2b + c = 3 \\ a + b + c = 6. \end{cases}$$

Work this problem here.

The solution yields $a = 2$, $b = 3$, $c = 1$. Thus the desired equation is $y = 2x^2 + 3x + 1$.

Example 5 Three men can complete a job together in 3 hours. Men A and B working together can finish in 5 hours. If men A and C work together, they can complete it in 4½ hours. How long would it take each man working alone to complete the job?

Remember to read the problem carefully.

Solution We let a represent the rate at which man A works (that is, the amount of the job he completes in 1 hour). The same is done for the others.

$$a = \text{rate for man } A$$
$$b = \text{rate for man } B$$
$$c = \text{rate for man } C$$

This gives us three unknowns.

Since it takes all three men 3 hours to complete the job, then in 1 hour they must complete one-third of the job. Thus

$$a + b + c = \frac{1}{3}.$$

Also

In one hour men A and B complete one-fifth of the job.

$$a + b = \frac{1}{5}$$

and

In one hour men A and C complete two-ninths of the job.

$$\frac{1}{4\frac{1}{2}} = \frac{2}{9}$$

$$a + c = \frac{2}{9}.$$

If we clear fractions, we have the system

$$\begin{cases} 3a + 3b + 3c = 1 \\ 5a + 5b \qquad = 1 \\ 9a \qquad + 9c = 2. \end{cases}$$

If man A does $\dfrac{4}{45}$ of the job in one hour, then it will take him $\dfrac{45}{4} = 11\dfrac{1}{4}$ hours to do the job.

Solving, we obtain $a = \dfrac{4}{45}$, $b = \dfrac{1}{9}$, $c = \dfrac{2}{15}$. Therefore man A can complete the job in 11¼ hours, man B in 9 hours, and man C in 7½ hours.

▼ **EXERCISE 9-6-1**

Solve each system.

1. $x + y + z = 6$
   $2x - y + z = 3$
   $x - y + 2z = 5$

2. $x + 2y + z = 0$
   $x - 3y - z = -2$
   $x + y - z = -2$

3. $2x + y + z = 0$
   $3x - 2y - z = -11$
   $x - y + 2z = 3$

4. $x - y + z = 8$
   $5x + 4y - z = 7$
   $2x + y - 3z = -7$

5. $x + 5y - 2z = 13$
   $6x + y + 3z = 4$
   $x - y + 2z = -5$

6. $x + y = 6$
   $2x - y + z = 7$
   $x + y - 3z = 12$

7. $3x + 4y + z = -2$
   $y + z = 1$
   $2x - y - z = -5$

8. $y - z = -3$
   $x + y = 1$
   $2x + 3y + z = 1$

9. A man has \$4.50 in nickels, dimes, and quarters. He has a total of 28 coins and the number of dimes is twice the number of nickels. How many of each type of coin does he have?

10. The equation of a circle can be put in the form $x^2 + y^2 + ax + by + c = 0$. If the circle contains the points $(-1,0)$, $(0,1)$, and $(1,3)$, find the values of a, b, and c and write the equation of the circle.

11. A chemist wishes to make 9 liters of a 30% acid solution by mixing three solutions of 5%, 20%, and 50%. How much of each must he use if he uses twice as much 50% solution as he does 5% solution?

12. Three women can complete a job in 2 hours. Women A and B, working together, can finish in $3\frac{1}{3}$ hours. Women B and C, working together, can complete it in 3 hours. How long would it take each woman working alone to complete the job?

 **9–7 EVALUATING SECOND- AND THIRD-ORDER DETERMINANTS**

Systems of equations are encountered in many different fields. In addition to the methods presented thus far, other procedures are available for solving these systems. Since computers accomplish long computations rapidly, they are used whenever possible. A procedure used by a computer usually depends on the concept of a matrix. In this section we will evaluate determinants of matrices in preparation for using them to solve systems of equations.

Any array of numbers arranged in rows and columns is called a **matrix.** The standard notation for a matrix encloses the array in brackets. The following are examples of matrices.

$$\begin{bmatrix} 2 & 3 & 1 \\ 5 & 4 & 2 \end{bmatrix} \qquad \begin{bmatrix} 2 & 0 & 4 \\ 3 & -1 & 6 \\ 2 & 11 & -5 \end{bmatrix} \qquad \begin{bmatrix} 1 & 2 \\ 5 & -6 \end{bmatrix}$$

Every *square* matrix (same number of rows and columns) has associated with it a unique real number called the *determinant* of the matrix. In this section we will discuss the determinant of two by two and three by three matrices and their uses in solving systems of first-degree equations in two or three unknowns.

O B J E C T I V E S

Upon completing this section you should be able to:

1. Evaluate a second order determinant.
2. Evaluate a third order determinant.

A two by two matrix is referred to as a second order matrix.

> The **determinant of a second-order matrix**
> $$\begin{bmatrix} a & c \\ b & d \end{bmatrix} \quad \text{represented by} \quad \begin{vmatrix} a & c \\ b & d \end{vmatrix}$$
> is the real number $ad - bc$.

Notice that the determinant of a matrix is represented by the array enclosed in straight lines rather than brackets.

First multiply (2)(4)

$$\begin{vmatrix} 2 & 3 \\ 1 & 4 \end{vmatrix}$$

Then subtract the product (1)(3)

$$\begin{vmatrix} 2 & 3 \\ 1 & 4 \end{vmatrix}$$

Example 1 Evaluate: $\begin{vmatrix} 2 & 3 \\ 1 & 4 \end{vmatrix}$

Solution $\begin{vmatrix} 2 & 3 \\ 1 & 4 \end{vmatrix} = (2)(4) - (1)(3) = 8 - 3 = 5$

Example 2 Evaluate: $\begin{vmatrix} 1 & -2 \\ 3 & 2 \end{vmatrix}$

Be careful of signs.

Solution $\begin{vmatrix} 1 & -2 \\ 3 & 2 \end{vmatrix} = (1)(2) - (3)(-2) = 2 + 6 = 8$

Example 3 Evaluate: $\begin{vmatrix} 4 & 7 \\ 3 & 2 \end{vmatrix}$

Solution $\begin{vmatrix} 4 & 7 \\ 3 & 2 \end{vmatrix} = (4)(2) - (3)(7) = 8 - 21 = -13$

▼ **EXERCISE 9–7–1**

Evaluate.

1. $\begin{vmatrix} 2 & 1 \\ 3 & 5 \end{vmatrix}$

2. $\begin{vmatrix} 3 & -2 \\ 1 & 2 \end{vmatrix}$

3. $\begin{vmatrix} 1 & 2 \\ 4 & 5 \end{vmatrix}$

4. $\begin{vmatrix} 2 & 4 \\ 3 & 6 \end{vmatrix}$

5. $\begin{vmatrix} 5 & -2 \\ 3 & 1 \end{vmatrix}$

6. $\begin{vmatrix} -2 & -3 \\ 5 & 4 \end{vmatrix}$

7. $\begin{vmatrix} 4 & 2 \\ -3 & -2 \end{vmatrix}$

8. $\begin{vmatrix} 4 & 0 \\ 6 & -3 \end{vmatrix}$

9. $\begin{vmatrix} -15 & -5 \\ 3 & 1 \end{vmatrix}$

10. $\begin{vmatrix} 3 & 7 \\ 0 & -2 \end{vmatrix}$

11. $\begin{vmatrix} 5 & 9 \\ 4 & -13 \end{vmatrix}$

12. $\begin{vmatrix} 0 & 0 \\ 3 & -6 \end{vmatrix}$

13. $\begin{vmatrix} 4 & 0 \\ 9 & 0 \end{vmatrix}$ **14.** $\begin{vmatrix} 8 & -4 \\ 6 & -3 \end{vmatrix}$ **15.** $\begin{vmatrix} 0 & -6 \\ 9 & 0 \end{vmatrix}$ **16.** $\begin{vmatrix} -2 & 0 \\ 0 & 3 \end{vmatrix}$

17. $\begin{vmatrix} -1 & 4 \\ -3 & 7 \end{vmatrix}$ **18.** $\begin{vmatrix} -2 & -5 \\ 4 & 3 \end{vmatrix}$ **19.** $\begin{vmatrix} -3 & -2 \\ -1 & -4 \end{vmatrix}$ **20.** $\begin{vmatrix} -4 & 6 \\ -2 & 3 \end{vmatrix}$

The determinant of a third-order matrix can be found in more than one way but we will use a general method that can be used to evaluate the determinant of a square matrix of any order.

> The **minor** of an element of a matrix is the determinant of the matrix that results when the row and column in which the element appears is eliminated.

Note that a minor is a determinant.

Example 4 In the matrix $\begin{bmatrix} 3 & 1 & 4 \\ 2 & 6 & 5 \\ 3 & -1 & 2 \end{bmatrix}$

the minor of 1 (first row, second column) is obtained by eliminating the first row and second column

$$\begin{bmatrix} 3 & 1 & 4 \\ 2 & 6 & 5 \\ 3 & -1 & 2 \end{bmatrix}$$

We only have four elements remaining.

giving

$$\begin{vmatrix} 2 & 5 \\ 3 & 2 \end{vmatrix}$$

and the minor of 5 (second row, third column) is

$$\begin{vmatrix} 3 & 1 \\ 3 & -1 \end{vmatrix}.$$

$$\begin{bmatrix} 3 & 1 & 4 \\ 2 & 6 & 5 \\ 3 & -1 & 2 \end{bmatrix}$$

> The **cofactor** of an element of a matrix is its minor multiplied by $(-1)^{i+j}$, where i is the number of the row and j is the number of the column of the element.

Notice that $(-1)^{i+j}$ will be positive if $(i + j)$ is an even number and negative if $(i + j)$ is an odd number.

This definition can be used to find the determinant of any order matrix.

The **determinant of a third-order matrix** is the sum of the products of each element of a row or column and its cofactor, that is, in

$$\begin{bmatrix} a_{11} & a_{12} & a_{13} \\ a_{21} & a_{22} & a_{23} \\ a_{31} & a_{32} & a_{33} \end{bmatrix}$$

the determinant is

$$a_{11} \cdot \text{cofactor } a_{11} + a_{21} \cdot \text{cofactor } a_{21} + a_{31} \cdot \text{cofactor } a_{31}.$$

Example 5 Evaluate: $\begin{vmatrix} 2 & 1 & 3 \\ 3 & -5 & 4 \\ 5 & 0 & 2 \end{vmatrix}$

Solution When we apply the definition to column one, we get

Note the cofactor of 2 is
$(-1)^{1+1}\begin{vmatrix} -5 & 4 \\ 0 & 2 \end{vmatrix}.$

$$\begin{vmatrix} 2 & 1 & 3 \\ 3 & -5 & 4 \\ 5 & 0 & 2 \end{vmatrix} = 2(-1)^{1+1}\begin{vmatrix} -5 & 4 \\ 0 & 2 \end{vmatrix} + 3(-1)^{2+1}\begin{vmatrix} 1 & 3 \\ 0 & 2 \end{vmatrix} + 5(-1)^{3+1}\begin{vmatrix} 1 & 3 \\ -5 & 4 \end{vmatrix}$$

$$= 2(1)(-10) + 3(-1)(2) + 5(1)(19)$$
$$= -20 - 6 + 95$$
$$= 69.$$

If we instead apply the definition to row three, we get

Notice there is less work using row three since the second term is zero.

$$\begin{vmatrix} 2 & 1 & 3 \\ 3 & -5 & 4 \\ 5 & 0 & 2 \end{vmatrix} = 5(-1)^{3+1}\begin{vmatrix} 1 & 3 \\ -5 & 4 \end{vmatrix} + 0(-1)^{3+2}\begin{vmatrix} 2 & 3 \\ 3 & 4 \end{vmatrix} + 2(-1)^{3+3}\begin{vmatrix} 2 & 1 \\ 3 & -5 \end{vmatrix}$$

$$= 5(1)(19) + 0(-1)(-1) + 2(1)(-13)$$
$$= 95 + 0 - 26$$
$$= 69.$$

The value of a determinant is a constant and can be found by applying the definition to any row or column. Notice that choosing a row or column containing a zero cuts the amount of computation necessary.

Example 6 Evaluate: $\begin{vmatrix} 3 & 0 & 5 \\ 2 & 1 & 6 \\ -4 & 3 & -2 \end{vmatrix}$

Solution   It is wise to choose either the first row or the second column since the zero will simplify the work. Using the second column gives

$$\begin{vmatrix} 3 & 0 & 5 \\ 2 & 1 & 6 \\ -4 & 3 & -2 \end{vmatrix} = 1(-1)^{2+2}\begin{vmatrix} 3 & 5 \\ -4 & -2 \end{vmatrix} + 3(-1)^{2+3}\begin{vmatrix} 3 & 5 \\ 2 & 6 \end{vmatrix}$$

We omitted the zero term.

$$= 1(1)(14) + 3(-1)(8)$$
$$= 14 - 24$$
$$= -10.$$

▼ **EXERCISE 9–7–2**

Evaluate the determinants.

1. $\begin{vmatrix} 1 & 2 & 5 \\ 3 & 1 & 4 \\ 2 & 0 & 3 \end{vmatrix}$

2. $\begin{vmatrix} 3 & 4 & 1 \\ 1 & 0 & -2 \\ 5 & -3 & 1 \end{vmatrix}$

3. $\begin{vmatrix} 4 & -2 & 1 \\ -2 & 3 & 5 \\ 1 & 0 & 2 \end{vmatrix}$

4. $\begin{vmatrix} 6 & -1 & 5 \\ 1 & 0 & 4 \\ -5 & -2 & 3 \end{vmatrix}$

5. $\begin{vmatrix} -1 & -2 & 4 \\ 3 & 1 & -1 \\ 5 & 6 & 5 \end{vmatrix}$

6. $\begin{vmatrix} 4 & 0 & 1 \\ -2 & 1 & 0 \\ 7 & 3 & 5 \end{vmatrix}$

7. $\begin{vmatrix} 3 & 5 & -3 \\ 4 & 1 & -2 \\ -2 & 6 & -4 \end{vmatrix}$

8. $\begin{vmatrix} 0 & 0 & 0 \\ -2 & 3 & -1 \\ 4 & 8 & -2 \end{vmatrix}$

9. $\begin{vmatrix} 4 & -2 & 0 \\ 3 & 4 & 0 \\ 1 & 3 & 0 \end{vmatrix}$

10. $\begin{vmatrix} -1 & 3 & 5 \\ -2 & 1 & -1 \\ 4 & 6 & 5 \end{vmatrix}$

11. $\begin{vmatrix} 3 & 4 & -1 \\ -1 & 0 & 3 \\ 2 & 1 & 5 \end{vmatrix}$

12. $\begin{vmatrix} 7 & -2 & 1 \\ -3 & 5 & 0 \\ 1 & 2 & 4 \end{vmatrix}$

13. $\begin{vmatrix} -4 & -1 & 0 \\ 3 & 2 & -5 \\ -2 & 1 & 4 \end{vmatrix}$

14. $\begin{vmatrix} 9 & 2 & 0 \\ 0 & 4 & 0 \\ -6 & 5 & 3 \end{vmatrix}$

15. $\begin{vmatrix} 2 & 6 & 1 \\ 3 & 5 & 1 \\ 1 & 9 & 1 \end{vmatrix}$

16. $\begin{vmatrix} 3 & -2 & 1 \\ -1 & 4 & 1 \\ 5 & 7 & 1 \end{vmatrix}$

17. $\begin{vmatrix} 5 & -2 & 8 \\ 0 & 4 & 5 \\ -1 & 3 & 1 \end{vmatrix}$

18. $\begin{vmatrix} 3 & -4 & 10 \\ 2 & 7 & 0 \\ 6 & 0 & -3 \end{vmatrix}$

19. Show that the equation of a line containing the points (x_1, y_1) and (x_2, y_2) may be expressed as
$$\begin{vmatrix} x & y & 1 \\ x_1 & y_1 & 1 \\ x_2 & y_2 & 1 \end{vmatrix} = 0.$$

    Hint: Evaluate the determinant on the left. Then show that the two-point form of the equation of a line is equivalent to this form.

20. Use the determinant form discussed in problem 19 to find the equation of a line through the points $(3,1)$ and $(2,3)$.

◣ 9–8 SOLVING SYSTEMS OF EQUATIONS BY DETERMINANTS

Now that we have learned to evaluate determinants, we will proceed to use them in solving systems of linear equations with two or three variables.

The standard form of a system of two linear equations is

$$\begin{cases} ax + cy = e \\ bx + dy = f. \end{cases}$$

If we solve this system, either by substitution or addition, the solution will be

$$x = \frac{de - cf}{ad - bc}$$

$$y = \frac{af - de}{ad - bc}.$$

From the definition of the value of the determinant of a second-order matrix we see that these values for x and y could be expressed as

$$x = \frac{\begin{vmatrix} e & c \\ f & d \end{vmatrix}}{\begin{vmatrix} a & c \\ b & d \end{vmatrix}}$$

$$y = \frac{\begin{vmatrix} a & e \\ b & f \end{vmatrix}}{\begin{vmatrix} a & c \\ b & d \end{vmatrix}}.$$

This gives us a method of solving a system of two linear equations by determinants. This method is often referred to as **Cramer's Rule.** In a sense this is a formula and, as all formulas, must come from the standard form. If an equation is not in standard form, then it must first be put in standard form before this method will apply.

Example 1   Solve by determinants: $\begin{cases} 3x - y = 5 \\ x + y = 3 \end{cases}$

Solution   The equations are already in standard form so

$$x = \frac{\begin{vmatrix} 5 & -1 \\ 3 & 1 \end{vmatrix}}{\begin{vmatrix} 3 & -1 \\ 1 & 1 \end{vmatrix}} = \frac{8}{4} = 2$$

$$y = \frac{\begin{vmatrix} 3 & 5 \\ 1 & 3 \end{vmatrix}}{\begin{vmatrix} 3 & -1 \\ 1 & 1 \end{vmatrix}} = \frac{4}{4} = 1.$$

The solution is (2,1).

OBJECTIVES

Upon completing this section you should be able to:
1. Use determinants to solve a system of two equations with two variables.
2. Use determinants to solve a system of three equations with three variables.

IMPORTANT!

Make sure that each number is entered in the proper place.

Some special patterns can be noted that make it easy to set up the determinants involved.

x's are replaced with constants.

$$x = \frac{\begin{vmatrix} 5 & -1 \\ 3 & 1 \end{vmatrix}}{\begin{vmatrix} 3 & -1 \\ 1 & 1 \end{vmatrix}}$$

coefficients of y
coefficients of x

When the equations are in standard form:

1. The denominator of each variable is the determinant of the matrix of coefficients.
2. The numerator is the same as the denominator *except* that the column of the variable being found is replaced by the column of constants.

This system is also in standard form.

Example 2   Solve by determinants: $\begin{cases} 4x + 3y = 2 \\ 3x + 4y = 5 \end{cases}$

Solution

x's are replaced by constants.

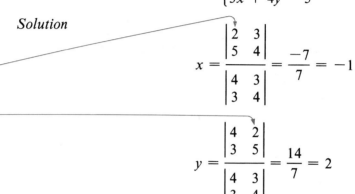

$$x = \frac{\begin{vmatrix} 2 & 3 \\ 5 & 4 \end{vmatrix}}{\begin{vmatrix} 4 & 3 \\ 3 & 4 \end{vmatrix}} = \frac{-7}{7} = -1$$

y's are replaced by constants.

$$y = \frac{\begin{vmatrix} 4 & 2 \\ 3 & 5 \end{vmatrix}}{\begin{vmatrix} 4 & 3 \\ 3 & 4 \end{vmatrix}} = \frac{14}{7} = 2$$

The solution is $(-1, 2)$.

These equations are not in standard form.

Example 3   Solve by determinants: $\begin{cases} y = 2x - 1 \\ \dfrac{x+1}{2} + \dfrac{3y}{5} = \dfrac{8}{5} \end{cases}$

Solution   Put the equations in standard form by rewriting the first one and clearing the fractions in the second.

The first equation becomes

$$2x - y = 1.$$

Multiplying both sides of the second equation by the least common denominator (10) and rearranging gives

$$5x + 6y = 11.$$

We now have the system

In standard form

$$\begin{cases} 2x - y = 1 \\ 5x + 6y = 11. \end{cases}$$

Hence

x's replaced by constants

$$x = \frac{\begin{vmatrix} 1 & -1 \\ 11 & 6 \end{vmatrix}}{\begin{vmatrix} 2 & -1 \\ 5 & 6 \end{vmatrix}} = \frac{17}{17} = 1$$

y's replaced by constants

$$y = \frac{\begin{vmatrix} 2 & 1 \\ 5 & 11 \end{vmatrix}}{\begin{vmatrix} 2 & -1 \\ 5 & 6 \end{vmatrix}} = \frac{17}{17} = 1.$$

The solution is $(1,1)$.

Example 4 Solve by determinants: $\begin{cases} x - 2y = 3 \\ 3x - 6y = 9 \end{cases}$

Solution

$$x = \frac{\begin{vmatrix} 3 & -2 \\ 9 & -6 \end{vmatrix}}{\begin{vmatrix} 1 & -2 \\ 3 & -6 \end{vmatrix}} = \frac{-18 + 18}{-6 + 6} = \frac{0}{0}$$

An expression having zero as the denominator has no meaning.

$$y = \frac{\begin{vmatrix} 1 & 3 \\ 3 & 9 \end{vmatrix}}{\begin{vmatrix} 1 & -2 \\ 3 & -6 \end{vmatrix}} = \frac{9 - 9}{-6 + 6} = \frac{0}{0}$$

Notice that *both* numerator and denominator are zero. These of course are undefined quantities and we seem to be at an impasse.

Try solving the system by the addition method. Do you see that the equations are dependent?

Multiply the first equation by 3 and add to the second equation.

We have thus discovered a characteristic of a system of dependent equations when solving by determinants.

If the determinants in both numerator and denominator are zero, the equations are dependent.

Example 5 Solve by determinants: $\begin{cases} 2x + 4y = 5 \\ x + 2y = 6 \end{cases}$

Solution

$$x = \frac{\begin{vmatrix} 5 & 4 \\ 6 & 2 \end{vmatrix}}{\begin{vmatrix} 2 & 4 \\ 1 & 2 \end{vmatrix}} = \frac{10 - 24}{4 - 4} = \frac{-14}{0}$$

Here again we obtain a denominator of zero.

$$y = \frac{\begin{vmatrix} 2 & 5 \\ 1 & 6 \end{vmatrix}}{\begin{vmatrix} 2 & 4 \\ 1 & 2 \end{vmatrix}} = \frac{12 - 5}{4 - 4} = \frac{7}{0}$$

Here again we see meaningless values for x and y since the denominators are zero.

Solve the system by the addition method. Do you see that the equations are inconsistent?

Multiply the second equation by -2 and add to the first equation.

Thus when solving a system by determinants, if the determinant of the matrix of coefficients is zero but the other determinants (numerators) are not, the equations are inconsistent.

▼ **EXERCISE 9–8–1**

Solve the systems by determinants.

1. $\begin{cases} x + y = 3 \\ 2x - y = 0 \end{cases}$

2. $\begin{cases} 2x + y = 7 \\ x + 2y = 11 \end{cases}$

3. $\begin{cases} x - y = -1 \\ 2x - y = 1 \end{cases}$

4. $\begin{cases} 2x + 3y = 10 \\ x - y = -5 \end{cases}$

5. $\begin{cases} 3x + y = 2 \\ 6x + 2y = 4 \end{cases}$

6. $\begin{cases} 3x + 2y = -8 \\ 2x - 5y = 1 \end{cases}$

7. $\begin{cases} 2x - 3y = 6 \\ x + 2y = 3 \end{cases}$

8. $\begin{cases} 2x + 8y = 1 \\ x + 4y = 3 \end{cases}$

9. $\begin{cases} x + 4y = 1 \\ 2x - 3y = 13 \end{cases}$

10. $\begin{cases} 2x + y = 0 \\ 4x + 3y = 6 \end{cases}$

11. $\begin{cases} 2x + y = 6 \\ 4x - 3y = -13 \end{cases}$

12. $\begin{cases} 5x + 6y = -1 \\ 2x + 3y = 0 \end{cases}$

13. $\begin{cases} x = 2y - 4 \\ 3y = 5 + x \end{cases}$

14. $\begin{cases} 2x = 3y + 12 \\ 2y + 8 = 3x \end{cases}$

15. $\begin{cases} 2x - 3y = 5 \\ \dfrac{4}{3}x - 2y = \dfrac{10}{3} \end{cases}$

16. $\begin{cases} x - 2y = 6 \\ \dfrac{x-1}{3} + \dfrac{y}{2} = \dfrac{1}{2} \end{cases}$

17. $\begin{cases} y = x + 3 \\ x = \dfrac{2y}{3} - 2 \end{cases}$

18. $\begin{cases} \dfrac{x+1}{5} + \dfrac{y-3}{2} = 1 \\ y = -5x \end{cases}$

19. A boat travels 15 miles per hour downstream and 9 miles per hour upstream. Find the speed of the current and the speed of the boat in still water.

20. How much of a 40% solution of alcohol and how much of an 80% solution should be mixed to give 40 liters of a 50% solution?

A system of three first-degree equations in three variables can be solved by following the same pattern used for two equations with two variables (Cramer's Rule).

If the equations are in standard form then:

1. The denominator of each variable is the same and is the determinant of the matrix of coefficients of the variables as they appear.
2. The numerator is the same as the denominator *except* that the column of the variable being found is replaced by the column of constants.

Example 6 Solve by determinants: $\begin{cases} 2x - 3y + z = 7 \\ x + y + z = 2 \\ 3x + 3y - z = -2 \end{cases}$

Solution The equations are in standard form so we set up the determinants to find x, y, and z as discussed.

x's are replaced with constants.

$$ x = \frac{\begin{vmatrix} 7 & -3 & 1 \\ 2 & 1 & 1 \\ -2 & 3 & -1 \end{vmatrix}}{\begin{vmatrix} 2 & -3 & 1 \\ 1 & 1 & 1 \\ 3 & 3 & -1 \end{vmatrix}} = \frac{-20}{-20} = 1 $$

These are coefficients of x, y, and z.

y's are replaced with constants.

$$y = \frac{\begin{vmatrix} 2 & 7 & 1 \\ 1 & 2 & 1 \\ 3 & -2 & -1 \end{vmatrix}}{\begin{vmatrix} 2 & -3 & 1 \\ 1 & 1 & 1 \\ 3 & 3 & -1 \end{vmatrix}} = \frac{20}{-20} = -1$$

z's are replaced with constants.

$$z = \frac{\begin{vmatrix} 2 & -3 & 7 \\ 1 & 1 & 2 \\ 3 & 3 & -2 \end{vmatrix}}{\begin{vmatrix} 2 & -3 & 1 \\ 1 & 1 & 1 \\ 3 & 3 & -1 \end{vmatrix}} = \frac{-40}{-20} = 2$$

Thus the solution is $(1, -1, 2)$.

WARNING

Since the denominator is the same for each, we should be extremely careful when it is evaluated.

Note that the denominator need not be evaluated each time since it is the same. Also it is possible to find x and y and then substitute these values in one of the equations to find z. Thus it is possible to solve the system by evaluating three third-order determinants and making one substitution.

Should the value of the denominator be zero, two possibilities exist:

1. If the numerators of all the variables are zero, the system is dependent.
2. If at the least one of the variables has a nonzero numerator, the system is inconsistent.

EXERCISE 9–8–2

Solve the systems by determinants.

1. $\begin{cases} x + y + z = 6 \\ 2x - y + 2z = 6 \\ 3x + 2y - z = 4 \end{cases}$

2. $\begin{cases} x - y + 2z = 3 \\ 3x + y + z = -1 \\ 2x - 3y + 5z = 8 \end{cases}$

3. $\begin{cases} x + 2y - z = 13 \\ 2x - y - 2z = 11 \\ 3x + y + z = 4 \end{cases}$

4. $\begin{cases} 2x + y - z = -2 \\ 3x + 2y + z = -4 \\ x - y + 3z = 13 \end{cases}$

5. $\begin{cases} 3x - y + z = -10 \\ 2x + 3y - 2z = -3 \\ x - 5y + 3z = -8 \end{cases}$

6. $\begin{cases} x + y = -2 \\ 3x - z = 1 \\ 2x + y + z = 1 \end{cases}$

7. $\begin{cases} x + 2y - 3z = -15 \\ 3x - y + z = 9 \\ 2x + 3y - z = -8 \end{cases}$

8. $\begin{cases} x - 3z = -15 \\ x - 2y = 2 \\ y + z = 4 \end{cases}$

9. $\begin{cases} x + 3y + 2z = 5 \\ 3x - y + z = 5 \\ 2x - 2y + 3z = 14 \end{cases}$

10. $\begin{cases} 4x - 3y + z = 5 \\ 2x + y + 3z = -5 \\ x - y - 2z = 8 \end{cases}$

11. $\begin{cases} x + y + 3z = 1 \\ 3x - y + 4z = 5 \\ 2x + 2y + 6z = 3 \end{cases}$

12. $\begin{cases} 2x - y + 3z = 1 \\ 3x - 2y - z = 7 \\ 6x - 3y + 9z = 3 \end{cases}$

13. $\begin{cases} 2x - 6y + 4z = 2 \\ x + 3y + 2z = 1 \\ 3x + 9y + 6z = 3 \end{cases}$

14. $\begin{cases} 2x - 4y + 8z = 4 \\ 3x - 6y + 12z = 6 \\ 5x - 10y + 20z = 10 \end{cases}$

15. $\begin{cases} 2x + 8y + 4 = 0 \\ x = 4 + 5z \\ 5y - 1 = -2z \end{cases}$

16. $\begin{cases} x - 16 = 3y - z \\ 4x + z = 15 - 3y \\ x = y + 6 \end{cases}$

17. The sum of three numbers is zero. Twice the first number added to the second is 11 less than the third. The third number is 17 more than the second. Find the numbers.

18. A man buys 11 gallons of three different kinds of paint, some at $8.00 a gallon, some at $7.00 a gallon, and some at $5.00 a gallon. He has twice as many gallons of $5.00 paint as he does of $7.00 paint. If his total bill for the paint is $67.00, how many gallons of $5.00 paint did he buy?

CHAPTER 9 SUMMARY

Key Words

Section 9–1

- A **system of equations** contains two or more equations for which we wish to find simultaneous solutions.

- **Independent equations** have unique solutions.

- **Inconsistent equations** have no solution.

- **Dependent equations** have infinitely many solutions.

Section 9–2

- A **system of linear inequalities** contains two or more inequalities for which we wish to find simultaneous solutions.

- **Linear programming** is a branch of mathematics that involves solutions of systems of inequalities.

Section 9–3

- A **contradiction** occurs when solving a system of inconsistent equations.

- An **identity** occurs when solving a system of dependent equations.

Section 9–6

- **Ordered triples** of numbers occur as solutions to systems of three equations with three variables.

Section 9–7

- A **matrix** is an array of elements arranged in rows and columns.

- Every square matrix has associated with it a unique real number called the **determinant** of the matrix.

Section 9–8

- **Cramer's Rule** is a method of solving systems of equations using determinants.

Procedures

Section 9–1

- To solve a system of equations in two variables by graphing, graph the equations carefully on the same coordinate system. Their point(s) of intersection will be the solution(s) of the system.

Section 9–2

- To solve a system of inequalities by graphing determine the region of the plane that satisfies both inequality statements.

Section 9–3

- A system of two linear equations is solved algebraically by using substitution or addition to eliminate one of the variables. The value of the remaining variable is then substituted into one of the original equations to obtain the value of the other variable. The ordered pair thus obtained must be a solution of each equation to be a solution of the system.

Section 9–6

- The algebraic solution of a system of three equations with three variables involves choosing any two of the three equations and eliminating one of the variables. Next choose any other two of the three equations and eliminate the same variable. The two equations with two variables thus obtained are solved and the solution is substituted into any one of the three original equations to obtain the value of the third variable. Again, the values obtained must be a solution of each of the three equations to be a solution of the system.

Section 9–7

- To find the determinant of a second-order matrix use the following definition:

$$\begin{vmatrix} a & c \\ b & d \end{vmatrix} = ad - bc$$

- To find the determinant of a third-order matrix use the following definition:

$$\begin{vmatrix} a & d & g \\ b & e & h \\ c & f & i \end{vmatrix} = a\begin{vmatrix} e & h \\ f & i \end{vmatrix} - b\begin{vmatrix} d & g \\ f & i \end{vmatrix} + c\begin{vmatrix} d & g \\ e & h \end{vmatrix}$$

Section 9–8

- When the equations of a system are in standard form, the values of the variables can be represented as the ratio of two determinants such that:
  1. The denominator of each ratio is the same and is the determinant of coefficients of the variables as they appear.
  2. The only way in which the numerator of each ratio differs from the denominator is that the column of the particular variable being represented is replaced by the column of constant terms.

- To solve a word problem with two (three) unknowns find two (three) equations that show a relationship between the unknowns. Then solve the system by any method. Always check the solution in the stated problem.

A N S W E R S

Solve the systems by graphing.

1. $\begin{cases} x - y = 3 \\ x + 2y = -6 \end{cases}$

2. $\begin{cases} x + y = -1 \\ 2x - y = 4 \end{cases}$

1. _____

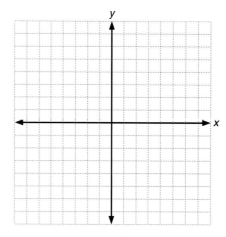
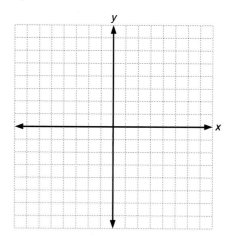

2. _____

3. $\begin{cases} 3x - y = 7 \\ 2x + y = 8 \end{cases}$

4. $\begin{cases} 2x + y = -3 \\ x - 2y = -4 \end{cases}$

3. _____

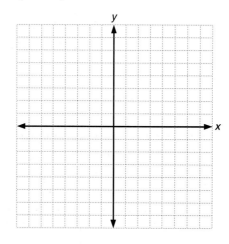
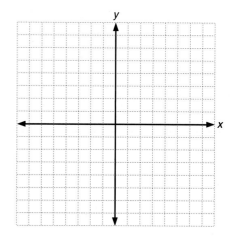

4. _____

5. $\begin{cases} x + y = -1 \\ y = x^2 - 1 \end{cases}$

6. $\begin{cases} x - y = 2 \\ y = x^2 + 3x - 5 \end{cases}$

5. _____

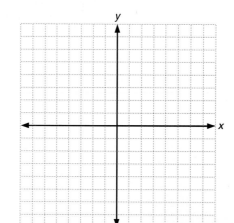
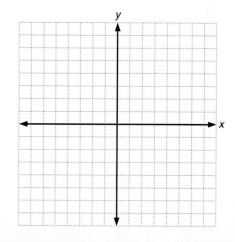

6. _____

7. $\begin{cases} x + y \le -4 \\ 5x - 3y \ge 10 \end{cases}$

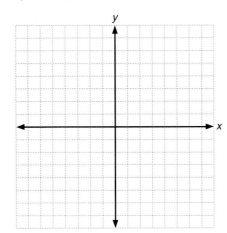

8. $\begin{cases} x - 2y \le 4 \\ 3x + y \ge 4 \end{cases}$

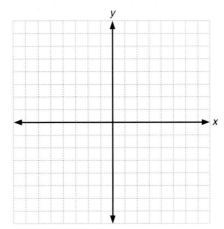

7. _____

9. $\begin{cases} x - y \ge -4 \\ x + y \ge -2 \\ 2x + 3y \le 6 \\ x \le 0 \end{cases}$

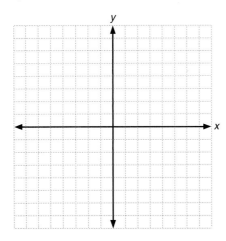

10. A man has a maximum of $12,000 to invest in two types of bonds. Bond A returns 8 percent and bond B returns 10 percent per year. His investment in bond B must not exceed 50 percent of his investment in bond A. How much should he invest at each rate in order to obtain a maximum profit?

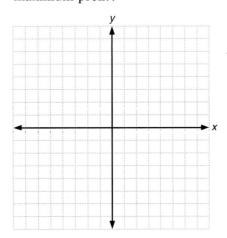

8. _____

9. _____

10. _____

A N S W E R S

Solve the systems by the substitution method.

11. $\begin{cases} x + 5y = 2 \\ 2x + 3y = -3 \end{cases}$ **12.** $\begin{cases} 2x + y = 4 \\ 3x - 2y = -1 \end{cases}$

11. _____

12. _____

13. $\begin{cases} 2x + 3y = 1 \\ 4x + 5y = 3 \end{cases}$ **14.** $\begin{cases} x + y = 5 \\ 2x - y = -5 \end{cases}$

13. _____

14. _____

Solve the systems by the addition method.

15. $\begin{cases} 2x - 3y = 5 \\ 2x + y = 9 \end{cases}$ **16.** $\begin{cases} 2x + y = 1 \\ x - y = 5 \end{cases}$

15. _____

17. $\begin{cases} 2x + 3y = 21 \\ 3x + 2y = 19 \end{cases}$ **18.** $\begin{cases} 2x + y = 1 \\ 3x - 2y = 5 \end{cases}$

16. _____

17. _____

Classify each system as independent, inconsistent, or dependent. If the system is independent, find its solution.

19. $\begin{cases} x + y = 1 \\ 2x - y = 5 \end{cases}$ **20.** $\begin{cases} x + 2y = 3 \\ 4x + 8y = 7 \end{cases}$

18. _____

19. _____

20. _____

21. $\begin{cases} 2x - y = 4 \\ 6x - 3y = 8 \end{cases}$

22. $\begin{cases} 2x + y = -5 \\ 3x + 2y = -10 \end{cases}$

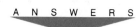

A N S W E R S

21. _____

23. $\begin{cases} 3x + 6y = 15 \\ 2x + 4y = 10 \end{cases}$

24. $\begin{cases} 2x + 3y = -5 \\ 5x - 2y = -22 \end{cases}$

22. _____

23. _____

24. _____

Change to standard form.

25. $3x = y - 2$

26. $\dfrac{1}{2}x + \dfrac{2}{3}y = 1$

25. _____

26. _____

27. $\dfrac{3}{5} - \dfrac{1}{2}x = 2y$

28. $4y = 3x - 5$

27. _____

Change to standard form when necessary and solve each system by any method.

28. _____

29. $\begin{cases} 3x = y + 2 \\ \dfrac{1}{3}x + \dfrac{1}{2}y = -1 \end{cases}$

30. $\begin{cases} 2x + y = 1 \\ x - \dfrac{1}{3}y = 3 \end{cases}$

29. _____

30. _____

31. $\begin{cases} \dfrac{2}{5}x + y = 0 \\ 2y = 3x - 19 \end{cases}$

32. $\begin{cases} x = 15 + 2y \\ x + \dfrac{1}{3}y = 1 \end{cases}$

31. _____

32. _____

A N S W E R S

33. _____

34. _____

35. _____

36. _____

37. _____

38. _____

39. _____

40. _____

41. _____

42. _____

43. _____

44. _____

Solve the systems algebraically.

33. $\begin{cases} x + y - z = 4 \\ 2x + y + z = 3 \\ 2x + 2y + z = 5 \end{cases}$ **34.** $\begin{cases} x - y + 2z = -1 \\ 3x + 2y - z = 14 \\ 2x + 3y + 5z = 7 \end{cases}$

35. $\begin{cases} x + y - z = -3 \\ x + y + z = 3 \\ 3x - y + z = 7 \end{cases}$ **36.** $\begin{cases} 2x - y + z = 5 \\ x + 2y - z = -2 \\ x + y - 2z = -5 \end{cases}$

37. $\begin{cases} 2x - 3y + z = 11 \\ x + y + 2z = 8 \\ x + 3y - z = -11 \end{cases}$ **38.** $\begin{cases} x + y = -2 \\ x - z = -1 \\ x + 2y - z = 1 \end{cases}$

Evaluate the determinants.

39. $\begin{vmatrix} 3 & -6 \\ 1 & 2 \end{vmatrix}$ **40.** $\begin{vmatrix} 4 & 5 \\ 3 & 7 \end{vmatrix}$ **41.** $\begin{vmatrix} -4 & 16 \\ 1 & 2 \end{vmatrix}$

42. $\begin{vmatrix} 2 & -3 \\ 8 & -12 \end{vmatrix}$ **43.** $\begin{vmatrix} 2 & 1 & 4 \\ 1 & 0 & 2 \\ -3 & 1 & -3 \end{vmatrix}$ **44.** $\begin{vmatrix} -2 & 1 & 0 \\ 0 & 6 & -3 \\ 4 & 2 & -5 \end{vmatrix}$

Solve the systems by determinants.

45. $\begin{cases} x - y = 6 \\ 3x + 2y = -7 \end{cases}$

46. $\begin{cases} 2x + 3y = 2 \\ x - 5y = 14 \end{cases}$

45. _____

46. _____

47. $\begin{cases} x + 2y = -6 \\ 2x - 3y = 9 \end{cases}$

48. $\begin{cases} 2x + 5y = 0 \\ 3x - 2y = 19 \end{cases}$

47. _____

48. _____

49. $\begin{cases} x + y + z = 2 \\ 2x - y + z = -1 \\ x - y + z = -2 \end{cases}$

50. $\begin{cases} 2x - y + z = 5 \\ x + 2y - z = -2 \\ x + y - 2z = -5 \end{cases}$

49. _____

50. _____

51. $\begin{cases} 2x - 3y + z = 11 \\ x + y + 2z = 8 \\ x + 3y - z = -11 \end{cases}$

52. $\begin{cases} x + y = -2 \\ x - z = -1 \\ x + 2y - z = 1 \end{cases}$

51. _____

52. _____

Solve the systems of equations using any method.

53. _____

53. $\begin{cases} x + 2y = 3 \\ 2x + 5y = 1 \end{cases}$

54. $\begin{cases} 2x + 3y = 6 \\ 3x - 4y = -8 \end{cases}$

54. _____

55. $\begin{cases} x + y - z = 0 \\ 3x - y + z = 12 \\ 2x + 3y + 2z = 7 \end{cases}$

56. $\begin{cases} 3x - 2y + z = 4 \\ 4x + 3z = 12 \\ x - 5y + 2z = 8 \end{cases}$

55. _____

56. _____

57. $\begin{cases} 3x + z = -2 \\ 5y - 3z = 23 \\ 6x + 7y = 26 \end{cases}$

58. $\begin{cases} x + 2y = 6 \\ 6y - 5z = 3 \\ 2x + 3z = 10 \end{cases}$

57. _____

Solve.

58. _____

59. The sum of two numbers is 147 and their difference is 19. Find the numbers.

60. An airliner took 6 hours to travel 3,864 kilometers from New York to Las Vegas. The return trip took 4 hours. Find the speeds of the plane and the wind.

59. _____

60. _____

61. A professor has 85 students enrolled in two classes. Forty-nine of these students are freshmen. If two-thirds of the first class and one-half of the second class are freshmen, how many students are in each class?

62. A boat travels 23 kilometers per hour downstream and 15 kilometers per hour upstream. Find the speed of the current and the speed of the boat.

61. _____

62. _____

63. Maria has $7.80 in nickels and quarters. If she has six more nickels than quarters, how many of each kind of coin does she have?

64. How many liters each of 12% salt solution and 20% salt solution must be mixed to produce twenty liters of 14% solution?

A N S W E R S

63. _____

64. _____

65. A man buys 11 liters of three different kinds of paint, some at $5.00 a liter, some at $4.00 a liter, and some at $3.00 a liter. He has twice as many liters of $3.00 paint as he does of $4.00 paint. If his total bill for the paint is $40.00, how many liters of $3.00 paint did he buy?

66. A man has 25 coins consisting of nickels, dimes, and quarters. The value of the coins is $3.30. If he has two more quarters than nickels, how many dimes does he have?

65. _____

66. _____

SCORE: _____

A   N   S   W   E   R   S

1. _____

2. _____

3. _____

4. _____

5. _____

6. _____

7. _____

8. _____

1. Solve the system by graphing.

$$\begin{cases} x + y = 4 \\ 2x - y = 5 \end{cases}$$

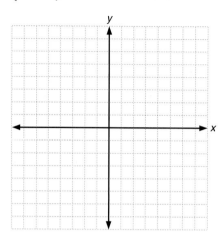

2. Solve the system of inequalities by graphing.

$$\begin{cases} 3x - 2y \le 8 \\ 2x + 3y \le 6 \end{cases}$$

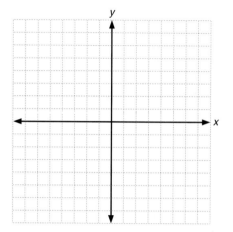

3. Solve the system algebraically:

$$\begin{cases} 2x + 3y = -2 \\ 3x + 2y = 7 \end{cases}$$

4. Solve algebraically:

$$\begin{cases} x + \dfrac{3}{4}y = 2 \\ 2y = 3x - 6 \end{cases}$$

5. Solve algebraically:

$$\begin{cases} x + 2y + z = 1 \\ 2x + 3y - z = 0 \\ x - 2y + 3z = 7 \end{cases}$$

6. Evaluate the determinants:

a. $\begin{vmatrix} 3 & -4 \\ 2 & 5 \end{vmatrix}$ **b.** $\begin{vmatrix} 2 & 3 & 0 \\ 4 & 1 & 2 \\ -1 & 5 & -1 \end{vmatrix}$

7. Solve the system by determinants:

$$\begin{cases} x + z = 4 \\ 2x - y = 3 \\ 3x + 2y - z = -2 \end{cases}$$

8. Solve the following using two unknowns:

A girl has $9.00 in dimes and quarters. If she has twice as many quarters as dimes, how many of each kind of coin does she have?

SCORE: _____

1. Combine like terms:
$8xy - 4x^2y - 9xy - 16x^2y$

2. Remove grouping symbols and simplify: $214a - 3\{2a - 5[2a - 3(5a - 4)]\}$

A N S W E R S

1. _____

3. Solve for x:
$3x - (4x + 3) =$
$\frac{2}{3}(x + 1) + 2$

4. Solve for x and graph the solution on the number line:
$|2x - 3| \geq 5$

-1 0   4

2. _____

3. _____

5. A stereo VCR is priced $274 more than a nonstereo model. If the total cost of both models is $924, find the cost of each.

6. Two cars start at the same time from the same point and travel in opposite directions. At the end of three hours car B stops and car A continues for another hour. At the end of four hours they were 345 miles apart. If car B traveled at a speed of 10 mph faster than car A, what was the average speed of each?

4. _____

5. _____

6. _____

7. Factor: $x^3 - 64$

8. Factor completely:
$6x^4 + 9x^3 - 15x^2$

7. _____

8. _____

9. Divide:
$\frac{2x^2 - 7x + 3}{2x^2 - 5x - 3} \div \frac{2x^2 + 9x - 5}{2x + 1}$

10. Solve for x:
$\frac{2}{3} + \frac{1}{x - 1} = \frac{x}{3x - 3}$

9. _____

10. _____

A N S W E R S

11. _____

12. _____

13. _____

14. _____

15. _____

16. _____

17. _____

18. _____

11. Simplify and rewrite using only positive exponents:
$$\left(\frac{x^2y^{-1}}{x^{-3}}\right)\left(\frac{x^5y^{-2}}{y^3}\right)^{-2}$$

12. Simplify: $\sqrt{\dfrac{2}{5}} + \sqrt{10}$

13. Solve for x: $3x^2 - 5x + 4 = 0$

14. Find all solutions for x:
$$x^4 - x^2 - 12 = 0$$

15. Solve for x:
$$\sqrt{x-1} + 1 = \sqrt{2x-1}$$

16. Write the equation in standard form of the line through the points $(2,-1)$ and $(-3,-5)$.

17. Write the equation of the circle $x^2 + y^2 - 2x + 10y + 10 = 0$ in standard form. Give the center and radius.

18. Give the vertex, focus, and directrix of the parabola $x^2 - 8y - 4x = 4$.

19. Solve by graphing:

$$\begin{cases} x - 3y \geq -5 \\ 2x + 3y < 3 \end{cases}$$

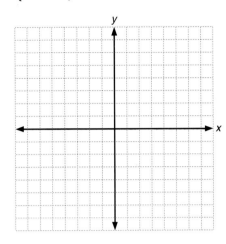

20. Solve by determinants:

$$\begin{cases} 2x - 3y = 16 \\ 5x + 2y = 2 \end{cases}$$

19. _____

20. _____

SCORE: _____

C   H   A   P   T   E   R

10

PRETEST

Before beginning this chapter, answer as many of the following questions as you can. When you have finished the chapter, take the practice test at the end of the chapter and compare the scores of the two tests to see how much you have learned.

A   N   S   W   E   R   S

1. Write in logarithmic form:
$2^3 = 8$

2. Write in exponential form:
$$\log_4 \frac{1}{16} = -2$$

1. _____

2. _____

3. Find a in each of the following:

   a. $\log_6 36 = a$

   b. $\log_3 a = 3$

   c. $\log_a 8 = 3$

   d. $\log_2 \frac{1}{8} = a$

   e. $\log 10^7 = a$

3a. _____

3b. _____

3c. _____

3d. _____

3e. _____

4. If $\log_b 2 = x$ and $\log_b 3 = y$, find the following:

   a. $\log_b 8$

   b. $\log_b 18$

4a. _____

4b. _____

5. Find the following logarithms:
  a. log 15.3

  b. log .000831

  c. log 9.05

  d. $\log_3 289$

6. Find *N:*
  a. log N = .9201

  b. log N = 4.0719

  c. log N = .8470 − 3

7. Evaluate using logarithms:
  a. $(24.7)^3$

  b. $\sqrt[5]{882}$

8. Solve for *x:*
  a. $2 \log(x + 1)$
     $= \log(5x + 11)$

  b. $2^{x+1} = 3^{2x+5}$

10

Logarithms

The Law of Natural Growth is given by the formula $A = A_0e^{rn}$, where A is the resulting amount, A_0 is the original amount, r is the rate of growth, and n is the time. If a city has a steady growth rate of 5 percent, in how many years will the population double?

The general topic of logarithms has been a part of the study of mathematics since their invention by John Napier (1550–1617). At that time their main use was to simplify certain numerical computations. They were used as the basis for early computers, and the slide rule is a classic example of the use of the logarithmic scale.

The advent of the modern electronic computer and the increasing popularity and availability of the hand-held calculator have diminished the need for the use of logarithms in arithmetic computations. However, the importance of a logarithm as a function is of eminent value in higher mathematics and many branches of the sciences make use of them.

In this chapter we will give an overview of the computational use of logarithms (this is important for an understanding of the theory of logarithms) with major emphasis on the implications of the definition of the logarithmic function.

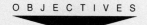 10-1 DEFINITION OF LOGARITHM

OBJECTIVES

Upon completing this section you should be able to:

1. Change an exponential statement of the form $x = b^y$ to logarithmic form.
2. Change a logarithmic statement of the form $\log_b x$ to exponential form.
3. Graph a logarithmic function.

The **logarithm** of a number x to the base b is the exponent y to which the base b is raised to obtain x. (b can represent any positive number except 1.) In symbols we may write

If $x = b^y$ then $y = \log_b x$.

We read $\log_b x$ as "the logarithm base b of x" or usually as "log base b of x." From the definition we see that

$$8 = 2^3 \quad \text{and} \quad \log_2 8 = 3$$

are equivalent statements. $8 = 2^3$ is the **exponential form** and $\log_2 8 = 3$ is the **logarithmic form.**

A sketch of the graph of the logarithmic curve might be helpful. We will sketch the graph of

$$y = \log_2 x.$$

If we first change from logarithmic form to exponential form, the equation becomes

$$x = 2^y.$$

We now set up a table of values choosing values for y and then finding the corresponding values for x.

Remember from the definition that a logarithm is an exponent.

x	$\dfrac{1}{8}$	$\dfrac{1}{4}$	$\dfrac{1}{2}$	1	2	4	8
y	-3	-2	-1	0	1	2	3

We then sketch the following curve.

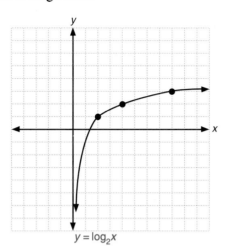

$y = \log_2 x$

The curve never actually touches the y-axis, which acts as an asymptote.

Some observations from the graph regarding the properties of $y = \log_2 x$:

1. $y = \log_2 x$ is a function since each value of x is associated with only one y value.
2. Each y value is associated with only one x value.
3. The point $(1,0)$ would be on the graph for the logarithm to any base.
4. Numbers less than or equal to zero have no logarithm. [The domain of the function is $(0, +\infty)$.]
5. Numbers greater than zero but less than 1 have negative logarithms.
6. Numbers greater than 1 have positive logarithms.
7. As the value of x increases, the value of y increases.

Take a few minutes to review these observations.

These facts would hold true for any base greater than 1. A different base would change the rate at which the curve changes direction but not its general shape.

▼ EXERCISE 10-1-1

Change to logarithmic form.

1. $3^2 = 9$ **2.** $5^2 = 25$ **3.** $5^3 = 125$ **4.** $49 = 7^2$

5. $27 = 3^3$ **6.** $64 = 2^6$ **7.** $\dfrac{1}{8} = 2^{-3}$ **8.** $3^0 = 1$

9. $\dfrac{1}{16} = 2^{-4}$ **10.** $3^1 = 3$

Change to exponential form.

11. $\log_2 4 = 2$ **12.** $\log_3 9 = 2$ **13.** $\log_4 16 = 2$ **14.** $\log_3 27 = 3$

15. $\log_3 81 = 4$ **16.** $\log_2 \frac{1}{2} = -1$ **17.** $\log_5 125 = 3$ **18.** $\log_2 32 = 5$

Find the value of the variable.

19. $\log_3 x = 4$ **20.** $\log_3 \frac{1}{3} = y$ **21.** $\log_b 16 = 4$ **22.** $\log_2 \frac{1}{8} = y$

23. $\log_b 125 = 3$ **24.** $\log_2 \frac{1}{32} = y$ **25.** $\log_3 x = -2$ **26.** $\log_2 x = -4$

27. $\log_7 x = 2$ **28.** $\log_{10} 1 = y$ **29.** $\log_{10} 10 = y$ **30.** $\log_{10} 100 = y$

31. $\log_{10} 1,000 = y$ **32.** $\log_{10} \frac{1}{10} = y$ **33.** $\log_{10} \frac{1}{100} = y$ **34.** $\log_{10} \frac{1}{1,000} = y$

Sketch the graphs.

35. $y = \log_3 x$

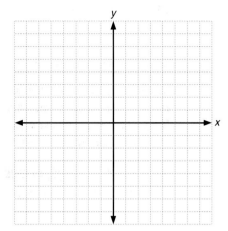

36. $y = \log_{10} x$

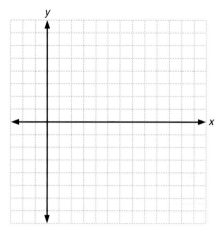

10-2 LAWS OF LOGARITHMS

From the definition of a logarithm we observe that *a logarithm is an exponent.* We then should expect that logarithms would have the properties of exponents and would follow the operational laws of exponents.

There are three laws of logarithms that concern us. They will allow us to multiply, divide, and find powers by logarithms.

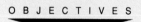

O B J E C T I V E S

Upon completing this section you should be able to:

1. State the three laws of logarithms.
2. Apply these laws to evaluate simple logarithmic expressions.

---

1. **First law of logarithms**   The logarithm of the product of two numbers is the sum of the logarithms of the two numbers. In symbols

$$\log_b(xy) = \log_b x + \log_b y.$$

2. **Second law of logarithms**   The logarithm of the quotient of two numbers (that is, a fraction) is the logarithm of the numerator minus the logarithm of the denominator. In symbols

$$\log_b \frac{x}{y} = \log_b x - \log_b y.$$

3. **Third law of logarithms**   The logarithm of a power of a number is the product of the exponent and the logarithm of the number. In symbols

$$\log_b(x)^n = n \log_b x.$$

Note that $\log_b b = 1$ since $b^1 = b$ and $\log_b 1 = 0$ since $b^0 = 1$.

These three laws are consequences of the laws of exponents. We will demonstrate how the first law is derived.

Let

$$\log_b x = k$$

and

$$\log_b y = n.$$

We change logarithmic form to exponential.

Then $\log_b x = k$ may be written as $b^k = x$ and $\log_b y = n$ may be written as $b^n = y$. Then

$$xy = b^k b^n$$
$$= b^{k + n}.$$

But $xy = b^{k + n}$ may be written as $\log_b xy = k + n$. And since $\log_b x = k$ and $\log_b y = n$

$$\log_b xy = \log_b x + \log_b y.$$

The student may wish to construct a similar proof for the other two laws. We now look at some applications of these laws.

Example 1 If $\log_b 5 = x$ and $\log_b 3 = y$, find $\log_b 15$.

Solution

First law of logarithms

$$\log_b 15 = \log_b (5)(3)$$
$$= \log_b 5 + \log_b 3$$
$$= x + y$$

Example 2 If $\log_b 5 = x$ and $\log_b 3 = y$, find $\log_b \dfrac{3}{5}$.

Second law of logarithms

Solution

$$\log_b \frac{3}{5} = \log_b 3 - \log_b 5$$
$$= y - x$$

Example 3 If $\log_b 5 = x$, find $\log_b 125$.

Solution

Third law of logarithms

$$\log_b 125 = \log_b (5)^3$$
$$= 3 \log_b 5$$
$$= 3x$$

▼ **EXERCISE 10-2-1**

Use the laws of logarithms to find an expression for the following if $\log_b 5 = x$ and $\log_b 3 = y$.

1. $\log_b 9$

2. $\log_b 25$

3. $\log_b \dfrac{3}{25}$

4. $\log_b 75$

5. $\log_b 45$ **6.** $\log_b \dfrac{5}{9}$ **7.** $\log_b \dfrac{5}{3}$ **8.** $\log_b \dfrac{9}{25}$

9. $\log_b 225$ **10.** $\log_b 375$

Using the facts that $\log_{10} 1 = 0$ and $\log_{10} 10 = 1$, evaluate.

11. $\log_{10} 100$ **12.** $\log_{10} 1{,}000$ **13.** $\log_{10} 10{,}000$ **14.** $\log_{10} \dfrac{1}{10}$

15. $\log_{10} .01$ **16.** $\log_{10} .001$ **17.** $\log_{10} \dfrac{1}{100}$ **18.** $\log_{10} 10^4$

19. $\log_{10} 10^5$ **20.** $\log_{10} 10^n$

Using the fact that $\log_{10} 3.56 = .5514$, evaluate.

21. $\log_{10}(3.56)(10^2)$ **22.** $\log_{10}(3.56)(10^{-1})$ **23.** $\log_{10}(3.56)(10^{-2})$

24. $\log_{10}(3.56)^2$ **25.** $\log_{10}(3.56)(10^{-3})$ **26.** $\log_{10} 3{,}560$

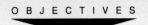 **10-3 TABLE OF LOGARITHMS**

O B J E C T I V E S

Upon completing this
section you should be able
to:

1. Understand the terms
   mantissa and
   characteristic.
2. Use a table to find the
   logarithm of a given
   number.
3. Find the antilogarithm of
   a number by using the
   table of logarithms.

You may want to review section
6–3.

In chapter 6 we showed that any decimal number can be written in scientific notation. Recall that a number is expressed in scientific notation if it is written as a product of two factors, one of which is a number equal to or greater than 1 but less than 10, and the other is a power of 10.

We therefore now have three facts that will aid us in finding logarithms of numbers from a table.

1. Every decimal number can be written in scientific notation.
2. $\log_{10}10^n = n$
3. $\log_{10}xy = \log_{10}x + \log_{10}y$

These facts help us recognize that a logarithmic table (base 10) need only contain the logarithms of numbers from 1 to 10. For instance, if we wish to find the logarithm of 368 (that is, the power of 10 that will yield 368), we first write 368 in scientific notation.

$$368 = 3.68 \times 10^2$$

We now apply the first law of logarithms.

$$\log_{10}(3.68 \times 10^2) = \log_{10}3.68 + \log_{10}10^2$$

Then since we know that $\log_{10}10^2 = 2$ we may write

$$\log_{10}(3.68 \times 10^2) = \log_{10}3.68 + 2.$$

Now if we can find the value of $\log_{10}3.68$ in the table, we will add 2 and have $\log_{10}368$. For this we need to be able to read the table of logarithms located in appendix A.

The first column of the table is headed N. In this column you will find the numbers 1 to 10. Notice that no decimals occur in the table. This is for the ease of printing a table, and you are expected to supply the decimals. The entry in column N of 36 is actually 3.6.

The other ten columns supply the second decimal or hundredths place.

To find the logarithm of 3.68 you look across the row from 3.6 until you reach the column headed 8 and find that

$$\log_{10}3.68 = .5658.$$

Notice that the decimal is also omitted in the logarithm columns. It is understood since for $1 \le N \le 10$ we have $0 \le \log_{10}N < 1$ that each entry is preceded by a decimal point.

So

$$\begin{aligned}
\log_{10}368 &= \log_{10}3.68 + 2 \\
&= .5658 + 2 \\
&= 2.5658.
\end{aligned}$$

Look at the table as you read this.

From this point on we will omit the base number for base 10. If another base is desired, the base number must be written. Logarithms, base 10, are referred to as **common logarithms.** Log 368 will imply base 10.

We do this for ease of writing.

Remember these facts about the table.

  1. Every entry in column N represents a number from 1 to 10.
  2. Every logarithm given is a number from 0 to 1.

An entry in the table (logarithm of a number from 1 to 10) is referred to as the **mantissa.** The whole number part of the logarithm (the exponent of 10 when the number is in scientific notation) is called the **characteristic.**

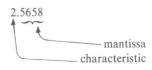

A step-by-step method of finding the logarithm of any number follows.

Step 1 Write the number in scientific notation.

Step 2 Use the table to find the mantissa.

Step 3 Add the characteristic (exponent of 10).

Example 1 Find log 5,280.

Solution
$$\log 5{,}280 = \log(5.28 \times 10^3)$$
$$= \log 5.28 + 3$$
$$= .7226 + 3$$

What is the mantissa? What is the characteristic?

This answer may be left as .7226 + 3 or written as 3.7226.

Example 2 Find log .0184.

Solution
$$\log .0184 = \log (1.84 \times 10^{-2})$$
$$= \log 1.84 - 2$$
$$= .2648 - 2$$

Notice a negative exponent is needed here.

This answer may be left as .2648 − 2 or written as −1.7352. Usually it is more convenient to leave it as .2648 − 2. For those with calculators the answer will appear as −1.7352 in the display window.

Example 3 Find log .008.

Solution
$$\log .008 = \log(8.0 \times 10^{-3})$$
$$= \log 8.0 - 3$$
$$= .9031 - 3$$

You may wish to also work these examples using a calculator that has the logarithmic function.

Example 4 Find log 93,400.

Solution
$$\log 93{,}400 = \log(9.34 \times 10^4)$$
$$= \log 9.34 + 4$$
$$= .9703 + 4$$

▼   **EXERCISE 10-3-1**

Find the logarithms of the numbers. Check your answers using a calculator having the logarithmic function.

1. 2.35   **2.** 6.82   **3.** 2.3   **4.** 7.9

5. 8.0   **6.** 3.0   **7.** 28.3   **8.** 37.1

9. 10   **10.** 38   **11.** 468   **12.** 329

13. 8,040   **14.** 3,180   **15.** 32,700   **16.** 14,900

17. .136   **18.** .014   **19.** .000508   **20.** .000682

21. .0000085   **22.** .000001   **23.** .00213   **24.** .00909

25. 683,000   **26.** 216,000

> The number having a certain logarithm is called the **antilogarithm** of the given logarithm.

In other words, find the number whose logarithm is given.

The problem "Find the antilogarithm of .8235" is the same as the problem "Find N if log N = .8235."

To find the antilogarithms we must reverse the procedure for finding logarithms. A step-by-step procedure follows.

> **Step 1** Separate the logarithm into mantissa and characteristic. (Remember that all mantissas are positive and between 0 and 1.)
> **Step 2** Use the table to find the antilogarithm of the mantissa.
> **Step 3** Use the characteristic as the exponent of 10 and multiply by the antilogarithm of the mantissa found in step 2.

Example 5 Find the antilogarithm of 3.4393.

Solution
$$3.4393 = .4393 + 3$$

The antilogarithm of .4393 is 2.75 (from the table), therefore   *Check this in the table.*

$$\text{antilog } 3.4393 = 2.75 \times 10^3 \qquad \text{(characteristic is 3)}$$
$$= 2,750.$$

Example 6 Find N if $\log N = 1.6365$.

Solution
$$1.6365 = .6365 + 1$$
$$\text{antilog } .6365 = 4.33 \qquad\qquad\qquad\qquad \textit{Again, check the table.}$$

Therefore

$$N = 4.33 \times 10^1$$
$$= 43.3.$$

Example 7 Find N if $\log N = .3284 - 3$.

Solution
$$\text{antilog } .3284 = 2.13 \qquad\qquad\qquad\qquad \textit{From the table}$$

Therefore

$$N = 2.13 \times 10^{-3} \qquad\qquad \textit{We must move the decimal point}$$
$$= .00213. \qquad\qquad\qquad\quad \textit{three places to the left.}$$

Example 8 Find the antilogarithm of -2.2984.

Solution Step 1 is a little more difficult in this case since the given logarithm is negative. It is necessary to rename -2.2984 so that the mantissa is positive. This can be accomplished by adding and then subtracting 3. Thus

$$-2.2984 = (-2.2984 + 3) - 3 \qquad\qquad \textit{Be very careful when performing}$$
$$= .7016 - 3. \qquad\qquad\qquad\qquad\quad \textit{this step.}$$

Then

$$\text{antilog } (.7016) = 5.03. \qquad\qquad\qquad \textit{From the table}$$

Therefore

$$\text{antilog } (.7016) - 3 = 5.03 \times 10^{-3}$$
$$= .00503.$$

▼ **EXERCISE 10–3–2**

Find the antilogarithm of each. Check answers using a calculator having the antilog or inverse log function.

1. .6117	**2.** .7033	**3.** .0607	**4.** .0899
5. .5888 + 1	**6.** .8306 + 1	**7.** .8698 + 2	**8.** .2625 + 2
9. .8756 − 1	**10.** .4871 − 1	**11.** .7810 − 3	**12.** .9440 − 3
13. 3.8797	**14.** 5.4843	**15.** 1.6435	**16.** 2.5911
17. −1.3468	**18.** −2.4989	**19.** −3.0958	**20.** −.2426

◣◥ 10–4 COMPUTATIONS USING LOGARITHMS

O B J E C T I V E S
▼

Upon completing this section you should be able to:

1. Use logarithms to multiply two numbers.
2. Use logarithms to divide two numbers.
3. Use logarithms to find the power of a number.
4. Use logarithms to find a root of a number.

We will use these properties again in section 10–6.

In the "pre-calculator age" (only a little more than a quarter of a century ago), complicated computations utilized logarithms or a *slide rule,* which is based on logarithms. Although the solutions were not exact, they were accurate enough for the desired purpose at the time.

All of the following examples and exercises can be solved more easily using a calculator. We shall use logarithms only for the purpose of reinforcing the laws of logarithms. It may be of interest to check your solutions with a calculator to see just how accurate these answers are. The logarithms for the examples and answers to the exercises are all taken from the table of logarithms in appendix A.

Two important properties will be used in much of our work with logarithms.

1. If $M = N$ (both M and N are positive), then $\log_b M = \log_b N$.
2. If $\log_b M = \log_b N$, then $M = N$.

Example 1 Evaluate using logarithms: $(5.280)(361.0)$

Solution Since the log of a product is equal to the sum of the logs, we may write

$$
\begin{aligned}
\log(5.280)(361.0) &= \log 5.28 + \log 361 \\
&= \log 5.28 + \log(3.61 \times 10^2) \\
&= .7226 + 2.5575 \\
&= 3.2801 \\
&= .2801 + 3.
\end{aligned}
$$

First law of logarithms

We now refer to the table and note that .2801 is not an entry. However, it is between the entries .2788 and .2810. If we were stressing accuracy, rather than just reinforcing the laws, we would resort to a process called *interpolation.* But for our purposes we will use the nearest entry, which is .2810, giving the antilog 1.91. Thus .2801 + 3 = log 1,910.

Which entry is closest to .2801?

We now have the statement

$$
\log(5.280)(361.0) = \log 1{,}910.
$$

If we use the second property, we may write

$$
(5.280)(361.0) = 1{,}910
$$

which is the desired result.

If the logs of two numbers are equal, then the numbers are equal.

Example 2 Evaluate using logarithms: $\dfrac{.0065}{.0732}$

Solution Using the second law of logarithms, we write

$$
\begin{aligned}
\log \frac{.0065}{.0732} &= \log .0065 - \log .0732 \\
&= \log(6.5 \times 10^{-3}) - \log(7.32 \times 10^{-2}) \\
&= (.8129 - 3) - (.8645 - 2) \\
&= -.0516 - 1 \\
&= .9484 - 2 \text{ (by adding and subtracting 1)} \\
&= \log(8.88 \times 10^{-2}) \\
&= \log .0888.
\end{aligned}
$$

The logarithm of a quotient is the difference of the logarithms.

Therefore,

$$
\frac{.0065}{.0732} = .0888.
$$

▼ **EXERCISE 10–4–1**

Evaluate using logarithms.

1. (3.2)(7.5)

2. (8.5)(2.4)

3. (2.3)(5.2)

4. (4.3)(6.4)

5. (5.19)(9.01)

6. (3.71)(6.03)

7. $\dfrac{8.4}{2.5}$

8. $\dfrac{6.8}{4.5}$

9. $\dfrac{1.7}{5.3}$

10. $\dfrac{3.2}{8.1}$

11. $\dfrac{3.57}{2.11}$

12. $\dfrac{9.13}{2.38}$

13. (25)(42)

14. (64)(35)

15. (23)(58)

16. (21)(56)

17. $\dfrac{86}{25}$

18. $\dfrac{78}{15}$

19. $(593)(26.8)$ **20.** $(13.2)(406)$ **21.** $\dfrac{29.3}{2.69}$

22. $\dfrac{.0154}{.0023}$ **23.** $\dfrac{.254}{34.6}$ **24.** $\dfrac{.314}{5.29}$

The third law of logarithms states

$$\log x^n = n \log x.$$

This law is used where n is any rational number.

Example 3   Evaluate using logarithms: $(.0432)^8$

Solution
$$\begin{aligned}
\log(.0432)^8 &= 8 \log .0432 \\
&= 8(.6355 - 2) \\
&= 5.0840 - 16 \\
&= .0840 - 11 \\
&= \log(1.21 \times 10^{-11}) \\
&= \log .0000000000121
\end{aligned}$$

Note that .0840 is between .0828 and .0864 but is closest to .0828.

Therefore

$$(.0432)^8 = .0000000000121.$$

Example 4   Evaluate using logarithms: $\sqrt[5]{125}$

Solution
$$\begin{aligned}
\log \sqrt[5]{125} &= \log(125)^{1/5} \\
&= \frac{1}{5}\log 125 \\
&= \frac{1}{5}(2.0969) \\
&= .4194 \\
&= \log 2.63
\end{aligned}$$

Recall that $\sqrt[n]{x} = x^{1/n}$.

Note that .4194 is nearest to .4200.

Therefore

$$\sqrt[5]{125} = 2.63.$$

◥ EXERCISE 10-4-2

Evaluate using logarithms. Check answers using a calculator with a power function (remember you may convert roots to fractional powers).

1. $(2.5)^2$

2. $(1.5)^2$

3. $(3.16)^2$

4. $(7.04)^2$

5. $(2.15)^3$

6. $(4.26)^3$

7. $(16.3)^2$

8. $(23.1)^2$

9. $(.054)^2$

10. $(.038)^2$

11. $(.21)^4$

12. $(.64)^4$

13. $(15.3)^5$

14. $(12.6)^5$

15. $(.053)^7$

16. $(.075)^7$

17. $\sqrt{6.76}$

18. $\sqrt{9.61}$

19. $\sqrt{256}$

20. $\sqrt{324}$

21. $\sqrt[3]{29.6}$

22. $\sqrt[3]{42.7}$

23. $\sqrt{2,416}$

24. $\sqrt{1,090}$

25. $\sqrt[5]{2,230}$

26. $\sqrt[5]{536}$

27. $\sqrt{.0951}$

28. $\sqrt{.0483}$

10-5 LOGARITHMS TO DIFFERENT BASES

As mentioned earlier, logarithms to the base 10 are referred to as common logarithms. We now wish to introduce other bases and use common logarithms to compute logarithms to these other bases.

Changing the base of a logarithm is accomplished by the following formula.

$$\log_b x = \frac{\log_a x}{\log_a b}$$

The formula is obtained as follows.

Let

$$\log_b x = N$$

then

$$b^N = x.$$

OBJECTIVES

Upon completing this section you should be able to:

1. State the formula for finding the logarithm of a number to any desired base.
2. Find logarithms of numbers to various bases using the base 10 table.
3. Find logarithms to base e.

If we then take the log of each side to the base *a,* we obtain

$$\log_a b^N = \log_a x$$

or

$$N \log_a b = \log_a x.$$

Then solving for *N,* we obtain the desired result.

$$N = \frac{\log_a x}{\log_a b}$$

Example 1   Find $\log_5 45$.

Solution   It is not necessary to have a table of logarithms to base 5 to find this logarithm. We may use our base-10 table by applying the preceding formula.

Using the formula,

$$\log_5 45 = \frac{\log 45}{\log 5}.$$

We find the values of log 45 and log 5 in our base-10 table, obtaining

$$\log_5 45 = \frac{1.6532}{.6990}.$$

Hence

$$\log_5 45 = 2.365.$$

Note that the original problem, find $\log_5 45$, could have been stated

$$5^x = 45, \text{ find } x$$

since

$$\log_5 45 = 2.365$$

in exponential form is

$$5^{2.365} = 45.$$

The number *e* (an irrational number) is often used in higher mathematics as a base for logarithmic tables. *e* correct to four digits is 2.718 and

$$\log e = .4343.$$

With this value we can change from base 10 (common logarithms) to base *e* (natural logarithms). Since natural logarithms are used so much in mathematics, we use special notation to represent them. ln *x* means "the logarithm of *x* to the base *e*." Thus ln $x = \log_e x$.

WARNING

Do not make the mistake of subtracting these numbers. $\dfrac{\log 45}{\log 5}$ has no relationship to log 45 − log 5.

A logarithm is an exponent.

This is an approximation.

Some calculators have an "ln *x*" button that will give you a logarithm to base *e*.

Example 2   Find ln 25.

Solution
$$\ln 25 = \frac{\log 25}{\log e}$$
$$= \frac{1.3979}{.4343}$$

Hence

$$\ln 25 = 3.219.$$

▼ **EXERCISE 10–5–1**

Find the logarithms.

1. $\log_2 8$

2. $\log_2 16$

3. $\log_3 5.6$

4. $\log_3 8.4$

5. $\log_7 8.92$

6. $\log_8 3.68$

7. $\log_2 29$

8. $\log_2 75$

9. $\log_5 35$

10. $\log_5 16$

11. $\log_4 94.7$

12. $\log_6 20.9$

13. $\log_9 613$

14. $\log_7 537$

15. $\log_{11} 155$

16. $\log_{13} 483$

17. $\log_3 .035$

18. $\log_5 .061$

19. $\log_{24} 793$

20. $\log_3 817$

21. $\ln 8.19$

22. $\ln 4.36$

23. $\ln 346$

24. $\ln 595$

25. $\ln 50.1$

26. $\ln 24.1$

27. $\ln .0082$

28. $\ln .0368$

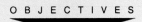 **10–6 EXPONENTIAL AND LOGARITHMIC EQUATIONS**

In certain equations the unknown appears as an exponent or in a logarithmic expression. Many such equations may be solved using the methods in the following examples.

Example 1   Solve for x: $\log(x + 1) = \log x + \log 3$

Solution   In general, when the unknown appears in a logarithmic expression, we can find a solution by taking the antilog of both sides of the equation. This requires that we apply the laws of logarithms in reverse.

$\log x + \log 3 = \log 3x.$

$$\log(x + 1) = \log x + \log 3$$
$$\log(x + 1) = \log 3x$$
$$x + 1 = 3x$$
$$x = \frac{1}{2}$$

The solutions for equations such as this should be checked by substituting the value back into the original equation.

Checking a solution is a very important step.

$$\log(x + 1) = \log\left(\frac{1}{2} + 1\right) = \log\frac{3}{2}$$

Also,

$$\log x + \log 3 = \log\frac{1}{2} + \log 3 = \log\left(\frac{1}{2}\right)(3) = \log\frac{3}{2}.$$

Therefore, $\frac{1}{2}$ is a solution to the equation.

Example 2   Solve for x: $\log(2x - 1) = \log(x + 2) + \log(x - 2)$

If the logs of two quantities are equal, then the quantities must be equal.

Solution
$$\log(2x - 1) = \log(x + 2)(x - 2)$$
$$2x - 1 = (x + 2)(x - 2)$$
$$2x - 1 = x^2 - 4$$
$$x^2 - 2x - 3 = 0$$
$$(x - 3)(x + 1) = 0$$
$$x = 3 \text{ or } x = -1$$

We see here why checking is important.

Checking our answers, we see that 3 is a solution. However, if we substitute $x = -1$ in the last term of the equation, $\log(x - 2)$, we obtain $\log(-3)$. But we cannot find the logarithm of a negative number. Thus, -1 is not a solution of the original equation.

Therefore, the solution is 3.

Example 3   Solve for x: $3^x = 5$ This is an exponential equation.

Solution   In this equation the unknown occurs in the exponent. In such cases it is usually necessary to take the log of both sides of the equation.

$$3^x = 5$$
$$\log 3^x = \log 5$$ If two quantities are equal, then their logs are equal.
$$x \log 3 = \log 5$$
$$x = \frac{\log 5}{\log 3}$$ This answer is an exact value.

We will leave the solution in terms of logarithms. To find a numerical answer would require only that we use the table to find log 5 and log 3 and then divide.

Example 4   Solve for x: $2^{x+1} = 8$

Solution   Taking the log of each side, we obtain

$$\log 2^{x+1} = \log 8$$
$$\log 2^{x+1} = \log 2^3$$ 8 can be written as 2^3.
$$(x + 1)\log 2 = 3 \log 2.$$

Dividing each side by log 2 yields

$$x + 1 = 3$$
$$x = 2.$$ Check this value in the original equation.

Example 5   Solve for x: $3^{x-4} = 2^{x+1}$

Solution   Taking the log of each side, we obtain

$$\log 3^{x-4} = \log 2^{x+1}$$
$$(x - 4)\log 3 = (x + 1)\log 2$$
$$x \log 3 - 4 \log 3 = x \log 2 + \log 2$$ We wish to get the x terms on one side of the equation.
$$x \log 3 - x \log 2 = \log 2 + 4 \log 3$$
$$x(\log 3 - \log 2) = \log 2 + 4 \log 3$$
$$x = \frac{\log 2 + 4 \log 3}{\log 3 - \log 2}.$$

Applications of logarithms are found in areas of business and science. The following example is an illustration of the use of logarithms in biology.

Example 6   The number N of bacteria present in a certain culture at the end of t hours is given by $N = N_0 \times 10^{.01t}$, where N_0 is the original number of bacteria. How many hours will it take for the number of bacteria to double?

We must solve for t when $N = 2N_0$.

Solution   When the number of bacteria doubles, the value of N will be twice that of N_0. Thus, the formula will read

$$2N_0 = N_0 \times 10^{.01t}.$$

Dividing each side by N_0, we obtain

When the variable is in the exponent, take the log of each side of the equation.

$$2 = 10^{.01t}.$$

We now take the logarithm of each side, obtaining

$$\log 2 = \log 10^{.01t}.$$

Using the third law of logarithms, we have

$$\log 2 = .01t \log 10.$$

Since $\log 10 = 1$, we may write

$$\log 2 = .01t$$

Find log 2 from the table.   or $$t = \frac{\log 2}{.01}.$$

Evaluating log 2 yields

$$t = \frac{.3010}{.01} = 30 \text{ hours.}$$

▼ EXERCISE 10–6–1

Solve the equations.

1. $\log(x + 2) = \log x + \log 2$ **2.** $\log(x + 3) = \log x + \log 5$

3. $\log(5x - 3) = \log(2x + 1) + \log 2$ **4.** $\log(3x + 1) = \log(x + 2) + \log 4$

5. $\log(x + 3) = \log(2x - 1) - \log 3$

6. $\log(2x - 4) = \log(x - 2) + \log(x + 3)$

7. $\log(x + 5) - \log 2 = \log(5x + 2)$

8. $2 \log x = \log 9$

9. $2 \log(x + 3) = \log(x + 1) + \log(x + 4)$

10. $2 \log(x + 1) = \log(5x + 11)$

11. $\log(x - 1) + \log(x - 2)$
$= \log 2 + \log (2x - 5)$

12. $\log(x + 6) = \log(x + 20) - \log(x + 2)$

13. $3^{x-2} = 9$

14. $5^{2x-1} = 125$

15. $2^x = 7$

16. $6^{x-1} = 11$

17. $2^{3x+4} = 9$

18. $7^{2x-3} = 1$

19. $2^{x+1} = 3^{2x+5}$

20. $5^{2x-1} = 11^{x+5}$

21. $3^{x-2} = 15^{2x-1}$

22. $\log(x+3) + \log 5 = 2$

23. $\log 3 = 3 - \log(x-2)$

24. $\log x + \log(x+15) = 2$

25. In chemistry the pH (hydrogen potential) of a solution is defined to be pH $= -\log[\text{H}^+]$, where $[\text{H}^+]$ is the concentration of hydrogen ions in moles per liter. Find the pH of a solution in which the concentration of hydrogen ions is 3.0×10^{-4} moles per liter.

26. The number of bacteria N in a certain culture at the end of t hours is given by $N = 1000 \times 10^{0.1t}$. Find the number of bacteria present at the end of three hours.

27. A radioactive substance decomposes according to the formula $N = N_0 e^{-.001t}$, where N represents the number of milligrams remaining, N_0 represents the original number of milligrams, and t represents the number of years. Find the number of milligrams remaining if there were 2 grams of the substance 1,000 years ago.

28. The *half-life* of a radioactive substance is defined as the length of time it takes for half of the substance to decompose. Find the half-life of the substance referred to in problem 27.

29. The amount A present in an account that pays r percent annual interest for n years is given by the formula $A = p(1 + r)^n$, where p is the principle invested. If the annual interest rate is 8 percent, how many years will it take an investment to double in value?

30. The intensity I (in lumens) of a beam of light passing through a substance is given by the formula $I = I_0 e^{-at}$, where I_0 is the intensity of the light before passing through the substance, a is the absorption coefficient of the substance, and t is the thickness (in centimeters) of the substance. If the initial intensity is 1,000 lumens and the absorption coefficient of the substance is .2, what thickness is necessary to reduce the illumination to 600 lumens?

C H A P T E R
10   **SUMMARY**

Key Words

Section 10–1

- The **logarithm** of a number x to the base b is the exponent to which the base b is raised to obtain x $(b > 0, b \neq 1)$.

- The **exponential form** $x = b^y$ and the **logarithmic form** $y = \log_b x$ are equivalent.

- The **logarithmic function** has as its domain $(0, +\infty)$. Its range is $(-\infty, +\infty)$.

Section 10–2

- The **laws of logarithms** are derived from the laws of exponents.

Section 10–3

- **Common logarithms** are logarithms to the base 10. log a written without the base is understood to be $\log_{10} a$.

- The **characteristic** and **mantissa** are the two parts of the logarithm of a number.

- The **antilogarithm** of a logarithm is the number that has that logarithm.

Section 10–5

- The **natural logarithm** (written ln x) of a number x is $\log_e x$.

Section 10–6

- A **logarithmic equation** is an equation in which the variable appears in a logarithmic expression.

- An **exponential equation** is an equation in which the variable appears in an exponent.

Procedures

Section 10–2

- The laws of logarithms:
  1. $\log_b(xy) = \log_b x + \log_b y$
  2. $\log_b\left(\dfrac{x}{y}\right) = \log_b x - \log_b y$
  3. $\log_b(x)^n = n \log_b x$

Section 10–3

- To find the common logarithm of a number:
  Step 1 Write the number in scientific notation.
  Step 2 Use the table to find the mantissa.
  Step 3 Add the characteristic (exponent of 10).

- To find the antilogarithm of a given logarithm:
  Step 1 Separate the logarithm into mantissa and characteristic. (Remember that all mantissas are positive and between 0 and 1.)
  Step 2 Use the table to find the antilogarithm of the mantissa.
  Step 3 Use the characteristic as the exponent of 10 and multiply by the antilogarithm of the mantissa found in step 2.

Section 10–4

- Computations using logarithms depend on two properties:
  1. If $M = N$ (both M and N are positive), then $\log_b M = \log_b N$.
  2. If $\log_b M = \log_b N$, then $M = N$.

Section 10–5

- To change the base of a logarithm use the formula

$$\log_b(x) = \frac{\log_a x}{\log_a b}.$$

Section 10–6

- To solve a logarithmic equation take the antilog of both sides, then proceed to solve for the variable. The solution thus obtained must check in the original equation to be valid.

- To solve an exponential equation take the logarithm of both sides and solve the resulting equation.

Solve for the variable.

A N S W E R S

1. $\log_4 x = 3$ **2.** $\log_2 8 = y$ **3.** $\log_b 81 = 2$

1. _____

2. _____

4. $\log_a 125 = 3$ **5.** $\log_2 x = -3$ **6.** $\log_3 \dfrac{1}{27} = y$

3. _____

4. _____

5. _____

If $\log_b 2 = x$ and $\log_b 3 = y$, find the following.

6. _____

7. $\log_b 6$ **8.** $\log_b \dfrac{2}{3}$ **9.** $\log_b \dfrac{2}{9}$

7. _____

8. _____

10. $\log_b 8$ **11.** $\log_b 72$ **12.** $\log_b \dfrac{4}{27}$

9. _____

10. _____

11. _____

12. _____

Find the logarithms (base 10) of the following numbers.

13. 8.04   **14.** 38.6   **15.** 49,000

13. _____

14. _____

16. .803   **17.** .0062   **18.** .000156

15. _____

16. _____

17. _____

Find N.

18. _____

19. $\log N = .6405$ **20.** $\log N = .9474$ **21.** $\log N = 3.1761$

19. _____

20. _____

21. _____

585

A N S W E R S

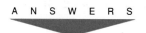

22. $\log N = 2.9581$ **23.** $\log N = .8681 - 2$ **24.** $\log N = -4.7471$

22. _____

23. _____

24. _____

25. _____

Evaluate using logarithms.

25. $(2.7)(8.1)$ **26.** $(25)(470)$ **27.** $(.013)(6.24)$

26. _____

27. _____

28. $(.152)(7.28)$ **29.** $\dfrac{.034}{.263}$ **30.** $\dfrac{381}{29.3}$

28. _____

29. _____

30. _____

31. $(38.7)^3$ **32.** $(.052)^3$ **33.** $(2.09)^4$

31. _____

32. _____

34. $\sqrt{529}$ **35.** $\sqrt[3]{71.4}$ **36.** $\sqrt[3]{.00342}$

33. _____

34. _____

35. _____

Find the logarithms.

37. $\log_4 104$ **38.** $\log_2 33.6$ **39.** $\log_5 2.27$

36. _____

37. _____

38. _____

40. $\ln 912$ **41.** $\ln .826$ **42.** $\ln 38.7$

39. _____

40. _____

41. _____

Solve.

42. _____

43. $\log(3x - 1)$ **44.** $\log(x - 6)$
$= \log 2 + \log(x + 3)$ $= \log(x + 4) - \log 3$

43. _____

44. _____

45. $\log(x + 1) + \log(x + 2)$
$= \log(7x + 23)$

46. $\log(4x + 1)$
$= \log(x - 5) + \log(x + 3)$

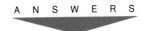

45. _____

47. $3^{x+1} = 7$

48. $2^{x-1} = 5^{2x+1}$

46. _____

47. _____

49. The Law of Natural Growth
is given by the formula
$A = A_0 e^{rn}$, where A is the
resulting amount, A_0 is the
original amount, r is the rate of
growth, and n is the time. If a
city has a steady growth rate of
5 percent, in how many years
will the population double?

50. The amount A of radioactive
carbon 14 remaining at any
time t is given the formula
$A = A_0(2)^{-t/5,750}$, where A_0 is
the amount that was originally
present. An animal's tooth was
tested and found to have lost 60
percent of its carbon 14 content.
How old is the animal?

48. _____

49. _____

50. _____

SCORE: _____

1. _____

2. _____

3a. _____

3b. _____

3c. _____

3d. _____

3e. _____

4a. _____

4b. _____

5a. _____

5b. _____

5c. _____

5d. _____

6a. _____

6b. _____

6c. _____

1. Write in logarithmic form:
$3^2 = 9$

2. Write in exponential form:
$\log_2 \dfrac{1}{4} = -2$

3. Find a:
   a. $\log_4 16 = a$ **b.** $\log_3 a = 2$

   c. $\log_a 27 = 3$ **d.** $\log_2 \dfrac{1}{16} = a$

   e. $\log 10^5 = a$

4. If $\log_b 2 = x$ and $\log_b 3 = y$, find:
   a. $\log_b 9$ **b.** $\log_b 12$

5. Find the logarithms:
   a. $\log 24.7$ **b.** $\log .000156$

   c. $\log 7.14$ **d.** $\log_4 316$

6. Find N:
   a. $\log N = .8331$ **b.** $\log N = 3.4742$

   c. $\log N = .7364 - 2$

7. Evaluate using logarithms:
  a. $(23.6)^3$
  b. $\sqrt{547}$

8. Solve for x:
  a. $\log(x + 2)$
     $= \log 5 + \log x$
  b. $3^{x-1} = 81$

7a. _____

7b. _____

8a. _____

8b. _____

SCORE: _____

PRETEST

Before beginning this chapter, answer as many of the following questions as you can. When you have finished the chapter, take the practice test at the end of the chapter and compare the scores of the two tests to see how much you have learned.

A N S W E R S

1. Use mathematical induction to prove
$$1 + \frac{3}{2} + 2 + \cdots + \frac{n+1}{2} = \frac{n(n+3)}{4}.$$

2. Write the first four terms of the sequence whose nth term is defined as $a_n = n^2 - 1$.

3. Evaluate: $\displaystyle\sum_{i=0}^{4} (3i - 4)$

4. Express the series $2 + 5 + 10 + 17$ in sigma notation.

5. If the first term of an arithmetic sequence is 30 and the constant difference is -6, find the 10th term.

6. The 3rd term of an arithmetic sequence is 15 and the 14th term is 59. Find the sum of the first eight terms.

7. In an arithmetic sequence if $a_1 = 2$, $a_n = 32$, and $S_n = 374$, find n.

8. Find the sum of the first six terms of the geometric sequence in which $a_1 = 4$ and $r = 2$.

1. _____

2. _____

3. _____

4. _____

5. _____

6. _____

7. _____

8. _____

9. Insert four geometric means between -4 and 128.

10. Find the sum of the infinite geometric series 40, $-10, \dfrac{5}{2}, \ldots$.

9. _____

10. _____

11. Find the eighth term of $(2x - 1)^{10}$.

12. Find the value of $(1.03)^5$ to three decimal places using the binomial expansion.

11. _____

12. _____

SCORE: _____

Sequences, Series, and the Binomial Theorem

A man borrows $10,000 for 10 years and agrees to pay back $1,000 principal each year plus simple interest at 8% on all principal left unpaid during the year. How much interest does he pay over the 10 year period?

M ost of our work in algebra involves the set of real numbers. We have also done some work involving the set of complex numbers. These sets of numbers were developed rather late on the time scale of the history of mathematics. The first set of numbers used by man was probably the set of counting numbers $\{1, 2, 3, \ldots\}$. In this chapter we will encounter mathematical ideas that center around the set of counting numbers and the set of nonnegative integers $\{0, 1, 2, \ldots\}$.

11-1 MATHEMATICAL INDUCTION

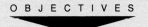

OBJECTIVES

Upon completing this section you should be able to:

1. Understand the principle of mathematical induction.
2. Prove theorems using mathematical induction.

The student is already aware of the fact that one exception to a mathematical rule or formula is all that is necessary to prove it to be false. This means that a statement, to be true about a set, must be true for every element within the set. Whenever a set is infinite, it is, of course, not possible to check the statement for every element of the set. **Mathematical induction** is a method of proof used when we have a statement concerning the infinite set of positive integers. This method is based on the following theorem.

> *The principle of mathematical induction:* If a statement P is true for the integer 1 and if the assumption that P is true for the integer k implies that P is true for $(k + 1)$, then P is true for the set of positive integers.

A theoretical example might make the principle more meaningful.

Suppose there is a stepladder with infinitely many steps. What two things would you need to know so that you could climb the ladder to any height? A little thought should produce the following two items.

1. You must know that you can at least get on the first step of the ladder.
2. You must know that if you are on any step of the ladder, you can get to the next step.

A little reflection will convince you that these two concepts are not only necessary but also sufficient for you to climb the ladder to any height. Do you see that they represent the statement of the principle of mathematical induction?

We now look at a mathematical example.

Example 1   Prove that the sum of integers 1 through n is given by the formula $\dfrac{(n^2 + n)}{2}$. In other words, show that

$$1 + 2 + 3 + \cdots + n = \frac{n^2 + n}{2}.$$

Solution   **Step 1** Is the formula true for $n = 1$?

$$\frac{1^2 + 1}{2} = \frac{1 + 1}{2} = 1$$

We substitute 1 for n.

Thus the formula holds true for $n = 1$.

Step 2 Assume the formula is true for $n = k$.

$$1 + 2 + 3 + \cdots + k = \frac{k^2 + k}{2}$$

We substitute k for n.

Step 3 We must now show that the assumption of step 2 will imply that the sum of the first $(k + 1)$ integers is $\dfrac{(k + 1)^2 + (k + 1)}{2}$. We know from step 2 that

$$1 + 2 + 3 + \cdots + k = \frac{k^2 + k}{2}. \longleftarrow$$

This is the form $\dfrac{n^2 + n}{2}$ where $n = k + 1$.

Now if we add $(k + 1)$ to each side, we obtain

$$1 + 2 + 3 + \cdots k + (k + 1) = \frac{k^2 + k}{2} + (k + 1)$$

We combine the terms on the right side into a single term.

$$= \frac{k^2 + k + 2k + 2}{2}$$

$$= \frac{k^2 + 3k + 2}{2}.$$

At this point we should always reflect on the form we desire and use any necessary algebraic manipulations to obtain it. The sum of the first $(k + 1)$ integers must fit the original formula, $\dfrac{(n^2 + n)}{2}$, with $n = (k + 1)$. We thus write $k^2 + 3k + 2$ as $k^2 + 2k + 1 + k + 1$ and factor by grouping the first three terms as $(k + 1)^2 + (k + 1)$. We now have the desired result

Recall what we want the final form to look like.

$$1 + 2 + 3 + \cdots + (k + 1) = \frac{(k + 1)^2 + (k + 1)}{2},$$

This is the desired form.

which shows that the formula holds for $n = k + 1$. Thus by the principle of mathematical induction the formula is true for n equal to any integer.

Any and all manipulative skills of algebra can be used to rearrange the right side of step 3 to obtain the proper form. A great deal of ingenuity is sometimes required.

Example 2 Prove $n < 2^n$ for all positive integers n.

Solution

Step 1 Show the statement is true for $n = 1$.

We see this is a true statement.

$$1 < 2^1 \quad \text{or} \quad 1 < 2$$

Step 2 Assume true for $n = k$.

$$k < 2^k$$

Again, this is the form we want where $n = k + 1$.

Step 3 Show that the assumption of step 2 implies that $k + 1 < 2^{k+1}$. From step 2 we have

$$k < 2^k.$$

We add the two inequalities.

We also know that $1 < 2^k$, so we may add this to the inequality $k < 2^k$ obtaining

$$k + 1 < 2^k + 2^k$$
$$k + 1 < 2^k(1 + 1)$$

When multiplying like bases, we add the exponents.

$$k + 1 < 2^k(2)$$
$$k + 1 < 2^{k+1}.$$

Therefore the statement is true for all n.

▼ **EXERCISE 11–1–1**

Prove the statements by using mathematical induction.

1. $1 + 3 + 5 + \cdots + (2n - 1) = n^2$

2. $2 + 6 + 10 + \cdots + (4n - 2) = 2n^2$

3. $1 + 4 + 7 + \cdots + (3n - 2) = \dfrac{3n^2 - n}{2}$

4. $1^3 + 2^3 + 3^3 + \cdots + n^3 = \dfrac{n^2(n + 1)^2}{4}$

5. $\dfrac{1}{1(2)} + \dfrac{1}{2(3)} + \dfrac{1}{3(4)} + \cdots$
$+ \dfrac{1}{n(n+1)} = \dfrac{n}{n+1}$

6. $3 + 3^2 + 3^3 + \cdots + 3^n = \dfrac{3(3^n - 1)}{2}$

7. $1(2) + 2(3) + 3(4) + \cdots + n(n+1)$
$= \dfrac{n(n+1)(n+2)}{3}$

8. $1 < x^n$ for all $x > 1$

9. $1 + 4n \le 5^n$

10. $1 + x + x^2 + \cdots + x^{n-1} = \dfrac{x^n - 1}{x - 1}$
$x \ne 1$

◣ 11-2 SEQUENCES AND SERIES

O B J E C T I V E S

Upon completing this section you should be able to:

1. Distinguish between a sequence and a series.
2. Find any term of a designated sequence.
3. Use sigma notation.

A second topic involving the nonnegative integers is that of sequences and series. A sequence can be defined as any array of numbers such that there is a first, second, third, and so forth. A more formal definition follows.

> A **sequence** is a function whose domain is the set of positive integers.

Note that the definition of a function would require one term for each nonnegative integer. $f(1)$ would be the first term, $f(2)$ the second term, and so on. Subscripts are generally used to designate the number of a term. Instead of $f(1)$, $f(2)$, $f(3)$, . . . , $f(n)$, it is generally customary to use a_1, a_2, a_3, . . . , a_n to designate a sequence with n terms.

> A sequence with a finite number of terms is called a **finite sequence,** and a sequence with an infinite number of terms is called an **infinite sequence.**

To avoid any misunderstanding about ordering we will designate the nth term (a_n) of a sequence by an algebraic formula.

Example 1   Write the first four terms and the tenth term of $a_n = \dfrac{n}{n + 1}$.

To find any given term we substitute the number of that term for n.

Solution

$$a_1 = \frac{1}{1 + 1} = \frac{1}{2}$$

$$a_2 = \frac{2}{2 + 1} = \frac{2}{3}$$

$$a_3 = \frac{3}{3 + 1} = \frac{3}{4}$$

$$a_4 = \frac{4}{4 + 1} = \frac{4}{5}$$

$$\vdots$$

$$a_{10} = \frac{10}{10 + 1} = \frac{10}{11}$$

A *recursive* definition of a sequence occurs when we give the first term and then give a method of finding the $(k + 1)$ term from the kth term.

Example 2   Find the first four terms of the sequence if $a_1 = 1$ and $a_{k+1} = 3a_k - 1$.

Solution

$$a_1 = 1$$
$$a_2 = 3a_1 - 1 = 3(1) - 1 = 2$$
$$a_3 = 3a_2 - 1 = 3(2) - 1 = 5$$
$$a_4 = 3a_3 - 1 = 3(5) - 1 = 14$$

We could, in this manner, find as many terms of the sequence as we wish. However, if we wanted to find the twentieth term, we would need to first find the first nineteen terms. It is desirable to have a formula that would give us the nth term without having to find all the preceding terms.

For the previous sequence the formula $a_n = \dfrac{1}{2}(3^{n-1} + 1)$ is valid for the terms we have obtained. To prove this formula for the entire sequence we must use mathematical induction.

Check this formula with the first four terms in example 2.

Step 1 Show the formula holds for $n = 1$.

$$a_1 = \frac{1}{2}(3^{1-1} + 1) = \frac{1}{2}(2) = 1$$

Step 2 Assume true for $n = k$.

$$a_k = \frac{1}{2}(3^{k-1} + 1)$$

Step 3 Show that the $(k + 1)$ term is $\dfrac{1}{2}[3^{(k+1)-1} + 1]$. We know from the recursive definition that the $(k + 1)$ term is $3a_k - 1$. Therefore using the value for a_k assumed in step 2, the $(k + 1)$ term is

$$3a_k - 1 = 3\left[\frac{1}{2}(3^{k-1} + 1)\right] - 1$$

$$= \frac{1}{2}(3^k + 3) - 1$$

$$= \frac{1}{2}(3^k + 3 - 2)$$

$$= \frac{1}{2}(3^k + 1)$$

$$= \frac{1}{2}[3^{(k+1)-1} + 1].$$

Again, notice the importance of knowing the form of the expression we are trying to obtain.

This last expression is our formula with $n = (k + 1)$, and hence by mathematical induction the formula holds for all nonnegative integers n.

A **series** is the indicated sum of a sequence.

Example 3   If the sequence is 1, 3, 5, 7, 9, then the series is $1 + 3 + 5 + 7 + 9$.

A special notation is sometimes used to indicate a series. It is especially useful when we have an expression for the general term of a sequence. The uppercase Greek letter sigma (Σ) is used to indicate a sum.

This says "find the values of $(2n - 1)$ when $n = 1, 2, 3, 4,$ and 5 and then find their sum."

Example 4   Find $\displaystyle\sum_{n=1}^{5} (2n - 1)$.

The notation is read as "Find the sum of the terms $2n - 1$ as n takes on successive integral values from 1 to 5."

Solution   We get

$$[2(1) - 1] + [2(2) - 1] + [2(3) - 1] + [2(4) - 1] + [2(5) - 1]$$
$$= 1 + 3 + 5 + 7 + 9$$
$$= 25.$$

Therefore

This is a neat and compact notation.

$$\sum_{n=1}^{5} (2n - 1) = 25.$$

Note that any letter may be used to indicate the variable.

Find this sum.

Example 5 $\displaystyle\sum_{i=1}^{8} i^2 = 1^2 + 2^2 + 3^2 + \cdots + 8^2$

Find this sum.

Example 6 $\displaystyle\sum_{k=1}^{4} (k + 1)^2 = 2^2 + 3^2 + 4^2 + 5^2$

▼ **EXERCISE 11-2-1**

Write the first four terms and the tenth term of the sequence whose nth term is defined as follows.

1. $a_n = 3n - 2$

2. $a_n = \dfrac{n}{2} + 1$

3. $a_n = n^2 + 3$

4. $a_n = 2n^2 - 5n + 1$

5. $a_n = \dfrac{\log n}{2n}$

Find the first four terms of each of the sequences defined recursively as follows.

6. $a_1 = 2,\ a_{k+1} = 3a_k - 4$

7. $a_1 = -4,\ a_{k+1} = 5a_k + 1$

8. $a_1 = \dfrac{1}{2},\ a_{k+1} = \dfrac{a_k}{2}$

9. $a_1 = 1,\ a_{k+1} = (a_k)^2 + 3$

10. $a_1 = 2,\ a_{k+1} = ka_k - 1$

Find the value.

11. $\displaystyle\sum_{n=1}^{4} 5n$

12. $\displaystyle\sum_{k=3}^{5} \dfrac{1}{2}k$

13. $\displaystyle\sum_{i=0}^{4} (2^i + 1)$

14. $\displaystyle\sum_{k=2}^{5} \frac{k}{k-1}$

15. $\displaystyle\sum_{j=1}^{5} \frac{j+2}{2^{j-1}}$

Write each series in sigma notation.

16. $1 + 2 + 3 + 4 + 5$

17. $4 + 9 + 16 + 25$

18. $\dfrac{1}{3} + \dfrac{1}{9} + \dfrac{1}{27} + \dfrac{1}{81}$

19. $1 - 1 + 1 - 1 + 1 - 1$

20. $1 + 6 + 11 + 16 + 21$

 11-3 ARITHMETIC SEQUENCES

Even though the definition of a sequence allows any list of numbers to qualify as a sequence, we have found that generally the terms are related in some way to preceding terms or to the counting numbers of the terms. In this section we will investigate one special type of sequence known as an *arithmetic sequence*.

OBJECTIVES

Upon completing this section you should be able to:

1. Identify an arithmetic sequence.
2. Find various terms of an arithmetic sequence.
3. Find the sum of the first *n* terms of an arithmetic sequence.

> An **arithmetic sequence** is a sequence in which each term is obtained by adding a constant (we will designate the constant by *d*) to the preceding term.

Note that the definition is recursive since $a_{k+1} = a_k + d$. Given the first term and *d*, we could write the first *n* terms.

Example 1 If the first term of an arithmetic sequence is 7 and $d = 4$, the first five terms are

$$7, 11, 15, 19, 23.$$

If we designate the first term of an arithmetic sequence by a and the constant difference by d, then the sequence is

$$a, a + d, (a + d) + d, [(a + d) + d] + d, \text{ and so on}$$

To obtain a term of the sequence we add d to the previous term.

or

$$a, a + d, a + 2d, a + 3d, \text{ and so on.}$$

An inspection of the sequence leads us to a formula for the nth term of an arithmetic sequence.

> If a is the first term and d is the constant difference in an arithmetic sequence, then the nth term is given by $a_n = a + (n - 1)d$.

A formal proof of this theorem could be given using mathematical induction.

Example 2 Find the tenth term of an arithmetic sequence where $a = 3$ and $d = 7$.

$$a_n = a + (n - 1)d$$
$$a_{10} = 3 + (10 - 1)7$$
$$= 66$$

Notice that by using this formula we do not need to find the first nine terms.

> The sum S_n of the first n terms of an arithmetic series is given by the formula $S_n = \dfrac{n}{2}(a_1 + a_n)$, where a_1 is the first term and a_n is the nth term.

Proof $S_n = a_1 + (a_1 + d) + (a_1 + 2d) + \cdots + [a_1 + (n - 1)d]$ by summing the sequence of n terms having the first term as a_1 and the constant difference of d. If we now write the series in reverse order, we obtain

$$S_n = a_n + (a_n - d) + (a_n - 2d) + \cdots + [a_n - (n - 1)d].$$

Note that $a_n - (n - 1)d = a_1$.

Adding the two expressions above gives

$$2S_n = (a_1 + a_n) + (a_1 + a_n) + (a_1 + a_n) + \cdots + (a_1 + a_n),$$

How many $(a_1 + a_n)$ terms are there?

which can be condensed to

$$2S_n = n(a_1 + a_n)$$

or

$$S_n = \frac{n}{2}(a_1 + a_n).$$

Since the nth term is $a_n = a_1 + (n-1)d$, we can rewrite the formula as

The other form is simpler however.

$$S_n = \frac{n}{2}[2a_1 + (n-1)d].$$

Example 3   If the first term of an arithmetic sequence is 7 and the constant difference is 4 find

    1. the third term,
    2. the fourteenth term,
    3. the sum of the first ten terms.

Solution   **1.** $a_3 = 7 + (3-1)4$
           $= 7 + (2)4$
           $= 15$
    2. $a_{14} = 7 + (14-1)4$
           $= 7 + (13)4$
           $= 59$

Find S_{10} using the formula $S_n = \frac{n}{2}(a_1 + a_n)$.

    3. $S_{10} = \frac{10}{2}[2(7) + (9)(4)]$
           $= 5(14 + 36)$
           $= 250$

▼ **EXERCISE 11–3–1**

1. Write the first six terms of the arithmetic sequence where $a = 3$ and $d = 5$.

2. Write the first seven terms of the arithmetic sequence where $a = 8$ and $d = -3$.

Find the indicated term in each arithmetic sequence.

3. $a = 2, d = 5$, 11th term

4. $a = 6, d = \frac{1}{2}$, 13th term

5. $a = 3, d = -2$, 6th term

6. $a = \frac{1}{2}, d = \frac{1}{3}$, 20th term

7. $a = -3, d = 4$, 50th term

8. $a = -4, d = -\dfrac{1}{3}$, 8th term

Find the sum of each arithmetic sequence.

9. $a = 2, d = 4, n = 10$

10. $a = 3, d = \dfrac{1}{2}, n = 9$

11. $a = 10, d = -4, n = 8$

12. $a = -\dfrac{1}{2}, d = 5, n = 10$

13. $a = 7, d = -\dfrac{3}{2}, n = 12$

Problems 14 to 20 all refer to arithmetic sequences.

14. If $a_1 = 7, a_n = 91$, and $S_n = 343$, find n.

15. If $a_{16} = -20$ and $S_{16} = -80$, find a_1.

16. If $a_n = 11$, $d = 2$, and $S_n = 32$, find n.

17. If $a_1 = 1$, $d = \dfrac{1}{2}$, and $S_n = 27$, find a_n.

18. Find the sum of the first 100 positive integers.

19. For what value of k is the sequence $2k + 4$, $3k - 7$, $k + 12$ an arithmetic sequence?

20. If the numbers a, b, c, d, e form an arithmetic sequence, then $b, c,$ and d are called *arithmetic means* between a and e. Find three arithmetic means between -12 and 60.

21. A sequence of numbers is called a *harmonic progression* if their reciprocals form an arithmetic sequence. The sequence $1, \dfrac{1}{3}, \dfrac{1}{6}, \dfrac{1}{9}, \dfrac{1}{12}$ is a harmonic progression since their reciprocals $1, 3, 6, 9, 12$ form an arithmetic sequence. Is the sequence $\dfrac{6}{7}, \dfrac{3}{5}, \dfrac{6}{13}, \dfrac{3}{8}$ a harmonic progression?

22. The *harmonic mean* of two numbers is the reciprocal of the arithmetic mean of the reciprocals of the two numbers. Find the harmonic mean of $\dfrac{1}{5}$ and $\dfrac{1}{8}$.

23. A man borrows $10,000 for 10 years and agrees to pay back $1,000 principal each year plus simple interest at 8% on all principal left unpaid during the year. How much interest does he pay over the 10-year period?

◥◣ 11–4 GEOMETRIC SEQUENCES

In section 11–3 we found that each term of an arithmetic sequence is found by adding a constant to the preceding term. In this section we introduce a sequence in which each term is found by multiplying the previous term by a constant. Such a sequence is called a *geometric sequence*.

> A **geometric sequence** is a sequence in which each term is obtained by multiplying the preceding term by a constant ratio (which we will denote by r).

Note that the definition is a recursive definition. If we denote the first term as a_1 and the constant ratio as r, we obtain the expansion of the geometric sequence as

$$a_1, a_1r, a_1r^2, a_1r^3, \ldots, a_1r^{n-1}.$$

> The formula for the nth term of a geometric sequence having a first term a_1 and a constant ratio r is $a_n = a_1r^{n-1}$.

Example 1 Write the first four terms and the seventh term of a geometric sequence if the first term is 3 and the constant ratio is 2.

Solution The first four terms are 3, 6, 12, 24.

The seventh term is

$$a_7 = 3(2)^6$$
$$= 3(64)$$
$$= 192.$$

$a_1 = 3$
$a_2 = 3(2)^{2-1} = 3(2)^1 = 6$
$a_3 = 3(2)^{3-1} = 3(2)^2 = 12$
$a_4 = 3(2)^{4-1} = 3(2)^3 = 24$

> The sum of the first n terms of a geometric sequence is
> $$S_n = \frac{a_1 - a_1r^n}{1 - r}.$$

Here again, a_1 is the first term and r is the constant ratio.

Proof $S_n = a_1 + a_1r + a_1r^2 + \cdots + a_1r^{n-1}$ by summing the terms of the sequence. If we multiply both sides of the equality by r, we obtain

$$rS_n = a_1r + a_1r^2 + a_1r^3 + \cdots + a_1r^{n-1} + a_1r^n.$$

$a_1 + a_1r + a_1r^2 + \cdots + a_1r^{n-1}$
$- (a_1r + a_1r^2 + \cdots + a_1r^{n-1}$
$+ a_1r^n) = a_1 - a_1r^n.$

Subtracting the last expression from the first gives

$$S_n - rS_n = a_1 - a_1r^n$$
$$S_n(1 - r) = a_1 - a_1r^n$$
$$S_n = \frac{a_1 - a_1r^n}{1 - r}.$$

Another form of this formula may be obtained by observing that

$$a_n = a_1r^{n-1}.$$

Multiplying by r gives

$$a_nr = a_1r^n.$$

If we replace a_1r^n by a_nr in the formula, we obtain

$$S_n = \frac{a_1 - a_nr}{1 - r}.$$

This form is easier to use if we already know a_n, since we do not need to raise r to the nth power.

Example 2   Find the sum of the first six terms of the geometric sequence in which $a_1 = \dfrac{2}{3}$ and $r = \dfrac{1}{2}$.

$a_6 = \dfrac{2}{3}\left(\dfrac{1}{2}\right)^{6-1} = \dfrac{2}{3}\left(\dfrac{1}{2}\right)^5$
$= \dfrac{2}{3}\left(\dfrac{1}{32}\right) = \dfrac{1}{48}$

Solution

$$S_6 = \frac{a_1 - a_6r}{1 - r}$$
$$= \frac{\dfrac{2}{3} - \left(\dfrac{1}{48}\right)\left(\dfrac{1}{2}\right)}{1 - \dfrac{1}{2}}$$
$$= \frac{\dfrac{2}{3} - \dfrac{1}{96}}{\dfrac{1}{2}}$$
$$= \frac{63}{48}$$

If the numbers $a_1, a_2, a_3, \ldots, a_n$ form a geometric sequence, the numbers $a_2, \ldots, a_{n-1}$ are called **geometric means** between a_1 and a_n.

$\dfrac{3}{48}$, ____ , ____ , ____ , 48
48 must be the fifth term.

Example 3   Insert three geometric means between 3 and 48.

Solution   We are given that $a_1 = 3$ and $a_5 = 48$. We know that

$$a_5 = a_1r^4$$

so

$$48 = 3r^4$$
$$r^4 = 16$$
$$r = \pm 2$$

Note there are two possibilities
for the value of r in this case.

hence the sequence is

3, 6, 12, 24, 48

or

3, -6, 12, -24, 48.

▼ **EXERCISE 11-4-1**

1. Write the first five terms of a geometric
   sequence where the first term is 2 and the
   constant ratio is 3.

2. Write the first four terms of a geometric
   sequence where the first term is 10 and the
   constant ratio is -2.

Find the indicated term in each geometric sequence.

3. $a_1 = 3$, $r = 5$, 6th term

4. $a_1 = 8$, $r = \dfrac{3}{2}$, 5th term

5. $a_1 = -3$, $r = -2$, 8th term

6. $a_1 = 4$, $r = -\dfrac{1}{2}$, 6th term

Problems 7 to 10 refer to geometric sequences.

7. If $a_1 = 1$, $r = -2$, and $S_n = -21$, find n.

8. If $a_7 = 192$ and $r = 2$, find S_7.

9. If $a_1 = 4$ and $S_4 = -80$, find r.

10. If $r = -3$, and $S_7 = 2{,}188$, find a_7.

11. If log 3 is the first term of a geometric sequence and log 9 is the second term, find the third and fourth terms.

12. For what value of k is the sequence $k - 2$, $k - 6$, $2k + 3$ a geometric sequence?

13. Insert four geometric means between 3 and 96.

14. Insert three geometric means between 16 and 625.

15. If you saved one cent the first day and added 2 cents the second day, 4 cents the third day, 8 cents the fourth day, and so on, how many days would it take to save 1 million dollars?

If $|r| > 1$, the successive terms of the sequence become farther and farther away from zero, but if $|r| < 1$, these terms get closer and closer to zero.

The formula for S_n gives us the sum of the first n terms in a geometric sequence. The question arises "Can we find the sum of the terms in an infinite geometric sequence?" The answer is "sometimes," and the determining factor is the size of the constant ratio r.

A study of limits is necessary to prove the following theorem, and such a study is beyond the scope of this text. We therefore state the theorem without proof and leave as an exercise an intuitive development of the formula.

> The sum S of an infinite geometric sequence $a_1, a_1 r, a_1 r^2, \ldots$ with $|r| < 1$ is given by the formula $S = \dfrac{a_1}{1 - r}$.

Example 4   Find the sum of the infinite geometric sequence $\dfrac{2}{3}$, $\dfrac{2}{9}$, $\dfrac{2}{27}$, ...

Solution   We see that $a_1 = \dfrac{2}{3}$ and $r = \dfrac{1}{3}$. Since $|r| < 1$ we may write

$$S = \frac{a_1}{1 - r}$$

$$= \frac{\dfrac{2}{3}}{1 - \dfrac{1}{3}}$$

$$= 1.$$

We can find the sum of an infinite geometric sequence only if $|r| < 1$.

▼ **EXERCISE 11-4-2**

Find the sum, if it exists, of each infinite geometric sequence.

1. $4 + 2 + 1 + \dfrac{1}{2} + \cdots$

2. $1 - \dfrac{1}{2} + \dfrac{1}{4} - \dfrac{1}{8} + \cdots$

3. $\sqrt{3} + 3 + \sqrt{27} + 9 + \cdots$

4. $15 - 5 + \dfrac{5}{3} - \dfrac{5}{9} + \cdots$

5. $4 + 2\sqrt{2} + 2 + \sqrt{2} + \cdots$

6. $\pi + .5 + .25 + .125 + \cdots$

7. $\displaystyle\sum_{i=1}^{\infty} \left(\frac{2}{3}\right)^i$

8. Given the decimal number $.3333 \cdots$
   a. Write this number as an infinite geometric series.
   b. Find a_1 and r.
   c. Find the sum of the series.

9. Repeat the directions in problem 8 for the following decimals:
   a. $.363636 \cdots$
   b. $.345345 \cdots$

10. Discuss what happens to the formula
$$S_n = \frac{a_1 - a_1 r^n}{1 - r} \text{ if } |r| < 1 \text{ and } n \text{ becomes very}$$
large.

11. A rubber ball is dropped from a height of 12 meters. Each time it strikes the ground it rebounds three-fourths of the height from which it fell. What is the approximate total distance traveled by the ball?

◣◢ 11-5 THE BINOMIAL THEOREM

Raising a binomial to a power occurs in many phases of mathematics, and the coefficients of this expansion have a special place in the study of statistics. In this section we will discuss and use the theorem for the binomial expansion.

Note the following expansions of $x + y$.

$$(x + y)^1 = x + y$$
$$(x + y)^2 = x^2 + 2xy + y^2$$
$$(x + y)^3 = x^3 + 3x^2y + 3xy^2 + y^3$$
$$(x + y)^4 = x^4 + 4x^3y + 6x^2y^2 + 4xy^3 + y^4$$

Several patterns are seen.

1. For $(x + y)^n$, there are $(n + 1)$ terms.
2. The exponents of x descend from n to zero.
3. The exponents of y ascend from zero to n.
4. The numerical coefficients are symmetrical with respect to the "middle" of the expansion.

If the pattern persists, $(x + y)^{10}$ would have eleven terms, the exponents of x would start at x^{10} in the first term and decrease to x^0 in the eleventh term, while the exponents of y would start with y^0 in the first term and increase to y^{10} in the eleventh term. The numerical coefficients of the first and eleventh, second and tenth, third and ninth, and so on, would be the same.

Finding the value of the numerical coefficients is the only part of a binomial expansion that involves any degree of difficulty.

At this point we wish to define factorial notation.

> If n is a positive integer, then the symbol $n!$ (read as "n factorial") is defined to be
>
> $$n! = n(n - 1)(n - 2) \cdots (3)(2)(1).$$

Another definition is necessary for consistency.

> $$0! = 1$$

Example 1 $3! = (3)(2)(1) = 6$

Example 2 $7! = (7)(6)(5)(4)(3)(2)(1) = 5{,}040$

OBJECTIVES

Upon completing this section you should be able to:

1. Expand a binomial using the binomial theorem.
2. Find any term of a binomial expansion.
3. Find powers and roots of numbers using the binomial expansion.

Notice that the coefficients are
expressed using factorial notation.

Binomial theorem: For any natural number n,

$$(a + b)^n = \sum_{r=0}^{n} \frac{n!}{(n - r)!r!} a^{n-r}b^r.$$

Notice in the theorem that n is the exponent or power of the binomial $(a + b)$ and r is the exponent of b or one less than the number of the term.

The proof is by mathematical induction and is a rather long exercise in algebraic manipulation. We will therefore leave this proof to the more enterprising student and proceed to use it for expanding binomials.

Example 3   Expand: $(x + y)^5$

Solution $(x + y)^5$

Notice that we need only find the first three coefficients.

$$= \sum_{r=0}^{5} \frac{5!}{(5 - r)!r!} x^{5-r}y^r$$

$$= \frac{5!}{5!0!}x^5 + \frac{5!}{4!1!}x^4y + \frac{5!}{3!2!}x^3y^2 + \frac{5!}{2!3!}x^2y^3 + \frac{5!}{1!4!}xy^4 + \frac{5!}{0!5!}y^5$$

$$= x^5 + 5x^4y + 10x^3y^2 + 10x^2y^3 + 5xy^4 + y^5$$

Example 4   Find the eighth term of the expansion of $(x + y)^{12}$.

r is always equal to one less than the number of the term.

Solution   Note that in the eighth term $r = 7$. Thus the eighth term is

$$\frac{12!}{5!7!}x^5y^7 = 792x^5y^7.$$

Example 5   Find the third term of $(2x + 3)^8$.

$a = 2x, b = 3$

Solution $(2x + 3)^8 = [(2x) + 3]^8$

Note that in the third term $r = 2$. Thus the third term is

$$\frac{8!}{6!2!}(2x)^6(3)^2 = 28(64x^6)(9)$$

$$= 16,128x^6.$$

Here is an example of a very useful application of the binomial expansion.

Example 6   Find the value of $(1.01)^7$ to three decimal places using the binomial expansion.

Solution   We may write $(1.01)^7$ as $(1 + .01)^7$. Then

$$(1 + .01)^7 = (1)^7 + 7(1)^6(.01) + 21(1)^5(.01)^2$$
$$+ 35(1)^4(.01)^3 + \cdots$$
$$= 1 + .07 + .0021 + .000035 + \cdots$$
$$\approx 1.072.$$

Notice that we do not need to go beyond the third term for three decimal place accuracy.

An alternative method of finding the coefficients of the terms of the binomial expansion is as follows:

This method is an alternative to finding the coefficients by using factorial notation.

1. The coefficient of the first term is 1.
2. The coefficient of the rth term can be found by multiplying the coefficient of the $(r - 1)$ term by the exponent of x and dividing this quantity by the number of the $(r - 1)$ term.

Example 7   Expand $(a + b)^6$.

Solution   First note the pattern of the exponents of a and b discussed earlier and see that the first term is a^6. Next multiply 6 (the exponent of a) by 1 (the coefficient of a^6) and divide this quantity by 1 (the number of the term), giving the second term to be $6a^5b$. Now multiply 5 (the exponent of a) by 6 (the coefficient of the second term) and divide this product by 2 (the number of the term) giving $15a^4b^2$ as the third term. Proceeding this way gives

$$(a + b)^6 = a^6 + 6a^5b + 15a^4b^2 + 20a^3b^3 + 15a^2b^4 + 6ab^5 + b^6.$$

You would only need to find the coefficients of the first four terms.

The weakness of this method of expansion lies in the fact that to find the eighth term of $(x + y)^{12}$ we would first need to find the first seven terms.

In other words, the above method is recursive.

The strength of this method lies in the fact that the arithmetic is easier than expanding by the binomial theorem.

A third method of determining the coefficients of the terms of an expansion of a binomial uses an array of numbers called *Pascal's triangle*. The pattern of this array is indicated here.

$(a + b)^0$				1				
$(a + b)^1$			1		1			
$(a + b)^2$			1	2	1			
$(a + b)^3$		1	3		3	1		
$(a + b)^4$	1		4	6	4		1	
$(a + b)^5$	1	5	10	10	5	1		

The numbers in each row represent the coefficients of the terms in the expansion of the binomials at the left.

This pattern may be continued indefinitely. Notice that each row in the array begins and ends with a 1. The second number in any row is obtained by adding the first and second numbers in the preceding row. The third number is obtained by adding the second and third numbers in the preceding row, and so on.

The numbers in the nth row of the triangle are the coefficients of the binomial expansion of $(a + b)^{n-1}$. Notice from the triangle, for example, that the coefficients of the expansion $(a + b)^4$ are 1, 4, 6, 4, 1. Thus,

$$(a + b)^4 = a^4 + 4a^3b + 6a^2b^2 + 4ab^3 + b^4.$$

Example 8 Using Pascal's triangle, expand $(a + b)^7$.

Solution

Using this method we must find all the coefficients of all expansions up to $(a + b)^7$. However, the work involved is not difficult.

$(a + b)^0$								1							
$(a + b)^1$							1		1						
$(a + b)^2$						1		2		1					
$(a + b)^3$					1		3		3		1				
$(a + b)^4$				1		4		6		4		1			
$(a + b)^5$			1		5		10		10		5		1		
$(a + b)^6$		1		6		15		20		15		6		1	
$(a + b)^7$	1		7		21		35		35		21		7		1

Thus,

$$(a + b)^7 = a^7 + 7a^6b + 21a^5b^2 + 35a^4b^3 + 35a^3b^4 + 21a^2b^5 + 7ab^6 + b^7.$$

If we make the assumption that the binomial theorem is valid for rational values of n, we see that the expansion becomes an infinite series in some cases. This can be very useful in creating series that evaluate roots.

Example 9 Use the binomial expansion to find $\sqrt{29}$ correct to three decimal places.

Solution

We write 29 as $(25 + 4)$ since 25 is a perfect square.

$$\sqrt{29} = (29)^{1/2} = (25 + 4)^{1/2}$$

$$= (25)^{1/2} + \frac{1}{2}(25)^{-1/2}(4) - \frac{1}{8}(25)^{-3/2}(4)^2 + \frac{1}{16}(25)^{-5/2}(4)^3 - \cdots$$

$$= 5 + \left(\frac{1}{2}\right)\left(\frac{1}{5}\right)(4) - \left(\frac{1}{8}\right)\left(\frac{1}{125}\right)(16) + \left(\frac{1}{16}\right)\left(\frac{1}{3125}\right)(64) - \cdots$$

$$= 5 + .4 - .016 + .0013 - \cdots$$

$$\approx 5.385 \text{ (to three decimal places)}$$

Notice that the expansion is infinite but we only needed the first four terms to evaluate the radical to three decimal places.

The key to this type of example, of course, is to find the sum of two numbers, one of which is a perfect power of the desired root.

27 is a perfect cube.

Example 10 $\sqrt[3]{30} = \sqrt[3]{27 + 3} = (27 + 3)^{1/3}$

$32 = 2^5$

Example 11 $\sqrt[5]{38} = \sqrt[5]{32 + 6} = (32 + 6)^{1/5}$

By this method any root of any number can be found to the desired degree of accuracy.

▼ **EXERCISE 11–5–1**

Expand by the binomial theorem. Repeat the expansion using the alternative methods discussed in this section.

1. $(x + 2)^4$

2. $(1 + 2x)^6$

3. $(x - y)^6$

4. $(2x^2 - y)^5$

5. $\left(2a + \dfrac{b}{2}\right)^7$

Use the binomial theorem to find the indicated term or terms.

6. $(x + y)^{30}$, first four terms

7. $(a - 3b)^{100}$, first four terms

8. $\left(x - \dfrac{2}{x}\right)^{10}$, first four terms

9. $(2a^2 - b^2)^7$, fourth term

10. $(2x - \sqrt{y})^9$, seventh term

Find the exact value using the binomial expansion.

11. 102^4 [*Hint:* $102^4 = (100 + 2)^4$]

12. 99^6

Use the binomial expansion to evaluate to three decimal places. Check your answer by using a calculator or logarithms.

13. $(1.02)^8$

14. $(.97)^6$

15. $\sqrt{50}$ **16.** $\sqrt[3]{29}$

17. $\sqrt[4]{1.02}$ **18.** $(1.03)^{-6}$

19. $(1.99)^{-5}$

20. Use mathematical induction to prove the binomial theorem.

Key Words

Section 11–1

- **Mathematical induction** is a method of proof used when we have a statement concerning the infinite set of positive integers.

Section 11–2

- A **sequence** is a function whose domain is the set of positive integers.

- Sequences are classified as **finite** or **infinite.**

- A **series** is the indicated sum of a sequence.

Section 11–3

- An **arithmetic sequence** is a sequence in which each term is obtained by adding a constant to the preceding term.

Section 11–4

- A **geometric sequence** is a sequence in which each term is obtained by multiplying the preceding term by a constant ratio.

- The terms between a_1 and a_n in a geometric sequence are called **geometric means.**

Section 11–5

- The **binomial theorem** is used to expand any power of a binomial.

Procedures

Section 11–1

- To prove a theorem by mathematical induction follow these steps:
  Step 1 Show true for $n = 1$.
  Step 2 Assume true for $n = k$.
  Step 3 Show that the assumption of step 2 implies true for $n = k + 1$.

Section 11–3

- To find the nth term (a_n) of an arithmetic sequence use the formula

$$a_n = a + (n - 1)d.$$

- To find the sum (S_n) of the first n terms of an arithmetic sequence use the formula

$$S_n = \frac{n}{2}(a_1 + a_n).$$

Section 11–4

- To find the nth term of a geometric sequence use the formula

$$a_n = a_1 r^{n-1}.$$

- To find the sum (S_n) of the first n terms of a geometric sequence use the formula

$$S_n = \frac{a_1 - a_1 r^n}{1 - r}.$$

- To find the sum (S) of the terms of an infinite geometric sequence when $|r| < 1$ use the formula

$$S = \frac{a_1}{1 - r}.$$

Section 11–5

- To expand a binomial use the following:

$$(a + b)^n = \sum_{r=0}^{n} \frac{n!}{(n - r)!r!} a^{n-r} b^r.$$

NAME:

CLASS / SECTION: DATE:

Use mathematical induction to prove the statements.

 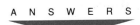
1. $\dfrac{1}{2}(1) + \dfrac{1}{2}(2) + \dfrac{1}{2}(3) + \cdots + \dfrac{1}{2}(n) = \dfrac{n^2 + n}{4}$

1. _____

2. $5 + 9 + 13 + \cdots + (4n + 1) = 2n^2 + 3n$

2. _____

3. $2 + 2^2 + 2^3 + \cdots + 2^n = 2^{n+1} - 2$

3. _____

4. $1 + 3 + 6 + \cdots + \dfrac{n(n + 1)}{2} = \dfrac{n(n + 1)(n + 2)}{6}$

4. _____

621

5. $1^2 + 3^2 + 5^2 + \cdots + (2n - 1)^2 = \dfrac{n(4n^2 - 1)}{3}$

5. _____

6. _____

6. $\dfrac{1}{1(3)} + \dfrac{1}{3(5)} + \dfrac{1}{5(7)} + \cdots + \dfrac{1}{(2n - 1)(2n + 1)} = \dfrac{n}{2n + 1}$

7. _____

7. Write the first four terms and the eighth term of the sequence whose nth term is given as $a_n = n^3 - 4$.

8. Write the first and the 112th term of the sequence whose nth term is $a_n = (-1)^n + 5$.

8. _____

9. _____

9. Find the first four terms of the sequence that is defined recursively as $a_1 = 1$, $a_{k+1} = k(a_k + 1)$.

10. Evaluate: $\displaystyle\sum_{i=1}^{5} 2^{i-1}$

10. _____

11. Express in expanded form: $\displaystyle\sum_{i=1}^{4} \frac{x^{i+2}}{4i}$

12. Express the series $2 + 5 + 10 + 17 + 26$ in sigma notation.

11. _____

12. _____

13. Express the series $1 + 5 + 9 + 13 + 17 + 21$ in sigma notation.

14. In an arithmetic sequence if $a = -8$ and $d = 5$, find the eighteenth term.

13. _____

14. _____

15. In an arithmetic sequence if $a = 8$ and $d = -\dfrac{1}{2}$, find the twenty-ninth term.

16. In an arithmetic sequence if $a = x + 2y$ and $d = 2x + y$, find the twelfth term.

15. _____

16. _____

17. _____

17. Find the sum of the first 100 positive odd integers.

18. If the sum of the first nine terms of an arithmetic sequence where $d = -3$ is 468, find the first term.

18. _____

19. _____

20. _____

21. _____

22. _____

23. _____

24. _____

25. _____

26. _____

19. An arithmetic sequence having ten terms has a first term of 18 and a sum of 45. Find the seventh term.

20. For what value of k is the sequence $k - 3$, $k + 5$, $2k - 1$ an arithmetic sequence?

21. Write the first four terms of a geometric sequence where the first term is $\sqrt{2}$ and the constant ratio is $\sqrt{2}$.

22. Find the eighth term of a geometric sequence where the first term is 6 and the constant ratio is 2.

23. Find the fifth term of a geometric sequence where the first term is -8 and the constant ratio is $-\dfrac{1}{2}$.

24. Find the sum of the first six terms of a geometric sequence where $a_1 = 16$ and $r = -\dfrac{1}{4}$.

25. For what values of k is the sequence $3k + 4$, $k - 2$, $5k + 1$ a geometric sequence?

26. Insert three geometric means between 81 and 16.

27. Using an infinite geometric series, find the common fraction that is equal to .6666 · · · .

28. Expand $(y^3 - 3a)^4$ by using the binomial theorem.

27. _____

28. _____

29. Find the fifth term of the expansion of $(y^2 - 2)^{10}$.

30. Find the middle term of the expansion of $\left(x - \dfrac{2}{x} \right)^8$.

29. _____

30. _____

31. _____

31. Evaluate $(.8)^5$ using the binomial expansion.

32. Evaluate $\sqrt[4]{19}$ to three decimal places using the binomial expansion.

32. _____

SCORE: _____

NAME: _____

CLASS / SECTION: _____ DATE: _____

A N S W E R S

1. _____

2. _____

3. _____

4. _____

5. _____

6. _____

7. _____

8. _____

1. Use mathematical induction to prove $2 + 4 + 6 + \cdots + 2n = n(n + 1)$.

2. Write the first four terms of the sequence whose nth term is defined as $a_n = 3n^2 - n - 2$.

3. Evaluate: $\displaystyle\sum_{i=2}^{7} (2i - 1)$

4. Express the series $-1 + 2 + 7 + 14$ in sigma notation.

5. If the first term of an arithmetic sequence is $\dfrac{1}{3}$ and the constant difference is $\dfrac{1}{2}$, find the ninth term.

6. The fourteenth term of an arithmetic sequence is 81 and the twentieth term is 117. Find the sum of the first four terms.

7. In an arithmetic sequence if $a_1 = 3$, $a_n = 27$, and $S_n = 150$, find n.

8. The first term of a geometric sequence is 625 and the last term is 1. If the sum of the terms is 521, find the fourth term.

9. Insert 3 geometric means between $\dfrac{2}{3}$ and 54.

10. Find the sum of the infinite geometric series $6 - 4 + \dfrac{8}{3} - \dfrac{16}{9} + \cdots$.

11. Find the tenth term of $\left(x + \dfrac{1}{2y}\right)^{12}$.

12. Use the binomial expansion to evaluate $\sqrt{15}$ to two decimal places.

A N S W E R S

9. _____

10. _____

11. _____

12. _____

SCORE: _____

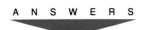

A N S W E R S

1. _____

2. _____

3. _____

4. _____

5. _____

6. _____

7. _____

8. _____

9. _____

10. _____

1. Evaluate $2x^3 - 3x^2y + 5xy^2$ if $x = -4$, $y = -2$.

2. In a collection of coins the ratio of nickels to dimes was 2 to 3. If there were 14 nickels, how many dimes were there?

3. Solve for x: $|3x - 1| = 5$

4. A collection of nickels, dimes, and quarters has a total value of $2.80. If there are 20 coins in all and there are two more quarters than nickels, find the number of each type of coin.

5. Factor: $6ax - 9x - 4ay + 6y$

6. Factor completely: $8x^3y - 2xy^3$

7. Subtract:
$$\frac{2x}{x - 1} - \frac{3x + 5}{x^2 + 2x - 3}$$

8. Simplify: $\dfrac{\dfrac{2}{x^2 - 1} - 1}{2 - \dfrac{1}{x + 1}}$

9. Write .00000581 in scientific notation.

10. Simplify: $\dfrac{1}{\sqrt{2} + \sqrt{3}}$

11. Solve for x: $2x^2 - 4x - 3 = 0$

12. Solve for x:
$$\sqrt{2x + 3} = \sqrt{x - 2} + 2$$

11. _____

13. Write the equation in standard form of the line through the point $(3, -1)$ and having slope of $\frac{2}{3}$.

14. Write the equation of the circle $x^2 + y^2 + 6x - 4y = 23$ in standard form. Give the center and radius.

12. _____

13. _____

15. Solve by graphing:
$$\begin{cases} x - 2y < -1 \\ 3x + 2y \geq 4 \end{cases}$$

16. Solve: $\begin{cases} 2x - y + 2z = -4 \\ 4x + 5y - z = 7 \\ x + 2y - 3z = 10 \end{cases}$

14. _____

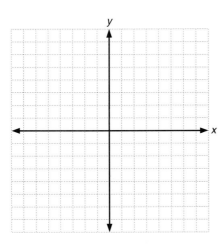

15. _____

16. _____

17. _____

17. Solve for x: $\log(x + 2)$ $= \log(x + 3) + \log(x - 2)$

18. Find the eleventh term of the arithmetic sequence where $a_1 = 2$ and $d = 6$.

18. _____

A N S W E R S

19. _____

20. _____

SCORE: _____

19. Find the sum of the infinite geometric series $4 + 2 + 1 + \dfrac{1}{2} + \cdots$.

20. Find the fourth term of $(x - 2)^7$.

Common Logarithms

N	0	1	2	3	4	5	6	7	8	9
10	0000	0043	0086	0128	0170	0212	0253	0294	0334	0374
11	0414	0453	0492	0531	0569	0607	0645	0682	0719	0755
12	0792	0828	0864	0899	0934	0969	1004	1038	1072	1106
13	1139	1173	1206	1239	1271	1303	1335	1367	1399	1430
14	1461	1492	1523	1553	1584	1614	1644	1673	1703	1732
15	1761	1790	1818	1847	1875	1903	1931	1959	1987	2014
16	2041	2068	2095	2122	2148	2175	2201	2227	2253	2279
17	2304	2330	2355	2380	2405	2430	2455	2480	2504	2529
18	2553	2577	2601	2625	2648	2672	2695	2718	2742	2765
19	2788	2810	2833	2856	2878	2900	2923	2945	2967	2989
20	3010	3032	3054	3075	3096	3118	3139	3160	3181	3201
21	3222	3243	3263	3284	3304	3324	3345	3365	3385	3404
22	3424	3444	3464	3483	3502	3522	3541	3560	3579	3598
23	3617	3636	3655	3674	3692	3711	3729	3747	3766	3784
24	3802	3820	3838	3856	3874	3892	3909	3927	3945	3962
25	3979	3997	4014	4031	4048	4065	4082	4099	4116	4133
26	4150	4166	4183	4200	4216	4232	4249	4265	4281	4298
27	4314	4330	4346	4362	4378	4393	4409	4425	4440	4456
28	4472	4487	4502	4518	4533	4548	4564	4579	4594	4609
29	4624	4639	4654	4669	4683	4698	4713	4728	4742	4757
30	4771	4786	4800	4814	4829	4843	4857	4871	4886	4900
31	4914	4928	4942	4955	4969	4983	4997	5011	5024	5038
32	5051	5065	5079	5092	5105	5119	5132	5145	5159	5172
33	5185	5198	5211	5224	5237	5250	5263	5276	5289	5302
34	5315	5328	5340	5353	5366	5378	5391	5403	5416	5428
35	5441	5453	5465	5478	5490	5502	5514	5527	5539	5551
36	5563	5575	5587	5599	5611	5623	5635	5647	5658	5670
37	5682	5694	5705	5717	5729	5740	5752	5763	5775	5786
38	5798	5809	5821	5832	5843	5855	5866	5877	5888	5899
39	5911	5922	5933	5944	5955	5966	5977	5988	5999	6010
40	6021	6031	6042	6053	6064	6075	6085	6096	6107	6117
41	6128	6138	6149	6160	6170	6180	6191	6201	6212	6222
42	6232	6243	6253	6263	6274	6284	6294	6304	6314	6325
43	6335	6345	6355	6365	6375	6385	6395	6405	6415	6425
44	6435	6444	6454	6464	6474	6484	6493	6503	6513	6522
45	6532	6542	6551	6561	6571	6580	6590	6599	6609	6618
46	6628	6637	6646	6656	6665	6675	6684	6693	6702	6712
47	6721	6730	6739	6749	6758	6767	6776	6785	6794	6803
48	6812	6821	6830	6839	6848	6857	6866	6875	6884	6893
49	6902	6911	6920	6928	6937	6946	6955	6964	6972	6981
50	6990	6998	7007	7016	7024	7033	7042	7050	7059	7067
51	7076	7084	7093	7101	7110	7118	7126	7135	7143	7152
52	7160	7168	7177	7185	7193	7202	7210	7218	7226	7235
53	7243	7251	7259	7267	7275	7284	7292	7300	7308	7316
54	7324	7332	7340	7348	7356	7364	7372	7380	7388	7396

Common Logarithms

N	0	1	2	3	4	5	6	7	8	9
55	7404	7412	7419	7427	7435	7443	7451	7459	7466	7474
56	7482	7490	7497	7505	7513	7520	7528	7536	7543	7551
57	7559	7566	7574	7582	7589	7597	7604	7612	7619	7627
58	7634	7642	7649	7657	7664	7672	7679	7686	7694	7701
59	7709	7716	7723	7731	7738	7745	7752	7760	7767	7774
60	7782	7789	7796	7803	7810	7818	7825	7832	7839	7846
61	7853	7860	7868	7875	7882	7889	7896	7903	7910	7917
62	7924	7931	7938	7945	7952	7959	7966	7973	7980	7987
63	7993	8000	8007	8014	8021	8028	8035	8041	8048	8055
64	8062	8069	8075	8082	8089	8096	8102	8109	8116	8122
65	8129	8136	8142	8149	8156	8162	8169	8176	8182	8189
66	8195	8202	8209	8215	8222	8228	8235	8241	8248	8254
67	8261	8267	8274	8280	8287	8293	8299	8306	8312	8319
68	8325	8331	8338	8344	8351	8357	8363	8370	8376	8382
69	8388	8395	8401	8407	8414	8420	8426	8432	8439	8445
70	8451	8457	8463	8470	8476	8482	8488	8494	8500	8506
71	8513	8519	8525	8531	8537	8543	8549	8555	8561	8567
72	8573	8579	8585	8591	8597	8603	8609	8615	8621	8627
73	8633	8639	8645	8651	8657	8663	8669	8675	8681	8686
74	8692	8698	8704	8710	8716	8722	8727	8733	8739	8745
75	8751	8756	8762	8768	8774	8779	8785	8791	8797	8802
76	8808	8814	8820	8825	8831	8837	8842	8848	8854	8859
77	8865	8871	8876	8882	8887	8893	8899	8904	8910	8915
78	8921	8927	8932	8938	8943	8949	8954	8960	8965	8971
79	8976	8982	8987	8993	8998	9004	9009	9015	9020	9025
80	9031	9036	9042	9047	9053	9058	9063	9069	9074	9079
81	9085	9090	9096	9101	9106	9112	9117	9122	9128	9133
82	9138	9143	9149	9154	9159	9165	9170	9175	9180	9186
83	9191	9196	9201	9206	9212	9217	9222	9227	9232	9238
84	9243	9248	9253	9258	9263	9269	9274	9279	9284	9289
85	9294	9299	9304	9309	9315	9320	9325	9330	9335	9340
86	9345	9350	9355	9360	9365	9370	9375	9380	9385	9390
87	9395	9400	9405	9410	9415	9420	9425	9430	9435	9440
88	9445	9450	9455	9460	9465	9469	9474	9479	9484	9489
89	9494	9499	9504	9509	9513	9518	9523	9528	9533	9538
90	9542	9547	9552	9557	9562	9566	9571	9576	9581	9586
91	9590	9595	9600	9605	9609	9614	9619	9624	9628	9633
92	9638	9643	9647	9652	9657	9661	9666	9671	9675	9680
93	9685	9689	9694	9699	9703	9708	9713	9717	9722	9727
94	9731	9736	9741	9745	9750	9754	9759	9763	9768	9773
95	9777	9782	9786	9791	9795	9800	9805	9809	9814	9818
96	9823	9827	9832	9836	9841	9845	9850	9854	9859	9863
97	9868	9872	9877	9881	9886	9890	9894	9899	9903	9908
98	9912	9917	9921	9926	9930	9934	9939	9943	9948	9952
99	9956	9961	9965	9969	9974	9978	9983	9987	9991	9996

Chapter 1 Pretest

The number in brackets after each answer refers to the section of the chapter that discusses that type of problem.

1. Associative property of addition [1–1]   **2.** $\frac{1}{2}$ [1–2]   **3.** -3.5 [1–2]   **4.** -3 [1–2]   **5.** 5 [1–2]   **6.** -9 [1–2]   **7.** 3 [1–2]

8. $-5°$ [1–3]   **9.** -7 [1–4]   **10.** 15 [1–4]   **11.** 48 [1–5]   **12.** -6 [1–5]   **13.** -24 [1–5]   **14.** $12x^7y^4$ [1–6]

15. 31 [1–4]   **16.** $2x^2 + 15xy$ [1–7]   **17.** $6x^3y - 8x^2y^2 + 10xy^4$ [1–8]   **18.** $-232x + 108$ [1–9]   **19.** -21 [1–10]

20. 161 [1–10]

1–1–1

1. a, b, c, d   **3.** c, d   **5.** b, c, d   **7.** a, b, c, d   **9.** e   **11.** d   **13.** d   **15.** e   **17.** d   **19.** d   **21.** Closure under addition

23. Additive identity   **25.** Associative property of multiplication   **27.** Commutative property of addition

29. Multiplicative identity   **31.** Distributive property of multiplication over addition

1–2–1

1. -3 **3.** $+4$ **5.** $-\frac{1}{2}$ **7.** $+5.3$ **9.** $-y$ **11.** $+b$ **13.** -7 **15.** $+4\frac{1}{3}$ **17.** 0   **19.** -100 **21.** $+7.38$ **23.** -4

25. -2.05 **27.** $-\frac{1}{2}$ **29.** $+\frac{3}{4}$ **31.** $A = -7, B = -3, C = +7$

33. **35.** **37.** **39.** **41.** -20 **43.** $+10$ **45.** -5

47. -13 **49.** -6 **51.** 5   **53.** 4   **55.** 0   **57.** 0   **59.** 3.14   **61.** -9

1–2–2

1. $-x - y$ **3.** $x - 5$ **5.** $-a + b + c$ **7.** $-x - y + 3$ **9.** $a - b + 8$ **11.** $a - c - 4$ **13.** $-a - b + c$

15. $-x + y + 15$ **17.** $-x - 4 + z$ **19.** $a - b$ **21.** $a - 2b + c$ **23.** $x - y$ **25.** $a + 6b - 4$

27. $-5x + y - 4z + 3$ **29.** $-a - 2b + 5c - 6$

1–3–1

1. $+7$ **3.** -2 **5.** $+2$ **7.** $+7$ **9.** -24 **11.** -4 **13.** $+4$ **15.** -8 **17.** $+53$ **19.** $+1$ **21.** $+10 - 3 = +7$

23. $+4 - 6 = -2$ **25.** $-8 + 10 = +2$ **27.** $+16 - 9 = +7$ **29.** $-6 - 18 = -24$ **31.** $+2 - 6 = -4$

33. $-8 + 12 = +4$ **35.** $-6 - 2 = -8$ **37.** $+65 - 12 = +53$ **39.** $+10 - 6 + 9 - 12 = +1$

1–3–2

1. $+8$ **3.** -7 **5.** $+6$ **7.** $+3$ **9.** $+5$ **11.** -6 **13.** 0   **15.** -7 **17.** $+7$ **19.** -9 **21.** $+9$ **23.** 0   **25.** 0

27. -14 **29.** 0   **31.** $+4$ **33.** $-3°$ **35.** $+7$ **37.** -8 **39.** $+23°$

1–4–1

1. $+12$ **3.** $+41$ **5.** $+17$ **7.** -8 **9.** -23 **11.** -21 **13.** $+61$ **15.** -32 **17.** -31 **19.** $+35$ **21.** $+25$ **23.** $+23$

25. -28 **27.** -30 **29.** $+17$ **31.** -24 **33.** -27 **35.** $+74$ **37.** $+29$ **39.** -38

1-4-2

1. +5   **3.** +4   **5.** +9   **7.** −6   **9.** −2   **11.** −14   **13.** 0   **15.** +1   **17.** 0   **19.** +3   **21.** −8
23. −2   **25.** −1
27. −9   **29.** +14   **31.** −33   **33.** 0   **35.** +25   **37.** −1   **39.** +26

1-4-3

1. 6   **3.** 0   **5.** 1   **7.** −10   **9.** 3   **11.** −6   **13.** 1   **15.** 1   **17.** 9   **19.** −1   **21.** −3   **23.** −2   **25.** 14   **27.** 25
29. −9   **31.** 4   **33.** 17   **35.** 8   **37.** 40   **39.** 9

1-4-4

1. 3   **3.** 6   **5.** 11   **7.** 12   **9.** 2   **11.** 3   **13.** 12   **15.** 27   **17.** −1   **19.** 0   **21.** 7   **23.** −6   **25.** 5   **27.** −6   **29.** 4

1-4-5

1. $6 + (−4) = 2$ **3.** $9 + (−12) = −3$ **5.** $13 + 2 = 15$ **7.** $−16 + 9 = −7$ **9.** $22 + 6 = 28$
11. $3 + (−5) = −2$ **13.** $−11 + 1 = −10$ **15.** $24 + (−24) = 0$ **17.** $−25 + 3 = −22$ **19.** $18 + (−18) = 0$
21. $8 + (−3) = 5$ **23.** $8 + (−21) = −13$ **25.** −4   **27.** 12   **29.** −23   **31.** −7

1-5-1

1. −40   **3.** −40   **5.** −21   **7.** −24   **9.** −42   **11.** −4   **13.** $−\dfrac{3}{5}$ **15.** −7.5   **17.** −11.73   **19.** $−\dfrac{1}{4}$ **21.** −2.1

23. −40   **25.** −48   **27.** −60   **29.** −35   **31.** 0   **33.** −6   **35.** $−9\dfrac{1}{2}$ **37.** −165.6   **39.** 0

1-5-2

1. $5x + 20$ **3.** $6a + 30$ **5.** $6a + 12$ **7.** $2x − 18$ **9.** $6x + 8$ **11.** $−3a − 21$ **13.** $2x − 6$ **15.** $−6x − 15$
17. $42x − 28$ **19.** $−16x − 40$ **21.** $4a + 4b + 16$ **23.** $2x + 6y + 8$ **25.** $6a + 12b + 6$ **27.** $10x − 5y + 15$
29. $−2x − 6y − 22$ **31.** $−12a − 8b − 4c$

1-5-3

1. 6   **3.** 8   **5.** 21   **7.** 45   **9.** 15   **11.** $\dfrac{1}{5}$ **13.** −40   **15.** −21   **17.** 6.0   **19.** 60   **21.** −12   **23.** 60   **25.** −1
27. 0   **29.** 150   **31.** 30   **33.** 168   **35.** −300   **37.** −80   **39.** −37

1-5-4

1. 3   **3.** −3   **5.** 5   **7.** −5   **9.** −4   **11.** 7   **13.** 4   **15.** −27   **17.** −7   **19.** 4   **21.** $−\dfrac{14}{15}$ **23.** −2   **25.** $\dfrac{7}{2}$
27. $−\dfrac{1}{21}$ **29.** $−\dfrac{5}{2}$ **31.** −2   **33.** 12   **35.** −2   **37.** 2   **39.** −3

1-6-1

1. 9   **3.** 8   **5.** 625   **7.** 243   **9.** $\dfrac{1}{8}$ **11.** 50   **13.** 7,776   **15.** 405   **17.** 35   **19.** 77

1-6-2

1. x^7 **3.** y^{11} **5.** a^5 **7.** a^3b^4 **9.** $x^{13}y^3$ **11.** $a^6b^3c^4$ **13.** $12x^8$ **15.** $-21x^5$ **17.** $-66a^{10}$ **19.** $32a^2b^3$ **21.** $35x^2y^2$
23. $-6x^3y^4$ **25.** $30a^5bc$ **27.** $-30x^{11}$ **29.** $42a^4b^4$ **31.** $-48x^3y^6$ **33.** $-165a^3b^3c^4$ **35.** $120x^3y^9z^3$ **37.** $-60x^2y^2z$
39. $60a^2b^3c^5d^2$

1-6-3

1. x^6 **3.** a^{12} **5.** x^8 **7.** x^9 **9.** w^{16} **11.** x^6y^9 **13.** $a^{20}b^5$ **15.** x^6y^{12} **17.** $x^5y^{10}z^{15}$ **19.** $a^{27}b^{45}c^{36}$ **21.** $-8x^9y^{12}$
23. $625a^4b^{12}$ **25.** $32x^{15}y^{25}$ **27.** $81a^8b^4c^{16}$ **29.** $36x^4y^{16}z^6$

1-6-4

1. x **3.** $\dfrac{1}{a^5}$ **5.** $\dfrac{x^2}{y^4}$ **7.** $\dfrac{x^2z^2}{y^3}$ **9.** $\dfrac{xy}{z^2}$ **11.** $2x^2$ **13.** $2x^6$ **15.** $\dfrac{b^5}{3}$ **17.** $\dfrac{3}{5a^8}$ **19.** $5xy$ **21.** $-\dfrac{4}{x^2}$ **23.** $\dfrac{a^2}{3b^3}$ **25.** $\dfrac{3x^2}{4y^2}$

27. $\dfrac{1}{a}$ **29.** $\dfrac{x^7}{4}$ **31.** $-\dfrac{7y^4}{2x^3}$ **33.** $-\dfrac{4y^4}{x^3}$ **35.** 1 **37.** $-\dfrac{5x}{7z}$ **39.** $-\dfrac{3wx^2}{4}$

1-7-1

1. $11x$ **3.** $4x$ **5.** $4a$ **7.** $8x^2$ **9.** $6a^3$ **11.** $16xy$ **13.** xy **15.** $4a + 6b$ **17.** $9a^2b + 11ab^2$ **19.** $22abc + ab$
21. $ab + ac$ **23.** $3a^2 + 8a$ **25.** $8x^3y^2$ **27.** $7x^2y - 5xy^2 + 9x^2y^2$ **29.** $3m^2n + 6mn^2$

1-8-1

1. $3x + 6y$ **3.** $-4x - 2y$ **5.** $x^2 + xy$ **7.** $6x^2 + 15xy$ **9.** $-2x + 7y$ **11.** $2x^2 - 10xy$ **13.** $10x + 5y - 15z$
15. $-6x^2 + 18xy$ **17.** $-2x^2 + 4xy$ **19.** $-6x^2 + 4xy + 2xz$ **21.** $-3x - 5y + z$ **23.** $6x^2y - 9xy^2 - 15xyz$
25. $14y^2 - 7xy + 21yz$ **27.** $-2x^3 + 6x^2 + 2x$ **29.** $6x + 15x^2 - 12x^3$ **31.** $20y^3 + 10y^2 - 5y$
33. $-14x^3y^4 - 21x^3y^2 + 35x^2y^3$ **35.** $6x^3y^2 - 4x^2y^3 + 2x^2y^2$ **37.** $2x^3y^2z - 10xy^3z^2 + 18x^2yz^2$
39. $-x^3y^3z^2 + 10x^2yz^2 - 3xy^3z^4$ **41.** $x + 3y$ **43.** $10x + 2$ **45.** $10x^2 - 2x$ **47.** $2x^3$ **49.** $x^2 + 7x$

1-9-1

1. 9 **3.** 2 **5.** 11 **7.** 8 **9.** 7 **11.** 18 **13.** 24 **15.** 20 **17.** 29 **19.** 14 **21.** 20 **23.** 26 **25.** 7 **27.** 17 **29.** 23
31. -9 **33.** 21 **35.** 2 **37.** 20 **39.** 22

1-9-2

1. 6 **3.** 40 **5.** $10x + 24$ **7.** $6 - 6x$ **9.** $12x + 44$ **11.** $12x^2 - 2xy$ **13.** $3x^3 - 12x^2 + 6x$ **15.** $-13a - 2$
17. $39x^2 + 27x$ **19.** $14x^2 - 8x^3 + 12x$ **21.** x^2 **23.** $8x^2 - 10x$ **25.** $90a^2 - 12a^3 - 5a$ **27.** $6x^4 - 12x^2$
29. $12xy^2 - 20x^2$

1-10-1

1. 15 **3.** -16 **5.** 9 **7.** -9 **9.** 9 **11.** -24 **13.** -216 **15.** -1 **17.** 28 **19.** 14
21. 8 **23.** -80 **25.** 17 **27.** 38 **29.** 32 **31.** 42.9 **33.** 220 **35.** $\$4,320$ **37.** 154 **39.** 14
41. 42.875 **43.** 90 **45.** 400 **47.** $1,390$ **49.** 80

Chapter 1 Review

The number in brackets after each answer refers to the section of the chapter that discusses that type problem.
1. Counting number, whole number, integer, rational number [1–1] **3.** Rational number [1–1] **5.** Irrational number [1–1]
7. Rational number [1–1] **9.** Commutative property of multiplication [1–1] **11.** Commutative property of addition [1–1]

13. Distributive property [1–1]   **15.** Additive inverse [1–1]   **17.** -14 [1–2]   **19.** $+\dfrac{1}{3}$ [1–2]   **21.** [1–2]

23. [1–2]   **25.** 8 [1–2]   **27.** -4 [1–2]   **29.** $-x + 4$ [1–2]   **31.** $-3x + 2y - 5$ [1–2]   **33.** -8 [1–3]

35. 6 [1–3]   **37.** -2 [1–4]   **39.** $+\dfrac{1}{3}$ [1–4]   **41.** -7 [1–4]   **43.** -1 [1–4]   **45.** 2 [1–4]   **47.** -2 [1–4]   **49.** 22 [1–4]

51. -48 [1–5]   **53.** $\dfrac{3}{14}$ [1–5]   **55.** 280 [1–5]   **57.** 12 [1–5]   **59.** $-\dfrac{2}{3}$ [1–5]   **61.** 3 [1–5]   **63.** -60 [1–5]   **65.** -6 [1–5]

67. x^8 [1–6]   **69.** $-8a^6$ [1–6]   **71.** $12x^6y^7$ [1–6]   **73.** x^8 [1–6]   **75.** $16a^4b^{12}c^8$ [1–6]   **77.** $-5x^3$ [1–6]   **79.** $-\dfrac{a^2c}{3b^2}$ [1–6]

81. $2x + 3y$ [1–7]   **83.** $12x^2y - 2xy - 7xy^2$ [1–7]   **85.** $-6x + 3y$ [1–8]   **87.** $6x^3y - 4x^2y^2 + 8x^2y^4$ [1–8]

89. $-15a^6b^4 + 21a^3b^4c - 9a^5b^3c^2$ [1–8]   **91.** 27 [1–9]   **93.** 11 [1–9]   **95.** $x - 12$ [1–9]   **97.** $9x + 3$ [1–9]

99. $17 - 6x$ [1–9]   **101.** $14 - 7a$ [1–9]   **103.** $23x + 20$ [1–9]   **105.** $4a - 12$ [1–9]   **107.** $-3a - 20$ [1–9]

109. -375 [1–10]   **111.** 45 [1–10]   **113.** -51 [1–10]   **115.** 7 [1–10]   **117.** 16 [1–10]   **119.** 21 [1–10]

Chapter 1 Test

The number in brackets after each answer refers to the section of the chapter that discusses that type problem.

1. Commutative property of multiplication [1–1]   **2.** $-\dfrac{4}{5}$ [1–2]   **3.** 8.3 [1–2]   **4.** -4 [1–2]   **5.** 3 [1–2]   **6.** -5 [1–2]

7. 11 [1–2]   **8.** $-\$15$ [1–3]   **9.** 9 [1–4]   **10.** -5 [1–4]   **11.** -36 [1–5]   **12.** 12 [1–5]   **13.** 120 [1–5]

14. $-30a^6b^7$ [1–6]   **15.** 18 [1–4]   **16.** $4ab - 3a^2$ [1–7]   **17.** $12x^3y^3 - 20x^5y + 16x^5y^2$ [1–8]   **18.** $64x - 48$ [1–9]

19. 28 [1–10]   **20.** 87.5 [1–10]

Chapter 2 Pretest

The number in brackets after each answer refers to the section of the chapter that discusses that type problem.

1. Identity [2–1]   **2.** Yes [2–1]   **3.** $x = -7$ [2–2]   **4.** $x = \dfrac{7}{2}$ [2–3]   **5.** $x = \dfrac{5}{2}$ [2–3]   **6.** $x = -6$ [2–5]

7. $x = -22$ [2–4]   **8.** $x = -4$ [2–5]   **9.** 18 boys [2–4]   **10.** Height = 8 in. [2–5]   **11.** $x = \dfrac{5c}{b}$ [2–6]

12. $x \le 4$ [2–7]   **13.** $-6, 9$ [2–8]   **14.** [2–8]

2-1-1

1. Identity   **3.** Conditional   **5.** Identity   **7.** Identity   **9.** Conditional   **11.** 3   **13.** 11   **15.** 4   **17.** 10   **19.** 6   **21.** $\dfrac{3}{5}$

23. $\dfrac{1}{3}$ **25.** 0   **27.** 7   **29.** 6   **31.** Equivalent   **33.** Not equivalent   **35.** Not equivalent   **37.** Not equivalent   **39.** Equivalent

2-2-1

1. $x = 12$ **3.** $x = -6$ **5.** $x = 3$ **7.** $x = \dfrac{1}{3}$ **9.** $x = -\dfrac{1}{2}$ **11.** $x = 3$ **13.** $x = -3$ **15.** $x = \dfrac{2}{5}$

17. $x = -\dfrac{1}{4}$ **19.** $x = -14$ **21.** $x = 12$ **23.** $x = 2$ **25.** $w = 17$ m   **27.** $r = 69\dfrac{2}{5}$ km/hr   **29.** 14%   **31.** 45%

2-3-1

1. 2   **3.** -5 **5.** 9   **7.** 8   **9.** -9 **11.** 16   **13.** $3\frac{13}{20}$ **15.** -3 **17.** $\frac{19}{15}$ **19.** 0   **21.** $\frac{3}{13}$ **23.** 16.1   **25.** 7   **27.** 9

29. 8   **31.** 8   **33.** 9   **35.** 1   **37.** 9   **39.** -1 **41.** 2   **43.** 1   **45.** -6 **47.** 1   **49.** 7

51. $-\frac{7}{11}$ **53.** $\frac{7}{11}$ **55.** 4   **57.** -6 **59.** $\frac{5}{2}$ **61.** 8   **63.** 11   **65.** -5 **67.** 12.4 cm   **69.** \$38.75   **71.** $30°\,C$

73. \$13.89

2-4-1

1. 12   **3.** -12 **5.** 15   **7.** -25 **9.** $7\frac{1}{2}$ **11.** 48   **13.** $10\frac{2}{3}$ **15.** $-16\frac{4}{5}$ **17.** $\frac{15}{16}$ **19.** $\frac{2}{5}$ **21.** $\frac{4}{5}$ **23.** $-\frac{4}{5}$

25. -9.76 **27.** $4\frac{2}{3}$ **29.** $1\frac{3}{7}$ **31.** 48   **33.** $5\frac{1}{17}$ **35.** $\frac{1}{4}$ **37.** $1\frac{1}{8}$ **39.** 8 cm   **41.** $100°$

2-4-2

1. 9   **3.** 4   **5.** 6   **7.** 60   **9.** $\frac{5}{2}$ **11.** 10   **13.** 14   **15.** $62\frac{1}{2}$ miles   **17.** Answers will vary   **19.** 7 gal   **21.** 12 tsp

23. 219 games   **25.** 16.69 cm

2-5-1

1. $8\frac{1}{2}$ **3.** -3 **5.** -2 **7.** $\frac{1}{2}$ **9.** $-\frac{2}{3}$ **11.** 5   **13.** $1\frac{19}{21}$ **15.** -10 **17.** $1\frac{3}{5}$ **19.** 2   **21.** 19   **23.** $3\frac{1}{3}$ **25.** $13\frac{1}{2}$

27. $\frac{51}{64}$ **29.** -4 **31.** $2\frac{1}{3}$ **33.** 140   **35.** $-\frac{3}{16}$ **37.** $-5\frac{3}{5}$ **39.** 30   **41.** $5\frac{3}{4}$ **43.** 5   **45.** 36   **47.** 0   **49.** $\frac{2}{15}$

51. $22\frac{8}{13}$ **53.** $-12\frac{3}{5}$ **55.** $34\frac{2}{7}$ **57.** 83   **59.** $17\frac{1}{2}$ **61.** -5 **63.** $1\frac{1}{45}$ **65.** $10\frac{1}{2}$ **67.** $-5\frac{5}{7}$ **69.** $-\frac{1}{5}$ **71.** 75 ft

73. \$8,526   **75.** \$67.50

2-6-1

1. $2y$ **3.** $2y$ **5.** $-6a$ **7.** $2y$ **9.** $2a$ **11.** $\frac{22bc}{a}$ **13.** $\frac{5a+3y}{3ab}$ **15.** $\frac{10b-3a}{9}$ **17.** $\frac{a-75}{29}$ **19.** $\frac{2a+3b}{2}$

21. $h=\frac{2A}{b}$ **23.** $g=\frac{2s}{t^2}$ **25.** $p=\frac{365I}{rD}$; \$4,506.17

2-7-1

1. $6<10$ **3.** $-3<3$ **5.** $4>1$ **7.** $-2>-3$ **9.** $0<7$

2-7-2

17. $x>-4$ **19.** $-4 \le x \le 4$

2-7-3

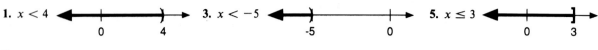

1. $x < 4$

3. $x < -5$

5. $x \leq 3$

7. $x > -17$

9. $x \geq -5$

2-7-4

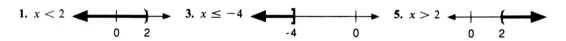

1. $x < 2$

3. $x \leq -4$

5. $x > 2$

7. $x < 4$

9. $x > 6$

11. $x \leq 3$

13. $x > 2$

15. $x \geq -2$

17. $x \leq -6$

19. $x > 3$

21. $x \leq 4\frac{1}{2}$

23. $x \geq 2$

25. $x < \frac{1}{2}$

27. $x \leq -20$

29. $x > -4$

31. $x \leq \frac{7}{3}$

33. $x < 140$

35. $x \leq -\frac{3}{16}$

37. $x > -5\frac{3}{5}$

39. $x > 3$

41. $x \leq 5\frac{3}{4}$

43. $x < 5$

45. $x > 36$

47. $x < 0$

49. $x < -6$

2-8-1

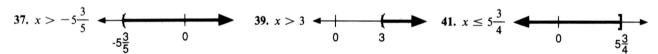

1. $-2, 2$ **3.** $-9, 9$ **5.** $-4, 2$ **7.** $3, 5$ **9.** 0 **11.** $-3, 2$ **13.** $-\frac{2}{3}, 2$ **15.** $-20, 4$ **17.** 5 **19.** $-12, -\frac{20}{3}$

2-8-2

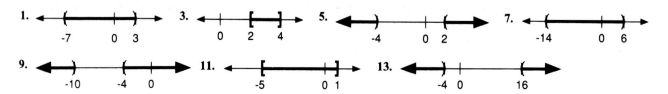

1.

3.

5.

7.

9.

11.

13.

15. (number line: closed bracket at -4, open bracket at 7) **17.** (number line: open bracket at $-\frac{15}{2}$, closed bracket at $\frac{9}{2}$) **19.** (number line: open paren at 12, open bracket at 18)

Chapter 2 Review

The number in brackets after each answer refers to the section of the chapter that discusses that type problem.

1. Conditional [2–1]   **3.** Identity [2–1]   **5.** Conditional [2–1]   **7.** Equivalent [2–1]   **9.** Not equivalent [2–1]

11. Equivalent [2–1]   **13.** 8 [2–2]   **15.** 19 [2–2]   **17.** -5 [2–3]   **19.** 39 [2–2]   **21.** 75 [2–2]   **23.** 3 [2–3]   **25.** 4 [2–3]

27. 6 [2–4]   **29.** $8\frac{1}{5}$ [2–2]   **31.** $4\frac{1}{2}$ [2–3]   **33.** $2\frac{4}{9}$ [2–4]   **35.** $2\frac{1}{7}$ [2–5]   **37.** 6 [2–4]   **39.** $\frac{19}{35}$ [2–4]   **41.** $4\frac{1}{5}$ [2–5]

43. 12 [2–3]   **45.** $4\frac{1}{4}$ [2–3]   **47.** $2\frac{2}{9}$ [2–4]   **49.** $\frac{45}{62}$ [2–5]   **51.** 2 [2–5]   **53.** $\frac{5}{12}$ [2–4]   **55.** 0 [2–3]   **57.** $1\frac{5}{21}$ [2–4]

59. $1\frac{31}{35}$ [2–5]   **61.** $1\frac{5}{9}$ [2–4]   **63.** $8\frac{2}{5}$ [2–5]   **65.** $\frac{6-z}{3y}$ [2–6]   **67.** $\frac{7y-7}{2}$ [2–6]   **69.** $\frac{6-2y}{7}$ [2–6]   **71.** $6-y$ [2–6]

73. $b = \frac{2A - ha}{h}$ [2–6]   **75.** $66\frac{2}{3}$ [2–4]   **77.** 14 [2–4]   **79.** (number line: closed bracket at 0, closed bracket at 5) [2–7]   **81.** $-11, 1$ [2–8]

83. $-3, \frac{11}{3}$ [2–9]   **85.** $-9, -6$ [2–8]   **87.** $x < 4$ (number line: open paren at 4) [2–7]

89. $x \geq -5$ (number line: closed bracket at -5) [2–7]   **91.** $x \geq -3$ (number line: closed bracket at -3) [2–7]   **93.** $x \leq -2$ (number line: closed bracket at -2) [2–7]

95. $x > \frac{11}{6}$ (number line: open paren at $\frac{11}{6}$) [2–7]   **97.** (number line: open paren at -1, open paren at 7) [2–8]   **99.** (number line: closed bracket at -4, closed bracket at -1) [2–8]

101. (number line: closed bracket at -12, closed bracket at 6) [2–8]   **103.** 18 m [2–2]   **105.** 18 m [2–5]   **107.** \$474 [2–4]   **109.** \$5,500 [2–2]

111. 84.2 cm [2–3]

Chapter 2 Test

The number in brackets after each answer refers to the section of the chapter that discusses that type problem.

1. Conditional [2–1]   **2.** Yes [2–1]   **3.** -4 [2–2]   **4.** $1\frac{1}{2}$ [2–3]   **5.** -2 [2–3]   **6.** $22\frac{1}{2}$ [2–4]   **7.** $-2\frac{5}{11}$ [2–5]

8. 140 [2–4]   **9.** 8 [2–5]   **10.** $w = 5, \ell = 12\frac{1}{2}$ [2–5]   **11.** $b = \frac{2A - ha}{h}$ [2–6]

12. $x \leq \frac{14}{3}$ (number line: closed bracket at $\frac{14}{3}$) [2–7]   **13.** $-\frac{8}{5}, 2$ [2–8]   **14.** (number line: closed bracket at $-\frac{1}{3}$, closed bracket at 3) [2–8]

Chapter 3 Pretest

The number in brackets after each answer refers to the section of the chapter that discusses that type problem.

1. a. $2x - 9$ **b.** $\frac{y}{3} + 1$ **c.** $.173x$ **d.** $10d$ [3–1]   **2.** 6, 12, 8 [3–2]   **3.** 10 [3–3]   **4.** 4 hours [3–4]   **5.** 1.4 liters [3–5]

3-1-1

1. $x + 5$ **3.** $x + 5$ **5.** $x - 9$ **7.** $x + 10$ **9.** $2x$ **11.** $2x + 2$ **13.** $6x - 4$ **15.** $\frac{1}{5}x$ **17.** $x - 3$ **19.** $8a$

21. $\frac{5x}{2}$ **23.** $\frac{b + 7}{2}$ **25.** $3(x - 9)$ **27.** $.15x$ **29.** $.1x$ **31.** $.095x$ **33.** $10d$ **35.** $10d + 25q$ **37.** $12y$ **39.** $\frac{d}{7}$

3-1-2

The following represent one way of answering each problem.

1. $x + 8$ = first number
   x = second number

3. x = height of World Trade Center
   $x + 104$ = height of Sears Tower

5. x = units produced first week
   $x + 612$ = units produced second week

7. x = number students in 7:00 A.M. class
   $x + 8$ = number students in 9:00 A.M. class

9. x = population of Long Beach
   $2x$ = population of San Francisco

11. x = width
   $x + 5$ = length

13. x = width
   $2x + 3$ = length

15. x = first number
   $40 - x$ = second number

17. x = first number
   $x + 18$ = second number

19. x = first investment
   $10,000 - x$ = second investment

21. x = last year's price
   $x + .05x$ = this year's price

23. $x + 4$ = first number
   x = second number
   $x + 13$ = third number

25. x = first number
   $2x$ = second number
   $\frac{x}{3}$ = third number

27. x = first even number
   $x + 2$ = second even number
   $x + 4$ = third even number

29. x = cost in 1970
   $\frac{3}{2}x$ = cost in 1976
   $3x$ = cost now

31. $x + 5$ = number dollars Jane has
   x = number dollars Bob has
   $x + 18$ = number dollars Jim has

33. $x + 4$ = mileage of Buick
   x = mileage of Chevrolet
   $x + 12$ = mileage of Toyota

35. x = original amount
   $.14x$ = interest
   $x + .14x$ = amount in the account

37. x = length of first piece
   $100 - x$ = length of second piece

39. x = number of nickels
   $2x$ = number of dimes
   $2x + 2$ = number of quarters

3-2-1

1. First number = 15, second number = 7   **3.** Sears Tower = 1,454 ft; World Trade Center = 1,350 ft
5. First week = 8,902; second week = 9,514   **7.** 7:00 A.M. class = 38, 9:00 A.M. class = 46   **9.** 716,000
11. Length = 14 m, width = 9 m   **13.** 65 cm   **15.** First number = 14, second number = 26
17. First number = 32, second number = 50   **19.** First investment = $6,628; second investment = $3,372
21. First number = 26, second number = −11   **23.** First number = 15, second number = 11, third number = 24
25. Coach ticket = $109, first-class ticket = $218, economy ticket = $79
27. Buick = 22 mpg, Chevrolet = 18 mpg, Toyota = 30 mpg   **29.** First side = 32 cm, second side = 16 cm, third side = 36 cm

3-3-1

1. 39, 44   **3.** Jane's age = 23, Louis' age = 18   **5.** 12, 63   **7.** Mark's age = 31, Mario's age = 17
9. Completions in 1978 = 231, completions in 1979 = 267   **11.** Length = 66 cm, width = 46 cm   **13.** 8, 24   **15.** 112, 121
17. 11, 29   **19.** First number = 48, second number = 36, third number = 68   **21.** 28, 30, 32
23. Jim's age = 19, Sue's age = 17, Hugh's age = 23   **25.** Selling price = $26.80, sales tax = $1.34
27. Population of town A = 25,795; population of town B = 23,450
29. Alice's height = 167 cm, Jane's height = 152 cm, Diane's height = 170 cm
31. Jim's weight = 72 kg, Tom's weight = 77 kg, Bob's weight = 74 kg   **33.** Saxon copies = 20, Canon copies = 12
35. Cyclist A = 77 km, cyclist B = 72 km, cyclist C = 74 km   **37.** Length = 10, width = 5   **39.** 8   **41.** 5,400
43. 12.5%   **45.** 14%

3-4-1

1. 264 miles   **3.** 2 hours   **5.** 19 mph   **7.** 11.69 m/sec   **9.** 238,080 miles   **11.** 6 hours   **13.** 2 hours   **15.** 2 hours

17. 2 hours   **19.** $\frac{1}{2}$ hour   **21.** 7 hours   **23.** 2,289 km   **25.** Walker = 6 mph, cyclist = 15 mph

27. Joan's rate = 6 mph, Sally's rate = 4 mph   **29.** Car A's rate = 77 km/hr, car B's rate = 85 km/hr

31. 2 hours 45 minutes   **33.** 390 km/hr   **35.** 170 mph   **37.** 40 km/hr   **39.** $4\frac{2}{7}$ km/hr

3-5-1

1. 10 lb caramels, 20 lb creams   **3.** 12.5 lb peanuts, 7.5 lb cashews   **5.** 18 nickels, 57 quarters   **7.** 21 dimes, 14 quarters
9. 25 bushels oranges, 10 bushels grapefruit   **11.** 30 kg pecans, 20 kg cashews   **13.** 63 nickels, 21 quarters   **15.** 15 liters
17. 50 children, 160 adults   **19.** 22 lb @ $1.90, 17 lb @ $2.30, 11 lb @ $2.60   **21.** 10 nickels, 25 dimes, 18 quarters

23. 13 nickels, 16 dimes, 8 quarters   **25.** 100 gal skim milk, 125 gal 3.6% milk   **27.** $3\frac{1}{3}$ liters

29. 24 first-class, 120 coach, 32 super saver

Chapter 3 Review

(The number in brackets after each answer refers to the section of the chapter that discusses that type of problem.)

1. $2x - 7$ [3–1]   **3.** $x + 5$ [3–1]   **5.** $\frac{y}{100}$ [3–1]   **7.** $x, 84 - x$ [3–1]   **9.** $2x, 181 - 3x, x$ [3–1]

11. $x + (x + 5) = 23$ [3–2]   **13.** $3x - (x - 7) = 13$ [3–2]   **15.** $.09x + x = 1.26$ [3–2]   **17.** 1, 8 [3–2]
19. 25.7 kg [3–5]   **21.** 12, 24, 28 [3–3]   **23.** 4 nickels, 8 dimes, 10 quarters [3–5]   **25.** A: 15 kg, B: 20 kg, C: 15 kg [3–5]
27. 126 km [3–4]   **29.** 80 km/hr [3–4]   **31.** 4.5 m, 13.5 m, 22 m [3–3]   **33.** 30 nickels, 24 dimes, 53 quarters [3–5]

35. 1.2 liters [3–5]   **37.** $\frac{1}{2}$ hour [3–4]

Chapter 3 Test

(The numbers in brackets after each answer refer to the section of the chapter that discusses that type of problem.)

1. a. $3x + 6$ **b.** $\frac{y}{2} - 7$ **c.** $.094x$ **d.** $\frac{d}{7}$ [3–1]   **2.** 5 nickels, 8 dimes, 16 quarters [3–2]   **3.** 12m, 24m, 27m [3–3]
4. 3 km/hr [3–4]   **5.** 5 liters [3–5]

Chapters 1–3 Cumulative Test

(The number in brackets after each answer refers to the chapter and section that discusses that type of problem.)
1. +1.5 [1–2]   **2.** −20 [1–2]   **3.** +14 [1–4]   **4.** 30 [1–5]   **5.** $40x^5y^6$ [1–6]   **6.** $19a - 8$ [1–9]   **7.** 43 [1–10]

8. Identity [2–1]   **9.** −7 [2–2]   **10.** 4 [2–3]   **11.** 28 [2–4]   **12.** $\frac{2x + 5y}{3}$ [2–6]   **13.** $x \leq 2\frac{9}{10}$ [2–7]

14. $-\frac{2}{3} < x < 4$ [2–8]   **15.** $10x$ [3–1]   **16.** $(8x - 3) - 5x = 6$ [3–2]

17. Length = 8 ft, width = 3 ft [3–2]   **18.** 12, 13, 14 [3–3]   **19.** 35 mph [3–4]   **20.** 10 liters [3–5]

Chapter 4 Pretest

The number in brackets after each answer refers to the section of the chapter that discusses that type problem.
1. a. $2x^2 + 13x - 7$ [4–1]   **b.** $2x^3 - 9x^2 + 17x - 12$ [4–1]   **2. a.** $6a^2b^2 - ab - 3$ [4–2]
b. $x^3 - 3x^2 + 8x - 24, R\ 76$ [4–2]   **3.** $7x(5x - 4)$ [4–3]   **4.** $2x^2y^2(3xy - 4x + 5y)$ [4–3]   **5.** $(x - y)(a - 1)$ [4–3]
6. $(x - 11)(x + 11)$ [4–7]   **7.** $(5x - 2)^2$ [4–7]   **8.** $(a - b)(x + 3)$ [4–4]   **9.** $(x + 10)^2$ [4–7]
10. $(x + 4)(3x + 5)$ [4–5]   **11.** $(2x - 1)(3x + 5)$ [4–5]   **12.** $(x + 2y)(x^2 - 2xy + 4y^2)$ [4–7]
13. $(6x - 11)(6x + 11)$ [4–7]   **14.** $(x - 9)(x + 7)$ [4–5]   **15.** $(a - 2)(a + 2)(x - 3)$ [4–7]   **16.** $3(2x - 5)(x + 7)$ [4–8]
17. $17x(x - 1)(x + 1)$ [4–8]   **18.** $3x(x - 5)(2x + 1)$ [4–8]   **19.** $(4x - 3)(2x + 5)$ [4–5]
20. $(2x - y - 1)(2x + y - 1)$ [4–9]

4–1–1

1. $9x^2 - 5x + 6$ **3.** $3a^2 + 1$ **5.** $x^2 - 5x + 7$ **7.** $2a^3 - a^2 + 13$ **9.** $5x^2 - 5x + 10$ **11.** $6x^2 - 2x$
13. $2x^3 - 4x^2 + 2x$ **15.** $6x^4 + 3x^3 - 15x^2$ **17.** $-10a^3 + 5a^2 - 35a$ **19.** $8a^6 - 16a^5 + 12a^4 - 20a^3$

4–1–2

1. $x^3 + 8x^2 + 16x + 5$ **3.** $x^3 - 5x^2 + 7x - 3$ **5.** $2x^3 - 9x^2 + 10x - 3$ **7.** $15a^3 + 8a^2b - 19ab^2 - 4b^3$
9. $8x^2 - 9y^2 - 6xy + 4xz + 3yz$ **11.** $x^3 + 27$ **13.** $2x^3 - 13x^2 + 7x - 6$ **15.** $x^3 - 1$ **17.** $x^4 + 5x^3 + 3x^2 + 2x + 3$
19. $x^3 - 27$ **21.** $a^2 + b^2 + c^2 + 2ab + 2ac + 2bc$ **23.** $x^2 - y^2 + x + 7y - 12$ **25.** $2a^2 - 2b^2 - 7a - 9b + 3ab + 5$
27. $5x^2 - 3y^2 - 3z^2 + 2xy - 14xz + 10yz$ **29.** $x^3 + 5x^2 + x - x^2y^2 + 4xy + 5 + x^2y - x^3y$

4–2–1

1. $2x^2$ **3.** $2x^6$ **5.** $\dfrac{b^5}{3}$ **7.** $\dfrac{3}{5a^8}$ **9.** $5xy$ **11.** $-\dfrac{4}{x^2}$ **13.** $\dfrac{a^2}{3b^3}$ **15.** $\dfrac{3x^2}{4y^2}$ **17.** $\dfrac{1}{a}$ **19.** $\dfrac{x^7}{4}$ **21.** $-\dfrac{7y^4}{2x^3}$ **23.** $-\dfrac{4y^4}{x^3}$ **25.** 1

27. $-\dfrac{5x}{7z}$ **29.** $-\dfrac{3wx^2}{4}$

4–2–2

1. $a + 5b$ **3.** $2x^2 + 5y^2$ **5.** $ab + b^2$ **7.** $a + 3b$ **9.** $2 - 3x^2$ **11.** $b + ab^2$ **13.** $5x - 6y$ **15.** $x + y$ **17.** $x + y$
19. $1 - 6xy^2$ **21.** $3b - 7ab^3$ **23.** $2y + 5xy^2z^3$ **25.** $8xy - 12y^2 + 2$ **27.** $-x - 2y$ **29.** $2b^3 - 4a^3 + 1$
31. $10 - 7x - 2y$ **33.** $-4x - 3y$ **35.** $9y - 7$ **37.** $8ab - 1 + 2b$ **39.** $7y^2 + 5x - 1$

4–2–3

1. $(x + 2)$ **3.** $(x - 2)$ **5.** $(x - 2)$ **7.** $(x - 15)$ R 5   **9.** $(x - 2)$ **11.** $(x - 11)$ **13.** $(x + 4)$ **15.** $(x^2 - 3x + 4)$
17. $(2x^2 + 4x + 7)$, R 27   **19.** $(x^2 - 6x + 54)$, R 14   **21.** $(x^2 + 11x + 120)$, R 200   **23.** $(x^2 + 3x + 1)$, R -12
25. $(x^2 + 7x - 1)$, R -6 **27.** $(x^3 - 3x^2 + 8x - 24)$ **29.** $(x^3 - x^2 + x - 1)$

4–3–1

1. $2(6a + 5b)$ **3.** $4(3x - 5y)$ **5.** $a(a + 5)$ **7.** $x(x - y)$ **9.** $2a(a + 3)$ **11.** $5x(3x - 7)$ **13.** $3x(3x - 1)$
15. $3a(2a + 1)$ **17.** $4xy(2x + 3)$ **19.** $x^2y(x - y)$ **21.** $2ab(5a - b + 3)$ **23.** $5x^2y(3x^2 - 4y + 1)$ **25.** Prime
27. $3(2x + 7y - 9z)$ **29.** $7a(2ab - 5b - 9)$ **31.** $(x + 4)(2x + 3)$ **33.** $(a - 1)(a + 5)$ **35.** $(y + 4)(6x - 7)$
37. $(x - 2)(3x + 1)$ **39.** $(a - 4)(7a^2 - 28a - 15)$ **41.** $(x + 3)(x + 3)$ **43.** $x(x + 1)(3x + 2)$
45. $a(a + b)(a^2 + ab + 4)$ **47.** $2a(a + 4)(4a + 1)$ **49.** $6x^2(x - 5)(2x^2 - 10x - 3)$

4–4–1

1. $(x + y)(a + 3)$ **3.** $(b + c)(a - 2)$ **5.** $(2x + y)(a + 3)$ **7.** $(x - y)(a + 2)$ **9.** $(x + 2y)(a + 5)$
11. $(a - b)(c - 5)$ **13.** $(x - y)(3a + 2)$ **15.** $(x - y)(a + 3)$ **17.** $(a + 1)(3x - y)$ **19.** $(x + y)(2a - 1)$
21. $(3x + y)(2a - 1)$ **23.** $(2x + 3)(2a + c)$ **25.** $(a + 3)(a - b)$ **27.** $(x - 2)(y + 3)$ **29.** $(x + 5)(y - 2)$

4–5–1

1. $a^2 + 5a + 6$ **3.** $x^2 + 7x + 10$ **5.** $a^2 - 5a + 6$ **7.** $a^2 - 8a + 15$ **9.** $x^2 + 4x - 5$ **11.** $a^2 + 4a - 45$
13. $6x^2 + 7x + 2$ **15.** $6a^2 + 5a - 4$ **17.** $6x^2 - 19x + 15$ **19.** $10a^2 - 5ab + 6a - 3b$ **21.** $4a^2 - 20a + ac - 5c$
23. $x^2 + 10x + 25$ **25.** $x^2 - 14x + 49$ **27.** $x^2 - 9$
29. $4x^2 - 49$

4-5-2

1. $(x + 1)(x + 3)$ **3.** $(x + 2)(x + 3)$ **5.** $(x + 1)(x + 9)$ **7.** $(x + 4)(x + 4)$ **9.** $(x + 2)(x + 12)$
11. $(x - 1)(x - 3)$ **13.** $(x - 1)(x - 2)$ **15.** $(x - 3)(x - 7)$ **17.** $(x - 3)(x - 5)$ **19.** $(x - 4)(x - 4)$
21. $(x + 1)(x - 3)$ **23.** $(x + 2)(x - 4)$ **25.** $(x + 3)(x - 2)$ **27.** $(x + 4)(x - 5)$ **29.** $(x + 7)(x - 3)$
31. $(x + 40)(x + 1)$ **33.** $(x + 30)(x - 1)$ **35.** $(x + 8)(x - 6)$ **37.** $(x + 9)(x - 2)$ **39.** $(x - 6)(x - 2)$
41. $(x + 35)(x - 1)$ **43.** $(x + 12)(x - 3)$ **45.** $(x - 8)(x - 3)$ **47.** $(x + 16)(x - 3)$ **49.** $(x + 10)(x - 4)$
51. $(x + 10)(x + 5)$ **53.** $(x + 50)(x - 1)$ **55.** $(x + 4)(x + 18)$ **57.** $(x - 8)(x + 3)$ **59.** $(x - 3)(x + 5)$
61. $(x + 6)(x + 8)$ **63.** $(x + 3)(x + 28)$ **65.** $(x - 1)(x + 84)$ **67.** $(x - 8)(x + 5)$ **69.** $(x - 1)(x + 22)$
71. $(x - 6)(x + 14)$ **73.** $(x - 14)(x + 2)$ **75.** $(x + 5)(x + 26)$ **77.** $(x - 10)(x + 13)$ **79.** $(x - 48)(x - 2)$

4-5-3

1. $(x + 2)(2x + 1)$ **3.** $(2x + 3)(x + 1)$ **5.** $(x - 3)(2x - 1)$ **7.** $(3x + 2)(2x + 1)$ **9.** $(2x - 5)(x + 4)$
11. $(3x + 5)(x - 2)$ **13.** $(4x - 3)(2x + 1)$ **15.** $(6x + 1)(x + 5)$ **17.** $(5x + 3)(x + 1)$ **19.** $(8x + 1)(2x - 1)$
21. $(x + 2)(4x - 9)$ **23.** $(2x + 1)(2x + 1)$ **25.** $(2x - 15)(3x - 2)$ **27.** $(3x - 2)(3x - 2)$ **29.** $(3x + 1)(2x - 1)$
31. $(2x + 3)(3x + 2)$ **33.** $(5x + 2)(3x - 5)$ **35.** $(6x - 5)(2x - 1)$ **37.** $(2x - 3)(3x + 5)$ **39.** $(3x + 4)(5x - 6)$

4-6-1

1. 6   **3.** -30 **5.** -40 **7.** 60   **9.** -16 **11.** 40   **13.** -21 **15.** 180   **17.** 24   **19.** -150

4-6-2

1. $(x + 2)(2x + 3)$ **3.** $(x + 4)(2x + 5)$ **5.** $(3x + 2)(x + 4)$ **7.** $(3x - 2)(x + 2)$ **9.** $(x - 5)(2x + 1)$
11. $(5x - 1)(x + 4)$ **13.** $(x - 5)(3x - 2)$ **15.** $(4x - 3)(x - 2)$ **17.** $(2x + 1)(3x + 2)$ **19.** $(2x + 1)(3x - 2)$
21. $(4x + 3)(x - 9)$ **23.** $(4x - 1)(2x + 5)$ **25.** $(x - 6)(6x - 5)$ **27.** $(2x - 1)(4x + 7)$ **29.** $(3x + 2)(6x - 5)$

4-6-3

1. $(x + 4)(x + 5)$ **3.** $(x - 3)(x + 2)$ **5.** $(x - 6)(x - 2)$ **7.** $(2x + 1)(x + 3)$ **9.** $(x + 1)(3x + 2)$
11. $(2x + 3)(x - 4)$ **13.** $(5x - 1)(x + 2)$ **15.** $(x - 3)(2x - 5)$ **17.** $(x - 4)(3x - 2)$ **19.** $(x - 8)(3x - 4)$
21. $(x + 3)(4x - 3)$ **23.** $(x + 4)(x + 4)$ **25.** $(2x + 1)(2x + 3)$ **27.** $(2x - 3)(3x + 4)$ **29.** $(4x - 3)(4x + 1)$

4-7-1

1. $(x - 2)(x + 2)$ **3.** $(x - 1)(x + 1)$ **5.** $(a - 9)(a + 9)$ **7.** $(x - 10)(x + 10)$ **9.** $(y - 11)(y + 11)$
11. $(2x - 3)(2x + 3)$ **13.** $(4x - 1)(4x + 1)$ **15.** $(2x - y)(2x + y)$ **17.** $(2x - 5y)(2x + 5y)$
19. $(4a - 11b)(4a + 11b)$ **21.** $(x + y - 8)(x + y + 8)$ **23.** $(x^2 - 3)(x^2 + 3)$ **25.** $(x^3 - 6)(x^3 + 6)$
27. $(5a^2 - 7)(5a^2 + 7)$ **29.** $[(x + y)^2 - z][(x + y)^2 + z]$

4-7-2

1. $x^2 + 2xy + y^2$ **3.** $x^2 + 12x + 36$ **5.** $a^2 - 4a + 4$ **7.** $4x^2 + 28x + 49$ **9.** $9a^2 + 6a + 1$ **11.** $4x$ **13.** $20x$
15. $2x$ **17.** $4xy$ **19.** $14xy$ **21.** $(x + 3)^2$ **23.** $(x - 7)^2$ **25.** $(2x + 3)^2$ **27.** $(3x + 5)^2$ **29.** $(6x - 5y)^2$

4-7-3

1. $(x + y)(x^2 - xy + y^2)$ **3.** $(x - y)(x^2 + xy + y^2)$ **5.** $(x - 1)(x^2 + x + 1)$ **7.** $(3a + 1)(9a^2 - 3a + 1)$
9. $(2x - y)(4x^2 + 2xy + y^2)$ **11.** $(3a + 2b)(9a^2 - 6ab + 4b^2)$ **13.** $(5a - 3b)(25a^2 + 15ab + 9b^2)$
15. $(x - y^2)(x^2 + xy^2 + y^4)$ **17.** $(2a^2 + b)(4a^4 - 2a^2b + b^2)$ **19.** $(a - 1)(a^2 + a + 1)(a^6 + a^3 + 1)$

4-8-1

1. $2(2x + 1)(x + 3)$ **3.** $(3x + 1)(2x + 1)$ **5.** $3(x + 8)(x - 6)$ **7.** $(3x + 1)(x - 3)$ **9.** $3(x - 2)(x + 2)$
11. $4(x + 3)^2$ **13.** Prime **15.** $x(x + 4)(3x - 2)$ **17.** $7(3x + 1)(2x - 1)$ **19.** $5(x + 4)(x - 3)$
21. $x(x - 6)(x - 2)$ **23.** $5(2x - 1)(2x + 1)$ **25.** Prime **27.** $(2x + 1)(7x + 2)$ **29.** $2(2x - 1)^2$
31. $3x(x - 8)(x - 5)$ **33.** $(x - 2)(x + 7)$ **35.** Prime **37.** $(2x - 3)(2x + 1)$ **39.** $(x - 5)(x + 7)$
41. $3x(3x - 5)(3x + 5)$ **43.** $2(x^2 + 2x + 3)$ **45.** $(x - 5)(x + 6)$ **47.** $3(x + 3)(x + 28)$ **49.** Prime
51. $2a(a - 4)(a^2 + 4a + 16)$ **53.** $3y(3x + 1)(9x^2 - 3x + 1)$ **55.** $a^2b(ab - 1)(a^2b^2 + ab + 1)$

4-9-1

1. $(x - y + 2)(x + y + 2)$ **3.** $(x + y - 5)(x + y + 5)$ **5.** $(2a - b - 3)(2a - b + 3)$ **7.** $(x - y - 2)(x + y + 2)$
9. $(6 - a - 2b)(6 + a + 2b)$ **11.** $(5 - 3x + y)(5 + 3x - y)$ **13.** $(x - y - 1)(x + y + 3)$
15. $(a - b - 2)(a + b - 8)$ **17.** $(2x^2 - 2x + 1)(2x^2 + 2x + 1)$ **19.** $(x^2 - x + 1)(x^2 + x + 1)$

Chapter 4 Review

The number in brackets after each answer refers to the section of the chapter that discusses that type problem.
1. $x^2 + 18x + 81$ [4–1] **3.** $6x^2 - 17x + 5$ [4–1] **5.** $x^3 - 3x^2 - 6x + 8$ [4–1] **7.** $a^2 - 2b^2 + ab + 3a + 3b + 2$ [4–1]
9. $\dfrac{2}{x^4}$ [4–2] **11.** $2x^2 - 3y^2 + 5xy$ [4–2] **13.** $x + 8$ [4–2] **15.** $x - 8, R -10$ [4–2] **17.** $x^2 - x + 1, R\ 2$ [4–2]
19. $x^2 - 3x + 9, R -26$ [4–2] **21.** $3(3x + 2y)$ [4–3] **23.** $a(a - 1)$ [4–3] **25.** $2x^2y(3x + 5y)$ [4–3]
27. $3ab(2a + 3b - 2)$ [4–3] **29.** $(a + b)(x - 2)$ [4–3] **31.** $(x + y)(c + d)$ [4–4] **33.** $(x + y)(a - 3)$ [4–4]
35. $(x + 2)(a - 2)(a + 2)$ [4–7] **37.** $(x + 1)(x + 3)$ [4–5] **39.** $(x - 5)(x + 2)$ [4–5] **41.** $(2a + 3)(b - 2)$ [4–4]
43. $(x + 3)^2$ [4–7] **45.** $(5x - 7)(5x + 7)$ [4–7] **47.** $(x + 4)^2$ [4–7] **49.** $(3x - 2)^2$ [4–7]
51. $(a + 3)(a^2 - 3a + 9)$ [4–7] **53.** Prime [4–5] **55.** $(3x - y + 1)(3x + y + 1)$ [4–9] **57.** $(x - 2)(x + 7)$ [4–5]
59. $5(x - 3)(x + 3)$ [4–8] **61.** $(x - 8)(x + 9)$ [4–5] **63.** $(a - 2)(a - b)$ [4–4] **65.** $(2a - b - 2)(2a + b + 2)$ [4–9]
67. $(3x - 2)(x + 5)$ [4–5] **69.** $(3x - 8)(3x + 8)$ [4–7] **71.** $3(x - 2)(x^2 + 2x + 4)$ [4–8] **73.** $(3x - 2)(2x - 1)$ [4–5]
75. $(x - y - 1)(x + y - 5)$ [4–9] **77.** $2x(x + 2)(3x - 2)$ [4–8] **79.** Prime [4–5]
81. $(x^2 - 2x - 2)(x^2 + 2x - 2)$ [4–9]

Chapter 4 Test

The number in brackets after each answer refers to the section of the chapter that discusses that type problem.
1. a. $x^2 - 3x - 40$ [4–1] **b.** $2a^3 - a^2b - ab^2 + 2ab + b^2$ [4–1] **2. a.** $5x^2y + xy^3 - 4$ [4–2]
b. $x^3 + 4x^2 + 8x + 16, R\ 64$ [4–2] **3.** $11a(2a + 3)$ [4–3] **4.** $2xy(5x^2 - 3xy + y^2)$ [4–3] **5.** $(a - 1)(2x - 3)$ [4–3]
6. $(x + 9)(x - 9)$ [4–7] **7.** $(x - 11)^2$ [4–7] **8.** $(a + b)(x + 3)$ [4–4] **9.** $(x - 7)^2$ [4–7] **10.** $(x - 4)(4x - 3)$ [4–5]
11. $(2x + 3)(x + 5)$ [4–5] **12.** $(3a - 2b)(9a^2 + 6ab + 4b^2)$ [4–7] **13.** $(5x - 12)(5x + 12)$ [4–7]
14. $(x - 9)(x + 3)$ [4–5] **15.** $(x - 3)(x + 3)(a + 3)$ [4–7] **16.** $2(3x + 1)(x - 2)$ [4–8] **17.** $8x(2x - 3)(2x + 3)$ [4–8]
18. $3x(x + 2)(2x - 3)$ [4–8] **19.** $(3x - 5)(2x + 3)$ [4–5] **20.** $(x - y - 7)(x + y + 3)$ [4–9]

Chapter 5 Pretest

The number in brackets after each answer refers to the section of the chapter that discusses that type problem.
1. $\dfrac{x + 3}{x(2x + 5)}$ [5–1] **2.** 1 [5–2] **3.** $\dfrac{x - 4}{x + 5}$ [5–3] **4. a.** $(x + 4)(x + 6)(3x - 2)$ [5–4] **b.** $2x^2 - 13x + 21$ [5–5]
5. $\dfrac{x^2 + 12x + 7}{(x - 2)(x + 4)(x - 1)}$ [5–6] **6.** $\dfrac{x + 2}{x^2 - 1}$ [5–7] **7.** $-\dfrac{1}{x + 3}$ [5–8] **8.** $x = -11$ [5–9]

5-1-1

1. $\dfrac{2}{3}$ **3.** $\dfrac{3}{4}$ **5.** $\dfrac{2}{5}$ **7.** $\dfrac{3}{5}$ **9.** $\dfrac{1}{x}$ **11.** $\dfrac{2y}{3z}$ **13.** $\dfrac{3}{4}$ **15.** $\dfrac{x + 3}{x - 2}$ **17.** $\dfrac{a}{b}$ **19.** $\dfrac{2a - 1}{a + 4}$ **21.** -1 **23.** $-\dfrac{2}{3}$

25. $-\dfrac{a}{4}$ **27.** In simplified form **29.** $\dfrac{x}{3x - 2}$ **31.** $\dfrac{2x - 3}{a - 2}$ **33.** $\dfrac{x + 1}{x + 2}$ **35.** $x - 3$ **37.** $\dfrac{x - 1}{(x - 3)(x + 1)}$ **39.** $\dfrac{4 - x}{x}$

41. $\dfrac{x}{x+3}$ **43.** $\dfrac{x-2}{a+7}$ **45.** $\dfrac{2x-1}{4x^2-2x+1}$ **47.** $\dfrac{x}{x+1}$ **49.** $\dfrac{3a+2}{2a-3}$

5-2-1

1. $\dfrac{3}{5}$ **3.** $\dfrac{1}{6}$ **5.** $\dfrac{z^2}{2}$ **7.** $\dfrac{3}{x+2}$ **9.** 1   **11.** 1   **13.** $\dfrac{2x+4}{x-1}$ **15.** $\dfrac{x-1}{x-2}$ **17.** $\dfrac{3}{x-1}$ **19.** x^2+3x+2 **21.** $\dfrac{4-2x}{x+5}$

23. $\dfrac{2x+2}{x+2}$ **25.** $\dfrac{2x+4}{4x-3}$ **27.** $\dfrac{x+6}{(x+2)(x+4)}$ **29.** $\dfrac{2(3x-2)(x+1)}{5(x+4)(x+3)}$

5-3-1

1. $\dfrac{1}{4}$ **3.** $\dfrac{4}{3}$ **5.** 1   **7.** $\dfrac{ab}{6}$ **9.** $\dfrac{x+1}{x-1}$ **11.** $-\dfrac{1}{2}$ **13.** $x+2$ **15.** 1   **17.** $\dfrac{1}{x+2}$ **19.** $\dfrac{4x^2-10x+25}{2x-5}$

21. $\dfrac{x+7}{(x+9)(x-4)}$ **23.** $\dfrac{2x-10}{x+1}$ **25.** $\dfrac{2x^2+13x+15}{(x+1)(2x-3)}$ **27.** $\dfrac{2x-1}{x+2}$ **29.** $-\dfrac{1}{2x+3}$

5-4-1

1. 15   **3.** 12   **5.** 24   **7.** 30   **9.** 240   **11.** xyz **13.** $4x^3y^2$ **15.** $x(x+2)$ **17.** $(x+1)(x+3)$ **19.** $6x^2(x^2-2)$
21. $(a-3)(a+3)$ **23.** $(x+3)(x-4)$ **25.** $(x-3)(x-3)$ **27.** $(a-2)(a+2)(a-5)$ **29.** $(x+1)(x+6)$
31. $(x-1)(x-1)(x-5)$ **33.** $(x+3)(x+2)(x-4)$ **35.** $(2x+3)(2x-3)(x+4)$ **37.** $20(x+1)(x-4)$
39. $(x+3)(x-2)(x-5)$ **41.** $(x-6)(x+3)(x+6)(x+2)$

5-5-1

1. 3   **3.** 12   **5.** 105   **7.** $x-3$ **9.** x^2-x-2 **11.** $8x+12$ **13.** x^2+x-12 **15.** $2x^2-13x+6$
17. $6x^2-7x-20$ **19.** $2x^2+14x+24$ **21.** $12x^2+60x+63$ **23.** $x^2-3x-10$ **25.** $2x^2+28x+98$
27. $6x^2-5x+6$ **29. a.** $\dfrac{x^2+5x+4}{(x-3)(x+4)}$ **b.** $\dfrac{x^2-5x+6}{(x-3)(x+4)}$ **31. a.** $\dfrac{2x^2+x-3}{2(x-4)(2x+3)}$ **b.** $\dfrac{4x^2-14x-8}{2(x-4)(2x+3)}$

33. a. $\dfrac{3x^2+14x+8}{(x-1)(x+3)(x+4)}$ **b.** $\dfrac{x^2+x-6}{(x-1)(x+3)(x+4)}$

5-6-1

1. $\dfrac{5}{7}$ **3.** $\dfrac{1}{2}$ **5.** $\dfrac{x+7}{x-4}$ **7.** 2   **9.** $\dfrac{x+y}{xy}$ **11.** $\dfrac{xy+xz+yz}{xyz}$ **13.** $\dfrac{x+11}{3(x+5)}$ **15.** $\dfrac{3x+2}{x(x+1)}$ **17.** $\dfrac{x-1}{(x-3)(x+2)}$

19. $\dfrac{6}{x+3}$ **21.** $\dfrac{3x-8}{(x-5)(x+2)}$ **23.** $\dfrac{8}{x+4}$ **25.** $\dfrac{16x-30}{5(x-5)(x+2)}$ **27.** $\dfrac{6}{x+1}$ **29.** $\dfrac{x^2+8x+11}{(x-6)(x+4)}$

31. $\dfrac{3x^2+8x+11}{(x-2)(x+1)(x+4)}$ **33.** $\dfrac{29}{3x}$ **35.** $\dfrac{4}{x-2}$ **37.** $\dfrac{3x+1}{(x-4)(x-2)}$ **39.** $\dfrac{1}{4x+1}$

5-7-1

1. $\dfrac{2}{5}$ **3.** $\dfrac{1}{12}$ **5.** $\dfrac{9-x}{3x}$ **7.** $\dfrac{2y-4x}{xy}$ **9.** $\dfrac{1-3x}{4(x+1)}$ **11.** $\dfrac{x-2}{3(x+1)}$ **13.** $\dfrac{x-20}{6(x-4)}$ **15.** $\dfrac{12-2x}{x(x+3)}$ **17.** $\dfrac{1}{(x+2)(x+3)}$

19. $\dfrac{x^2-14}{(x+2)(x+4)}$ **21.** $\dfrac{x^2-7x+5}{(x-2)(2x+1)}$ **23.** $\dfrac{x^2-x-4}{(x-1)(x+1)}$ **25.** $\dfrac{1}{x+2}$ **27.** $\dfrac{1}{x-2}$ **29.** $-\dfrac{x^2+2x+2}{2(x-3)(x+2)}$

31. $\dfrac{6x+12}{(x-2)(x+1)(x+3)}$ **33.** $\dfrac{3x+51}{(x+5)(x-4)(x+3)}$ **35.** 0   **37.** $-\dfrac{1}{5}$ **39.** $\dfrac{4}{x+5}$

5-8-1

1. a **3.** $\dfrac{1}{xy}$ **5.** $\dfrac{a+b}{a-b}$ **7.** $\dfrac{1}{ab}$ **9.** 3 **11.** $\dfrac{a+1}{a^2 b}$ **13.** $\dfrac{1}{x}$ **15.** $\dfrac{x^3 + x^2}{2x+1}$ **17.** $\dfrac{a-b}{ab}$ **19.** $y^2 - x^2$ **21.** $\dfrac{x+2}{x-3}$

23. $\dfrac{2a+b}{a^2 + 2a + ab}$ **25.** $\dfrac{xy - x^2}{y}$ **27.** $-\dfrac{1}{x+5}$ **29.** $\dfrac{x^2 + 5x + 1}{x^2 + 1}$

5-9-1

1. 4 **3.** $\dfrac{1}{2}$ **5.** $-\dfrac{9}{8}$ **7.** No solution   **9.** 4 **11.** -2 **13.** $\dfrac{1}{2}$ **15.** No solution   **17.** 3 **19.** 0 **21.** 3 **23.** -2

25. 5 **27.** 9 **29.** $\dfrac{9}{21}$ **31.** 18 cm   **33.** 20 hours

Chapter 5 Review

The numbers in brackets after each answer refer to the section of the chapter that discusses that type problem.

1. $\dfrac{x+15}{4}$ [5–1]   **3.** $\dfrac{2}{3}$ [5–1]   **5.** $\dfrac{x+3}{x+8}$ [5–1]   **7.** $\dfrac{x-7}{x+7}$ [5–1]   **9.** $\dfrac{3}{5(x-1)}$ [5–1]   **11.** $x(x+2)$ [5–4]

13. $x(x+3)$ [5–4]   **15.** $x(x-5)(x+1)$ [5–4]   **17.** $(x-2)(x+1)(x+2)$ [5–4]   **19.** $(x-2)(x-1)(x+1)$ [5–4]
21. $2x+20$ [5–5]   **23.** $x^2 + 10x + 21$ [5–5]   **25.** $8x^2 - 2x$ [5–5]   **27.** $2x^2 - 11x - 21$ [5–5]

29. $2x^2 + 19x + 24$ [5–5]   **31.** $\dfrac{3}{2(x+1)}$ [5–2]   **33.** $\dfrac{6x-20}{x(x-5)}$ [5–6]   **35.** $\dfrac{5}{7}$ [5–2]   **37.** $\dfrac{1}{x+1}$ [5–3]

39. $\dfrac{5x+6}{x(x+2)}$ [5–6]   **41.** $\dfrac{1}{x-4}$ [5–3]   **43.** $\dfrac{15-x}{4(x+5)}$ [5–7]   **45.** $\dfrac{x-2}{x+2}$ [5–2]   **47.** $\dfrac{x+2}{x+4}$ [5–3]

49. $\dfrac{x^2 + 2x + 3}{x(x-5)(x+1)}$ [5–6]   **51.** $\dfrac{17x+25}{(x+2)(x+1)(7-x)}$ [5–7]   **53.** $\dfrac{(2x+3)(x+2)}{2x+1}$ [5–2]   **55.** $\dfrac{2x+3}{2x+5}$ [5–3]

57. $\dfrac{6x^2 + 40x + 61}{(x-2)(x+3)(x+4)}$ [5–6]   **59.** $\dfrac{x+6}{x-2}$ [5–6]   **61.** $\dfrac{3}{x-1}$ [5–7]   **63.** $\dfrac{x^2 + 6x + 9}{(x-3)(3x+1)}$ [5–2]

65. $\dfrac{2x^2 + 9x + 2}{(2x-3)(x+2)(x+3)}$ [5–7]   **67.** $\dfrac{2x-5}{x+4}$ [5–2]   **69.** $\dfrac{2x-3}{(x-2)(x-1)}$ [5–6]   **71.** $\dfrac{1}{3}$ [5–8]   **73.** $\dfrac{x-2}{2}$ [5–8]

75. $\dfrac{1}{a}$ [5–8]   **77.** $3x+5$ [5–8]   **79.** $-x^2 + x + 6$ [5–8]   **81.** 1 [5–9]   **83.** $\dfrac{7}{6}$ [5–9]   **85.** No solution [5–9]

87. $\dfrac{1}{3}$ [5–9]   **89.** -1 [5–9]

Chapter 5 Test

The number in brackets after each answer refers to the section of the chapter that discusses that type problem.

1. $\dfrac{x+3}{x+5}$ [5–1]   **2.** $\dfrac{x+5}{x-1}$ [5–2]   **3.** $\dfrac{2}{3}$ [5–3]   **4. a.** $(x+6)(x-6)(x+4)$ [5–4]   **b.** $2x^2 - 9x + 10$ [5–5]

5. $\dfrac{x^2 - x + 12}{(x+2)(x+3)(x-3)}$ [5–6]   **6.** $\dfrac{3x^2 + 11x - 3}{(x+2)(x+4)}$ [5–7]   **7.** $x^2 - 3x$ [5–8]   **8.** $x = 2$ [5–9]

Chapter 6 Pretest

The number in brackets after each answer refers to the section of the chapter that discusses that type problem.

1. a. $x^4 y^8$ [6–1]   **b.** 2 [6–1]   **2. a.** $\dfrac{x^2}{y^6}$ [6–2]   **b.** $\dfrac{1}{\left(\dfrac{1}{a} + \dfrac{1}{b}\right)^2}$ or $\dfrac{a^2 b^2}{(b+a)^2}$ [6–2]   **3.** 2.73×10^6 [6–3]

4. 3.4×10^9 [6–3]   **5.** -4 [6–4]   **6.** 9 [6–5]   **7. a.** $x^3 y^2$ [6–6]   **b.** $-2xy^3 \sqrt[5]{2x^3 y}$ [6–6]   **8. a.** $\sqrt{15} + \sqrt{10} + \sqrt{6}$ [6–7]
b. -163 [6–7]   **9. a.** $\dfrac{3\sqrt[3]{x^2}}{x}$ [6–8]   **b.** $\dfrac{4(\sqrt{5} - \sqrt{2})}{3}$ [6–8]   **10. a.** $\dfrac{2x^{2/3} y^{1/2}}{xy}$ [6–9]   **b.** $\dfrac{11\sqrt{5}}{5}$ [6–9]

6-1-1

1. x^{10} **3.** $x^8 y^4$ **5.** $\dfrac{x^6}{y^9}$ **7.** $16x^8$ **9.** $\dfrac{8}{x^6 y^3}$ **11.** $\dfrac{y^2}{x^2}$ **13.** $8x^6 y^6$ **15.** $64x^{18}y^6$ **17.** $-3x^2$ **19.** $-\dfrac{y^3}{5x}$ **21.** $2x^8 y^4$

23. $-4x^8 y^9$ **25.** $288x^{22}y^{16}$ **27.** $-\dfrac{2}{x}$ **29.** $-\dfrac{1}{10xy^4}$

6-2-1

1. $\dfrac{1}{x^3}$ **3.** $\dfrac{1}{x^2 y^5}$ **5.** $\dfrac{1}{(ab)^3}$ **7.** $\dfrac{1}{8}$ **9.** 9 **11.** $\left(\dfrac{b}{a}\right)^5$ **13.** 1 **15.** $\dfrac{1}{ab}$ **17.** $\dfrac{1}{a}+\dfrac{1}{b}$ **19.** $\dfrac{1}{x^6}$ **21.** $\dfrac{1}{x^8}$ **23.** $\dfrac{1}{x^{15}}$

25. $\dfrac{1}{8x^6}$ **27.** $\dfrac{1}{8}$ **29.** $\dfrac{1}{y^7}$

6-2-2

1. $\dfrac{1}{x^{16}}$ **3.** $-\dfrac{9}{8x^{11}}$ **5.** $27x^{12}$ **7.** $\dfrac{x^8}{y^{12}}$ **9.** $\dfrac{x^6}{y^2}$ **11.** $\dfrac{x^{14}}{1,024y^2}$ **13.** $\dfrac{y^3}{x^3}$ **15.** $\dfrac{x^{25}}{y^{10}}$ **17.** $\dfrac{x+y}{x^2 y^2}$ **19.** $\dfrac{x^2 y^2}{(x+y)^2}$ **21.** $\dfrac{8x^6 y^3}{x+y}$

6-3-1

1. Yes **3.** Yes **5.** Yes **7.** No **9.** Yes **11.** 5×10^3 **13.** 2.35×10^{-7} **15.** 5.2×10^{-9} **17.** 6.8×10^1
19. 7.28×10^5 **21.** 320,100 **23.** 623,000,000,000,000,000,000,000 **25.** 50,200,000,000 **27.** 537.62 **29.** 3.6
31. 4.9×10^7 miles **33.** 6,000,000,000,000 miles **35.** .00000001 cm **37.** 1×10^{-3} cm **39.** 4.3×10^{19}

6-3-2

1. 4.8×10^9 **3.** 7.8×10^2 **5.** 1.7×10^7 **7.** 9.35×10^{-7} **9.** 3.78×10^{-6} **11.** 9.246×10^7 **13.** 2.1805×10^{-19}
15. 2.34×10^6 **17.** 4×10^{-11} **19.** 5×10^0

6-4-1

1. 2 **3.** 3 **5.** -1 **7.** 10 **9.** 2 **11.** -2 **13.** 15 **15.** 12 **17.** 25 **19.** -1

6-4-2

1. 3.873 **3.** 2.646 **5.** 10.100 **7.** 2.881 **9.** 4.626 **11.** 10.502 **13.** 14.832 **15.** 34.337 **17.** 32.787 **19.** 16.876

6-5-1

1. $\sqrt[3]{x}$ **3.** $\sqrt[4]{a^3}$ **5.** $\dfrac{1}{\sqrt[3]{x^2}}$ **7.** $\sqrt[7]{(-3)^2}$ **9.** $\dfrac{1}{\sqrt[3]{(ab)^2}}$ **11.** $x^{1/2}$ **13.** $a^{3/4}$ **15.** $x^{-1/2}$ **17.** $(ab)^{2/3}$ **19.** $x^{4/7}$ **21.** 2 **23.** 9
25. 16 **27.** 4 **29.** 1 **31.** 9 **33.** 1 **35.** 125 **37.** -2 **39.** -21

6-6-1

1. $2\sqrt{2}$ **3.** $3\sqrt{3}$ **5.** $2\sqrt[3]{2}$ **7.** $x\sqrt{x}$ **9.** $x^2 y^3$ **11.** $3x^2\sqrt{2x}$ **13.** $2y^3\sqrt{2x}$ **15.** $2x\sqrt[3]{2x^2}$ **17.** $2x^2\sqrt[5]{2x}$
19. $3x^2 y\sqrt{2xy}$ **21.** $3x^2 y\sqrt[4]{2xy}$ **23.** $-2x^2\sqrt[5]{2y^2}$ **25.** $3x^4 y^2\sqrt[3]{3x^2 y}$

6-6-2

1. $\sqrt[3]{x}$ **3.** $y\sqrt[3]{y}$ **5.** $y\sqrt{x}$ **7.** $\sqrt[5]{x^3 y}$ **9.** $\sqrt[6]{x^5 y^3}$ **11.** $\sqrt[4]{2x^2}$ **13.** $\sqrt[3]{3x}$ **15.** $y^2\sqrt{2x}$ **17.** $y^2\sqrt[3]{5xy}$ **19.** $x\sqrt[3]{3xy^2}$

6–7–1

1. $4\sqrt{2}$ **3.** $7\sqrt{x}$ **5.** $\sqrt{3}$ **7.** $9\sqrt{2}$ **9.** $3\sqrt{3}$ **11.** $7\sqrt{5} - 3\sqrt{3}$ **13.** $-3\sqrt[3]{2}$ **15.** $\sqrt{2}$ **17.** $\sqrt[3]{3} - \sqrt[3]{2}$ **19.** $\sqrt{2} - \sqrt[3]{2}$

6–7–2

1. $\sqrt{10}$ **3.** $6\sqrt{14}$ **5.** $8a^2$ **7.** $x\sqrt[4]{x}$ **9.** $-40x^2\sqrt{10}$ **11.** $\sqrt{6} + \sqrt{10}$ **13.** $6\sqrt{15} - 9\sqrt{21}$ **15.** $30\sqrt{2} + 15\sqrt{5}$
17. $12\sqrt{2} - 12 + 12\sqrt{6}$ **19.** $\sqrt{21} + \sqrt{10}$ **21.** $6\sqrt{3} - 8\sqrt{5} - 24$ **23.** $4\sqrt[3]{6} - 12$ **25.** $\sqrt{6} + \sqrt{15} + 2 + \sqrt{10}$
27. $6\sqrt{15} + \sqrt{10} - 15 - 2\sqrt{6}$ **29.** -2 **31.** -167 **33.** $6\sqrt{2} - 3\sqrt{30} + 2\sqrt{6}\sqrt[3]{2} - 3\sqrt{10}\sqrt[3]{2}$ **35.** $47 + 6\sqrt{10}$
37. $187 - 20\sqrt{21}$

6–8–1

1. $\dfrac{\sqrt{2}}{2}$ **3.** $\dfrac{\sqrt{7}}{7}$ **5.** $\dfrac{\sqrt{3}}{6}$ **7.** $\dfrac{\sqrt[3]{4}}{2}$ **9.** $\dfrac{\sqrt[5]{4}}{2}$ **11.** $\dfrac{\sqrt[3]{x}}{x}$ **13.** $\dfrac{3\sqrt[3]{4x^2}}{2x}$ **15.** $\dfrac{\sqrt[3]{2xy^2}}{x}$ **17.** $\dfrac{\sqrt{2}}{2}$ **19.** $\dfrac{\sqrt{2}}{4}$ **21.** $\dfrac{\sqrt{x+3}}{x+3}$

23. $\dfrac{2\sqrt{2x-4}}{x-2}$ **25.** $\sqrt{x+y}$

6–8–2

1. $\sqrt{3} - \sqrt{2}$ **3.** $\dfrac{3 - \sqrt{2}}{7}$ **5.** $\sqrt{2} + 1$ **7.** $\dfrac{\sqrt{x} - \sqrt{y}}{x - y}$ **9.** $\dfrac{\sqrt{5} - 1}{2}$ **11.** $\dfrac{6\sqrt{x} - 6y}{x - y^2}$ **13.** $3\sqrt{3} - 3\sqrt{2}$

15. $-\sqrt{2} - \sqrt{3}$ **17.** $2 - \sqrt{2}$ **19.** $2 + \sqrt{3}$ **21.** $\sqrt{a} + 1$ **23.** $\sqrt{7} + \sqrt{2}$ **25.** $5 + 2\sqrt{6}$ **27.** $\dfrac{23 - 17\sqrt{3}}{26}$

29. $\dfrac{12 + 7\sqrt{6}}{-10}$

6–9–1

1. $x^{5/6}$ **3.** $x^{3/10}$ **5.** $x^{7/6} + x^{11/12}$ **7.** $x^{1/8} - x^{11/12}$ **9.** $x^{4/3} - y^{4/3}$ **11.** $\dfrac{x + y}{xy}$ **13.** $\dfrac{1}{x}$ **15.** $\dfrac{x^{1/2}}{x}$ **17.** $\dfrac{2x^{1/2}}{x^2}$ **19.** $\dfrac{2x^{1/6}}{x}$

21. $\dfrac{2x^{2/3}y^{2/5}}{xy}$ **23.** $\dfrac{x(x^{1/2} - 5)}{x - 25}$ **25.** $\dfrac{3\sqrt{2}}{2}$ **27.** $\dfrac{5\sqrt{2}}{4}$ **29.** $\dfrac{-17\sqrt{3}}{6}$ **31.** $\dfrac{-17\sqrt{3}}{6}$ **33.** $\dfrac{11\sqrt{6}}{180}$ **35.** $\dfrac{4\sqrt{6}}{3}$

37. $\dfrac{16\sqrt{6}}{15}$ **39.** $\dfrac{\sqrt{15}}{6}$

Chapter 6 Review

The number in brackets after each answer refers to the section of the chapter that discusses that type problem.

1. $32x^{10}y^{15}$ [6–1]   **3.** $\dfrac{8x^9}{27y^3}$ [6–1]   **5.** $-\dfrac{1}{2x^4}$ [6–1]   **7.** x^6 [6–2]   **9.** $\dfrac{1}{x^{18}}$ [6–2]   **11.** $\dfrac{x + y}{x - y}$ [6–2]   **13.** 3.7623×10^7 [6–3]

15. $41{,}200$ [6–3]   **17.** 2.0×10^{10} [6–3]   **19.** 4 [6–4]   **21.** 3 [6–4]   **23.** 20.199 [6–4]   **25.** 9 [6–5]   **27.** $\dfrac{4}{25}$ [6–5]

29. 36 [6–5]   **31.** $5\sqrt{2}$ [6–6]   **33.** $xy^2\sqrt[5]{xy^2}$ [6–6]   **35.** $-2xy^2\sqrt[3]{5y}$ [6–6]   **37.** $61a\sqrt{2a} - 6a$ [6–7]

39. $30 - 2\sqrt{15}$ [6–7]   **41.** $\dfrac{\sqrt{2}}{6}$ [6–8]   **43.** $\sqrt{a - b}$ [6–8]   **45.** $\dfrac{9 - 5\sqrt{3}}{3}$ [6–8]   **47.** $x - 2(3x)^{1/2} + 3$ [6–9]

49. $\dfrac{-151\sqrt{2}}{40}$ [6–9]

Chapter 6 Test

The number in brackets after each answer refers to the section of the chapter that discusses that type problem.

1. a. $x^{24}y^{36}$ [6–1]   **b.** $\dfrac{1}{4x^7y}$ [6–1]   **2. a.** $\dfrac{1}{x^4y^2}$ [6–2]   **b.** $\dfrac{b}{ab+1}$ [6–2]   **3.** .0000256 [6–3]   **4.** 2.1×10^6 [6–3]

5. -6 [6–4]   **6.** $\dfrac{5}{7}$ [6–5]   **7. a.** $xy^2\sqrt[5]{xy^2}$ [6–6]   **b.** $-3b^2\sqrt[3]{ab^2}$ [6–6]   **8. a.** $3\sqrt{15}+\sqrt{10}-\sqrt{6}$ [6–7]

b. -29 [6–7]   **9. a.** $\dfrac{\sqrt[3]{4}}{2}$ [6–8]   **b.** $\dfrac{5\sqrt{2}+5\sqrt{x}}{2-x}$ [6–8]   **10. a.** $\dfrac{x^{1/3}y^{4/5}}{xy}$ [6–9]   **b.** $\dfrac{2\sqrt{2}+\sqrt{6}}{4}$ [6–9]

Chapters 1–6 Cumulative Test

The number in brackets after each answer refers to the chapter and section that discusses that type of problem.

1. 14 [1–2]   **2.** $8a^5b^3 - 2a^3b^5 + 12a^5b^2$ [1–8]   **3.** 78 [1–10]   **4.** 1 [2–5]   **5.** $x > -1$ ⟵(++++++++⟶   [2–7]
                                                                                                                                                    -1 0

6. $-2, 7$ [2–8]   **7.** $x, 48 - x$ [3–1]   **8.** 22 [3–3]   **9.** 6 liters of 20% solution, 4 liters of 45% solution [3–5]
10. $6x^3 - 11x^2 + 15x - 4$ [4–1]   **11.** $(x - y)(3 - a)$ [4–4]   **12.** $(2x - y)(4x - 3y)$ [4–5]
13. $(x + 2 + y)(x + 2 - y)$ [4–9]   **14.** $\dfrac{2x}{x+4}$ [5–1]   **15.** $\dfrac{2x+1}{x+1}$ [5–7]   **16.** $x + 1$ [5–8]   **17.** 6 [5–9]
18. 2.45×10^7 [6–3]   **19.** $\sqrt{5} - \sqrt{2}$ [6–8]   **20.** $\dfrac{2x^{1/2}y^{4/5}}{xy}$ [6–9]

Chapter 7 Pretest

The number in brackets after each answer refers to the section of the chapter that discusses that type problem.

1. $\left\{-4, \dfrac{1}{3}\right\}$ [7–1]   **2. a.** $\left\{0, \dfrac{5}{3}\right\}$ [7–2]   **b.** $\{\pm\sqrt{2}\}$ [7–2]   **3.** $\left\{\dfrac{10 \pm \sqrt{85}}{5}\right\}$ [7–3]   **4.** $\left\{\dfrac{-4 \pm \sqrt{31}}{3}\right\}$ [7–4]

5. a. 1, unequal, rational [7–5]   **b.** -3, no real roots [7–5]   **6.** $\left\{\dfrac{1 \pm 3i}{2}\right\}$ [7–6]   **7.** $\left\{\dfrac{\pm\sqrt{6}}{3}, \pm\sqrt{2}\right\}$ [7–7]   **8.** $\{-1\}$ [7–8]

9. 4 and 9, or -4 and -9 [7–9]

7-1-1

1. $x^2 + 5x + 6 = 0$ **3.** $x^2 - 3x + 2 = 0$ **5.** $6x^2 - 5x + 1 = 0$ **7.** $2x^2 - 5x + 1 = 0$ **9.** $5x^2 + 7x + 3 = 0$

11. $\{-2, -1\}$ **13.** $\{-7, -1\}$ **15.** $\{-4, 2\}$ **17.** $\{3, 6\}$ **19.** $\{-7, 3\}$ **21.** $\{-3, 1\}$ **23.** $\{-1\}$ **25.** $\{-4, 0\}$ **27.** $\left\{0, \dfrac{1}{2}\right\}$

29. $\{-5, -2\}$ **31.** $\left\{-\dfrac{3}{2}, -1\right\}$ **33.** $\left\{-\dfrac{3}{2}, -\dfrac{2}{5}\right\}$ **35.** $\left\{-\dfrac{1}{3}, \dfrac{5}{2}\right\}$ **37.** $\left\{-\dfrac{5}{2}, \dfrac{3}{5}\right\}$ **39.** $\left\{-2, -\dfrac{5}{6}\right\}$

7-2-1

1. $\{-3, 0\}$ **3.** $\left\{0, \dfrac{3}{2}\right\}$ **5.** $\{0, 8\}$ **7.** $\left\{0, \dfrac{5}{3}\right\}$ **9.** $\{0, 2\}$ **11.** $\{\pm 2\}$ **13.** $\{\pm\sqrt{5}\}$ **15.** $\{\pm 2\sqrt{5}\}$ **17.** $\{\pm\sqrt{14}\}$ **19.** $\{\pm 6\}$

21. $\{\pm 4\}$ **23.** $\{\pm\sqrt{2}\}$ **25.** $\{\pm 2\sqrt{2}\}$ **27.** $\{\pm\sqrt{3}\}$ **29.** No real solution

7-3-1

1. (4); $(x + 2)^2$ **3.** (9); $(x - 3)^2$ **5.** (16); $(x + 4)^2$ **7.** (121); $(x + 11)^2$ **9.** (256); $(x - 16)^2$ **11.** $\left(\dfrac{1}{4}\right)$; $\left(x + \dfrac{1}{2}\right)^2$

13. (144); $(x - 12)^2$ **15.** (1); $(x - 1)^2$ **17.** $\left(\dfrac{25}{4}\right)$; $\left(x - \dfrac{5}{2}\right)^2$ **19.** $\left(\dfrac{49}{4}\right)$; $\left(x + \dfrac{7}{2}\right)^2$

7-3-2

1. $\{-2,-1\}$ **3.** $\{-3,2\}$ **5.** $\{1,6\}$ **7.** $\{-3\}$ **9.** $\{-2 \pm \sqrt{7}\}$ **11.** $\{-5,1\}$ **13.** $\{-1,3\}$ **15.** No real solution
17. $\left\{\dfrac{5 \pm \sqrt{37}}{2}\right\}$ **19.** No real solution   **21.** $\left\{\dfrac{-4 \pm \sqrt{7}}{3}\right\}$ **23.** $\left\{\dfrac{5 \pm \sqrt{57}}{4}\right\}$ **25.** $\left\{-2,\dfrac{1}{2}\right\}$ **27.** $\left\{\dfrac{-3 \pm \sqrt{21}}{3}\right\}$
29. No real solution

7-4-1

1. $\{-5,3\}$ **3.** $\left\{\dfrac{-5 \pm \sqrt{13}}{2}\right\}$ **5.** $\left\{\dfrac{5 \pm \sqrt{21}}{2}\right\}$ **7.** $\left\{2,-\dfrac{3}{5}\right\}$ **9.** No real solution   **11.** $\left\{-1,-\dfrac{2}{3}\right\}$ **13.** $\left\{\dfrac{1 \pm \sqrt{13}}{6}\right\}$
15. $\left\{\dfrac{-3 \pm \sqrt{5}}{2}\right\}$ **17.** $\left\{\dfrac{1}{2},1\right\}$ **19.** $\left\{-\dfrac{1}{3},1\right\}$ **21.** $\left\{-1,-\dfrac{2}{3}\right\}$ **23.** $\left\{\dfrac{3 \pm \sqrt{3}}{2}\right\}$ **25.** $\{5\}$ **27.** $\left\{-\dfrac{2}{3}\right\}$ **29.** $\left\{1,\dfrac{1}{5}\right\}$

7-5-1

1. 0, real, equal   **3.** 72, irrational, unequal   **5.** 1, rational, unequal   **7.** 192, irrational, unequal   **9.** -11, no real roots
11. 61, irrational, unequal   **13.** 0, real, equal   **15.** 20, irrational, unequal   **17.** 225, rational, unequal   **19.** 0, real, equal
21. 25, unequal, rational   **23.** 0, real, equal   **25.** 1, unequal, rational   **27.** -3, no real roots   **29.** 17, irrational, unequal

7-6-1

1. $3i$ **3.** $i\sqrt{2}$ **5.** $5i$ **7.** $i\sqrt{7}$ **9.** $i\sqrt{30}$ **11.** $2i\sqrt{2}$ **13.** $10i$ **15.** $3i\sqrt{2}$ **17.** $5i\sqrt{3}$ **19.** $4i\sqrt{5}$

7-6-2

1. $7 + 7i$ **3.** $7 - i$ **5.** 13   **7.** $-3 - i$ **9.** $8 + 3i$ **11.** $5 + 16i$ **13.** $-10 + 11i$ **15.** $x^2 + y^2$ **17.** $9 + 40i$
19. $11 - 60i$ **21.** $\dfrac{-4 + 7i}{5}$ **23.** $\dfrac{4 - 3i}{5}$ **25.** $\dfrac{2 + 12i}{37}$ **27.** $1 - 3i$ **29.** $\dfrac{14 - 5i}{13}$

7-6-3

1. $\{1 \pm i\}$ **3.** $\{2 \pm 2i\}$ **5.** $\{-1 \pm 2i\}$ **7.** $\{-2 \pm 3i\}$ **9.** $\left\{\dfrac{1 \pm i}{2}\right\}$ **11.** $\left\{\dfrac{1 \pm i}{4}\right\}$ **13.** $\{-1 \pm 2i\}$ **15.** $\left\{\dfrac{-1 \pm i\sqrt{3}}{2}\right\}$
17. $\left\{\dfrac{-1 \pm i\sqrt{7}}{4}\right\}$ **19.** $\left\{\dfrac{3 \pm i\sqrt{7}}{4}\right\}$ **21.** $\{1 \pm i\sqrt{3}\}$ **23.** $\{2 \pm i\sqrt{2}\}$ **25.** $\left\{\dfrac{-1 \pm i\sqrt{11}}{6}\right\}$ **27.** $\left\{\dfrac{1 \pm i\sqrt{14}}{5}\right\}$
29. $\left\{\dfrac{-3 \pm i\sqrt{7}}{2}\right\}$

7-7-1

1. $\{\pm 1,\pm 2\}$ **3.** $\{\pm \sqrt{2},\pm 2\}$ **5.** $\{\pm \sqrt{2},\pm \sqrt{5}\}$ **7.** $\left\{\dfrac{\pm \sqrt{6}}{3},\pm \sqrt{2}\right\}$ **9.** $\{\pm 3,\pm i\sqrt{2}\}$ **11.** $\left\{1,2,\dfrac{-1 \pm i\sqrt{3}}{2},-1 \pm i\sqrt{3}\right\}$
13. $\left\{-1,-3,\dfrac{1 \pm i\sqrt{3}}{2},\dfrac{3 \pm 3i\sqrt{3}}{2}\right\}$ **15.** $\{625\}$ **17.** $\{4,16\}$ **19.** $\{-27,1\}$

7-8-1

1. $\{4\}$ **3.** $\{9\}$ **5.** $\{5\}$ **7.** $\{14\}$ **9.** $\{7\}$ **11.** $\{6\}$ **13.** $\{0,4\}$ **15.** $\{2\}$ **17.** $\{5\}$ **19.** $\{2\}$ **21.** $\{1\}$ **23.** $\{-1\}$
25. $\{-3\}$ **27.** $\{8\}$ **29.** $\{2\}$

7-9-1

1. 9 and 16   **3.** $\frac{2}{5}$ **5.** 24 rows, 30 seats per row   **7.** $h = 7, b = 10$ **9.** 11 and 16   **11.** $\frac{3}{4}, \frac{4}{3}$ **13.** 8, 10   **15.** 4 and 13

17. 6 cm, 8 cm   **19.** 2 m   **21.** 12   **23.** 1 sec going up, 2 sec coming down   **25.** 5   **27.** 5 m

Chapter 7 Review

The number in brackets after each answer refers to the section of the chapter that discusses that type problem.

1. $\{-7, -3\}$ [7–1]   **3.** $\{3, 8\}$ [7–1]   **5.** $\left\{-\frac{3}{2}, \frac{1}{3}\right\}$ [7–1]   **7.** $\{\pm\sqrt{17}\}$ [7–2]   **9.** $\{\pm 4i\}$ [7–2]   **11.** $\{-5, -3\}$ [7–3]

13. $\left\{\frac{-3 \pm \sqrt{13}}{2}\right\}$ [7–3]   **15.** $\left\{\frac{-5 \pm i\sqrt{7}}{4}\right\}$ [7–3, 7–6]   **17.** $\{1, 3\}$ [7–4]   **19.** $\{-5 \pm i\}$ [7–6]   **21.** 17, irrational, unequal [7–5]

23. 0, real, equal [7–5]   **25.** 1, rational, unequal [7–5]   **27.** $\{\pm\sqrt{2}, \pm i\sqrt{5}\}$ [7–7]   **29.** $\{16, 25\}$ [7–7]   **31.** $\{9\}$ [7–8]

33. $\{8\}$ [7–8]   **35.** $\{-6\}$ [7–8]   **37.** -7 and -5 or 5 and 7 [7–9]   **39.** $\frac{2}{5}, 5$ [7–9]

Chapter 7 Test

The number in brackets after each answer refers to the section of the chapter that discusses that type problem.

1. $\left\{-3, \frac{1}{2}\right\}$ [7–1]   **2. a.** $\left\{0, \frac{7}{2}\right\}$ [7–2]   **b.** $\{\pm\sqrt{10}\}$ [7–2]   **3.** $\left\{\frac{2 \pm \sqrt{5}}{3}\right\}$ [7–3]   **4.** $\left\{\frac{-3 \pm \sqrt{41}}{4}\right\}$ [7–4]

5. a. 17, unequal, irrational [7–5]   **b.** 0, equal, real [7–5]   **6.** $\left\{\frac{3 \pm i\sqrt{7}}{2}\right\}$ [7–6]   **7.** $\{\pm\sqrt{5}, \pm i\sqrt{2}\}$ [7–7]   **8.** $\{1\}$ [7–8]

9. 7 meters [7–9]

Chapter 8 Pretest

The number in brackets after each answer refers to the section of the chapter that discusses that type problem.
1. a. $D: (-\infty, +\infty)$, $R: [2, +\infty)$, yes [8–2]   **b.** $D: [-4, 4]$, $R: [-4, 4]$, no [8–2]

2.

x	2	4	5
y	-3	1	3

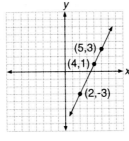

[8–3]

3. $x + y = 2$ [8–3]   **4.**

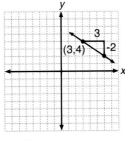

[8–3]

5. $(x + 2)^2 + (y - 4)^2 = 25$,
center: $(-2, 4)$,
radius: 5 [8–4]

6.

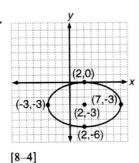

[8–4]

7.

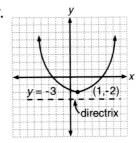

[8–4]

8.

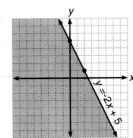

[8–5]

8-1-1

1. a. $(-\infty,+\infty)$ **b.** $(-1,+\infty)$ **c.** Yes **3. a.** $(-\infty,0) \cup (0,+\infty)$ **b.** $(-\infty,0) \cup (0,+\infty)$ **c.** Yes **5. a.** $[-3,+\infty)$
b. $(-\infty,+\infty)$ **c.** No **7. a.** $(-\infty,+\infty)$ **b.** $(-\infty,+\infty)$ **c.** Yes **9. a.** $(-\infty,+\infty)$ **b.** $(-\infty,+\infty)$ **c.** Yes
11. a. $[-3,3]$ **b.** $[-3,3]$ **c.** No **13. a.** $[-3,3]$ **b.** $[0,3]$ **c.** Yes **15. a.** $(-\infty,1) \cup (1,+\infty)$
b. $(-\infty,0) \cup (0,+\infty)$ **c.** Yes **17. a.** $[0,+\infty)$ **b.** $(-\infty,+\infty)$ **c.** No **19. a.** $(-\infty,+\infty)$ **b.** $[3,+\infty)$ **c.** Yes
21. a. $(-\infty,-3] \cup [3,+\infty)$ **b.** $[0,+\infty)$ **c.** Yes **23. a.** $(2,+\infty)$ **b.** $(0,+\infty)$ **c.** Yes

8-2-1

1.
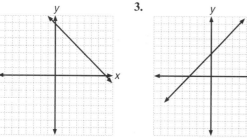

3. Quadrant II, $(-,+)$, quadrant III, $(-,-)$, quadrant IV, $(+,-)$

8-3-1

1.
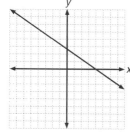

3.
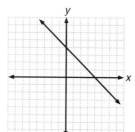

5.
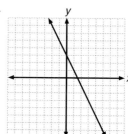

7.

9.
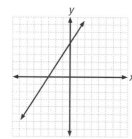

11.
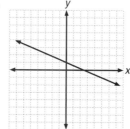

13.

15.

17.

19.

8-3-2

1. $8x + 5y = 21$ **3.** $x + 2y = -2$ **5.** $5x + 12y = 20$ **7.** $y = 2$ **9.** $4x + 11y = -43$

11. $3x - y = -4$

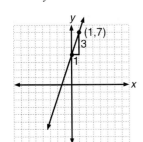

13. $x - 2y = -2$

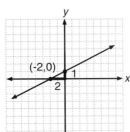

15. $8x + y = -31$

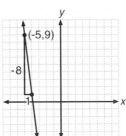

17. $x = -2$

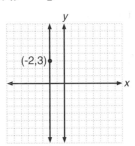

19. $y = 5$

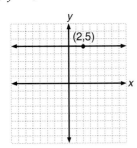

21. $y = -2x + 5$ **23.** $y = -\dfrac{3}{4}x + \dfrac{1}{4}$ **25.** $y = \dfrac{2}{5}x - \dfrac{9}{5}$ **27.** $y = \dfrac{5}{2}x$

29. $y = \dfrac{1}{8}x - \dfrac{1}{8}$ **31.** $7x - 2y = -14$ **33.** $2x - y = 5$ **35.** $5x - 2y = 27$

37. $y = 1$ **39.** $x = 4$ **41.** $3x + 2y = 13$ **43.** $5x - y = -9$

8-4-1

1. $(x + 2)^2 + (y - 3)^2 = 36$ **3.** $(x + 1)^2 + (y - 6)^2 = 81$ **5.** $(x - 3)^2 + (y - 2)^2 = 49$ **7.** $(x + 2)^2 + (y + 5)^2 = 1$

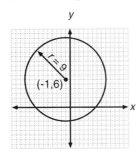

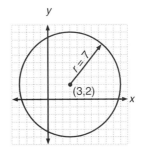

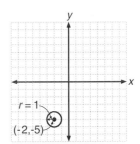

9. $\left(x - \dfrac{1}{2}\right)^2 + (y + 4)^2 = 25$

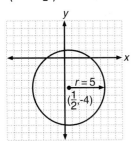

8-4-2

1.

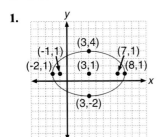

3.

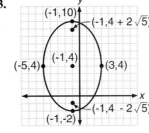

5.

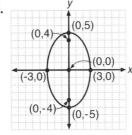

7.

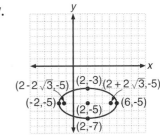

9. $\dfrac{(x + 3)^2}{49} + \dfrac{(y - 2)^2}{9} = 1$

8-4-3

1. 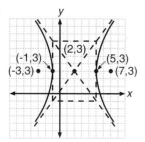   **3.** 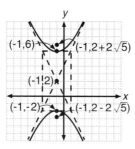   **5.** 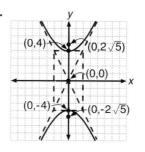   **7.**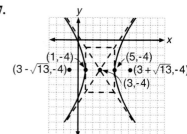

8-4-4

1. 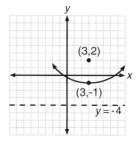   **3.** 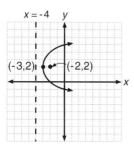   **5.** 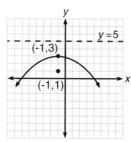   **7.**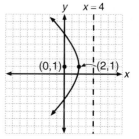

9. $\dfrac{(x-5)^2}{9} + \dfrac{(y+2)^2}{1} = 1$, horizontal ellipse   **11.** $(x-4)^2 = 8(y+1)$, vertical parabola

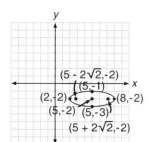   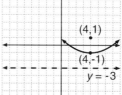

13. $(x+4)^2 + (y-3)^2 = 30$, circle   **15.** $\dfrac{(x+4)^2}{16} - \dfrac{(y-3)^2}{9} = -1$, vertical hyperbola

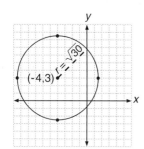

8-5-1

1.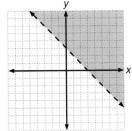

3.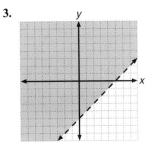

5.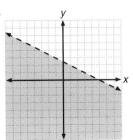

7.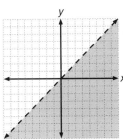

9.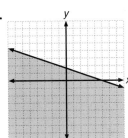

11.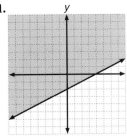

13.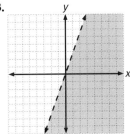

15.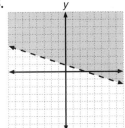

17.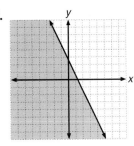

19.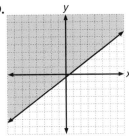

Chapter 8 Review

The number in brackets after each answer refers to the section of the chapter that discusses that type of problem.

1. (2,1) [8–1]   **3.** (−3,2) [8–1]   **5.** (−4,0) [8–1]   **7.** (−2,−5) [8–1]   **9.** (−2,−2) [8–1]

11. *D:* (−∞,+∞), *R:* [0,+∞), yes [8–2]   **13.** *D:* [1,+∞), *R:* (−∞,+∞), no [8–2]   **15.** *D:* [−2,2], *R:* [0,2], yes [8–2]

17. *D:* (−∞,−1) ∪ (1,+∞), *R:* (0,+∞), yes [8–2]

19.

x	−1	0	1
y	−3	−1	1

21.

x	0	2	3
y	7	1	−2

23. $11x + 7y = 2$ [8–3]   **25.** $3x + 5y = -39$ [8–3]

27. $y = \dfrac{2}{5}x - 2$ [8–3]   **29.** $x - 2y = -6$ [8–3]

[8–3]   [8–3]   **31.** [8–3]   **33.** $3x + 8y = 50$ [8–3]

35. $(x - 3)^2 + (y + 1)^2 = 25$, center: (3,−1), radius: 5 [8–4]

37. $\dfrac{(x + 3)^2}{9} + \dfrac{(y - 5)^2}{25} = 1$, vertical ellipse [8–4]

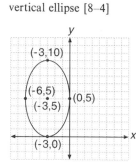

39. $\dfrac{(x - 2)^2}{25} - \dfrac{(y + 1)^2}{16} = -1$, vertical hyperbola [8–4]

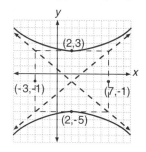

41.

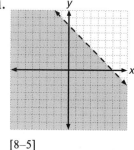

[8–5]

43.

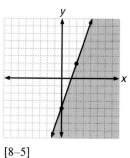

[8–5]

Chapter 8 Test

The number in brackets after each answer refers to the section of the chapter that discusses that type problem.

1. a. $D: \left[\dfrac{5}{2}, +\infty\right)$, $R: [0, +\infty)$, yes [8–2]   **b.** $D: [-3,3]$, $R: [-3,3]$, no [8–2]

2.

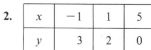

x	-1	1	5
y	3	2	0

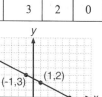

[8–3]

3. $x + 2y = 9$ [8–3]

4.

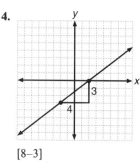

[8–3]

5. $(x - 5)^2 + (y + 2)^2 = 36$, center: $(5, -2)$, radius: 6 [8–4]

6.

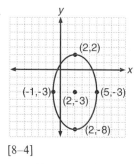

[8–4]

7.

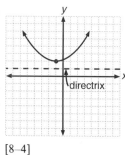

[8–4]

8.

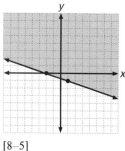

[8–5]

Chapter 9 Pretest

The number in brackets after each answer refers to the section of the chapter that discusses that type problem.

1.

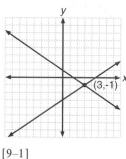

[9–1]

2.

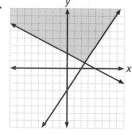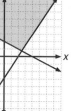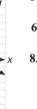

[9–2]

3. $(-2,1)$ [9–3]   **4.** $(-3,4)$ [9–4]   **5.** $(-2,1,0)$ [9–6]

6. a. -21 [9–7]   **b.** 43 [9–7]   **7.** $(1,-3,2)$ [9–8]

8. 150 ml of 6% solution, 50 ml of 10% solution [9–5]

9–1–1

1.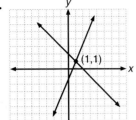

3.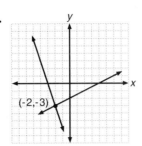

5.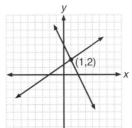

7.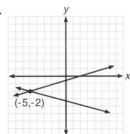

9.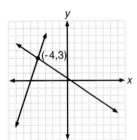

11.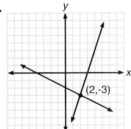

13.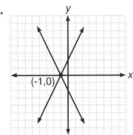

15.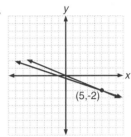

17.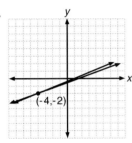

19.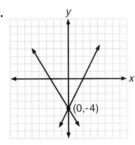

21.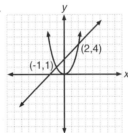

23.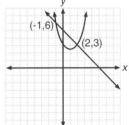

25. One point, two points, no points

9–2–1

1.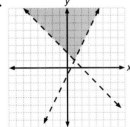

3.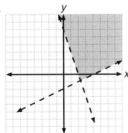

5.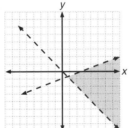

7.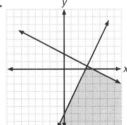

9.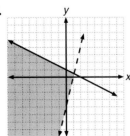

11.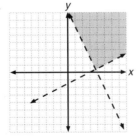

13.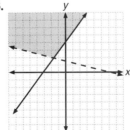

15.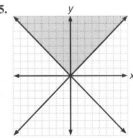

17. 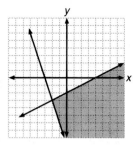   **19.** 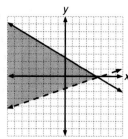   **21.** 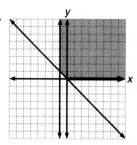   **23.**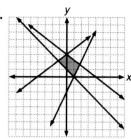

25. 40 half minute, 60 one minute

9–3–1

1. (2,1)   **3.** (5,0)   **5.** (1,−1)   **7.** (−3,1)   **9.** $\left(\frac{1}{2},4\right)$ **11.** (−1,−3)   **13.** $\left(8,3\frac{1}{2}\right)$ **15.** (5,−6)   **17.** (3,−10)
19. (−13,−9)

9–3–2

1. (3,−1)   **3.** (2,4)   **5.** (−1,4)   **7.** (10,−2)   **9.** (2,1)   **11.** (−4,5)   **13.** (2,−8)   **15.** $\left(\frac{1}{2},4\right)$ **17.** (−2,5)
19. (−1,1)

9–3–3

1. Independent, (2,−3)   **3.** Independent, (1,4)   **5.** Independent, (0,−2)   **7.** Dependent   **9.** Independent, (−3,7)
11. Independent, $\left(\frac{1}{2},3\right)$ **13.** Dependent   **15.** Independent, (3,−8)

9–4–1

1. $10x + 3y = 45$ **3.** $2x + 4y = 9$ **5.** $15x − 20y = 6$ **7.** (3,0)   **9.** (−2,−1)   **11.** (0,3)

9–5–1

1. 18, 39   **3.** 27 men, 21 women   **5.** length = 150 ft, width = 75 ft
7. Golden Gate Bridge = 4,200 ft; Brooklyn Bridge = 1,595 ft   **9.** 25°, 65°   **11.** Length = 223 ft, width = 77 ft
13. $7,000 at 11%; $8,000 at 14%   **15.** Rate of the boat is 9 km/hr, rate of the current is 3 km/hr
17. 16 nickels, 18 dimes   **19.** 60 kg of 40% alloy, 120 kg of 70% alloy

9–6–1

1. (1,2,3)   **3.** (−2,1,3)   **5.** (2,1,−3)   **7.** (−2,1,0)   **9.** 5 nickels, 10 dimes, 13 quarters
11. 2 liters of 5%, 3 liters of 20%, 4 liters of 50%

9–7–1

1. 7   **3.** −3   **5.** 11   **7.** −2   **9.** 0   **11.** −101   **13.** 0   **15.** 54   **17.** 5   **19.** 10

9–7–2

1. −9   **3.** 3   **5.** 81   **7.** 46   **9.** 0   **11.** 36   **13.** −50   **15.** 2   **17.** −13

9–8–1

1. (1,2)   **3.** (2,3)   **5.** Dependent   **7.** (3,0)   **9.** (5,−1)   **11.** $\left(\dfrac{1}{2},5\right)$ **13.** (−2,1)   **15.** Dependent   **17.** (0,3)

19. Current: 3 mph, boat: 12 mph

9–8–2

1. (1,2,3)   **3.** (2,3,−5)   **5.** (−3,1,0)   **7.** (1,−2,4)   **9.** (0,−1,4)   **11.** Inconsistent   **13.** Dependent   **15.** (−6,1,−2)
17. (3,−10,7)

Chapter 9 Review

The number in brackets after each answer refers to the section of the chapter that discusses that type problem.

1. 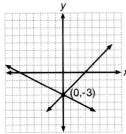   **3.** 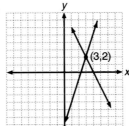   **5.** 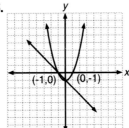   **7.**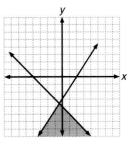

9.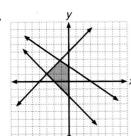
11. (−3,1) [9–3]   **13.** (2,−1) [9–3]   **15.** (4,1) [9–3]   **17.** (3,5) [9–3]   **19.** (2,−1) [9–3]
21. Inconsistent [9–3]   **23.** Dependent [9–3]   **25.** $3x - y = -2$ [9–4]   **27.** $5x + 20y = 6$ [9–4]
29. (0,−2) [9–4]   **31.** (5,−2) [9–4]   **33.** (1,2,−1) [9–6]   **35.** (1,−1,3) [9–6]
37. (0,−2,5) [9–6]   **39.** 12 [9–7]   **41.** −24 [9–7]   **43.** −3 [9–7]   **45.** (1,−5) [9–8]
47. (0,−3) [9–8]   **49.** (1,2,−1) [9–8]   **51.** (0,−2,5) [9–8]   **53.** (13,−5) [9–3, 9–8]
55. (3,−1,2) [9–6, 9–8]   **57.** $\left(-\dfrac{1}{3},4,-1\right)$ [9–6, 9–8]   **59.** 64, 83 [9–5]

61. 39 students in the first class, 46 students in the second class [9–5]   **63.** 31 nickels, 25 quarters [9–5]
65. 6 liters [9–6, 9–8]

Chapter 9 Test

The number in brackets after each answer refers to the section of the chapter that discusses that type problem.

1. 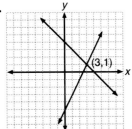   **2.**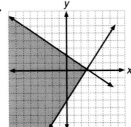
3. (5,−4) [9–3]   **4.** (2,0) [9–4]   **5.** (2,−1,1) [9–6]
6. a. 23 [9–7]   **b.** −16 [9–7]   **7.** (1,−1,3) [9–8]
8. 15 dimes, 30 quarters [9–5]

Chapters 1–9 Cumulative Test

The number in brackets after each refers to the chapter and section that discusses that type of problem.

1. $-xy - 20x^2y$ [1–7]   **2.** $13a + 180$ [1–9]   **3.** $-\dfrac{17}{5}$ [2–5]   **4.** $x \le -1$ or $x \ge 4$ [2–8]

5. Stereo: $599, nonstereo: $325 [3–2]   6. Car A: 45 mph, car B: 55 mph [3–4]   7. $(x - 4)(x^2 + 4x + 16)$ [4–7]

8. $3x^2(2x + 5)(x - 1)$ [4–8]   9. $\dfrac{1}{x + 5}$ [5–3]   10. -1 [5–9]   11. $\dfrac{y^9}{x^5}$ [6–2]   12. $\dfrac{6\sqrt{10}}{5}$ [6–9]   13. $\dfrac{5 \pm i\sqrt{23}}{6}$ [7–6]

14. $\{\pm 2, \pm i\sqrt{3}\}$ [7–7]   15. $\{1,5\}$ [7–8]   16. $4x - 5y = 13$ [8–3]

17. $(x - 1)^2 + (y + 5)^2 = 16$, center: $(1,-5)$, radius: 4 [8–4]

18. Vertex: $(2,-1)$, focus: $(2,1)$, directrix: $y = -3$ [8–4]

19. 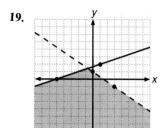   20. $(2,-4)$ [9–8]

[9–2]

Chapter 10 Pretest

The number in brackets after each answer refers to the section of the chapter that discusses that type problem.

1. $\log_2 8 = 3$ [10–1]   2. $4^{-2} = \dfrac{1}{16}$ [10–1]   3. a. 2 [10–1]   b. 27 [10–1]   c. 2 [10–1]   d. -3 [10–1]   e. 7 [10–1]

4. a. $3x$ [10–2]   b. $x + 2y$ [10–2]   5. a. 1.1847 [10–3]   b. $.9196 - 4$ [10–3]   c. .9566 [10–3]   d. 5.1580 [10–5]

6. a. 8.32 [10–3]   b. 11,800 [10–3]   c. .00703 [10–3]   7. a. 15,100 [10–4]   b. 3.88 [10–4]   8. a. 5 [10–6]

b. $\dfrac{5 \log 3 - \log 2}{\log 2 - 2 \log 3}$ [10–6]

10-1-1

1. $\log_3 9 = 2$ 3. $\log_5 125 = 3$ 5. $\log_3 27 = 3$ 7. $\log_2 \dfrac{1}{8} = -3$ 9. $\log_2 \dfrac{1}{16} = -4$ 11. $2^2 = 4$ 13. $4^2 = 16$

15. $3^4 = 81$ 17. $5^3 = 125$ 19. 81   21. 2   23. 5   25. $\dfrac{1}{9}$ 27. 49   29. 1   31. 3   33. -2

35.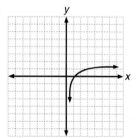

10-2-1

1. $2y$ 3. $y - 2x$ 5. $x + 2y$ 7. $x - y$ 9. $2x + 2y$ 11. 2   13. 4   15. -2 17. -2 19. 5   21. $.5514 + 2$

23. $.5514 - 2$ 25. $.5514 - 3$

10-3-1

1. .3711   3. .3617   5. .9031   7. $.4518 + 1$ 9. 1   11. $.6702 + 2$ 13. $.9053 + 3$ 15. $.5145 + 4$ 17. $.1335 - 1$

19. $.7059 - 4$ 21. $.9294 - 6$ 23. $.3284 - 3$ 25. $.8344 + 5$

10-3-2

1. 4.09 **3.** 1.15 **5.** 38.8 **7.** 741 **9.** .751 **11.** .00604 **13.** 7,580 **15.** 44 **17.** .045 **19.** .000802

10-4-1

1. 24 **3.** 12 **5.** 46.8 **7.** 3.36 **9.** .321 **11.** 1.69 **13.** 1,050 **15.** 1,330 **17.** 3.44 **19.** 15,900 **21.** 10.9
23. .00734

10-4-2

1. 6.25 **3.** 9.99 **5.** 9.94 **7.** 266 **9.** .00292 **11.** .00195 **13.** 838,000 **15.** .00000000118 **17.** 2.6 **19.** 16 **21.** 3.09
23. 49.2 **25.** 4.67 **27.** .308

10-5-1

1. 3.000 **3.** 1.568 **5.** 1.125 **7.** 4.858 **9.** 2.209 **11.** 3.282 **13.** 2.921 **15.** 2.103 **17.** -3.052 **19.** 2.101
21. 2.103 **23.** 5.846 **25.** 3.914 **27.** -4.804

10-6-1

1. 2 **3.** 5 **5.** No solution **7.** $\dfrac{1}{9}$ **9.** No solution **11.** 3,4 **13.** 4 **15.** $\dfrac{\log 7}{\log 2}$ **17.** $\dfrac{\log 9 - \log 16}{\log 8}$

19. $\dfrac{5 \log 3 - \log 2}{\log 2 - 2 \log 3}$ **21.** $\dfrac{\log 5 - \log 3}{\log 3 + 2 \log 5}$ **23.** $\dfrac{1,006}{3}$ **25.** 3.5229 **27.** 736 mg **29.** 9 years

Chapter 10 Review

The number in brackets after each answer refers to the section of the chapter that discusses that type problem.

1. 64 [10–1] **3.** 9 [10–1] **5.** $\dfrac{1}{8}$ [10–1] **7.** $x + y$ [10–2] **9.** $x - 2y$ [10–2] **11.** $3x + 2y$ [10–2] **13.** .9053 [10–3]
15. 4.6902 [10–3] **17.** -2.2076 [10–3] **19.** 4.37 [10–3] **21.** 1,500 [10–3] **23.** .0738 [10–3] **25.** 21.9 [10–4]
27. .0811 [10–4] **29.** .129 [10–4] **31.** 58,000 [10–4] **33.** 19.1 [10–4] **35.** 4.15 [10–4] **37.** 3.350 [10–5]
39. .5093 [10–5] **41.** -0.1911 [10–5] **43.** 7 [10–6] **45.** 7 [10–6] **47.** $\dfrac{\log 7 - \log 3}{\log 3}$ [10–6] **49.** About 14 years [10–6]

Chapter 10 Test

The number in brackets after each answer refers to the section of the chapter that discusses that type problem.

1. $\log_3 9 = 2$ [10–1] **2.** $2^{-2} = \dfrac{1}{4}$ [10–1] **3. a.** 2 [10–1] **b.** 9 [10–1] **c.** 3 [10–1] **d.** -4 [10–1] **e.** 5 [10–1]

4. a. $2y$ [10–2] **b.** $2x + y$ [10–2] **5. a.** 1.3927 [10–3] **b.** .1931 $- 4$ [10–3] **c.** .8537 [10–3] **d.** 4.1516 [10–5]

6. a. 6.81 [10–3] **b.** 2,980 [10–3] **c.** .0545 [10–3] **7. a.** 13,100 [10–4] **b.** 23.4 [10–4] **8. a.** $x = \dfrac{1}{2}$ [10–6]

b. $x = 5$ [10–6]

Chapter 11 Pretest

The number in brackets after each answer refers to the section of the chapter that discusses that type problem.

1. If $n = 1$, then $\dfrac{1(1 + 3)}{4} = \dfrac{4}{4} = 1$. Assume true for $n = k$, then $1 + \dfrac{3}{2} + 2 + \cdots + \dfrac{k + 1}{2} = \dfrac{k(k + 3)}{4}$.

For $n = k + 1$ we have $\dfrac{k(k + 3)}{4} + \dfrac{(k + 1) + 1}{2} = \dfrac{k(k + 3) + 2(k + 2)}{4} = \dfrac{k^2 + 5k + 4}{4} = \dfrac{(k + 1)[(k + 1) + 3]}{4}$. [11–1]

2. $a_1 = 0, a_2 = 3, a_3 = 8, a_4 = 15$ [11–2]   **3.** 10 [11–2]   **4.** $\sum_{i=1}^{4} (i^2 + 1)$ [11–2]

5. -24 [11–3]   **6.** 168 [11–3]   **7.** 22 [11–3]   **8.** 252 [11–4]   **9.** 8, $-16, 32, -64$ [11–4]

10. 32 [11–4]   **11.** $-960x^3$ [11–5]   **12.** 1.159 [11–5]

11-1-1

1. Step 1 If $n = 1$, then $1^2 = 1$.
Step 2 Assume true for $n = k$, then $1 + 3 + 5 + \cdots + (2k - 1) = k^2$.
Step 3 For $n = k + 1$ we have $k^2 + [2(k + 1) - 1] = k^2 + 2k + 1 = (k + 1)^2$.

3. Step 1 If $n = 1$, then $\dfrac{3(1)^2 - 1}{2} = 1$.

Step 2 Assume true for $n = k$, then $1 + 4 + 7 + \cdots + (3k - 2) = \dfrac{3k^2 - k}{2}$.

Step 3 For $n = k + 1$ we have $\dfrac{3k^2 - k}{2} + [3(k + 1) - 2]$

$$= \frac{3k^2 + 5k + 2}{2} = \frac{3k^2 + 6k + 3 - k - 1}{2} = \frac{3(k + 1)^2 - (k + 1)}{2}.$$

5. Step 1 If $n = 1$, then $\dfrac{1}{1 + 1} = \dfrac{1}{2} = \dfrac{1}{1(2)}$.

Step 2 Assume true for $n = k$, then $\dfrac{1}{1(2)} + \dfrac{1}{2(3)} + \dfrac{1}{3(4)} + \cdots + \dfrac{1}{k(k + 1)} = \dfrac{k}{k + 1}$.

Step 3 For $n = k + 1$ we have $\dfrac{k}{k + 1} + \dfrac{1}{(k + 1)[(k + 1) + 1]} = \dfrac{k(k + 2) + 1}{(k + 1)(k + 2)} = \dfrac{(k + 1)^2}{(k + 1)(k + 2)} = \dfrac{k + 1}{(k + 1) + 1}$.

7. Step 1 If $n = 1$, then $\dfrac{1(1 + 1)(1 + 2)}{3} = 2 = 1(2)$.

Step 2 Assume true for $n = k$, then $1(2) + 2(3) + 3(4) + \cdots + k(k + 1) = \dfrac{k(k + 1)(k + 2)}{3}$.

Step 3 For $n = k + 1$ we have $\dfrac{k(k + 1)(k + 2)}{3} + (k + 1)[(k + 1) + 1]$

$$= \frac{k(k + 1)(k + 2) + 3(k + 1)(k + 2)}{3} = \frac{(k + 1)[(k + 1) + 1][(k + 1) + 2]}{3}.$$

9. Step 1 If $n = 1$, then $1 + 4(1) \leq 5^1$.
Step 2 Assume true for $n = k$, then $1 + 4k \leq 5^k$.
Step 3 From step 1 we can see that $1 \leq 5^k$. Multiplying each side by 4 gives $4 \leq 4(5^k)$. Now add $1 + 4k \leq 5^k$ to $4 \leq 4(5^k)$ obtaining
$\quad 1 + 4k + 4 \leq 5^k + 4(5^k)$
$\quad 1 + 4k + 4 \leq 5^k(1 + 4)$
$\quad 1 + 4(k + 1) \leq 5^{k + 1}$.

11-2-1

1. $a_1 = 1, a_2 = 4, a_3 = 7, a_4 = 10, a_{10} = 28$ **3.** $a_1 = 4, a_2 = 7, a_3 = 12, a_4 = 19, a_{10} = 103$
5. $a_1 = 0, a_2 = .0753, a_3 = .0795, a_4 = .0753, a_{10} = .05$ **7.** $a_1 = -4, a_2 = -19, a_3 = -94, a_4 = -469$
9. $a_1 = 1, a_2 = 4, a_3 = 19, a_4 = 364$ **11.** 50   **13.** 36   **15.** $\dfrac{119}{16}$ **17.** $\sum_{i=1}^{4} (i + 1)^2$ **19.** $\sum_{i=1}^{6} (-1)^{i + 1}$

11-3-1

1. 3, 8, 13, 18, 23, 28   **3.** 52   **5.** -7 **7.** 193   **9.** 200   **11.** -32 **13.** -15 **15.** 10   **17.** 5   **19.** 10   **21.** Yes
23. \$4,400

11-4-1

1. 2, 6, 18, 54, 162   **3.** 9,375   **5.** 384   **7.** 6   **9.** -3 **11.** log 81; log 6,561   **13.** 6, 12, 24, 48   **15.** 27 days

11-4-2

1. 8   **3.** Sum does not exist   **5.** $4(2 + \sqrt{2})$ **7.** 2   **9. a.** $a_1 = .36, r = .01, S = \dfrac{4}{11}$ **b.** $a_1 = .345, r = .001, S = \dfrac{115}{333}$

11. 84 meters

11-5-1

1. $x^4 + 8x^3 + 24x^2 + 32x + 16$ **3.** $x^6 - 6x^5y + 15x^4y^2 - 20x^3y^3 + 15x^2y^4 - 6xy^5 + y^6$

5. $128a^7 + 224a^6b + 168a^5b^2 + 70a^4b^3 + \dfrac{35}{2}a^3b^4 + \dfrac{21}{8}a^2b^5 + \dfrac{7}{32}ab^6 + \dfrac{1}{128}b^7$

7. $a^{100}; -300a^{99}b; 44{,}550a^{98}b^2; -4{,}365{,}900a^{97}b^3$ **9.** $-560a^8b^6$ **11.** 108,243,216   **13.** 1.172   **15.** 7.071

17. 1.005   **19.** .032

Chapter 11 Review

The number in brackets after each answer refers to the section of the chapter that discusses that type problem.

1. Step 1 If $n = 1$, then $\dfrac{1^2 + 1}{4} = \dfrac{1}{2} = \dfrac{1}{2}(1)$

Step 2 Assume true for $n = k$, then $\dfrac{1}{2}(1) + \dfrac{1}{2}(2) + \cdots + \dfrac{1}{2}(k) = \dfrac{k^2 + k}{4}$

Step 3 For $n = k + 1$ we have $\dfrac{k^2 + k}{4} + \dfrac{1}{2}(k + 1) = \dfrac{k^2 + 3k + 2}{4} = \dfrac{(k + 1)^2 + (k + 1)}{4}$ [11–1]

3. Step 1 If $n = 1$, then $2^{1+1} - 2 = 4 - 2 = 2$

Step 2 Assume true for $n = k$, then $2 + 2^2 + \cdots + 2^k = 2^{k+1} - 2$

Step 3 For $n = k + 1$ we have $2^{k+1} - 2 + 2^{k+1} = 2^{k+1}(1 + 1) - 2$

$\qquad = 2^{k+1}(2) - 2$

$\qquad = 2^{(k+1)+1} - 2$ [11–1]

5. Step 1 If $n = 1$, then $\dfrac{1[4(1)^2 - 1]}{3} = 1 = 1^2$

Step 2 Assume true for $n = k$, then $1^2 + 3^2 + \cdots + (2k - 1)^2 = \dfrac{k(4k^2 - 1)}{3}$

Step 3 For $n = k + 1$ we have $\dfrac{k(4k^2 - 1)}{3} + [2(k + 1) - 1]^2 = \dfrac{4k^3 + 12k^2 + 11k + 3}{3}$

$\qquad = \dfrac{4k^3 + 12k^2 + 12k + 4 - k - 1}{3} = \dfrac{4(k + 1)^3 - (k + 1)}{3}$

$\qquad = \dfrac{(k + 1)[4(k + 1)^2 - 1]}{3}$ [11–1]

7. $a_1 = -3, a_2 = 4, a_3 = 23, a_4 = 60, a_8 = 508$ [11–2]   **9.** $a_1 = 1, a_2 = 2, a_3 = 6, a_4 = 21$ [11–2]

11. $\dfrac{x^3}{4} + \dfrac{x^4}{8} + \dfrac{x^5}{12} + \dfrac{x^6}{16}$ [11–2]   **13.** $\displaystyle\sum_{i=1}^{6} (4i - 3)$ [11–2]   **15.** -6 [11–3]   **17.** 10,000 [11–3]   **19.** 0 [11–3]

21. $\sqrt{2}, 2, 2\sqrt{2}, 4$ [11–4]   **23.** $-\dfrac{1}{2}$ [11–4]   **25.** $-\dfrac{27}{14}$ or 0 [11–4]   **27.** $\dfrac{2}{3}$ [11–4]   **29.** $3{,}360\, y^{12}$ [11–5]   **31.** .32768 [11–5]

Chapter 11 Test

The number in brackets after each answer refers to the section of the chapter that discusses that type problem.

1. Step 1 If $n = 1$, then $1(1 + 1) = 2$

Step 2 Assume true for $n = k$, then $2 + 4 + \cdots + 2k = k(k + 1)$.

Step 3 For $n = k + 1$ we have $k(k + 1) + 2(k + 1) = (k + 1)(k + 2)$

$\qquad = (k + 1)[(k + 1) + 1]$ [11–1]

2. $a_1 = 0, a_2 = 8, a_3 = 22, a_4 = 42$ [11–2]   **3.** 48 [11–2]   **4.** $\displaystyle\sum_{i=1}^{4} (i^2 - 2)$ [11–2]   **5.** $4\dfrac{1}{3}$ [11–3]   **6.** 48 [11–3]

7. 10 [11–3]   **8.** -5 [11–4]   **9.** 2, 6, 18 or $-2, 6, -18$ [11–4]   **10.** $\dfrac{18}{5}$ [11–4]   **11.** $\dfrac{55x^3}{128y^9}$ [11–5]   **12.** 3.87 [11–5]

Chapters 1–11 End of Book Test

The number in brackets after each answer refers to the chapter and section that discusses that type of problem.

1. -112 [1–10]   **2.** 21 [2–4]   **3.** $-\dfrac{4}{3}, 2$ [2–8]   **4.** 5 nickels, 8 dimes, 7 quarters [3–5]   **5.** $(2a - 3)(3x - 2y)$ [4–4]

6. $2xy(2x - y)(2x + y)$ [4–8]   **7.** $\dfrac{2x + 5}{x + 3}$ [5–7]   **8.** $\dfrac{3 - x^2}{(x - 1)(2x + 1)}$ [5–8]   **9.** 5.81×10^{-6} [6–3]   **10.** $\sqrt{3} - \sqrt{2}$ [6–8]

11. $\dfrac{2 \pm \sqrt{10}}{2}$ [7–6]   **12.** $\{3,11\}$ [7–8]   **13.** $2x - 3y = 9$ [8–3]

14. $(x + 3)^2 + (y - 2)^2 = 36$, center: $(-3,2)$, radius: 6 [8–4]

15.

[9–2]

16. $(1,0,-3)$ [9–6]   **17.** $2\sqrt{2}$ [10–6]   **18.** 62 [11–3]   **19.** 8 [11–4]   **20.** $-280x^4$ [11–5]

INDEX

A

Absolute value, 8
  equations involving, 147
  inequalities involving, 150
Addition method of solving a
  system, 514
Addition of signed numbers, 22
Additive identity, 5
Additive inverse, 5, 253
Algebraic expression, 49
Algebraic fractions, 274
  addition of, 297
  division of, 282
  equations containing, 311
  multiplication of, 279
  simplifying, 274
  subtraction of, 303
Algorithm, 229
Antilogarithm, 568
Area formulas
  circle, 66
  rectangle, 65, 93
  square, 65
  trapezoid, 63, 67, 128
  triangle, 65, 79, 93
Arithmetic means, 606
Arithmetic sequences, 602
Associative property of addition, 5
Associative property of
  multiplication, 5
Axes, 435
Axioms, 5

B

Base, 40
Binary operation, 31
Binomials, 52
  multiplying, 239
Binomial theorem, 614
Braces, 59
Brackets, 59, 431

C

Cartesian coordinate system, 434
Celsius temperature, 66, 102, 131
Characteristic of a logarithm, 567

Circle
  area of, 66
  center, 453
  circumference of, 66, 91
  radius, 453
  standard form, 453
Circumference of a circle, 66, 91
Closure, 5
Coefficient, 40
  numerical, 40
  zero, 231
Cofactor, 531
Combining like terms, 49
Combining signed numbers, 13
  rules for, 21
  using the number line, 16
Common denominator, 286
Common factors, 234
Common logarithms, 567
Commutative property of addition,
  5
Commutative property of
  multiplication, 5, 243
Completely factored, 234
Completing the square, 388
Complex fractions, 307
  simplifying, 307
Complex numbers, 401
  operations on, 403
  as solutions to quadratic
    equations, 405
Conditional equation, 84
Cone
  right circular, 67
Conic sections, 452
  circle, 453
  ellipse, 457
  hyperbola, 463
  parabola, 467
Constraints on a variable, 505
Contradiction, 518
Counting numbers, 4
Cramer's rule, 535
Cube, volume of, 67
Cylinder, volume of, 67

D

Dependent equations, 495, 518
Descartes, René, 434
Determinant, 529
  of a second-order matrix, 530
  solving equations using, 535
  of a third-order matrix, 532
Difference
  of two cubes, 257
  of two perfect squares, 253
Discriminant, 398
Distance-rate-time problems, 193
Distributive property, 5, 53, 224
Dividend, 230
Division of signed numbers, 38
Divisor, 230
Domain, 431

E

Ellipse, 457
  center, 458
  foci, 458
  horizontal, 457
  standard form, 457
  vertical, 457
  vertices, 458
Equations, 84
  absolute value, 147
  conditional, 84
  containing radicals, 410
  dependent, 495, 518
  equivalent, 86
  exponential, 579
  having algebraic fractions, 311
  identity, 84
  inconsistent, 495, 518
  independent, 495, 518
  linear, 438, 494, 511
  literal, 126, 495
  logarithmic, 578
  quadratic, 382
  quadratic in form, 408
  root of, 84
  solving, 84
  system of three equations, 524
  system of two equations, 494
Equivalent equations, 86

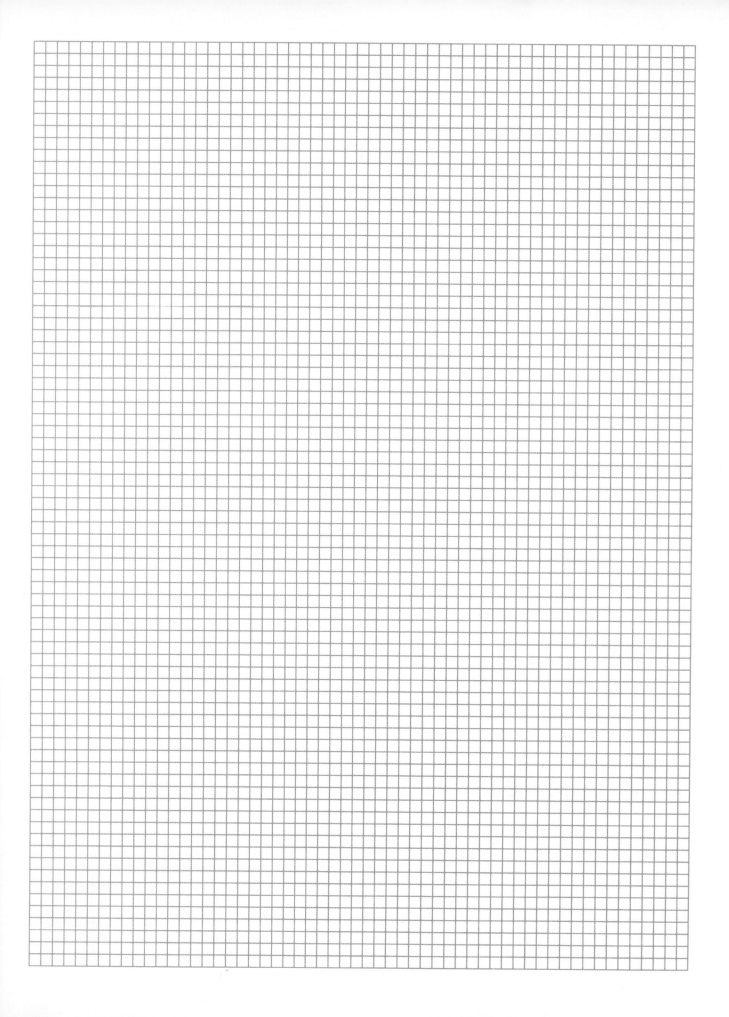

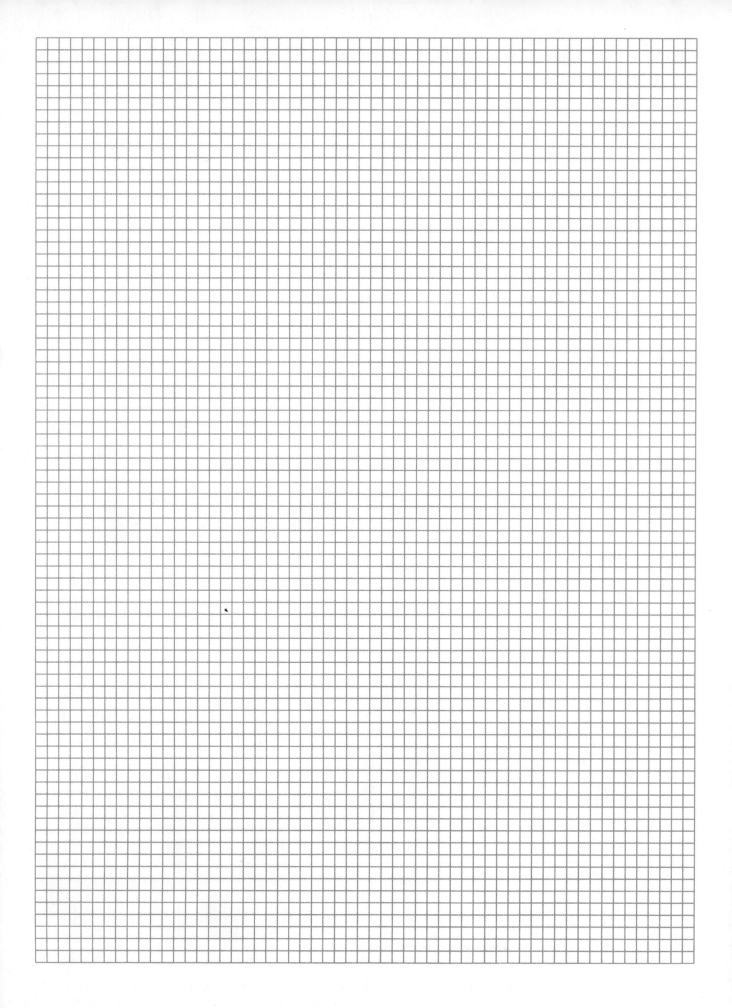